Control of Breathing and Its Modeling Perspective

AF787647

Control of Breathing and Its Modeling Perspective

Edited by

Yoshiyuki Honda

Chiba University
Chiba, Japan

Yoshimi Miyamoto

Yamagata University
Yonezawa, Japan

Kimio Konno

Tokyo Women's Medical College
Tokyo, Japan

and

John G. Widdicombe

St. George's Hospital Medical School
London, United Kingdom

Springer Science+Business Media, LLC

Library of Congress Cataloging-in-Publication Data

Control of breathing and its modeling prespective / edited by
Yoshiyuki Honda ... [et al.].
p. cm.
"Proceedings of the Fifth Oxford Conference on Control of
Breathing and Its Modelling Prespective, held September 17-19, 1991,
in Fuji, Japan"--T.p. verso.
Includes bibliographical references and index.
ISBN 978-1-4757-9849-4
1. Respiration--Regulation--Congresses. 2. Respiration-
-Regulation--Animal models--Congresses. I. Honda, Yoshiyuki, 1926-
. II. Oxford Conference on Control of Breathing and its Modelling
Perspective (1991 : Fuji-shi, Japan)
[DNLM: 1. Models, Biological--congresses. 2. Respiration-
-physiology--congresses. 3. Respiration Disorders--physiopathology-
-congresses. WF 102 C7623 1991]
QP123.C657 1992
612.2--dc20
DNLM/DLC
for Library of Congress 92-49245
CIP

Proceedings of the Fifth Oxford Conference on
Control of Breathing and Its Modelling Perspective,
held September 17-19, 1991, in Fuji, Japan

ISBN 978-1-4757-9849-4 ISBN 978-1-4757-9847-0 (eBook)
DOI 10.1007/978-1-4757-9847-0

© 1992 Springer Science+Business Media New York
Originally published by Plenum Press, New York in 1992
Softcover reprint of the hardcover 1st edition 1992

All rights reserved

No part of this book may be reproduced, stored in a retrieval system, or transmitted
in any form or by any means, electronic, mechanical, photocopying, microfilming,
recording, or otherwise, without written permission from the Publisher

PREFACE

The fifth Oxford Conference was held on September 17th-19th, 1991, at the Fuji Institute of Training in Japan - the first time that the meeting has taken place in the Asian area.

The facts that only a relatively few Japanese had attended previous Oxford Conferences and that Japan is far from other regions with possible participants made the organizers anticipate a small attendance at the meeting. However, contrary to our expectations, 198 active members (72 foreign and 126 domestic participants) submitted 146 papers from 15 countries. This was far beyond our preliminary estimate and could have caused problems in providing accommodation for the participants and in programming their scientific presentations. These difficulties, however, were successfully overcome by using nearby hotels, by telecasting presentations into a second lecture room and by displaying a substantial number of poster presentations during the whole period of the meeting.

The meeting had two types of sessions: regular and current topics. The first paper in each session represented a short overview or introduction so as to make it easier for the audience to comprehend the problems at issue. Because of the large number of papers submitted, carefully selected speakers (mostly well-known scholars) made excellent presentations that were followed by lively discussions. In this way, the conference laid a foundation on which to base its continued scientific success.

Of the papers submitted at the meeting, those dealing with the problem of hypoxic ventilatory depression were the most numerous. This issue was discussed from a number of perspectives such as rhythm genesis, neurochemicals, central and peripheral chemoreceptor activities, brain blood flow, accompanying depression of heart rate, and modelling.

The most vigorous discussion was developed in the session "Role of potassium ion in exercise hyperpnea"; the whole audience was much absorbed by a stimulating and exciting debate on the mechanisms of excess ventilation above the anaerobic threshold. We were also most grateful to Dr. S. Yamashiro, who dedicated a special lecture entitled "Modelling pioneer, Fred S. Grodins (1915-1989)."

We are unable to publish at full length all the excellent contributions to the conference. However, with the cooperation of the authors we are happy to include many reports and papers, sometimes brief, that represent the extent and high quality of all the research described at the meeting.

All attending the conference appreciated its beautiful setting at the foot of Mount Fuji. Although the latter was often obscured by clouds, the scientific discourse was clear and sparkling. We were pleased that the many distinguished foreign visitors not only made excellent scientific contributions, but also enjoyed and valued the discussions and dialogue with Japanese research workers. We were especially glad to have so many Japanese participants at the meeting. This is an indication of the rapidly growing interest in the control of breathing in Japan. We believe that the exchange of ideas and the good communication with foreign workers at this conference has encouraged and benefitted both groups, with good prospects for progress and success in the future.

Finally, we are grateful to Mike Mussell for coordinating this book's publication, and for financial support from: The Japanese Ministry of Education, The commemorative Association for Japan, World Exposition (1990), and numerous other associations and companies.

Y. Honda
Y. Miyamoto
K. Konno
J. G. Widdicombe

CONTENTS

PERIPHERAL CHEMOSENSITIVITY

CENTRAL CHEMOSENSITIVITY

INTEGRATIVE AND BEHAVIORAL CONTROL

ROLE OF NEUROCHEMICALS AND HORMONES

MECHANISMS OF EXERCISE HYPERPNEA

HYPOXIC VENTILATORY DEPRESSION

SLEEP DISORDERS

MECHANICAL ASPECTS IN THE CONTROL OF BREATHING

CELLULAR AND MOLECULAR ASPECTS

OPTIMIZATION HYPOTHESIS

ROLE OF POTASSIUM ION IN EXERCISE HYPERPNEA

ADAPTION OF PERIPHERAL CHEMORECEPTORS

MODELLING PIONEER: FRED S. GRODINS (1915-1989)

Stanley M. Yamashiro

Biomedical Engineering Department
University of Southern California
Los Angeles, CA 90089-1451
U.S.A.

INTRODUCTION

Fred S. Grodins was a pioneer in the application of control theory to study biological control systems. He was born in Chicago, Illinois and obtained all of his college training at Northwestern University. This consisted of a B.S. in Chemistry in 1937, a M.S. in Physiology in 1940, a M.D. in 1942, and Ph. D. in Physiology in 1944. Fred interned at Michael Reese Hospital in Chicago. This was followed by two years of active duty in the U.S. Army Air Force. After the war, Fred returned to Chicago to become Assistant Professor of Physiology at the University of Illinois College of Medicine. In 1947 he returned to Northwestern to become Associate Professor of Physiology and Abbott Professor of Physiology in 1951. Fred remained at Northwestern until 1967 when he left to become the founding chairman of the Biomedical Engineering Department at the University of Southern California in Los Angeles, California. In 1986 Fred retired from this position and became Emeritus Professor until his death. Summarized below are some of the major contributions Fred made to modeling and control of breathing. Personal recollections of this remarkable individual are also given.

Early Research Activities

The first scientific paper published by Fred was on traumatic shock in 1941[1]. This early work clearly demonstrated his command and special integrative skills for reviewing research literature. Shown in Figure 1 is a photograph of Fred at this phase of his career showing the unusual way he held his pen and the trademark cigarette dangling from his mouth. During this period Fred also became interested in electrical stimulation of muscle for therapeutic purposes and published several papers on this subject [2,3]. This work was the precursor to the use of this technique to study cardiorespiratory responses to exercise which will be discussed later.

World War II

World war II had a profound influence on Fred Grodins. He was drafted into the Air Force and sent to Randolph Field, Texas where the Army Air Force School of Aviation Medicine was located. Here he joined John Gray and other physiologists in work on

Control of Breathing and Its Modeling Perspective, Edited by
Y. Honda *et al.*, Plenum Press, New York, 1992

Figure 1. Fred Grodins in the early 1940's.

Figure 2. Fred Grodins as a member of the U.S. Air Force

respiratory physiology focusing on altitude and other Air Force related problems. One outcome of this effort was the Multiple Factor Theory of respiratory regulation[4]. Another was a paper on acid-base changes during asphyxia and resusitation which was Fred's first publication dealing with respiration[5]. John Gray's influence on this latter work was clearly acknowledged. Figure 2 shows Fred during his Air Force days.

Northwestern Years

Following the war, Fred continued his work in respiratory physiology. He began with a survey of literature which was so good that it became a much quoted publication on control of breathing during exercise[6]. Evident in this paper was the high regard for engineering control theory which had been developed during the war. In this work the approach of the Multiple Factor Theory was extended to cover exercise. Viewing respiratory control in terms of a CO_2 response curve as:

$$V_a = a\, P_{aCO2} - b \qquad (1)$$

where V_a = alveolar ventilation and P_{aCO2} = arterial CO_2 tension, the required conditions for isocapnic hyperpnea can be derived by consideration of the alveolar equation. The assumed form of the alveolar equation was:

$$P_{aCO2} = P_{ICO2} + 0.862\ MR_{CO2} / V_a \qquad (2)$$

where P_{ICO2} = inspired CO_2 tension and MR_{CO2} = metabolic CO_2 production rate.

Figure 3. Toshiro Sato, Mrs. Sato, and Fred Grodins in Japan

Combining Equations (1) and (2) led to at least two possible forms for an isocapnic controller when $P_{ICO2} = 0$:

$$a = b / P_{ISO} + 0.862 \, MR_{CO2} / P_{ISO}^2 \quad (3)$$

or

$$b = a \, P_{ISO} - 0.862 \, MR_{CO2} / P_{ISO} \quad (4)$$

where P_{ISO} = isocpnic CO_2 level. Equation (3) defines a multiplicative form where the controller gain changes with metabolic rate, while Equation (4) defines an additive form where the gain is fixed but the intercept point changes with metabolic rate instead. Available data in 1950 favored the additive hypothesis. However, in more recent times it is clear that this issue is still far from being settled[7,8].

This theoretical exploration was followed by experimental work on exercise hyperpnea based on electrical stimulation of muscle. Fred Kao was a student of Fred Grodins and started a fruitful series of cross-circulation experiments[9] based on electrical muscle stimulation while at Northwestern.

In 1954 Fred published his landmark paper on dynamic responses to CO_2 inhalation[10]. This was truly a remarkable achievement at the time. It was the first documented simulation model of a biological control system studied with the aid of an analog computer and included experimental verification. John Gray and engineers at Northwestern were collaborators in this effort. This type of endeavor is quite respectable today, but was quite radical in 1954. The application of control theory was extended to cardiovascular system analysis[11] in 1959. In 1965 Fred published the first book on applying control theory to biological control systems[12]. This book has had a seminal influence on the development of biological control modeling. It has been translated into Japanese by one of Fred's postdoctoral fellows, Toshiro Sato, who is shown with his wife and Fred in Figure 3 in a picture taken during a trip to Japan.

In the late 1960's Fred became acquainted with the mathematician Richard Bellman and spent one summer at the Rand Corporation in Santa Monica, California working on a digital computer model of respiratory control. This led to a publication still regarded as one of the most complete models of respiratory control[13]. This collaboration was significant becuse it eventually led Fred to leave his beloved city of Chicago and associated wind and snow for sunny Southern California!

Figure 4. Sylvia Grodins

Biomedical Engineering

For many years, Fred was convinced that engineering training in especially control theory was the ideal graduate preparation for research he was interested in. He finally had a chance to explore this conviction when he became the founding chairman of the Biomedical Engineering Department at the University of Southern California. The realization of his vision involved a lot of hard work because initially available research facilities consisted of several empty rooms. By recruiting key faculty and securing several training and research grants, a viable program was initiated and continues to this day. The current health of this department is living testament to the effectiveness of Fred's efforts.

Personal Notes

The professional success Fred Grodins achieved has to be credited to a large measure to the support he received over his entire career from his wife, Sylvia, shown in Figure 4 below. They met while Fred was still a graduate student and she a nurse. She eventually took up interior decorating and ran a highly successful business with offices in Chicago and Los Angeles.

In 1963 Sylvia bought Fred a Corvette sports car which became one of his prized possessions. He is shown driving this car in Figure 5. Fred enjoyed this and a newer model Corvette which most of his former students associated him with.

Figure 5. Fred Grodins in his 1963 Sting Ray.

Finally, the author would like to mention the great privilege it has been to be associated with Fred Grodins for over 20 years as student and colleague. Especially appreciated are the personal lesson learned about the value of a lifelong commitment to integrity.

REFERENCES

1. F.S. Grodins and S. Freeman, Traumatic shock, Int. Abstr. Surg. 72:1 (1941).

2. F.S. Grodins, S.L. Osborne, F.R, Johnson, S. Arana. and A.C. Ivy, Electrical stimulation and atrophy of denervated muscle, Am. J. Physiol. 142:222 (1944).

3. S. L. Osborne, F.S. Grodins, E. Mittlemann, and W.S. Milne, Rationale for electrodiagnosis and electrical stimulation in denervated muscle, Arch. Phys. Therapy 25:338 (1944).

4. Gray, J.S., Pulmonary ventilation and its physiological regulation, Charles C. Thomas, Springfield, (1950).

5. F.S. Grodins, A. Lein, and H.F. Adler, Changes in blood acid-base balance during asphyxia and resusitation, Am. J. Physiol. 147:433 (1946).

6. F.S. Grodins, Analysis of factors concerned in regulation of breathing in exercise. Physiol. Rev. 30: 220 (1950).

7. C.S. Poon, Ventilatory control in hypercapnia and exercise: optimization hypothesis. J. Appl. Physiol. 62: 2447 (1987).

8. G.D. Swanson and P.A. Robbins, Optimal respiratory controller structures, I.E.E.E. Trans. B.M.E. 33: 677 (1986).

9. F.F. Kao, An experimental study of the pathways involved in exercise hyperpnea employing cross-circulation techniques, In: The Regulation of Human Respiration, eds. D.J.C. Cunningham and B.B. Lloyd, Blackwell, Oxford, 461 (1963).

10. F.S. Grodins, J.S. Gray, K.R. Schroeder, A.L. Norins, and R.W. Jones, Respiratory responses to CO_2 inhalation, a theoretical study of a nonlinear biological regulator, J. Appl. Physiol. 7:283 (1954).

11. F.S. Grodins, Integrative cardiovascular physiology, a mathematical synthesis of cardiac and blood vessel hemodynamics, Q. Rev. Biol. 34:93 (1959).

12. F.S. Grodins, Control theory and biological systems, Columbia University Press, New York, (1963).

13. F.S. Grodins, J. Buell, and A.J. Bart, Mathematical analysis and digital simulation of the respiratory control system, J. Appl. Physiol 22:260 (1967).

MAINTENANCE OF THE RESPIRATORY RHYTHM DURING NORMOXIA AND HYPOXIA

Diethelm W. Richter, Mark C. Bellingham, and Christian Schmidt

II. Physiology Institute
University of Göttingen
Humboldtallee 23
Göttingen, W-3400, Germany

INTRODUCTION

The generation and maintenance of respiratory rhythm is currently the subject of some controversy, with several theories available for its' origin, based on experimental results from *in vivo*[1] and *in vitro* preparations.[2-5] While there appear to be some age-dependent differences in rhythm generation mechanisms,[5-8] there is little question that synaptic interconnection between different neuronal groups in the medulla oblongata plays a vital part in generating and shaping respiratory pattern in adult mammals.[9] Such synaptic interconnections are likely to be mainly inhibitory in nature.[10-12] Intrinsic membrane properties of certain types of respiratory neurons[2,6,13-17] may also underlie key mechanisms for phase transitions. Together the two processes constitute the core rhythm generator for respiration.[18,19]

THE GENERATION OF RESPIRATORY RHYTHM UNDER NORMOXIA

The Basic Types of Respiratory Neurons

Our present knowledge of medullary respiratory neurons and their possible interconnections have been recently reviewed.[20-22] Six broadly defined patterns of activity are seen, when phrenic nerve activity is used as a timing marker for rhythm generation:

1) early inspiratory neurons (Early-I) - these are propriobulbar neurons showing steep depolarisation and peak discharge slightly before onset of inspiratory phrenic nerve activity, with gradual membrane hyperpolarisation and declining discharge rates during the remainder of inspiratory phrenic nerve activity.

2) ramp inspiratory neurons (Ramp-I) - these are propriobulbar or bulbospinal neurons showing steady depolarisation and augmenting discharge rates during inspiratory phrenic nerve activity.

3) late inspiratory neurons (Late-I) - these are propriobulbar neurons which show delayed depolarisation during inspiratory phrenic nerve activity, but which do not begin to discharge until inspiratory phrenic nerve activity is more than half of maximal levels.

4) postinspiratory neurons (Post-I) - these are propriobulbar/bulbospinal neurons which show maximal depolarisation and discharge rate at the transition from peak inspiratory phrenic nerve activity to the declining afterdischarge in phrenic nerve activity, followed by gradual membrane hyperpolarisation and declining discharge rates throughout the remainder of the expiratory interval.

5) *stage 2 expiratory neurons (E2)* - these are propriobulbar and/or bulbospinal neurons which show gradual membrane depolarisation during the postinspiratory afterdischarge of phrenic nerve activity, reaching peak membrane depolarisation when phrenic nerve activity is completely silent, and firing at increasing rate through this second stage of expiration.

Temporal correlations of membrane potential hyperpolarisation in the inactive phases of these types of respiratory neurons with the firing patterns of concurrently active types led to a model of respiratory rhythmogenesis based on the action of inhibitory synaptic interconnections and intrinsic membrane properties.[1,9] Deficiencies in rhythmic output of computer models using the proposed connectivity scheme[23,24] has pointed to the significance of a widely ignored type of respiratory neuron, which has been found in both *in vivo*[25-27] and *in vitro*[5] preparations:

6) *pre-inspiratory neurons (Pre-I)*[5] - these are propriobulbar neurons which show peak membrane depolarisation and firing rate quite some time before the onset of inspiratory phrenic nerve activity, and then discharge at declining rates during inspiration.

Respiratory Rhythmogenesis

The rhythm generator seems to operate by the following general mechanism:

1) The decrementing neuron types (Early-I or Post-I) constitute an antagonistic half-oscillator. Their membrane potential depolarises rapidly due to activation of low voltage activated (LVA gCa)[15,17-19,28] and possibly also TTX-sensitive (gCa(TTX))[29] calcium currents with consequent rapid firing. It is possible that this rapid depolarisation may be further enhanced by activation of persistent sodium currents.[30] The action potential discharge then activates a high voltage activated calcium current[14,31] and possibly a non-specific cation conductance.[32] Intracellular accumulation of calcium leads to activation of a calcium-dependent potassium current,[13,33] causing adaptation of firing and gradual repolarisation of the membrane potential.
2) Activity of these decrementing neuron types produces widespread synaptic inhibition of all other respiratory neuron types. Their decrementing firing rate allows the gradual depolarisation of the augmenting neuron type for the respiratory phases, a inhibitory sub-set (propriobulbar Ramp-I and Bötzinger E2) of which then becomes the primary source of inhibitory output to the neurons of the antagonistic phase.
3) The augmenting neuron types differ in the means by which their membrane depolarisation is generated. There is evidence that both Ramp-I and E2 types possess low voltage activated calcium conductances (LVA gCa, gCa(TTX)), which results in an initial small, rapid depolarisation at the onset of the phases. In the case of Ramp-I neurons, this initial activity may be necessary to transiently overcome inhibition from Early-I neurons, and start a cascade of recurrent excitation between Ramp-I neurons, which generates depolarisation and produces the characteristic augmenting inspiratory output. E2 neurons probably do not possess such extensive excitatory interconnections; their gradual depolarisation is more likely to be due to decreasing synaptic inhibition from the decrementing pattern of postinspiratory neurons, controlling the rate of rise (and hence onset of firing) of the augmenting E2 neuron.
4) The late onset neuron types (Late-I and Pre-I) are strongly inhibited during the early part of their respiratory phase by the decrementing neurons active during the phase. They then rapidly reach peak depolarisation and firing rate at the end of the respiratory phases. The outputs of these neurons seem to be inhibitory to Ramp-I[11] and E2 neurons (unpublished data of this laboratory).
5) The cessation of firing in the decrementing Early-I and Post-I neurons, and the suppression of firing in the augmenting Ramp-I and E2 neurons, results in disinhibition of the decrementing neuron type of the antagonistic phase. The period of inhibition of the decrementing neuron of the antagonistic phase has de-inactivated the LVA gCa and the gCa(TTX), so that disinhibition results in rapid membrane depolarisation, and phase switching occurs.

The model of respiratory rhythmogenesis proposed above is thus a hybrid of synaptic interconnection and intrinsic properties of neurons.

DISTURBANCE OF RHYTHMOGENESIS DURING HYPOXIA

An essential property of any model is the ability to predict the response of the real system to a test condition. One important test condition that is likely to be encountered by the respiratory control system is that of hypoxia. A vital feature of the proposed model are phasic synaptic inputs which are largely inhibitory. These inputs are responsible for shaping pattern generation in either respiratory phase and for "priming" the LVA gCa and TTX-sensitive gCa by hyperpolarisation of the decrementing neuron types. Progressive suppression of inhibitory synaptic transmission should thus result in increasing destabilization of normal respiratory pattern generation and phase switching, and ultimately stop rhythmogenesis altogether. A hypoxia-induced decrease of the efficiency of synaptic transmission[34,35] to respiratory neurons should parallel the progressive disruption of respiratory patterns.

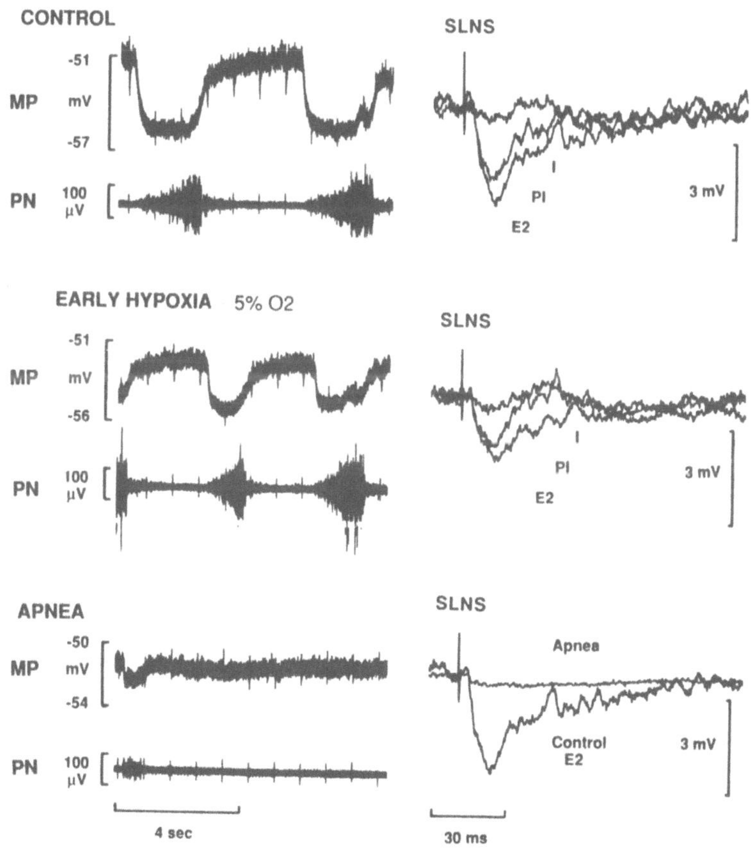

Figure 1. Continuous intracellular recording from an E2 neuron, showing membrane potential (MP) and phrenic nerve activity (PN) on the left, and responses to stimulation (1 V, 0.75 Hz) of the ipsilateral superior laryngeal nerve, averaged (n = 30-50) during inspiration (I), postinspiration (PI) and stage 2 of expiration (E2) on the right. Ventilation with 5% O_2 in N_2 reduced respiration-related MP shifts and IPSP amplitudes in early hypoxia (0-2 minutes), followed by loss of respiratory rhythm and almost complete abolition of the evoked IPSP during apnea (3-4 minutes).

Disturbance of synaptic transmission during hypoxia - effects and causes

All neurons recorded showed a characteristic sequence of changes in membrane potential pattern during hypoxia;[36] Figure 1 shows typical excerpts of membrane potential patterns from a single E2 neuron. The level of hyperpolarisation during inspiration became less negative (Fig 1) and irregular, in parallel with depression and irregularity of phrenic nerve activity; inspiratory hyperpolarisation dwindled further, and ceased within 1-2 respiratory cycles of the cessation of phrenic nerve activity (Fig 1). Loss of phrenic nerve activity and respiration-related membrane potential shifts usually occurred within 2 minutes after the beginning of severe hypoxia. This resulted in a slight overall depolarisation.

Excitatory (EPSPs) and inhibitory (IPSPs) postsynaptic potentials averaged immediately after the cessation of respiration-related membrane potential shifts (i.e., during central apnea) showed a significant decrease in amplitude.[37] At the same time, a primary increase in input resistance was seen, which turned into a secondary decrease later on. During this stage of prolonged hypoxia, some IPSPs were completely suppressed.[36] A sudden and potentially critical decrease in the efficiency of synaptic transmission thus occurs, coincident with the cessation of normal respiratory cycling. The inference to draw from this is that depression of synaptic transmission, particularly of synaptic inhibition, may be responsible for the loss of regular respiratory cycling and onset of apnea.

The secondary decrease in input resistance in the postsynaptic neuron cannot fully explain the profound depression of PSPs. It appears that depression of synaptic transmission occurs at some site presynaptic to the neuron recorded; in the *in vivo* situation it is difficult to locate the site of depression with certainty, because the pathways activated involve more than one synapse.

The role of adenosine in hypoxia-induced depression of synaptic transmission

The depression of synaptic transmission during hypoxia is unlikely to be due to a failure of axonal conduction, as axonal conduction is highly resistant to hypoxia, and also unlikely to be due to failure of action potential initiation, as respiratory neurons remain capable of firing action potentials after PSP depression has occurred during hypoxia.[36]

Hypoxia is known to induce changes in the levels of several neuromodulators, such as adenosine,[38,39] serotonin[40] and opioids,[41] and of neurotransmitters, such as GABA[42] and excitatory amino acids.[43] Alteration in the level of these substances can change the activity of neurons, or alter the release of neurotransmitters from presynaptic terminals. In particular, adenosine has to be considered, as it is produced by the breakdown of ATP and AMP when the energy supply/demand ratio is low, and is transported across the neuronal membrane into the extracellular space.[44] Adenosine is known to be a potent depressor of synaptic transmission in central nervous systems[45] reducing the number of quanta released from presynaptic terminals.[46] Taken together, these findings suggest that raised adenosine levels during hypoxia are responsible for depression of ventilation, and that this depression may be due to inhibitory modulation of synaptic transmission between respiratory neurons. We therefore tested whether adenosine could depress synaptic transmission in respiratory neurons. As adenosine is easily transported across the neuronal membrane,[44] intracellular injection of adenosine enabled us to mimic the effects of hypoxia in a single neuron, while avoiding other hypoxia-induced changes.

Intracellular injection of adenosine reduced IPSP and EPSP amplitudes (Figure 2), followed by recovery over 1-10 minutes.[47] IPSPs were depressed to a greater extent than EPSPs. Respiration-related membrane potential shifts were also diminished following adenosine injection, and recovery occurred over a similar time scale. These effects could be blocked by prior systemic administration of a specific antagonist of the A1 receptor type, 8-cyclopentyl-1,3-dipropylxanthine (DPCPX, 0.01 mg/kg i.v.), indicating that they were mediated by activation of this type of adenosine receptor (Figure 2). As we assume that any increase in extracellular adenosine levels induced by its' intracellular injection must be quite localized, it is reasonable to assume that the activated A1 receptor must be located on the presynaptic terminals contacting the neuron, or on the postsynaptic membrane of the neuron recorded. The postsynaptic conductance changes are, however, too small to solely account for the observed decrease in PSPs. Preliminary results also indicate that hypoxia-induced depression of evoked PSPs can also be largely blocked by administration of the A1 receptor blocker DPCPX (Figure 2).

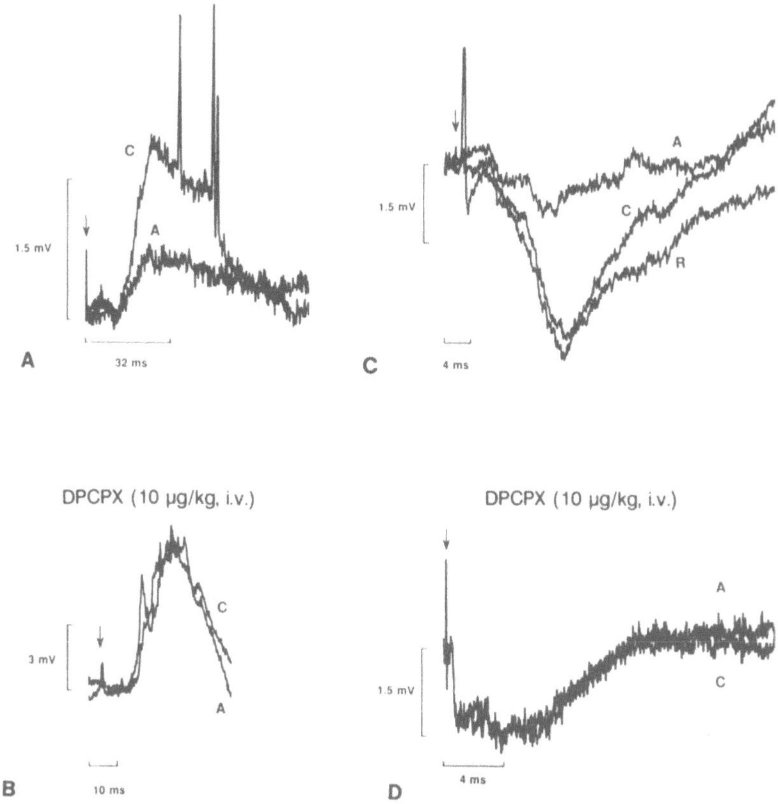

Figure 2. Averaged records of IPSPs evoked by superior laryngeal nerve stimulation (1-1.5 V, 0.5-1 Hz), recorded intracellularly from different E2 neurons. Upper records show the response of IPSPs to intracellular injection of adenosine (435 nA x sec) or 2 minutes of hypoxia (5% O_2 in N_2), while the lower records show responses to the same conditions after injection of DPCPX (0.01 mg/kg, i.v.)

Based on these findings, we would like to advance the following overview and hypothesis: In glomectomized adult mammals, hypoxia results in large increases in extracellular adenosine levels, due to its intracellular production from an enhanced breakdown of ATP/AMP and rapid transportation across the membrane. These increased levels of adenosine depress the release of neurotransmitters from presynaptic terminals, reducing synaptic modulation of postsynaptic membrane potential. As medullary respiratory neurons receive a greater proportion of inhibitory inputs than of excitatory inputs, this results in reduction of respiration-phased inhibitory inputs from other active respiratory neurons, first manifested as a reduction of respiration-related membrane potential shifts. As hypoxia continues, this progressive reduction in phasic inhibitory synaptic transmission reaches critical levels, shown as irregular respiratory cycling and then as complete loss of respiratory rhythm.

However, systemic hypoxia is a global pathophysiological condition, with multiple concurrent sequelae. While the hypothesis we have advanced can account for the loss of synaptic inputs apparent during hypoxia, it is probable that other critical factors in respiratory rhythmogenesis may be disturbed in parallel. Investigation of the responses to hypoxia of other vital components of our hypothesis for the generation of respiratory rhythm are necessary.

Acknowledgements

This work was supported by the SFB 330

REFERENCES

1. D.W. Richter, D. Ballantyne and J.E. Remmers, How is the respiratory rhythm generated? A model, *NIPS* 1:109 (1986).
2. H. Onimaru and I. Homma, Respiratory rhythm generator neurons in medulla of brainstem-spinal cord preparation from newborn rat, *Brain Res.* 403:380 (1987).
3. H. Onimaru, A. Arata and I. Homma, Primary respiratory rhythm generator in the medulla of brainstem-spinal cord preparation from newborn rat, *Brain Res.* 445:314 (1988).
4. J.L. Feldman and J.C. Smith, Cellular mechanisms underlying modulation of breathing pattern in mammals, *Ann. N. Y. Acad. Sci.* 563:114 (1989).
5. J.C. Smith, J.J. Greer, G. Liu and J.L. Feldman, Neural mechanisms generating respiratory pattern in mammalian brain stem-spinal cord in vitro. I. Spatiotemporal patterns of motor and medullary neuron activity, *J. Neurophysiol.* 64:1149 (1990).
6. J.C. Smith, H.H. Ellenberger, K. Ballanyi, D.W. Richter and J.L. Feldman, A brainstem region that may generate respiratory rhythm in mammals, *Science* 254:726 (1991).
7. D.W. Richter and K.M. Spyer, Cardio-respiratory control, in: "Central Regulation of Autonomic Functions", A.D. Loewy and K.M. Spyer, eds., Oxford University Press, Oxford (1990).
8. D.W. Richter, K.M. Spyer, M.P. Gilbey, E.E. Lawson, C.R. Bainton and Z. Wilhelm, On the existence of a common cardiorespiratory network, in: "Cardiorespiratory and Motor Coordination", H.P. Koepchen and T. Huopaniemi, eds., Springer-Verlag, Munich (1991).
9. D.W. Richter, Generation and maintenance of the respiratory rhythm, *J. Exp. Biol.* 100:93 (1982).
10. D.W. Richter, H. Camerer, M. Meesmann and N. Röhrig, Studies on the synaptic interconnection between bulbar respiratory neurones of cats, *Pflügers Arch.* 380:245 (1979).
11. D. Ballantyne and D.W. Richter, Postsynaptic inhibition of bulbar inspiratory neurones in the cat, *J. Physiol. (Lond.)* 348:67 (1984).
12. D. Ballantyne and D.W. Richter, The non-uniform character of expiratory synaptic activity in expiratory bulbospinal neurones of the cat, *J. Physiol. (Lond.)* 370:433 (1986).
13. S.W. Mifflin, D. Ballantyne, S.B. Backman and D.W. Richter, Evidence for a calcium activated potassium conductance in medullary respiratory neurones, in: "Neurogenesis of Central Respiratory Rhythm", A.L. Bianchi and M. Denavit-Saubie, eds., MTP Press, Lancaster (1985).
14. S.W. Mifflin and D.W. Richter, The effect of QX-314 on medullary respiratory neurones, *Brain Res.* 420:22 (1987).
15. J. Champagnat, T. Jacquin and D.W. Richter, Voltage-dependent currents in neurones of the nuclei of the solitary tract of rat brainstem slices, *Pflügers Arch.* 406:372 (1986).
16. M.S. Dekin and P.A. Getting, In vitro characterization of neurons in the ventral part of the nucleus tractus solitarius. II. Ionic basis for repetitive firing patterns, *J. Neurophysiol.* 58:215 (1987).
17. D.W. Richter, J. Champagnat and S.W. Mifflin, Membrane properties of medullary respiratory neurones of the cat, in: "Respiratory Muscles and their Neuromotor Control", G.C. Sieck, S.C. Gandevia and W.E. Cameron, eds., Alan R. Liss, New York (1987).
18. D.W. Richter, D. Ballantyne and S.W. Mifflin, Interaction between postsynaptic activities and membrane properties in medullary respiratory neurones, in: "Neurogenesis of Central Respiratory Rhythm", A.L. Bianchi and M. Denavit-Saubie, eds., MTP Press, Lancaster (1985).
19. D.W. Richter, J. Champagnat and S.W. Mifflin, Membrane properties involved in respiratory rhythm generation, in: "Neurobiology of the Control of Breathing", C.von Euler and H. Lagercrantz, eds., Raven Press, New York (1986).
20. S. Long and J. Duffin, The neuronal determinants of respiratory rhythm, *Prog. Neurobiol.* 27:101 (1986).
21. J. Duffin and D. Aweida, The propriobulbar respiratory neurons in the cat, *Exp. Brain Res.* 81:213 (1990).
22. K. Ezure, Synaptic connections between medullary respiratory neurons and considerations on the genesis of respiratory rhythm, *Prog. Neurobiol.* 35:429 (1990).
23. S.M. Botros and E.N. Bruce, Neural network implementation of the three-phase model of respiratory rhythm generation, *Biol. Cybern.* 63:143 (1990).
24. A.I. Pack and D.W. Richter, Modelling cardio-respiratory activities, *Eur. J. Neurosci.* Suppl. 3:172 (1990).
25. R. Nesland and F. Plum, Subtypes of medullary respiratory neurons, *Exp. Neurol.* 12:337 (1963).
26. M.I. Cohen, Discharge patterns of brain-stem respiratory neurons during Hering-Breuer reflex evoked by lung inflation, *J. Neurophysiol.* 32:356 (1969).
27. S.W. Schwarzacher, J.C. Smith and D.W. Richter, Respiratory neurones in the pre-Bötzinger region of cats, *Pflügers Arch.* 418:R17 (1991).

28. E. Carbone and H.D. Lux, Kinetics and selectivity of a low-voltage-activated calcium current in chick and rat sensory neurones, *J. Physiol. (Lond.)* 386:547 (1987).
29. H. Meves and W. Vogel, Calcium inward currents in internally perfused giant axons, *J. Physiol. (Lond.)* 235:225 (1973).
30. C.E. Stafstrom, P.C. Schwindt, M.C. Chubb and W.E. Crill, Properties of persistent sodium conductance and calcium conductance of layer V neurons from cat sensorimotor cortex in vitro, *J. Neurophysiol.* 53:153 (1985).
31. R.W. Tsien, D. Lipscombe, D.V. Madison, K.R. Bley and A.P. Fox, Multiple types of neuronal calcium channels and their selective modulation, *TINS* 11:431 (1988).
32. L.D. Partridge and D. Swandulla, Calcium-activated non-specific cation channels, *TINS* 11:69 (1988).
33. A. Marty, Ca^{+}-dependent K^{+} channels with large unitary conductance, *TINS* 6:262 (1983).
34. N. Fujiwara, H. Higashi, K. Shimoji and M. Yoshimura, Effects of hypoxia on rat hippocampal neurones in vitro, *J. Physiol. (Lond.)* 384:131 (1987).
35. E. Cherubini, Y. Ben-Ari and K. Krnjevic, Anoxia produces smaller changes in synaptic transmission, membrane potential, and input resistance in immature rat hippocampus, *J. Neurophysiol.* 62:882 (1989).
36. D.W. Richter, A. Bischoff, K. Anders, M.C. Bellingham and U. Windhorst, Hypoxia induced changes of the respiratory network of cats, *J. Physiol. (Lond.)* 443:1 (1991).
37. M.C. Bellingham, C. Schmidt, U. Windhorst and D.W. Richter, Effects of hypoxia on postsynaptic potentials in medullary respiratory neurons of the cat, *Soc. Neurosci. Abstr.* 17:104 (1991).
38. T. Hedner, J. Hedner, P. Wessberg and J. Jonason, Regulation of breathing in the rat: indications for a role of central adenosine mechanisms, *Neurosci. Lett.* 33:147 (1982).
39. M. Runold, H. Lagercrantz, N.R. Prabhakar and B.B. Fredholm, Role of adenosine in hypoxic ventilatory depression, *J. Appl. Physiol.* 67:541 (1989).
40. D.E. Millhorn, F.L. Eldridge and T.G. Waldrop, Prolonged stimulation of respiration by endogenous central serotonin, *Resp. Physiol.* 42:171 (1980).
41. M.M. Grunstein, T.A. Hazinski and H.A. Schlueter, Respiratory control during hypoxia in newborn rabbits: implied action of endorphins, *J. Appl. Physiol.* 51:122 (1981).
42. K. Iverson, T. Hedner and P. Lundborg, GABA concentrations and turn-over in neonatal rat brain during asphyxia and recovery, *Acta Physiol. Scand.* 118:91 (1983).
43. B.S. Meldrum, Excitatory amino acids and anoxic/ischaemic brain damage, *TINS* 8:47 (1985).
44. T.W. Stone, Physiological roles for adenosine and adenosine 5'-triphosphate in the nervous system, *Neurosci.* 6:523 (1981).
45. T.V. Dunwiddie, The physiological role of adenosine in the central nervous system, *Int. Rev. Neurobiol.* 27:63 (1985).
46. T.V. Dunwiddie, C.R. Lupica and W.R. Proctor, Unitary EPSPs measured by whole-cell recording are reduced by adenosine in rat hippocampal CA1 pyramidal neurons in vitro, *Soc. Neurosci. Abstr.* 17:1548 (1991).
47. C. Schmidt, M.C. Bellingham and D.W. Richter, Effects of intracellular injection of adenosine in medullary respiratory neurons of cat, *Pflügers Arch.* 420, Suppl. 1: R130 (1992).

RAPHE MAGNUS-INDUCED INHIBITION OF MEDULLARY AND SPINAL RESPIRATORY ACTIVITIES IN THE CAT

Mamoru Aoki and Yoshimi Nakazono

Department of Physiology
Sapporo Medical College
Sapporo 060, Japan

INTRODUCTION

Electrical stimulation of the medullary raphe complex results in significant changes in the discharge patterns of respiratory neurons and phrenic motoneurons. Recent studies by us[1,2] and others[3,4] have demonstrated that stimulation in the nucleus raphe magnus (NRM) produced marked inhibitory effects on respiratory activities in cats and other animals. Previous studies[5-7] have also provided evidence for the involvement of several putative transmitter substances such as serotonin (5-hydroxytryptamine) and other transmitters in raphe induced responses.

In the present study, we first investigated if the raphe magnus-induced inhibitory effects could be produced by injecting excitatory amino acid L-glutamate into the NRM, in order to determine the relative contribution of cell bodies to the responses elicited by electrical stimulation.[8] Secondly, we examined the possible involvement of putative transmitters such as γ-aminobutyric acid (GABA), bicuculline, glycine, serotonin and opiates, by observing the actions of receptor antagonists on the responses to stimulation of the NRM.[9-15] In addition, an attempt was made to determine if the raphe magnus-induced inhibition is attributable to direct and/or indirect actions on phrenic motoneurons.[1,16,17] Brief reports of some of the results have been published.[18,19]

METHODS

The experiments were carried out on 34 adult cats either anesthetized with sodium pentobarbital (20 mg/kg, i.v. initially, additional i.v. when required) or decerebrated at the precollicular level in some experiments. The animals were paralyzed with pancuronium-bromide and artificially ventilated. End-tidal CO_2 concentrations were held at 4-5 %. Rectal temperature was maintained at 37°C. The animals were placed in a stereotaxic head holder and fixed in a spinal frame. The spinal cord was exposed with a laminectomy from C_1 to C_3 segments, and in some preparations up to C_5 segments. The

Control of Breathing and Its Modeling Perspective, Edited by
Y. Honda *et al.*, Plenum Press, New York, 1992

medulla oblongata was exposed by an occipital craniotomy and by suction of part of the cerebellum. The phrenic nerve was exposed unilaterally in most preparations and the C_5 phrenic nerve discharges were recorded as an indicator of central respiratory outflow. A bilateral pneumothorax was made in most experiments. For electrical stimulation of the NRM, tungsten microelectrodes, insulated except for the tip, were inserted to the region of the NRM at P_4-P_8 levels. For chemical stimulation of the NRM, glass microelectrodes were used to inject a small amount of L-glutamate (500 μM, 1-4 μl).

In an initial series of experiments, the effects of several antagonist drugs of the putative transmitters on the sensitivity to the raphe magnus-induced inhibition were examined. The drugs intraveneously administered are as follows: picrotoxin (0.8-1.25 mg/kg) and (+)-bicuculline (Sigma, 0.25-0.5 mg/kg) as GABA receptor antagonists; strychnine (0.2 mg/kg) as a glycine antagonist; methysergide (2 mg/kg) as a serotonin

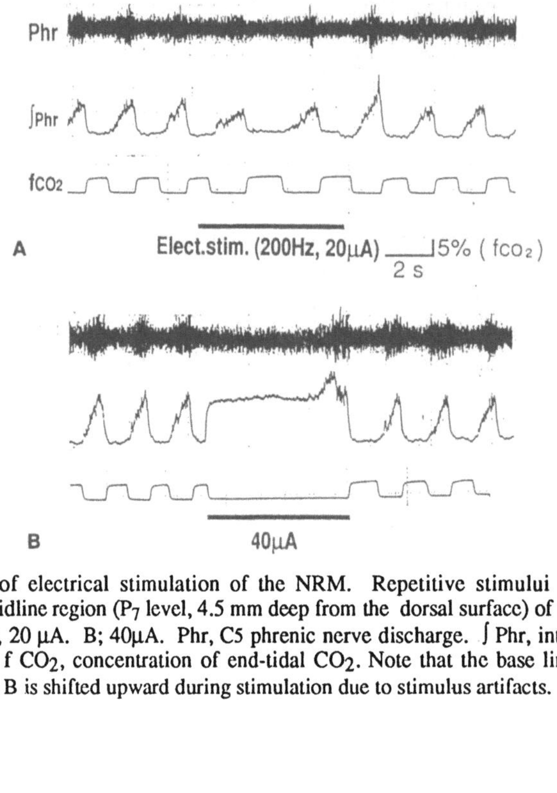

Fig. 1. Effects of electrical stimulation of the NRM. Repetitive stimului (200 Hz) were delivered to the midline region (P_7 level, 4.5 mm deep from the dorsal surface) of the medulla. A; stimulus intensity, 20 μA. B; 40μA. Phr, C5 phrenic nerve discharge. ∫ Phr, integrated phrenic nerve discharge. f CO_2, concentration of end-tidal CO_2. Note that the base line of integrated phrenic activity in B is shifted upward during stimulation due to stimulus artifacts.

antagonist; and naloxone (1-2 mg/kg) as an antagonist of opiates. These drugs were dissolved in a 165 mM NaCl solution. In some experiments, a small amount of picrotoxin (100 μM, 1-2 μl) was injected using glass micropipettes placed in the phrenic motor nucleus at the C_5 segment. In a later series of experiments, the recordings of extracellular unit activities of respiratory neurons and iontophoretic application of picrotoxin were made using a double-barrel glass microelectrode. The one recording pipette contained a tungsten wire, with its tip exposed while the other pipette was filled with picrotoxin solution which was administered iontophoretically with currents of 100-300 nA.

At the end of each experiment, recording and/or stimulating sites were marked with electrolytic lesions and later identified histologically.

RESULTS

Electrical and Chemical Stimulation of the NRM

Effects of Electrical Stimulation. Repetitive electrical stimuli (100~200 Hz, pulses of 0.2 msec duration, 20-100 μA) delivered to the region of the NRM produced significant inhibitory effects on the respiratory neural activities; depression or abolition of phrenic nerve discharge and slight prolongation of the respiratory cycle. The degree of inhibition was intensity dependent. As shown in Fig. 1, inspiratory phrenic nerve discharges were reduced, or abolished, during stimulation as a function of stimulus

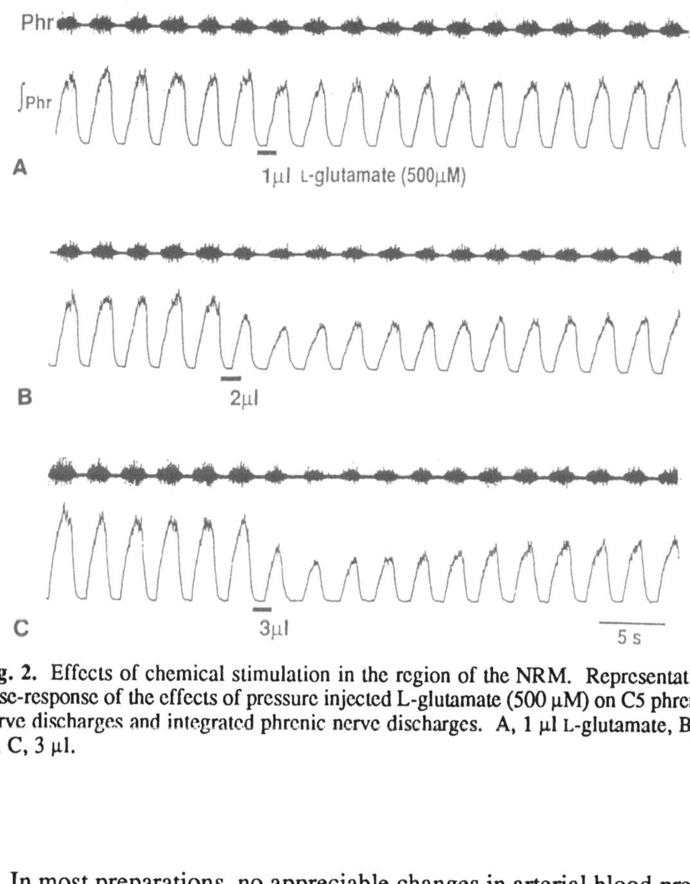

Fig. 2. Effects of chemical stimulation in the region of the NRM. Representative dose-response of the effects of pressure injected L-glutamate (500 μM) on C5 phrenic nerve discharges and integrated phrenic nerve discharges. A, 1 μl L-glutamate, B, 2 μl, C, 3 μl.

intensities. In most preparations, no appreciable changes in arterial blood pressure were induced by stimulation of the NRM, although blood pressure was slightly decreased with stimulus intensities above 100 μA. The effective sites of stimulation for producing such inhibitory effects were restricted to the midline region corresponding to the NRM (P_4-P_8). When the stimulation sites were varied by 0.5 to 1 mm lateral to the midline, little or no such inhibitory effects could be observed.

When a short stimulus train (200 Hz, 40 pulses) was delivered during the inspiratory phase, an inspiratory 'off-switch' effect could be produced as previously reported.

Chemical Stimulation Experiments. Selective activation of NRM neurons was performed in order to assess the relative contribution of cell bodies and axons of passage

induced inhibition of the C_5 phrenic nerve discharge ipsilateral to the injection side was significantly blocked (Fig. 4, A). The time course of this blocking action is illustrated in Fig. 4, B. The blocking action began immediately after microinjection of picrotoxin, reached its peak in 1 to 2 min and persisted for 20-30 min. In contrast, raphe magnus induced inhibition of the C_5 phrenic nerve discharge on the contralateral side was not affected. These results clearly demonstrate that the direct raphe spinal pathway is at least in part involved in the raphe magnus-induced respiratory inhibition.

Iontophoretic Administration. In another series of experiments, we further examined whether or not the raphe magnus-induced inhibition of the respiratory neuron discharges in the upper cervical cord[20-23] and the medulla could be blocked by iontophoretically applied picrotoxin. In Fig. 5, is a typical result obtained from an inspiratory neuron activity recorded at C_1 segment. In a control before picrotoxin application, both phrenic nerve activity and C_1 inspiratory neuron discharges were strongly depressed by electrical stimulation of the NRM. Picrotoxin was administered a few seconds before electrical stimulation iontophoretically (5 mM, 200 nA) near this neuron using a double-barrel pipette. It was a consistent finding that the raphe magnus-induced depression of C_1 neuron discharges was powerfully blocked and C_1 neuron discharges persisted during electrical stimulation, whereas the phrenic discharge was almost completely depressed (Fig. 5, lower set of records). Unit discharges of the ventral respiratory group (VRG) of neurons of the retroambigual nucleus were also recorded at the obex level and the effect of picrotoxin was examined. Electrical stimulation of the NRM induced strong suppression of the phrenic nerve discharge and the unit discharges of inspiratory type neurons. When picrotoxin was administered iontophoretically near the neuron, the depression of the unit discharge was markedly reduced, as in the case of cervical inspiratory neurons. Electrophoretically applied pirotoxin, with currents of 50-300nA, significantly reduced the raphe magnus-induced inhibition in the majority (19/27) of medullary and C_1 inspiratory neurons tested.

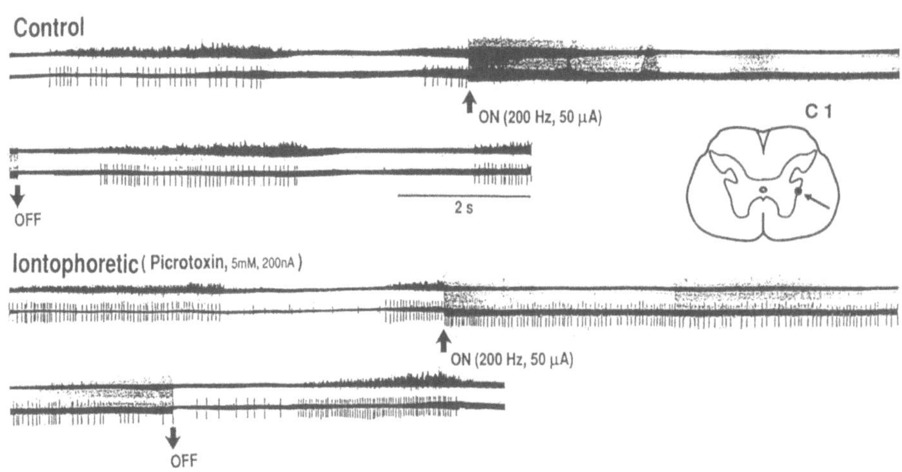

Fig. 5. Effects of iontophoretically applied picrotoxin on inhibition of unit discharge of a upper cervical inspiratory neuron and C_5 phrenic nerve discharge induced by stimulation of the NRM. Upper (Control) and lower (Iontophoretic) sets are continuous records. Upper traces are phrenic nerve discharges and lower traces are C_1 inspiratory neuron discharge (see inset, indicated by an arrow). Picrotoxin was continuously delivered (200 nA), (beginning a few seconds before electrical stimulation of the NRM and to the end of the trace) at the recording site with a double barreled electrode.

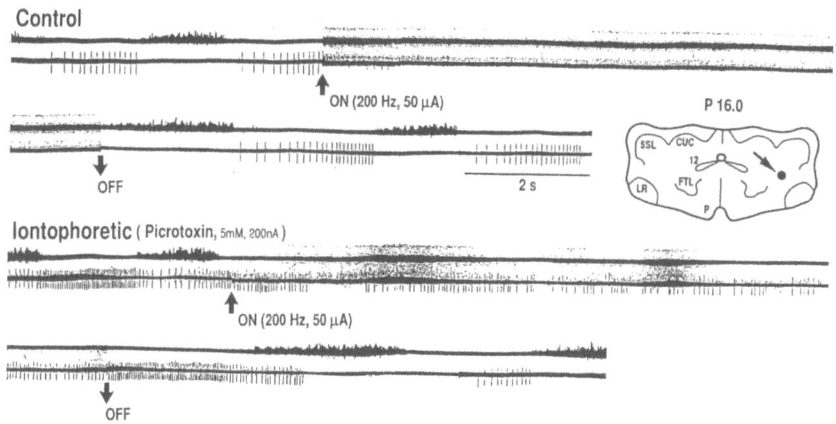

Fig. 6. Effects of iontophoretically applied picrotoxin on the raphe magnus-induced inhibition of unit discharge of an expiratory neuron. Format is the same with Fig. 5. Recording site corresponding to the nucleus retroambigualis is shown in an inset (shown by an arrow).

In unit discharges of an expiratory type neurons recorded at a few milimeters caudal to the obex, the raphe magnus-induced depressant effect was strongly blocked by an iontophoretic administration of picrotoxin (5 mM, 200 nA) as shown in Fig. 6. Similar results were obtained with other 5 expiratory neurons tested.

These results revealed that raphe magnus-induced depressive effects on both inspiratory and expiratory types of the VRG respiratory neurons are strongly blocked by picrotoxin. It is suggested that raphe magnus neurons project to the VRG neurons and exert inhibitory effects mediated by GABA.

DISCUSSION

It has been shown that électrical stimulation of the caudal raphe nuclei, consisting of the raphe obscurus, raphe pallidus, and raphe magnus, produce marked changes in respiratory nervous activities.[1-7] Thus, the medullary raphe nuclei may be involved in respiratory control. Previous studies[4-6] demonstrated that the responses to stimulation of raphe obscurus is excitatory, whereas a mixture of inhibitory and excitatory responses of lesser magnitude were produced by stimulation of raphe pallidus. The present experiments have clearly demonstrated that electrical stimulation of the NRM consistently produces short-latency inhibitory effects on phrenic nerve activity, as well as medullary and upper cervical respiratory neuron discharges. These inhibitory effects could be mimicked by chemical stimulation with microinjections of the excitatory amino acid L-glutamate into the NRM. Present results indicate that activation of cell bodies in the NRM, not axons of passage, is mainly responsible for the raphe magnus-induced inhibition.

There is now increasing evidence for the involvement of GABA as an inhibitory transmitter in central respiratory control.[9,10] Recently, it has been reported that GABA mediate postsynaptic inhibitions of the ventral respiratory group (VRG) neurons in the medulla. Recent anatomical and physiological studies have revealed that the caudal raphe nuclei project to several nuclei involved in respiratory control, such as the phrenic motor nucleus in the spinal cord, and the nucleus tractus solitarius where dorsal respiratory group (DRG) neurons exist. The present experiments have demonstrated that the raphe magnus-induced inhibition of VRG neurons and the C_5 phrenic nerve activity could be markedly blocked by systemic and topical application of picrotoxin. These results may support the view that GABAergic neurons in the NRM send their axonal projections to respiratory neurons in the medulla, as well as upper cervical respiratory neurons and phrenic motoneurons in the spinal cord. Both picrotoxin and bicuculline are $GABA_A$ receptor antagonists and they were effective in blocking the raphe magnus-induced inhibition. Sacrofen, a $GABA_B$ receptor antagonist, has been found to be ineffective (personal communication), it is strongly suggested that the raphe magnus-induced inhibition is the $GABA_A$ receptor mediated response.[24] In conclusion, this study has provided evidence for the presence of a GABAegic system, probably arising from the NRM. This system as well as a serotonergic system may play an important role in respiratory control.

REFERENCES

1. M. Aoki, Y. Fujito, Y. Kurosawa, H. Kawasaki, and I. Kosaka, Descending inputs to the upper cervical inspiratory neurons from the medullary respiratory neurons and the raphe nuclei in the cat, *in:* "Respiratory Muscles and Their Neuromotor Control, " G. C. Sieck, S. C. Gandevia, W. E. Cameron, eds., A. R. Liss, New York, pp 75-82 (1987).
2. M. Aoki, Y. Fujito, I. Kosaka and N. Kobayashi, Supraspinal descending control of propriospinal respiratory neurons in the cats. *in:*"Respiratory Control," G.D. Swanson, F.S. Grodins and R.L. Hughson, eds., Plenum, New York, pp 451-459 (1989).
3. B.J. Sessle, G.J. Ball, and G.E. Lucier, Suppressive influences from periaqueductal gray and nucleus raphe magnus on respiration and relared reflex activities and on solitary tract neurons, and effect of naloxone, *Brain Res.* 216:145-161 (1981).
4. P.M. Lalley, Responses of phrenic motoneurones of the cat to stimulation of medullary raphe nuclei, *J. Physiol.* 380:349-371 (1986).
5. P.M. Lalley, Serotonergic and non-serotonergic responses of phrenic motoneurones to raphe stimulation in the cat, *J. Physiol.* 308:373-385 (1986).
6. J.R. Holtman, Jr., T.E. Dick, and A.J. Berger, Involvement of serotonin in the excitation of phrenic motoneurons evoked by stimulation of the raphe obscurus, *J. Neurosci.* 6:1185-1193 (1986).
7. S.J. Fung and C.D. Barnes, Raphé-produced excitation of spinal cord motoneurons in the cat, *Neurosci. Lett.*, 103:185-190 (1989).
8. J.R. Haselton, R.W. Winters, D.R. Liskowsky, C.L. Haselton, P.M. McCabe, and N. Schneiderman, Cardiovascular responses elicited by electrical and chemical stimulation of the rostral medullary raphe of the rabbit, *Brain Res.* 453:167-175 (1988).
9. A. Haji, R. Takeda and J.E. Remmers, GABA-mediated inhibitory mechanisms in control of respiratory rhythm, *in:*"Control of Breathing and Dyspnea," T. Takishima and N.S. Cherniack, eds., Pergamon, Oxford-New York, pp 61-63 (1991).
10. K.A. Yamada, P. Hamosh, and R.A. Gillis, Respiratory depression produced by activation of GABA receptors in hindbrain of cat, *J. Appl. Physiol.*, 51:1278-1286 (1981).
11. D.R. Curtis, A.W. Duggan, D. Felix, G.A.R. Johnston and H. McLennan, Antagonism between bicuculline and GABA in the cat brain, *Brain Res.* 33:57-73 (1971).
12. J. Champagnat, M. Denavit-Saubié, S. Moyanova and G. Rondouin, Involvement of amino acids in periodic inhibitions of bulbar respiratory neurones, *Brain Res.* 237:351-365 (1982).
13. L. Grelot, S. Iscoe and A.L. Bianchi, Effects of amino acids on the excitability of respiratory bulbospinal neurons in solitary and para-ambigual regions of medulla in cat, Brain Res., 443:27-36 (1988).
14. H. Arita and M. Ochiishi, Opposing effects of 5-hydroxytryptamine on two types of medullary inspiratory neurons with distinct firing patterns, *J. Neurophysiol.*, 66:285-292 (1991).
15. M. Denavit-Saubié, J. Champagnat, and W. Zieglgänsberger, Effects of opiates and methionine-enkephalin on pontine and bulbar respiratory neurones of the cat, *Brain Res.* 155:55-67 (1978).
16. J. R. Holtman, Jr., W.P. Norman, and R. A. Gillis, Projection from the raphe nuclei to the phrenic motor nucleus in the cat, *Neurosci. Lett.*, 44:105-111 (1984).
17. R.M. Bowker, V.K. Reddy, S.J. Fung, J.Y.H. Chan and C.D. Barnes, Serotonergic and non-serotonergic raphe neurons projecting to the feline lumbar and cervical spinal cord: a quantitative horseradish peroxidase-immunocytochemical study, *Neurosci. Lett.* 75:31-37 (1987).
18. M. Aoki, Y. Nakazono and A. Mizuguchi, Inhibitory effects induced by electrical and chemical stimulation of the nucleus raphe magnus on respiratory activities, *Jap. J. Physiol.* 41: Suppl., S 234 (1991).
19. M. Aoki, Y. Nakazono and A. Mizuguchi, Respiratory inhibitom from nuclus naphe magnus and its reversal by a GABA antagonist, Abstr. 3rd IBRO World Congress of Neurosience, P123 (1991).

20. M. Aoki, S. Mori, K. Kawahara, H. Watanabe, and N. Ebata, Generation of spontaneous respiratory rhythm in high spinal cats, *Brain Res.*, 202:51-63 (1980).
21. M. Aoki, Respiratory-related neuron activities in the cervical cord of the cat. *in*: Proceedings of the International Symposium. "Central Neural Production of Periodic Respiratory Movements", J.R.Feldman, A.J. Berger, eds., Northwestern Univ., Chicago, pp 155-156 (1982).
22. J. Duffin and R. W. Hoskin, Intracellular recordings from upper cervical inspiratory neurons in the cat, *Brain Res.*, 435:351-354 (1987).
23. J. Lipski and J. Duffin, An electrophysiological investigation of ropriospinal inspiratory neurons in the upper cervical cord of the cat, *Exp. Brain Res.*, 61:625-637 (1986).
24. N.G. Bowery, A.L. Hudson, and G.W. Price, $GABA_A$ and $GABA_B$ receptor site distribution in the rat central nervous system, *Neurosci.* 20:365-383 (1987).

CRANIAL NERVE AND PHRENIC RESPIRATORY RHYTHMICITY DURING CHANGES IN CHEMICAL STIMULI IN THE ANESTHETIZED RAT

Yasuichiro Fukuda

Department of Physiology II
School of Medicine, Chiba University
1-8-1 Inoohana, 280 Chiba, Japan

INTRODUCTION

Central respiratory rhythmicity and/or timing has been determined by the trajectory of phrenic (Phr) nerve discharges. Cranial nerves innervating the upper airway muscles also display respiratory modulated activity synchronized with Phr activity. These cranial nerve motoneurons as well as spinal Phr motoneurons (or Phr driving medullary pre-motor neurons) are driven by a common rhythm generating mechanism. There are, however, significant differences in the discharge pattern and responses to chemical stimuli between Phr and cranial nerve respiratory activity[1]. The onset of inspiratory activity of the cranial nerves is much earlier than that of Phr nerve[2,3,4]. Changes in chemical stimuli (Pa_{O_2}, Pa_{CO_2}) initiate differential effects on I activity among various respiratory nerves, including Phr and cranial nerves[3,5,6,7]. On the other hand, cranial nerve respiratory activities are sensitive to anesthesia and/or sleep stage, and hence have not been considered as major output signals for observation of central respiratory rhythmicity[8,9,10]. In the present experiment we found that the glossopharyngeal (IX) nerve I activity showed much smaller suppression during hypocapnia or hypoxia than the Phr I discharge did. Furthermore a small ramp-like rhythmic IX activity even without Phr burst was seen during hypocapnic or hypoxic respiratory suppression.

MATERIALS AND METHODS

Eighteen male Wistar rats (300-350g body weight) were used. The animal was anesthetized with halothane (2.5% for induction, 1.2-1.5% during operation). The animal was tracheotomized and breathed spontaneously through a tracheal cannula during operation. The femoral artery and vein were cannulated for blood pressure measurement and for drug injection. after exposure of the phrenic, glossopharyngeal and vagal pharyngeal nerves, the animal was bilaterally vagotomized, immobilized with pancuronium bromide (Myoblock R, 0.2mg/h, iv), and artificially ventilated. The bilateral carotid sinus nerve, cervical sympathe-

Control of Breathing and Its Modeling Perspective, Edited by
Y. Honda *et al.*, Plenum Press, New York, 1992

tic and aortic nerves were sectioned. Halothane was discontinued and the anesesia was switched to urethane (1.0g/kg, iv) after completing all operations[11]. A small additional dose of urethane was injected (0.2-0.3g/kg, iv) repeatedly to maintain a constant level of anesthesia. End-tidal Po_2 and Pco_2 ($P_{ET}O_2$, $P_{ET}co_2$) were continuously monitored via sampling the tracheal gas with an expiratory gas analyzer (1H26, NEC San-Ei Instruments, Tokyo)[12,13]. Efferent I discharges were recorded from the cut central end of the Phr nerve and the glossopharyngeal (IX) nerve or its stylopharyngeal branch. In some experiment the efferent decremental expiratory (E) activity was recorded from the vagal pharyngeal nerve[14]. The animal was ventilated first with hyperoxic gas ($P_{ET}O_2 \geqq 200$ Torr), and $P_{ET}co_2$ was controlled at about 45 Torr (normocapnia) by adjusting

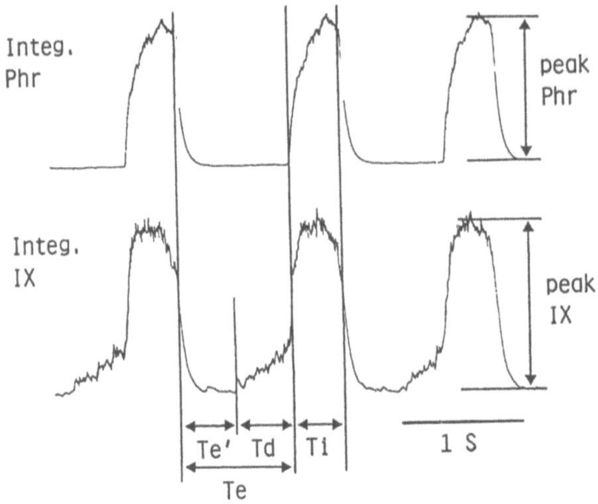

Figure 1. Inspiratory discharge of phrenic (Phr) and glossopharyngeal (IX) nerves. Integrated waves, measurements of respiratory timing and the height of peak integrated activites are shown. Ti, inspiratory time; Te, expiratory time; Td, time delay from the onset of IX I activity to the start of Phr discharge; Te', early expiratory time (= Te - Td).

the rate and depth of artificial ventilation (control condition). $P_{ET}co_2$ was reduced to about 30-35 Torr within 3 min by hyperventilating the animal in hyperoxia[3]. $P_{ET}O_2$ was decreased to about 30-40 Torr within 2 min by applying hypoxic gas in normocapnia. The reduction in $P_{ET}co_2$ or $P_{ET}O_2$ produced reversible cessation of respiratory rhythmicity which was determined by cessation of the phasic Phr activity (hypocapnic or hypoxic apnea). Inspiratory time (Ti, period of time from the onset to termination of Phr discharge), time delay between the onset of the IX I activity and the Phr onset (Td), and expiratory time (Te, from termination of Phr activity to the next Phr onset) were measured from integrated I activities (Fig.1). The early expiratory time (Te'= Te - Td) was calculated. The peak height of integrated I activity was compared between the Phr and IX nerves. Systolic blood pressure was always maintained above 70-80 Torr even during systemic hypoxia. $P_{ET}co_2$ or $P_{ET}O_2$ was quickly restored to control condition after cessation of rhythmic IX and Phr activities.

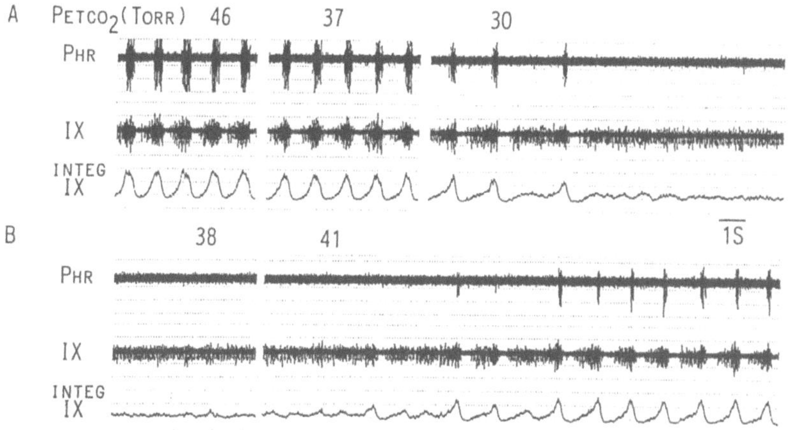

Figure 2. Effect of hypocapnia on the phrenic (PHR) and glossopharyngeal (IX) inspirtory activities. Values of $P_{ET}CO_2$ are shown at the top of each trace. A and B were continuous recordings obtained in one animal.

RESULTS

DISCHARGE PATTERN IN CONTROL HYPEROXIC AND NORMOCAPNIC CONDITION

The IX showed a gradual but earlier onset of I activity in the late expiratory phase (Fig. 1). Parameters of respiratory timing and peak height of integrated I activity obtained in control condition and at 5-10 s prior to hypocapnic or hypoxic apnea are summarized in Fig. 6. The vagal pharyngeal (Xphar) nerve showed decremental E activity immediately following the termination of Phr discharges (Fig. 3B).

EFFECT OF HYPOCAPNIA

A reduction in $P_{ET}CO_2$ in hyperoxia decreased respiratory frequency and attenuated the height of peak Phr activity (Fig. 2 and Fig. 6). The reduction in frequency was produced mainly by prolongation of Td, i.e., delayed onset of synchronized large IX and Phr I discharge following early start of gradual IX I activity. The magnitude of decrease in the IX I activity was small. Cessation of Phr discharge occurred at $P_{ET}CO_2$ about 30-35 Torr (apnea). However, several small ramp like rhythmic bursts persisted in the IX nerve even after the cessation of Phr discharge. Small rhythmic bursts became tonic and then disappeared with further reduction in $P_{ET}CO_2$. With elevating $P_{ET}CO_2$ small ramp like rhythmic IX activity appeared first. Large IX and Phr bursts appeared subsequently always at the peak of each small rhythmic activity. The Phr activity was small during early recovery phase. Appearance of Phr I activity was greatly delayed after prolonged hypocapnia or in deeply anesthetized condition (5-10 min after supplemental urethane) although clear rhythmic IX burst had already started (Fig. 3A). The decremental Xphar E activity decreased greatly during hypocapnia and disappeared much earier than the cessation of Phr discharge (Fig. 3B).

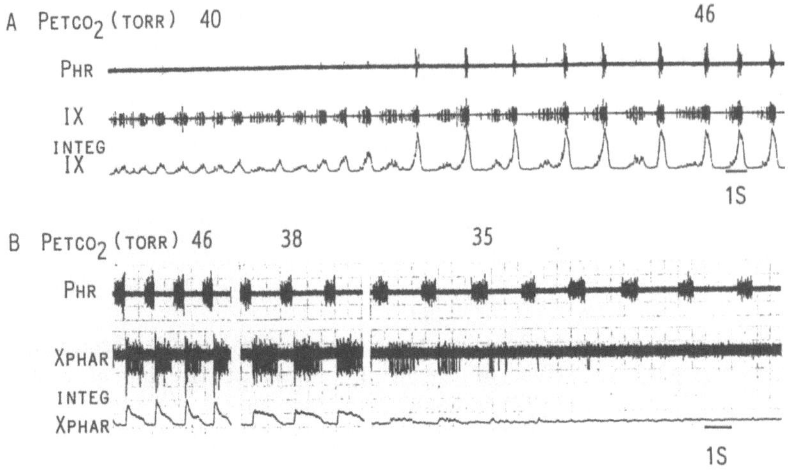

Figure 3. A Effect of prolonged hypocapnia on the recovery of phrenic (PHR) and glossopharyngeal (IX) nerve inspiratory activities. B Effect of hypocapnia on the vagal pharyngeal nerve (XPHAR) expiratory activity. $P_{ET}CO_2$ values are shown at the top of each trace. INTEG, integrated activity. A and B were obtained from different animals.

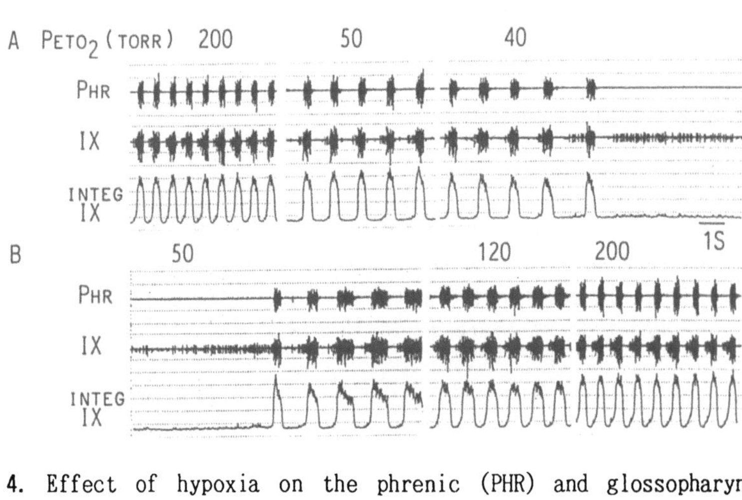

Figure 4. Effect of hypoxia on the phrenic (PHR) and glossopharyngeal (IX) nerve inspiratory activities. Value of $P_{ET}O_2$ is shown at the top each trace. INTEG, integrated activity. A and B were continuous recordings obtained in one animal.

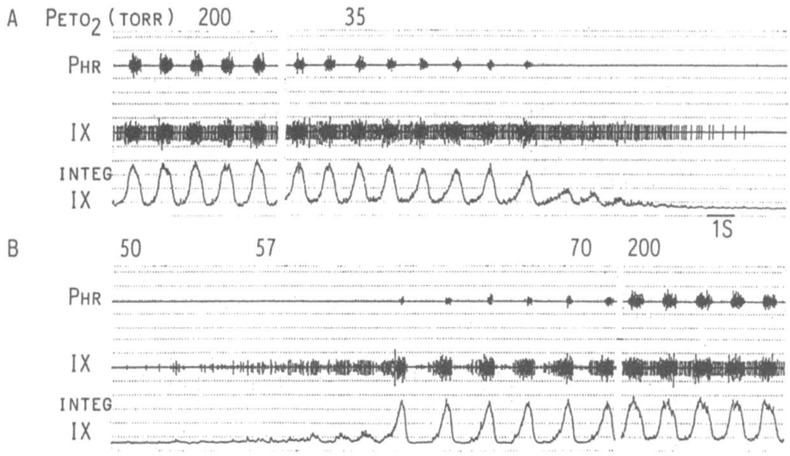

Figure 5. Effect of hypoxia on the phrenic (PHR) and glossopharyngeal (IX) nerve inspiratory activities (10 min after administration of supplemental dose of urethane). INTEG, integrated activity. $P_{ET}O_2$ values are shown at the top of each trace. A and B were continuous recordings obtained in one animal.

EFFECT OF HYPOXIA

The reduction in $P_{ET}O_2$ with maintained normocapnia prolonged Te' without significant change in Ti (Fig.4, Fig.6). The peak I activity remained unchanged in the IX and decreased slightly in the Phr nerve. The early and gradual onset of IX I activity became less remarkable, and respiratory rhythmicity was lost simultaneously in both nerves at $P_{ET}O_2$ about 30-40 Torr (hypoxic apnea). The IX activity became tonic and then silent after cessation of rhythmicity. During recovery phase, small tonic IX activity appeared first and large synchronized rhythmic Phr and IX bursts appeared when this small IX discharge increased. In some cases initial tonic IX activity turned out small ramp like rhythmic activity before appearance of synchronized phasic Phr and IX bursts. When hypoxia was applied shortly (5-10 min) after administration of supplemental dose of urethane (0.2-0.3g/kg, iv), however, the Phr I activity was greatly supperessed and disappeared earlier than the cessation of small ramp like rhythmic IX activity (Fig.5). Recovery of Phr I discharge was delayed. Initiation of synchronized Phr and IX bursts occurred always at the peak of each small ramp like rhythmic IX activity which was restored much earlier than the phasic Phr activity during elevation of $P_{ET}O_2$. The Xphar E activity was decreased greatly as in hypocapnia.

DISCUSSION

The present experiment showed two important differences in discharge responses to chemical stimuli between the Phr and IX I activities. The first is that the IX nerve I activity was less suppressed by hypocapnia and hypoxia, and second is that small ramp like rhythmic IX activity can be seen even in the absence of phasic Phr discharges. A possible persistence of intramedullary respiratory rhythmicity after Phr discharge has vanished during hypocapnia has been suggested in the observation of expiratory laryngeal activity of the cat[7]. Furthermore, the IX I activity was less sensitive to suppression by anesthesia in the rat[15]. There has been general agreement that the respiratory modulated activity in the cranial nerves are easily suppressed by anesthesia or sleep and by reduction in chemical stimuli[8,9,10]. However, present and recent findings

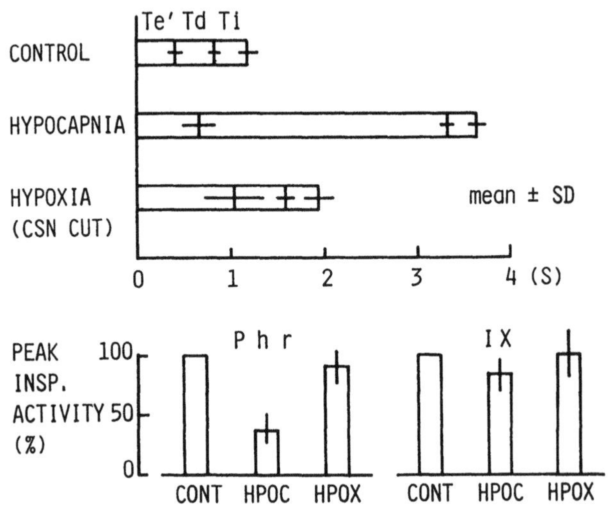

Figure 6. Summary of effects of hypocapnia and hypoxia on respiratory timing and peak inspiratory activities. Phr, phrenic nerve; IX, glossopharyngeal nerve; CONT, control condition; HPOC, hypocapnia; HPOX, hypoxia; CSN, carotid sinus nerve. Results were obtained in control condition and at 5-10 s prior to apnea. Values obtained at 5-10 min after administration of supplemental dose of urethane were not included.

indicated that significant spacial inhomogeneity in the responses of various cranial nerve motoneurons to reduced respiratory stimuli or to anesthetic agents. The mechanism of a higher'resistance' to hypocapnic suppression in the IX I motoneuron than in the Phr motoneurons was obscure. However, the net suppressive effects of hypocapnia on membrane properties of individual motoneuron, net drive inputs to its, and number of recruited neurons seem to be smaller in the IX I motoneurons than in the Phr motoneurons. The efferent IX nerve I activity, which represents overall activity of the rostral ambiguual or retrofacial IX (or stylopharyngeal) motoneurons[16,17], has not yet been analyzed systematically. Hypocapnia also inhibited a common rhythm generating mechanism resulting in cessation of rhythmicity. Hypoxia has direct or indirect effect on respiratory neurons especially when there is no compensatory stimulation from peripheral arterial chemoreceptors[18,19,20]. Hypoxia might have preferentially suppressed the rhythm generating mechanism because the reduction in the peak I activity was not large prior to apnea (Fig. 4, Fig. 6). However, in deeply anesthetized condition, the Phr I motoneurons activity became inhibited by hypoxia as in hypocapnia.

The gradual but early onset of cranial nerve I discharge preceding the start of Phr discharge is likely due to that cranial nerve I motoneurons have lower firing threshold (for similar inputs) or earlier and stronger inputs than the spinal Phr motoneuron[2,5,21]. This differnce in onset time occurs probably in the medulla oblongata[2,21]. The IX I activity starts earlier also in the isolated brain stem preparation of newborn rats[22]. The initial part of IX activity resembles temporally that of phase spanning E-I neurons in the rat medulla oblongata[23] or of pre-I neurons observed in in vitro brain stem reparation of newborn rats[22,24]. An interesting observation was that during recovery from hypocapnic or hypoxic apnea small ramp like rhythmic IX burst appeared first and synchronized large Phr and IX I bursts superimposed later always on the peak of small rhythmic IX activity. This evidence indicates that the IX or Phr motoneurons seem to be driven by inputs from following two temporally different mechanisms, i.e., 1) mechanism(s) generating initial small ramp like rhythmic activity (rRa) and 2) later synchronized I bursting mechanism (sIb). The transition from the rRa to the sIb in a triggering manner may be smooth in control condition. The IX I motoneurons are activated by both inputs due to their lower firing threshold (or stronger inputs) resulting in early onset of I activity. The Phr motoneurons or medullary premotor neurons may be activated by the sIb only because of their higher firing threshold (or weaker inputs). A small ramp like rhythmic activity was not obviously seen in other cranial nerves such as hypoglossal, superior or recurrent laryngeal nerves during hypocapnia or hypoxia in the vagotomized and peripherally chemodenervated rat. In conclusion central respiratory rhythmicity during hypocapnic or hypoxic suppression would be better recognized by discharge pattern of the IX nerve activity rather than that of the Phr activity.

ACKNOWLEDGMENTS

The author thanks Dr. D. W. Richter for valuable discussion.

REFERENCES

1. C. Von Euler, On the central pattern generator for the basic breathing rhythmicity. J. Appl. Physiol. 55:1647(1983).
2. M. I. Cohen, Phrenic and recurrent laryngeal discharge patterns and the Hering-Breuer reflex. Amer. J. Physiol. 228:1489(1975).
3. Y. Fukuda, and Y. Honda, Effects of hypocapnia on respiratory timing and inspiratory activities of the superior laryngeal, hypoglossal, and phrenic nerves in the vagotomized rat. Jpn. J. Physiol. 33:733(1983).
4. Y. Fukuda, and Y. Honda, Modification by chemical stimuli of temporal difference in the onset of inspiratory activity between vagal (superior laryngeal) or hypoglossal and phrenic nerves of the rat. Jpn. J. Physiol. 38:309 (1988).
5. S. D. Iscoe, Central control of the upper airway, *in*: "Respiratory function of the upper airway", O. P. Mathew and G. Sant'Ambriogio, eds., Marcel Dekker, New York-Basel (1988).

6. D.Weiner, J.Mitra, J.Salamone,and N.S.Cherniack, Effect of chemical stimuli on nerve supplying upper airway muscles. J.Appl.Physiol. 52:530(1982).
7. D.Zhou, Q.Huang, W.M.St.John, and D.Bartlett,Jr, Respiratory activities of intralaryngeal branches of the recurrent laryngeal nerve, J.Appl.Physiol. 67:1117(1989)
8. M.I.Cohen, Neurogenesis of respiratory rhythm in the mammal. Physiol.Rev. 59:1105(1979).
9. J.L.Feldman, Neurophysiology of breathing in mammals, *in*: "Handbook of physiology, sect. 1, The nervous system, Vol.I", V.B.Mountcastle, F.E.Bloom, and S.R.Geiger, eds., Amer.Physiol.Soc., Bethesda, (1986).
10. Y.Murakami, and J.I.Kirchner, Respiratory activity of the external laryngeal muscles: an electromyographic study in the cat, *in*:"Ventilatory and phonatory control", D.Wyke, ed., Oxford Univ.Press, London (1974).
11. Y.Fukuda, A.Sato, and A.Trzebski, Carotid chemoreceptor discharge responses to hypoxia and hypercapnia in normotensive and spontaneously hypertensive rats. J.Auton.Nerv.Syst. 19:1(1987).
12. Y.Fukuda, W.R.See, and Y.Honda, Effect of halothane anesthesia on end-tidal Pco_2 and pattern of respiration in the rat. Pflügers Arch. 392:244(1982).
13. H.Tojima, T.Kuriyama, and Y.Fukuda, Arterial to end-tidal Pco_2 differnce varies with different ventilatory conditions during steady state hypercapnia in the rat, Jpn.J.Physiol. 38:445(1988).
14. Y.Fukuda, Hypoxic inhibition of respiratory neural regulation in anesthetized rats, Jpn.J.Physiol. 41:893(1991).
15. Y.Fukuda, Differences in glossopharyngeal and phrenic inspiratory activities of rats during hypocapnia and hypoxia, Neurosci Lett.(1992)In press.
16. D.Bieger, and D.A.Hopkins, Viscerotopic representation of the upper alimentary tract in the medulla oblongata in the rat: The nucleus ambiguus, J. Comp.Neurol. 262:546(1987).
17. H.Ellenberger, and J.L.Feldman, Monosynaptic transmission of respiratory drive to phrenic motoneurons from brainstem bulbospinal neurons in rats, J.Comp.Physiol, 269:47(1988).
18. N.S.Cherniack, N.H.Edelman, and S.Lahiri, Hypoxia and hypercapnia as respiratory stimulants and depressants, Respir.Physiol. 11:113(1970/71).
19. R.Maruyama, A.Yoshida, and Y.Fukuda, Differential sensitivity to hypoxic inhibition of respiratory processes in the anesthetized rat, Jpn.J.Physiol. 39:857(1989).
20. J.A.Neubauer, J.E.Melton, and N.H.Edelman, Modulation of respiration during hypoxia, J.Appl.Physiol. 68:441(1990).
21. W.M.St.John, D.Bartlett,Jr, K.V.Knuth, and J.C.Hwang, Brain stem genesis of automatic ventilatory patterns independent of spinal mechanisms. J.Appl. Physiol. 51:204(1981).
22. J.C.Smith, J.J.Greer, G.Liu, and J.L.Feldman, Neural mechanisms generating respiratory pattern in mammalian brainstem-spinal cord in vitro.I Spatiotemporal patterns of motor and medullary neuron activity. J.Neurophysiol. 64:1149(1990).
23. S.W.Schwarzacher, Z.Wilhelm, K.Anders, and D.W.Richter, The medullary respiratory network in the rat. J.Physiol.(Lond.) 435:631(1991).
24. H.Onimaru, A.Arata, and I.Homma, Primary respiratory rhythm generator in the medulla of brainstem-spinal cord preparation from newborn rat. Brain Res. 445:314(1988).

POSSIBLE IMPLICATION OF N. PARABRACHIALIS IN OPIOID-MEDIATED RESPIRATORY SUPPRESSION INDUCED BY THIN-FIBER MUSCULAR AFFERENTS

Takao Kumazawa, Taijiro Hirano, Eiko Tadaki, Yasuko Kozaki, and Kunihiro Eguchi

Department of Neural Regulation
Research Institute of Environmental Medicine,
Nagoya University, Nagoya 464-01 Japan

INTRODUCTION

The great majority of thin-fiber muscular afferents are of the polymodal receptor type, signaling nociceptive information.[1] Arterial injection of various algesic substances into the gastrocnemius muscle of anesthetized, spontaneously ventilated dogs, causes an increase in minute respiratory volume similar to increases observed in discharge rates of muscular polymodal receptors in response to the same stimulus.[2] Such findings suggest an involvement of the muscular polymodal receptors in the respiratory response.

When the end-tidal gas concentration is kept constant in artificially ventilated dogs and cats, stimulation of the thin-fiber afferents causes post-stimulus suppression of the respiratory outputs as shown by phrenic discharges following respiratory facilitation.[3,4] The effects of naloxone and peptidase inhibitors indicate that the post-stimulus respiratory suppression was mediated by central opioidergic mechanisms.[5,6]

This opioid-mediated post-stimulus respiratory suppression can be observed even after intercollicular decerebration, but abolished by a slightly caudal transection.[7] Micro-stimulation of the parabrachial region as well as microinjection of morphine into the region caused naloxone-reversible respiratory suppression similar to that induced by muscular afferent stimulation.[8] These results suggest an involvement of the parabrachial region in the post-stimulus respiratory suppression.

To further investigate the role of the parabrachial region in the post-stimulus respiratory suppression induced by thin-fiber muscular afferents, the effects of a microinjection of kainic acid (KA), known as an excito-cytotoxic agent, into the parabrachial region was investigated in cats.

METHODS

Experiments were performed using cats anesthetized with chloralose-urethane (15 and 75 mg/kg iv, respectively) under preliminary anesthesia with sodium pentobarbital. Subsequent doses were given to maintain the state of anesthesia. Mean arterial pressure and heart rate were monitored

Control of Breathing and Its Modeling Perspective, Edited by
Y. Honda *et al.*, Plenum Press, New York, 1992

and the rectal temperature was maintained at 38±0.5°C throughout the experiment. The animals were bilaterally vagotomized and paralyzed before stimulation was started. They were artificially ventilated with the end-tidal CO_2 and O_2 kept constant. Phrenic nerve discharges were recorded, and the product of the peak amplitude of the integrated phrenic nerve activity and the instantaneous respiratory rate averaged over a 30-sec period was used as the neural respiratory output (RO). Gastrocnemius muscle nerve afferents were electrically stimulated for 1 min at 8 Hz, with the intensity set at multiples of the threshold (T) for the muscle contraction.

Kainic acid (0.48 or 1.91 nmol in 0.1 µl pontamine sky blue solution; pH 7.4) was injected into the N. parabrachialis (NPBL) and medialis (NPBM) with a needle connected to a Hamilton syringe which was inserted stereotaxically at the following coordinates according to Berman's atlas[9]: P 3.5; L 4; H -1.5 (NPBL) and H -2.5 (NPBM). A control injection of 0.1 µl of the solution containing pontamine sky blue alone into the regions did not induce any respiratory change (n=3). In most cases, KA was injected into 4 sites in each cat: in the NPBL and NPBM of the right and then left side. Enough time elapsed between the injections so that the effects on ongoing respiratory activities and the effects of thin-fiber muscular afferent stimulation could be observed. The effect of stimulation was tested before and after KA injection, the intensity being kept the same in each case throughout the series. The injected sites marked by pontamine sky blue were histologically identified.

RESULTS AND DISCUSSION

Effects of kainic acid injection into the parabrachial region on ongoing respiratory activities. Unilateral (on the right side) microinjections of a small dose (0.48 nmol) of KA into the dorsal part of NPBL caused facilitation of respiration within 30 min after injection in all 9 cats tested. Sample results are shown in the middle tracing of Figure 1. The time course facilitation to develop varied among cases, and in some cases the facilitation was preceded by slight and transient respiratory depression. More remarkable respiratory facilitation was induced at a faster rate with the larger dose (1.91 nmol) of KA: the averaged RO significantly increased 6 min after the injection in 7 of the cases. Effective injection sites were sufficiently restricted even in cases where the larger dose were used, since no respiratory facilitation could not seen within the 20-min observation period following the injection of the larger dose of KA into sites 2-3 mm dorsal to the NPBL.

A microinjection of the small dose of KA into the NPBM initially augmented respiration both in the 9 cats which had received the injection of KA into the NPBL of the same side (described above), and in 1 other cat which had not undergone any injections of KA. The degree of respiratory augmentation varied among the cases with maximal increases in RO ranging from 7 to 70% of the RO before injection. The time course for augmentation to develop also varied among the cases. In 5 cases augmentation continued until at least 15 min after the injection when muscle nerve stimulation was tested. In all 5 of these cases, electrical stimulation of the thin-fiber muscle nerve substantially suppressed respiration (see description below). In the remaining 5 cases including 1 case in which KA was first administered into the NPBM, RO began to decrease within 15 min after the injection. In fact, 2 cases became apneic at 2 and 8 min after the injection, respectively. The period for observing ongoing respiratory changes was extended in 1 case where RO gradually decreased to about half at 70 min after the injection. It is noteworthy that KA injected into the NPBM firstly as well as that subsequently to the injection into the NPBL similarly induced respiratory depression which was preceded by facilitation. Injections with the larger dose of KA caused more remarkable respiratory depression, but a recognizable degree of initial augmentation was observed in only 1 out of 5 cases tested. Averaged RO in 5 cases became significantly depressed 9 min after the injection. Phrenic activities in all cases almost completely disappeared within 30 min after the injection.

The above results on the effects of unilateral KA injection into the NPBL and NPBM on ongoing respiration showed that: 1) KA injection into the NPBL caused dose-dependent respiratory facilitation, while injection into the NPBM had the opposite effects; 2) Prior to the final effects as described above, the smaller dose of KA induced initial transient and slight respiratory depressions in some cases when KA was injected into the NPBL while it induced initial respiratory augmentation in all cases when injected into the NPBM. KA has been reported to cause excito-cytotoxic effects by acting on excitatory amino acid receptors of neuronal perikarya.[10] KA applied to respiratory-related neurons induces a rapid transient increase in discharges followed by irreversible cessation of firing[11] and also causes neuronal degeneration histologically.[12,13] Although a short interval before histological examination in the present study did not allow to reveal degenerative changes of the sites of KA injection, it is likely that neuronal activity became inactivated by the KA-induced sustained response as was reported in hippocampal cells.[14] Respiratory facilitation caused by KA injection into the NPBL and respiratory depression caused by injections into the NPBM are considered to be resulted from inactivation of the neurons in the NPBL and NPBM, indicating reciprocal respiratory action between the NPBL and NPBM. In other words, neurons in the NPBL exhibit a tonic inhibitory action on respiration while neurons in the NPBM have a tonic facilitatory action.

This reciprocal role between the NPBL and NPBM was repeatedly confirmed by observing the results of KA injection of KA into the NPBL and NPBM on the other side. In 5 cats, the respiratory activities of which had been depressed after the injection of KA into the NPBM on the right side, the subsequent injection of KA into the NPBL of the other side augmented ROs from 110% to 400% of the values before KA injection. The phrenic discharge pattern became quite regular after the injection into the NPBL; the variance of both the peak amplitude of the integrated phrenic discharges and respiratory rate became significantly low. A regular discharge pattern was recovered even from the apneic state in 1 case. Subsequent injections of KA into the remaining NPBM in a few cases substantially depressed respiratory activities and even induced apnea in 1 extreme case.

Neither large nor small doses of KA injected into the NPBL and NPBM induced any changes in arterial blood pressure. The absence of any circulatory changes further indicates the restrictiveness of sites for KA injections in producing the respiratory effects as observed in the present study, since it has been recently reported that the neurons in the NPBL, which locate in the slightly ventral sites, are capable of inducing circulatory responses accompanied by respiratory changes.[15]

Effects of kainic acid injection into the parabrachial region on post-stimulus respiratory suppression induced by thin-fiber muscular afferents. Stimulation of the gastrocnemius muscle nerve at the suprathreshold intensity to activate C-fiber afferents usually causes respiratory facilitation during stimulation and is followed by long-lasting, endogenous opioid-mediated respiratory suppression,[3,4,5] as the sample in the top tracing of Figure 1 shows. Figure 3A shows the time course for the response to the nociceptive muscular afferent stimulation before the injection of KA into the NPBL and is expressed as a percentage of the pre-stimulus RO averaged in 8 cats. When compared with the pre-stimulus value, RO observed 5 min after the cessation of stimulation showed a significant decrease. The post-

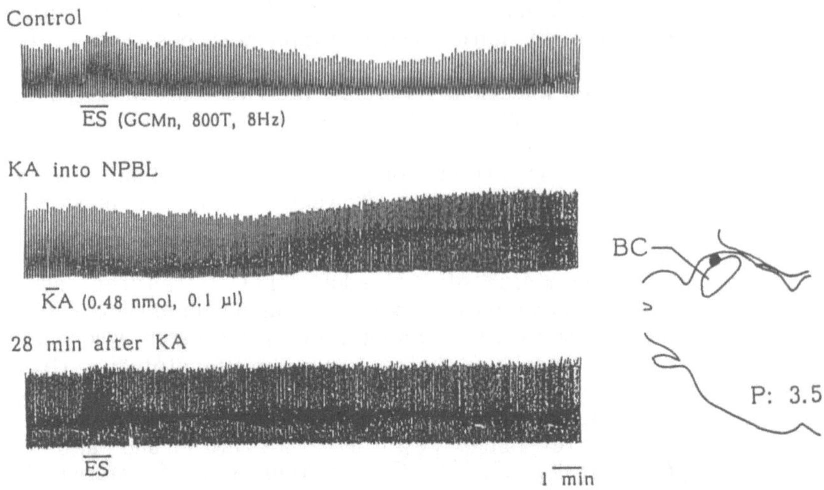

Figure 1. Effects of kainic acid injected into the dorsal part of the NPBL on integrated phrenic discharges of a cat.
KA: a small dose of kainic acid was injected into the dotted site shown at the right; ES: stimulation of the gastrocnemius muscle nerve (GCMn) at a suprathreshold intensity for C-fiber afferents; BC: brachium conjunctivum. Note that post-stimulus respiratory suppression disappeared after KA was injected into the NPBL.

stimulus respiratory suppression was recovered within 20-60 min in all cases.

As described above, the injection of the small dose of KA into the NPBL induced respiratory facilitation. Effects of muscle nerve stimulation was tested after respiratory facilitation reached a stable state in the 8 cats. As shown in the bottom tracing of Figure 1, muscle nerve stimulation did not induce any post-stimulus suppression after the injection. The averaged results for the 8 cases are shown in Figure 3B. After the injection of KA into the NPBL, the post-stimulus ROs showed no significant difference from the pre-stimulus value. The RO values between 4 and 14 min were significantly greater when compared with the corresponding values in the intact state as shown in Figure 3A. These findings indicate that post-stimulus respiratory suppression disappears after KA is injected into the NPBL.

For the case in Figure 2, post-stimulus respiratory suppression was not very clear in the intact state (1st tracing). The injection of KA into the NPBM caused a slight increase in phrenic activity, but muscle nerve stimulation induced marked post-stimulus suppression (3rd tracing) when compared with that observed in the intact state. Phrenic activities transiently recovered and then gradually decreased again to reach an apneic state 1 hr after the injection. The apneic state was partially reversed by naloxone (4th and 5th tracing). The same stimulation after naloxone did not cause any post-stimulus suppression (5th tracing). These findings indicate that KA injected into the NPBM caused potentiation of opioid-mediated post-stimulus suppression in addition to depression of ongoing respiratory activity. In 7 cases receiving KA injection into the NPBL, the averaged response to muscle nerve stimulation when KA was injected into the NPBM showed potentiation of post-stimulus respiratory suppression (Figures 3C and 3D). Marked potentiation of post-stimulus suppression developed to apnea 47 min after stimulation was observed in 1 cat in which KA was injected only into the NPBM and not the NPBL. Such a potentiation effect may result from

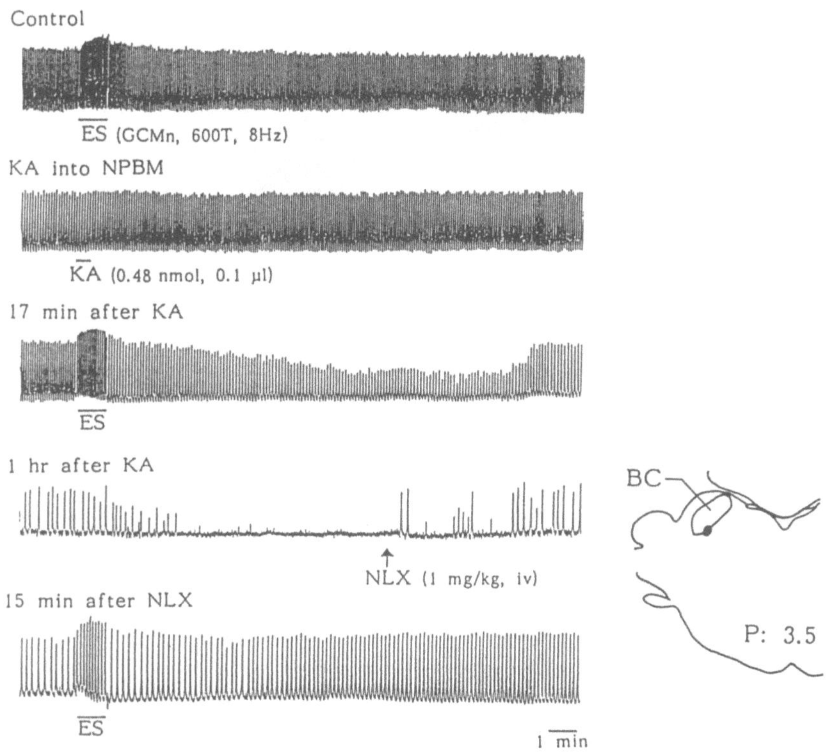

Figure 2. Effects of kainic acid injected into the NPBM.
NLX: naloxone. Note that after KA injection into the NPBM, post-stimulus respiratory suppression was induced more remarkably than in the intact state and that naloxone reversed the depressed ongoing phrenic activity seen after KA injection as well as the post-stimulus suppression.

inactivation of the neurons in the NPBM. It is presumed that elimination of the powerful, tonic respiratory facilitatory effect originating from neurons in the NPBM would potentiate the appearance of a reflexively induced inhibitory effect through the remaining NPBL in the other side.

Muscle nerve stimulation after the injection of KA into the NPBM resulted in apnea in 5 out of the 8 cases tested within 2 to 150 min after the cessation of stimulation. Together with the other 2 cases which became apneic shortly after the injection of KA into the NPBM without application of stimulation, a total of 7 out of 10 cases developed apnea after the injection of KA into the NPBM. In order to restore respiratory activity, the end-tidal CO_2 level was elevated or naloxone was applied during the apneic state. Elevation of the end-tidal CO_2 from 3.5% to 5% was effective in the tested 2 cases which became apneic after stimulation, but phrenic activity could not be recovered with elevations of up to 7% in 2 cases which became apneic shortly after the injection. Naloxone (1 mg/kg, iv) was effective in the other 2 cases which developed apnea after stimulation and in 1 of the 2 cases, which became apneic after the injection of KA alone but were resistant to recovery with elevations of CO_2. Naloxone was ineffective in the other case, however, even at a dose of 10 mg/kg.

Subsequent injections of KA into the NPBL on the other side induced regular, facilitated respiratory activity in 6 cases. In these cases, the

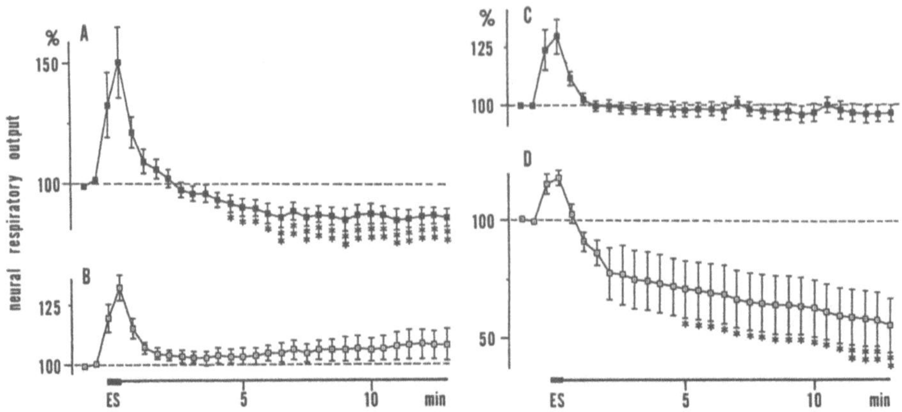

Figure 3. Time courses for the responses to thin-fiber muscular afferent stimulation before and after unilateral injections of the small dose of kainic acid into the NPBL and NPBM.

A: in the intact state, n=8; B: after kainic acid injection into the NPBL in the same cats in A; C: immediately before a subsequent injection of kainic acid into the NPBM in 7 out of the 8 cats in A; D: after kainic acid injection into the NPBM in the same cats in C. The same stimulation was tested in each cat throughout the series. Respiratory output was expressed as a percentage of the pre-stimulus value (mean $\pm$ S.E.). *, ⁑: significant differences ($p<0.05$, $p<0.01$, paired t test, respectively) when compared with the pre-stimulus value.

muscle nerve stimulation did not induced any signs of post-stimulus respiratory suppression, despite preservation of respiratory facilitation during the stimulating period as well as the facilitatory effect by elevation of CO_2 level. After subsequent injections of KA into the remaining NPBM in 5 cases, in contrast to the observation in the contralateral side, the same stimulation failed to potentiate the post-stimulus respiratory suppression except in 1 case where KA had not been injected into the contralateral NPBL. These findings further confirm implication of the NPBL in the mechanism of the post-stimulus respiratory suppression induced by nociceptive muscular afferents.

The dorsolateral region of the rostral pons including the parabrachial nucleus has been known to play an important role in regulating respiration, although such studies have focused mainly on the breath-by-breath switching effect.[16,17,18] Our study is the first to show that the NPBL and NPBM exhibit tonic and reciprocal modulatory effects on respiration.

The parabrachial regions receive projections of nociceptive inputs from the spinal cord and trigeminal nuclei[19,20,21] and opioid receptors and opioidergic neurons are densely distributed in this region.[22,23,24] Abolishment of the muscular afferent-induced respiratory suppression after an injection of KA indicates, therefore, that the neurons in the NPBL are implicated in reflexively induced respiratory suppression which is mediated through the endogenous opioidergic mechanism triggered by nociceptive muscular afferents.

Our previous study revealed that the great majority of thin-fiber afferents are of the polymodal receptor type which signal nociceptive information.[1] Stimulation of the polymodal receptor afferents induces intensity-dependent respiratory facilitation, followed by naloxone-reversible respiratory suppression, if the stimulus is strong enough.[3,4,5] Such findings suggest the existence of some type of negative feedback mechanism which works

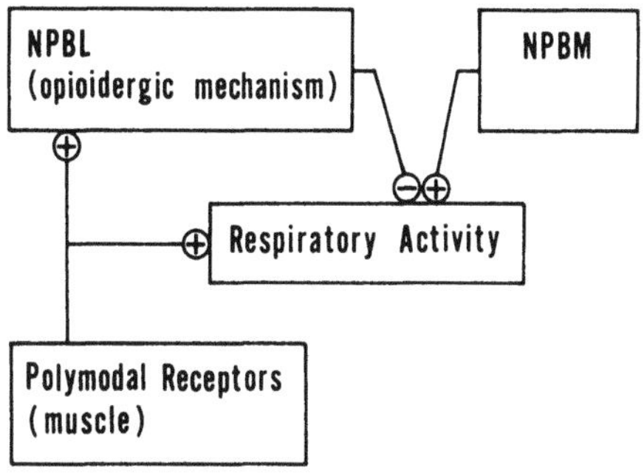

Figure 4. Schematic representation of functional roles of the NPBL and NPBM in spontaneous respiratory activities and the post-stimulus respiratory suppression induced by nociceptive muscular afferents, presumed from the present study.

against excessive excitation in respiration. Implication of the polymodal receptors in acupuncture analgesia which are mediated through the central opioidergic system has been proposed to act in a similar manner as a negative feedback mechanism in the pain system.[25] It has been reported that stimulation of the parabrachial area causes antinociception and that discharges concomitant with nocifensive withdrawal reflexes were recorded from some neurons in this area.[26,27] These findings suggest that similar negative feedback mechanisms exist in the modulation of respiration as well as that of pain. It may be quite interesting to explore the intimate relationship between the alarm-signaling system and regulation system for fundamental bodily functions.

In summary, as schematically shown in Figure 4,

1. Microinjection of KA into the dorsal part of the NPBL caused facilitation of respiration and eliminated the post-stimulus suppression, while an injection into the NPBM caused the opposite effects. These results suggest a reciprocal relationship between the NPBL and NPBM.
2. This is the first study to present evidences suggesting that the neurons in the NPBL exhibit tonic inhibition on spontaneous respiratory activities and are also implicated in the opioid-mediated post-stimulus respiratory suppression reflexively induced by nociceptive muscular afferents.

Acknowledgments

The authors thank Yoshiko Yamaguchi for production of the graphics and help in preparing this manuscript.

This work was supported in part by Grants-in-Aid for Scientific Research from the Ministry of education, Science and Culture, Japan.

REFERENCES

1. T.Kumazawa and K.Mizumura, Thin-fiber receptors responding to mechanical, chemical, and thermal stimulation in the skeletal muscle of the dog, J.Physiol.(Lond.) 273:179(1977).
2. K.Mizumura and T.Kumazawa, Reflex respiratory response induced by chemical stimulation of muscle afferents, Brain Res. 109:402(1976).
3. T.Kumazawa, E.Tadaki, K.Mizumura, and K.Kim, Post-stimulus facilitatory and inhibitory effects on respiration induced by chemical and electrical stimulation of thin-fiber muscular afferents in dogs, Neurosci.Lett. 35:283(1983).

4. T.Kumazawa and E.Tadaki, Two different inhibitory effects on respiration by thin-fiber muscular afferents in cats, Brain Res. 272:364(1983).
5. T.Kumazawa, E.Tadaki, and K.Kim, A possible participation of endogenous opiates in respiratory reflexes induced by thin-fiber muscular afferents, Brain Res. 199:244(1980).
6. Y.Kozaki, D.Simbulan, E.Tadaki, and T.Kumazawa, Effect of enkephalinase inhibitors on reflex respiratory suppression induced by thin-fiber muscular afferents, Environ.Med. 31:55(1987).
7. T.Kumazawa, K.Eguchi, and E.Tadaki, Naloxone-reversible respiratory inhibition induced by muscular thin-fiber afferents in decerebrated cats, Neurosci.Lett. 53:81(1985).
8. K.Eguchi, E.Tadaki, D.Simbulan,Jr., and T.Kumazawa, Respiratory depression caused by either morphine microinjection or repetitive electrical stimulation in the region of the nucleus parabrachialis of cats, Pflugers Arch. 409:367(1987).
9. A.L.Berman, "The Brain Stem of the Cat. A Cytoarchitectonic Atlas with Stereotaxic Coordinates," The University of Wisconsin Press(1968).
10. J.T.Coyle, M.E.Molliver, and M.J.Kuhar, In situ injection of kainic acid: a new method for selectively lesioning neuronal cell bodies while sparing axons of passage, J.Comp.Neurol. 180:301(1978).
11. M.Denavit-Saubie, D.Riche, J.Champagnat, and J.C.Velluti, Functional and morphological consequences of kainic acid microinjections into a pontine respiratory area of the cat, Neuroscience 5:1609(1980).
12. M.-P.Morin-Surun, J.Champagnat, E.Boudinot, and M.Denavit-Saubie, Differentiation of two respiratory areas in the cat medulla using kainic acid, Respir.Physiol. 58:323(1984).
13. A.J.Berger and K.A.Cooney, Ventilatory effects of kainic acid injection of the ventrolateral solitary nucleus, J.Appl.Physiol. 52:131(1982).
14. N.I.Kiskin, O.A.Krishtal, and A.Y.Tsyndrenko, Excitatory amino acid receptors in hippocampal neurons: kainate fails to desensitize them, Neurosci.Lett. 63:225(1986).
15. M.Miura and K.Takayama, Circulatory and respiratory responses to glutamate stimulation of the lateral parabrachial nucleus of the cat, J.Auton.Nerv.Syst. 32:121(1991).
16. M.I.Cohen, Switching of the respiratory phases and evoked phrenic responses produced by rostral pontine electrical stimulation, J.Physiol.(Lond.) 217:133(1971).
17. F.Bertrand and A.Hugelin, Respiratory synchronizing function of nucleus parabrachialis medialis: pneumotaxic mechanisms, J.Neurophysiol. 34:189(1971).
18. C.von Euler, I.Marttila, J.E.Remmers, and T.Trippenbach, Effects of lesions in the parabrachial nucleus on the mechanisms for central and reflex termination of inspiration in the cat, Acta Physiol.Scand. 96:324(1976).
19. J.K.Hylden, H.Hayashi, G.J.Bennett, and R.Dubner, Spinal lamina I neurons projecting to the parabrachial area of the cat midbrain, Brain Res. 336:195(1985).
20. D.G.Standaert, S.J.Watson, R.A.Houghten, and C.B.Saper, Opioid peptide immunoreactivity in spinal and trigeminal dorsal horn neurons projecting to the parabrachial nucleus in the rat, J.Neurosci. 6:1220(1986).
21. H.Hayashi and T.Tabata, Distribution of trigeminal sensory nucleus neurons projecting to the mesencephalic parabrachial area of the cat, Neurosci.Lett. 122:75(1991).
22. S.F.Atweh and M.J.Kuhar, Autoradiographic localization of opiate receptors in rat brain. II. The brain stem, Brain Res. 129:1(1977).
23. N.Sales, D.Riche, B.P.Roques, and M.Denavit-Saubie, Localization of mu- and delta-opioid receptors in cat respiratory areas: an autoradiographic study, Brain Res. 344:382(1985).

24. T.Ibuki, H.Okamura, M.Miyazaki, N.Yanaihara, E.A.Zimmerman, and Y.Ibata, Comparative distribution of three opioid systems in the lower brainstem of the monkey (Macaca fuscata), J.Comp.Neurol. 279:445(1989).
25. T.Kumazawa, Nociceptors and autonomic nervous control, Asian Med.J. 24:632(1981).
26. J.O.Dostrovsky, J.W.Hu, B.J.Sessle, and R.Sumino, Stimulation sites in periaqueductal gray nucleus raphe magnus and adjacent regions effective in suppressing oral-facial reflexes, Brain Res. 252:287(1982).
27. C.M.Haws, A.M.Williamson, and H.L.Fields, Putative nociceptive modulatory neurons in the dorsolateral pontomesencephalic reticular formation, Brain Res. 483:272(1989).

SYNAPTIC AND NON-SYNAPTIC CONTROL OF MEMBRANE POTENTIAL FLUCTUATIONS IN BULBAR RESPIRATORY NEURONS OF CATS

Ryuji Takeda and Akira Haji

Department of Pharmacology
Faculty of Medicine
Toyama Medical and Pharmaceutical University
2630 Sugitani
Toyama 930-01, Japan

INTRODUCTION

Bulbar respiratory neurons display rhythmic fluctuations of membrane potential in synchrony with the respiratory cycle. These undulatory potentials are largely dependent upon the periodically arriving synaptic activities[1]. Intrinsic membrane properties also play a certain important role in the generation or modulation of these potentials. Several types of ionic conductance have been demonstrated in the presumed respiration-related neurons in the brainstem slices[2,3]. However, since these neurons in tissue slices usually lack the spontaneous rhythmic modulation in membrane potential, the identification of the neuron type based upon the spontaneous patterns of firing and membrane potential fluctuations is yet uncertain. Moreover, it is hard to know at what timing any specific ionic conductance becomes active in the respiratory cycle change in membrane potential observed in different types of the respiratory neuron. Nonetheless, an intact brainstem preparation has also an inherent drawback to exclude the contamination of ionic currents mediated by action potentials and postsynaptic potentials[4,5]. The coaxial multi-barrelled microelectrode technique[6] can partly overcome these difficulties as it allows an in vivo intracellular recording of membrane potential in conjunction with an extracellular iontophoresis of drugs which could block action potentials and synaptic waves in the recorded neuron. The present study was aimed at elucidating the possible synaptic and non-synaptic mechanisms by which the periodic fluctuations of membrane potential are shaped in bulbar respiratory neurons of the ventral respiratory group.

METHODS

Experiments were carried out on decerebrate, paralyzed and artificially ventilated cats after vagotomy. The end-tidal concentration of CO_2 was kept at 4-5%. Arterial blood pressure was maintained at more than 80 mmHg, and rectal temperature at 37-39 °C. Efferent discharge of phrenic nerve was recorded to monitor the central respiratory rhythm. Two types of respiratory neurons, augmenting inspiratory (I) and decrementing expiratory or post-inspiratory (PI) neurons[1], were impaled in the ventral respiratory group (VRG) 0-2 mm rostral to the obex. The center recording pipette of a coaxial multi-barrelled microelectrode was filled with 2M potassium citrate and had a resistance of 20-30 MΩ. Tetrodotoxin (TTX; 0.5 mM in 165 mM NaCl solution at pH 6.5) and cadmium chloride (Cd^{2+}; 5 mM in 165 mM NaCl solution at pH 6) were contained in the peripheral drug pipettes, with their tips recessed 20-40 μm from the tip of the center recording pipette. Drugs were iontophoresed with positive currents, and a retaining current of -5 nA was applied between test periods. The pipette filled with 165 mM NaCl solution served both as a current sink and a drug control. Laryngeal motor (LM) and non-antidromically-activated (NAA) neurons were identified by antidromic stimulation of the vagus nerve and C2-C3 spinal cord[1]. No bulbospinal neuron was encountered. For current clamping and measurements of input resistance, intracellular currents were injected through the recording electrode using a high-frequency current injection and voltage sampling method.

RESULTS

Iontophoresis of tetrodotoxin

During the first 10-30 sec period of iontophoresis with a current strength of 25-50 nA, TTX completely blocked the firing of action potentials in all I (n=15) and PI neurons (n=20) examined. It eliminated excitatory postsynaptic potentials (EPSPs) and inhibitory postsynaptic potentials (IPSPs) evoked by stimulation of vagus nerve (Fig. 1A). An increase of input resistance was observed in a steady-state current-voltage relation which was taken after a 2 min iontophoresis of TTX (Fig. 1B,C). In addition, TTX produced a brief depolarization lasting for 30-40 sec and a late, sustained hyperpolarization in all respiratory neurons tested. During a 1-2 min iontophoresis of TTX, the respiratory oscillation of membrane potential progressively decreased and the membrane potential profile became relatively flat in a whole respiratory cycle. This was accompanied by an obvious decline of synaptic noise. These changes in membrane potential occurred similarly in NAA (n=20) and LM neurons (n=15). More prolonged iontophoresis of TTX (up to 5 min) produced no further change in the membrane potential trajectory. Figure 2 illustrates typical examples of the steady-state change in membrane potential of I and PI neurons. In the I neuron, TTX decreased the inspiratory ramp-depolarization and the rapid hyperpolarization

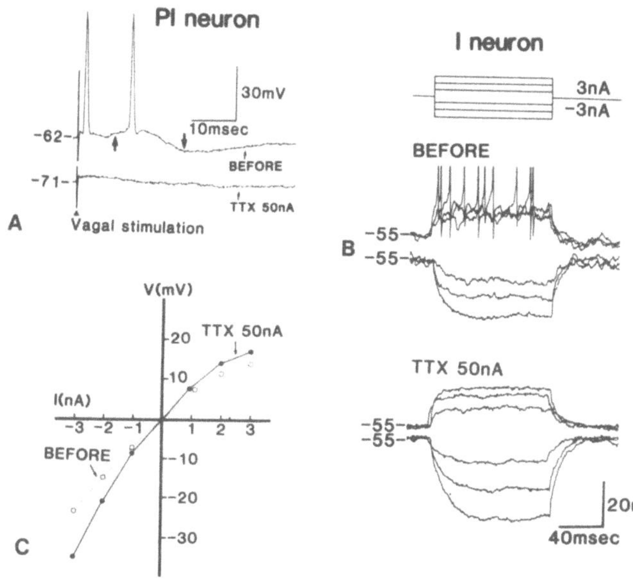

Figure 1. **A**: antidromic action potential and succeeding excitatory (upward arrow) and inhibitory postsynaptic potentials (downward arrow) of a PI-LM neuron evoked by stimulation of the ipsilateral vagus nerve with pulses of suprathreshold intensity (1.5 V, 0.2 msec). These were taken before (upper trace) and after (lower trace) iontophoresis of tetrodotoxin (TTX) with a 50 nA pulse for 1 min. Recordings were made 200 msec after the onset of inspiration (during the non-spiking phase). **B**: superimposed traces of membrane potential responses of an I-NAA neuron provoked by intracellular currents of various intensities (top traces). Sets of recordings were obtained before (upper) and after (bottom) iontophoresis of TTX (50 nA, 2 min). These were taken at 500 msec after the onset of stage 1 expiration and the membrane potential was current-clamped at the control potential (-55 mV) after TTX. **C**: current(I)-voltage(V) relation of the neuron shown in B obtained before (open circles) and after TTX (solid circles). Zero voltage axis corresponds to -55 mV.

during stage 1 expiration or post-inspiration[1] (Fig. 2A). In the PI neuron, the post-inspiratory depolarization and the inspiratory IPSP wave were severely depressed by TTX. Concomitantly, both types of neurons displayed a distinct hyperpolarization during a whole respiratory cycle (Fig. 2A,B). In the majority of cases, TTX did not completely eliminated the respiratory oscillations of membrane potential and left a small depolarization to occur during the most depolarized phase: late in inspiration for the I neuron (n=12) and early in post-inspiration for the PI neuron (n=17) (Fig. 2A,B). Complete abolition of the membrane potential fluctuations was observed after TTX in 2 I and 3 PI neurons.

Current-clamp experiments

Figure 3 shows the response of the TTX-resistant, residual depolarization to constant currents of various intensities (ranging from -6 nA to 5 nA). For the I neuron, the small inspiratory depolarization was decreased when the membrane was steadily hyperpolarized and increased when depolarized (Fig. 3A). For the PI neuron, hyperpolarization resulted in a decrease in the amplitude and the duration of the residual, post-inspiratory depolarization and depolarization caused a voltage-dependent increase of that wave (Fig. 3B). When depolarization exceeded a value of approximately -65 mV (ranging between -68 mV and -55 mV), a large depolarizing wave developed in an abrupt, non-linear fashion during stage 1 of expiration (Fig. 3B). This potential reached a rounded peak 200-250

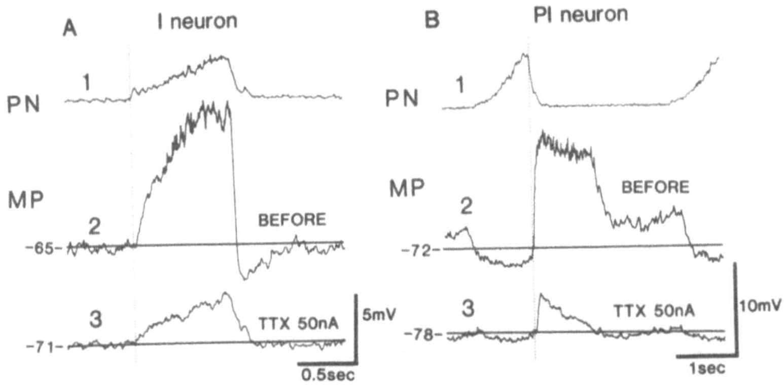

Figure 2. Alterations of membrane potential fluctuations after iontophoresis of tetrodotoxin (TTX) in an I-NAA (A) and a PI-NAA neuron (B). PN; integrated phrenic neurogram (trace 1), MP; membrane potential recordings, which were passed through a low-pass filter (bandwidth 0-100 Hz) to show a clear profile of the MP trajectory. Recordings were made before (trace 2) and after iontophoresis of TTX (trace 3) with a current strength of 50 nA for 2 min. Horizontal reference lines represent the membrane potential levels at which the inspiratory ramp-depolarization started in the I neuron and the post-inspiratory depolarization commenced in the PI neuron. Vertical dotted lines in A and B represent the onset of inspiration and of expiration, respectively.

msec after its abrupt onset, repolarized slowly in ensuing 500-700 msec and then declined rather rapidly in 500-600 msec to reach the resting potential (Fig. 3B, arrow). This depolarizing wave had a characteristic bell-shaped profile distinctive from that of the residual post-inspiratory wave observed at more hyperpolarized states after TTX application. Input resistance was lower during the bell-shaped depolarization than during the residual depolarization. In the particular cell shown in Fig. 3B, a small hyperpolarizing wave was still visible during inspiration, of which the polarity was reversed at a potential of more negative than -95 mV. This suggests the persistence of IPSPs during that phase of the respiratory cycle. However, similar bell-shaped waves were evoked in many PI cells where the inspiratory IPSPs were totally suppressed by TTX. For the I neuron, this kind of depolarization appeared to be less convincing. Excessive depolarization increased the residual depolarization, notably its rising slope (Fig. 3A, arrow).

Iontophoresis of cadmium

Iontophoresis of Cd^{2+} (25-50 nA, 30-60 sec) quickly halted the spontaneous spiking and progressively decreased the respiratory fluctuations in membrane potential of all I and PI neurons. It also caused a transient depolarization and a subsequent hyperpolarization during a whole respiratory cycle. These effects resembled those

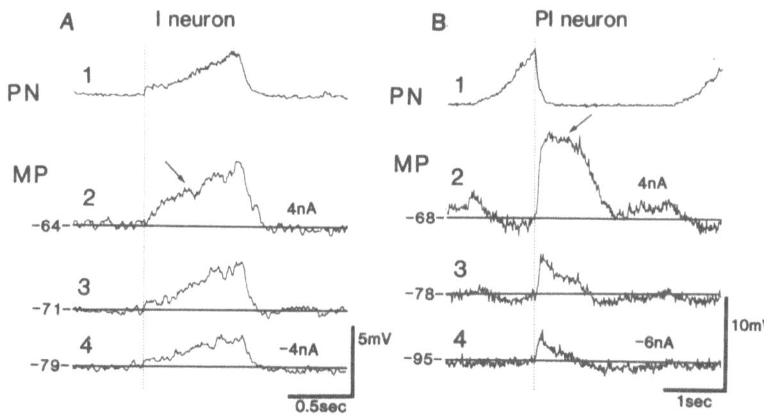

Figure 3. Responses in membrane potential of an I-NAA (A) and a PI-NAA neuron (B) to intracellular steady currents of various intensities after iontophoresis of TTX (50 nA, 2 min). PN; integrated phrenic neurogram (trace 1), MP; membrane potentials presented after passing through a low-pass filter (0-100 Hz). Recordings of MP for both neurons were made without current injection (trace 3), during a depolarizing current (trace 2) and during a hyperpolarizing current (trace 4). Horizontal solid lines and vertical dotted lines represent the reference MP and the delineation of the respiratory phase transition, respectively, as shown in Fig. 2.

induced by TTX. However, Cd^{2+} did not block action potentials evoked by antidromic stimulation or intracellular current injection even when the spontaneous firing had been suppressed. After a long (2 min) iontophoresis of TTX, Cd^{2+} produced no additional depressant effect on the depolarization which resided during inspiration for the I neuron and during post-inspiration for the PI neuron. However, in the presence of Cd^{2+}, the increase of the residual depolarization did not occur when the membrane was steadily depolarized. This was observed both in I (n=2) and PI neurons (n=4). More specifically, iontophoresed Cd^{2+} abolished the bell-shaped depolarization that had been evoked in the PI neuron by intensive depolarization.

DISCUSSION

The major finding derived from the present study is that the periodic membrane potential fluctuations in I and PI neurons consist of at least three components; (1) TTX- and Cd^{2+}-sensitive segments, (2) TTX-insensitive, Cd^{2+}-sensitive segments, and (3) the remaining TTX- and Cd^{2+}-insensitive segments. It is presumed that locally applied TTX, a specific blocker of the fast sodium channel, is capable of blocking synaptic inputs onto the impaled neuron by eliminating action potentials in pre- and postsynaptic elements. This can be supported by the following findings yielded by iontophoresis of TTX; namely, suppression of spontaneous and antidromically stimulated action potentials, abolition of the vagus nerve stimulated EPSPs and IPSPs, reduction of synaptic noise and an increase in input resistance. Thus, the segments of membrane potential fluctuations which were eliminated by iontophoresed TTX are assumed to be synaptically mediated. For the I neuron, they include a major part of the ramp-depolarization during inspiration and the rapid, large hyperpolarization during post-inspiration. For the PI neuron, they are the inspiratory IPSP wave and a dominant part of the post-inspiratory depolarization. In addition, TTX produced an initial depolarization and a late hyperpolarization prevailing in a whole respiratory cycle, suggesting that the VRG neurons receive not only phasic synaptic inputs but also tonic inhibitory and tonic excitatory synaptic drives[7,8]. Some cells totally lost the cyclic changes in membrane potential after TTX, implying that their membrane potential trajectory is formed exclusively by synaptic drives[8]. In addition, iontophoresis of Cd^{2+} without prior application of TTX had a similar effect as that of TTX described above. This can be explained by assuming that Cd^{2+} blocked the release of chemical mediators by preventing Ca^{2+} from entering into presynaptic nerve terminals and, consequently, eliminated synaptic inputs onto the recorded neuron.

Most I and PI neurons displayed residual depolarization during the most depolarized part of membrane potential fluctuations after local application of TTX. This potential was increased when the membrane was steadily depolarized. Moreover, for the PI neuron, a large, bell-shaped depolarization emerged when the membrane was depolarized to a potential exceeding approximately -65 mV. If the increased-conductance EPSP were involved in this depolarizing wave, depolarization would decrease its amplitude. However, this was not observed. Furthermore, the decreased-conductance EPSP presumably does not contribute to this depolarizing wave, because the neuron's input resistance remained low at that phase of the respiratory cycle. Therefore, it seems more likely that the TTX-resistant depolarizing wave is ascribed to an intrinsic membrane conductance. Since these waves were manifested or augmented by depolarization and suppressed by iontophoresed Cd^{2+}, they are attributable to the Ca^{2+}-conductance of the postsynaptic membrane[9]. It is known that two types of Ca^{2+}-conductance, the high- and low-threshold Ca^{2+} currents, can be postsynaptically induced in various central neurons[2,3,10]. The former is activated by intensive depolarization and decays at a relatively slow rate, contributing to repetitive firing. The latter remains inactivated at normal resting membrane potential and becomes apparent at depolarization when the inactivation is removed by prior hyperpolarization[10]. It is presumed that the residual depolarization and the bell-shaped depolarization are, respectively, due to the low- and the high-threshold Ca^{2+} conductance. However, electroresponsive features of the residual depolarization somewhat differ from those due to known types of Ca^{2+} conductance[2,3,10]; i.e., differences in the threshold potential for activation and the time course to decay.

Nevertheless, the possibility that synaptic mechanisms underlie the TTX-resistant, Cd^{2+}-sensitive depolarization cannot be excluded. In particular, because of its voltage-dependent appearance, the bell-shaped depolarization might be synaptically mediated through activation of the NMDA (N-methyl-D-aspartate) glutamate receptor subtype[11]. We observed that iontophoresis of MK-801, a potent antagonist for the NMDA receptor, lessened preferentially the post-inspiratory depolarization in the PI neurons (unpublished observation). Therefore, it is presumed that the active phase depolarization, especially its most depolarized segment, is partly due to synaptic excitation mediated by NMDA receptors. However, the latter possibility is still equivocal since Cd^{2+} is shown to be relatively ineffective to block the NMDA receptor-mediated excitation[11].

Finally, the remaining TTX- and Cd^{2+}-insensitive portions of depolarization are assumed to reflect TTX-resistant slow potentials or Ca^{2+} spikes at presynaptic axons or non-chemical synapses[6]. Iontophoresed Cd^{2+} may distribute less widely than TTX and hence the residual depolarization may be less vulnerable to Cd^{2+}. Alternatively, the sufficient concentration was not achieved for both TTX and Cd^{2+} at presynaptic nerve terminals or dendritic trees. Thus, it is possible that the residual depolarization could be due to synaptic potentials or dendritic spikes[10] generated remotely from the somatic penetration site.

The present study documents that synaptic activity and intrinsic membrane conductance cooperate to form the periodic membrane potential fluctuations in bulbar respiratory neurons in vivo. Both components can be identified based upon their pharmacological and electrophysiological properties characteristic of each type of the respiratory neuron. These two mechanisms, in conjunction with several other ionic currents reportedly found in the respiration-related neurons[2-5], may be involved in organizing the neuronal functions such as trigger, delay and off-switch of the central respiratory activity.

REFERENCES

1. D.W. Richter, Generation and maintenance of the respiratory rhythm, J. Exp. Biol., 100:93(1982).
2. J. Champagnat, T. Jacquin and D.W. Richter, Voltage-dependent currents in neurones of the nuclei of the solitary tract of rat brainstem slices, Pflugers Arch., 406:372(1986).

3. M.S. Dekin and P.A. Getting, In vitro characterization of neurons in the ventral part of the nucleus tractus solitarius. II. Ionic mechanisms responsible for repetitive firing activity, J. Neurophysiol., 58:215(1987).
4. D.W. Richter, D. Ballantyne and S. Mifflin, Interaction between postsynaptic activities and membrane properties in medullary respiratory neurones, in: "Neurogenesis of Central Respiratory Rhythm," A.L. Bianchi and M. Denavit-Saubie, ed., MTP, Lancaster, p172(1985).
5. D.W. Richter, J. Champagnat and S. Mifflin, Membrane properties involved in respiratory rhythm generation, in: "Neurobiology of the Control of Breathing," C. von Euler and H. Lagercrantz, ed., Raven, New York, p141(1987).
6. A. Haji, J.E. Remmmers, C. Connelley and R. Takeda, Effects of glycine and GABA on bulbar respiratory neurons in cats, J. Neurophysiol., 63:955(1990).
7. R. Takeda and A. Haji, Synaptic response of bulbar respiratory neurons to hypercapnic stimulation in peripherally chemodenervated cats, Brain Research, 561:307(1991).
8. R. Takeda, A. Haji, J.E. Remmers and T. Hukuhara, Respiratory pattern generation in the ventral respiratory group neurons, in: "Control of Breathing and Dyspnea," T. Takishima and N.S. Cherniack, ed., Pergamon, Oxford, p65(1991).
9. S. Hagiwara and L. Byerly, Calcium channel, Ann. Rev. Neurosci., 4:69(1981).
10. R. Llinas and Y. Yarom, Properties and distribution of ionic conductances generating electroresponsiveness of mammalian inferior olivary neurones in vitro, J. Physiol. (Lond.), 315:569(1981).
11. P.M. Headley and S. Grillner, Excitatory amino acids and synaptic transmission: the evidence for a physiological function, TIPS Special Report 1991, Elsevier, Cambridge, p30(1991).

NONLINEAR DYNAMICS OF A MODEL OF THE CENTRAL RESPIRATORY PATTERN GENERATOR

A. Gottschalk[1,2], K.A. Geitz[2], D.W. Richter[3], M.D. Ogilvie[2], and A.I. Pack[2]

[1]Department of Anesthesia
[2]Center for Sleep and Respiratory Neurobiology
Hospital of the University of Pennsylvania
Philadelphia, PA 19104
[3]II Department of Physiology
University of Goettingen, Humboldtallee 23, 34 Goettingen, FRG

INTRODUCTION

Recently, several models have computationally explored the network hypothesis of central respiratory rhythm generation.[1,2] One characteristic common to these models is the perfectly periodic nature of their outputs. This is not consistent with the impression that the respiratory rhythm is considerably more variable. This impression was recently quantified,[3] and the data supports the notion that lightly anesthetized vagotomized animals exhibit rhythmic behavior consistent with perfectly periodic limit cycle oscillations. However, animals with an intact vagus consistently displayed more irregular respiratory patterns. The dynamical features of these patterns were quantitated by computing the correlation dimension of the corresponding time series.[4] The vagotomized animals produced time series whose correlation dimension was equal to one, whereas the presence of an intact vagus, rather than stabilizing the rhythm, produced time series with non-integer correlation dimensions significantly greater than unity. The presence of a non-integer correlation dimension is consistent with a process exhibiting chaotic dynamics.[4] Thus, the breath by breath variability in the ventilatory pattern may be explainable as a fundamental component of the process generating the respiratory rhythm, and not the product of intrinsic or extrinsic noise, or overwhelming system dimensionality. We hypothesized that our model of the central respiratory pattern generator, when appropriately modified to include vagal feedback from the pulmonary stretch receptors, would exhibit chaotic dynamics.

The respiratory network of Fig. 1 represents the interactions of five types of neurons thought to be involved with respiratory rhythmogenesis,[1,5] and is the starting point of our study. As described,[1,6] each network element sums the inhibitory and

Control of Breathing and Its Modeling Perspective, Edited by
Y. Honda *et al.*, Plenum Press, New York, 1992

excitatory inputs it receives from the other elements in addition to a tonic excitatory input, and the dissipation of this sum, which is analogous to membrane potential, is regulated by a time constant. When the threshold of the neural element is reached, the firing rate of the neural element is its output and this is proportional to the membrane potential. Additionally, the firing rate is modulated by an adaptation time constant. In the absence of additional inputs or extreme modification of its parameters, this model replicates the three-phases of the central respiratory pattern generator in a perfectly periodic fashion.

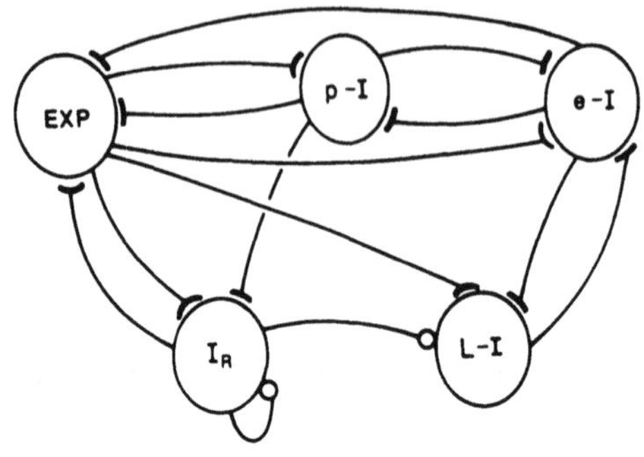

o - Inhibitory — - Excitatory

e-I: Early-Inspiratory
I_R: Ramp-Inspiratory
L-I: Late-Inspiratory
p-I: Post-Inspiratory
Exp: Expiratory

Fig. 1. Network model of the central respiratory pattern generator.

METHODS

Our approach is to first modify the model of Fig. 1 to include vagal feedback corresponding to the pulmonary stretch receptors, and then to determine the alteration in the dynamic behavior of the model as vagal feedback is varied. There is evidence of oligosynaptic vagal interactions with of all of the neurons in the model except for the ramp-inspiratory neurons.[7,8,9] Here, we will only consider excitation of the post-inspiratory neurons,[7] and inhibition of the early-inspiratory neurons.[8] Since the time course of the firing rate of the ramp-inspiratory neuron is very similar to that of the phrenic nerve, we can conceptualize the addition of vagal afferent activity to the model as the addition of feedback pathways from the ramp-inspiratory neuron to the post-inspiratory and early-inspiratory neurons. Along these feedback pathways is the lung, which, following others,[10] we model as a leaky integrator of ramp-inspiratory output. If the level of ventilatory drive is modulated by increasing the tonic drive to each of the network elements, then we can examine ventilatory timing as drive is increased in a manner analogous to the study of Clark and von Euler.[11] The ventilatory timing can then be determined from the outputs of the lung model.

To examine the dynamics of the model once feedback from the pulmonary stretch receptors has been included, we have employed a combination of techniques. These include estimation of the first Lyapunov exponent,[4] bifurcation diagrams,[12] first-return maps,[12] and estimation of the correlation dimension.[4] One difficulty which is encountered in this analysis is that the absolute level of vagal feedback, as well as the ratio of post-inspiratory to early-inspiratory feedback, is unknown. Consequently, we have found it necessary to examine the dynamics of all possible configurations.

RESULTS

For a large range of combinations of feedback to the post-inspiratory and early-inspiratory neurons, we were able to demonstrate that as ventilatory drive was modulated, we could replicate the hyperbolic tidal volume verses inspiratory period curves described by Clark and von Euler[11]. This motivated us to further evaluate the dynamics of the model using the dynamically oriented techniques listed above.

Fig. 2 summarizes the dynamics of our model when vagal feedback is present.

Here, the first Lyapunov exponent[4] is displayed as a function of feedback to the post-inspiratory and early-inspiratory neurons. Values of zero or below are shown in black. The presence of a positive first Lyapunov exponent demonstrates the exponential divergence of adjacent trajectories which is characteristic of chaotic processes.[4,12] Sporadic small islands of chaotic activity are present throughout Fig. 2 with the exception of the "tail" of activity in the upper portion of the figure.

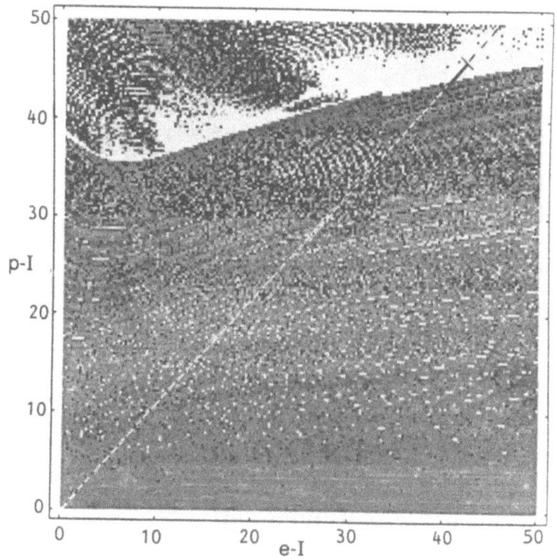

Fig. 2. Positive first Lyapunov exponent as a function of feedback to p-I (vertical axis) and e-I (horizontal axis) neural elements.

To better relate the data of Fig. 2 to the actual behavior of the model, bifurcation diagrams like those in the upper half of Fig. 3 were constructed. Here, we have depicted model behavior as a function of vagal feedback for a given ratio of post-inspiratory to early-inspiratory feedback. This corresponds to the behavior of the model along a straight line passing through the origin of Fig. 2. For each level of feedback in Fig. 3, we have successively plotted the period T_n of sequential breaths. Thus, for a given level of vagal feedback, perfectly periodic behavior produces only a single point (to left of fig.), whereas a chaotic process results in points dispersed over a range of values (to right of fig.). For clarity, we have plotted the corresponding first Lyapunov exponents in the lower half of Fig. 3.

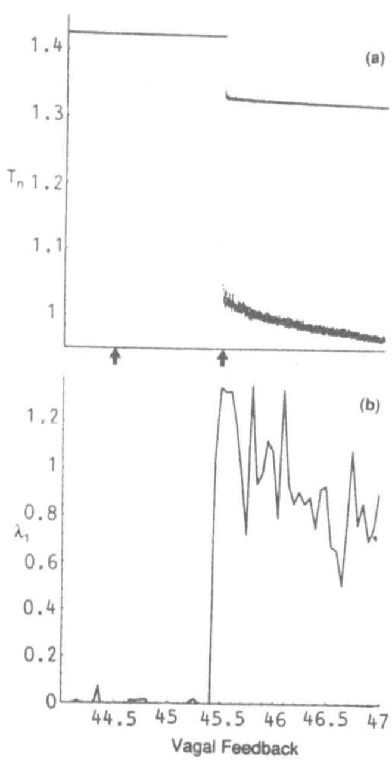

Fig. 3. (a)Bifurcation diagram and (b) first Lyapunov exponents as a function of vagal feedback along the solid portion of the line in Fig. 2.

To examine the dynamics of the model in even greater detail, the first-return maps[12] of Fig. 4 were constructed for the specific levels of vagal feedback marked by the arrows in Fig. 3. These are displayed along with their first Lyapunov exponent (λ_1) and correlation dimension (D_2). The first return map plots the current value of a system attribute (T_n) as a function of a previous value (T_{n-1}), and is used to reveal the organization of the chaotic attractor. In the case of periodic behavior (Fig. 4a), a single point emerges. More elaborate forms, which exhibit significant organization, are produced when chaotic behavior is present (Fig. 4b).

DISCUSSION

Our computational studies demonstrate that the dynamical features of respiratory pattern generation in our model of the central respiratory pattern generator parallels the physiology in its dependence on the

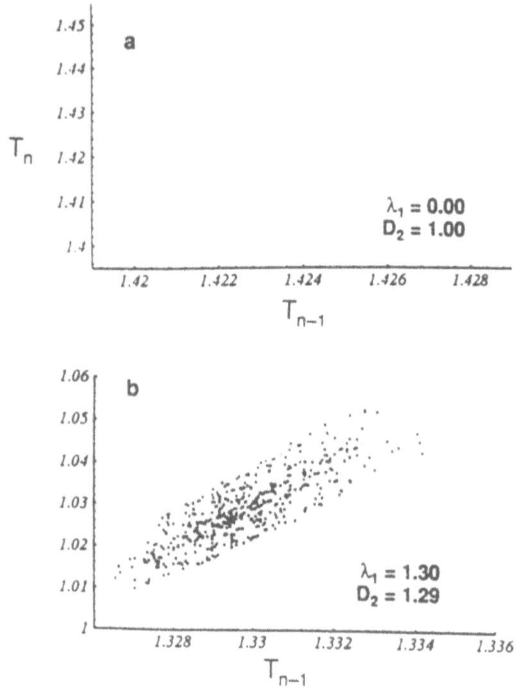

Fig. 4. First return maps for the points of Fig. 3 which are marked with arrows. For clarity, only a portion of the map is shown in Fig. 4b.

presence of feedback from the pulmonary stretch receptors for the emergence of its important dynamical features. Our correlation dimension estimates of about 1.3 are substantially below those observed physiologically,[3] suggesting that additional features need to be incorporated in the model. For example, an important feedback loop which is currently lacking is the ability of the model to adjust ventilatory drive in response to changes in alveolar ventilation, and hence blood gas tensions, nor have we considered all possible vagal feedback pathways. A more extensive comparison between the dynamical features of the physiological system and the model would require, in addition to examining first Lyapunov exponents and correlation dimensions, data demonstrating the same route to chaotic behavior as a system parameter, such as vagal feedback, is varied. Physiologic data of this type are presently unavailable, nor has the route to chaotic behavior in our model been fully characterized. We speculate that identifying the specific route to chaos from the many possibilities[13] will be an important indication of the state of the central respiratory pattern generator.

REFERENCES

1. M.D. Ogilvie, A. Gottschalk, K. Anders, D.W. Richter and A.I. Pack, A network model of respiratory rhythmogenesis. *Am. J. Physiol.* (In Review)
2. S.M. Botros and E.N. Bruce, Neural network implementation of the three-phase model of respiratory rhythm generation. *Biol. Cybern.* 63:143-153 (1990).
3. M. Sammon and E.N. Bruce, Pulmonary vagal afferent activity increases dynamical dimension of respiration in rats. *J. Appl. Physiol.* 70:1748-62 (1991).
4. G. Mayer-Kress (ed), "Dimensions and Entropies in Chaotic Systems," Springer Verlag, New York (1989).
5. D.W. Richter, D. Ballantyne and J.E. Remmers, How is the respiratory rhythm generated? A model. *NIPS* 1:109-112 (1986).
6. K. Matsuoka, Sustained oscillations generated by mutually inhibiting neurons with adaptation. *Biol. Cybern.* 52:367-376 (1985).
7. J.E. Remmers, D.W. Richter and D. Ballantyne, Reflex prolongation of state I of expiration. *Pflugers Arch.* 407:190-198 (1986).

8. C. von Euler, Brainstem mechanisms for generation and control of breathing pattern, *in:* "Handbook of Physiology. The Respiratory System II," Am. Physiol. Soc., Washington, (1986).
9. J.L. Feldman, Neurophysiology of breathing in mammals, *in:* "Handbook of Physiology. The Nervous System IV," Am. Physiol. Soc., Washington (1986).
10. S. Geman and M. Miller, Computer simulation of brainstem respiratory activity. *J. Appl. Physiol.* 41:931-38 (1976).
11. F.J. Clark and C. von Euler, On the regulation of depth and rate of breathing. *J. Physiol. (Lond.)* 222:267-95 (1972).
12. T.S. Parker and L.O. Chua, "Practical Numerical Algorithms for Chaotic Systems," Springer-Verlag, New York (1989).
13. P. Berge, Y. Pomeau and C. Vidal, "Order Within Chaos - Towards a Deterministic Approach to Turbulence," John Wiley & Sons, New York (1984).

PHASE-DEPENDENT TRANSIENT RESPONSES OF RESPIRATORY MOTOR ACTIVITIES FOLLOWING PERTURBATION OF THE CYCLE

Yoshitaka Oku, J.R. Romaniuk, and Thomas E. Dick

Departments of Medicine, Physiology, and Biomedical Engineering
Case Western Reserve University
Cleveland, OH 44106 USA

Dynamics of an oscillator can be characterized by its behavior following a brief perturbing stimulus. The transient behavior of the oscillator returning to its steady state provides insights regarding the intrinsic properties of the oscillator. In Figure 1, the phrenic nerve activity represents the respiratory oscillation. The timing of the oscillator may be altered following the perturbing stimulus relative to that predicted from unperturbed breaths. This shift in timing is referred to as phase-resetting. We define old phase as the time from the onset of phrenic activity to the onset of the stimulus, and phase shift as the amount of resetting. The relationship between old phase and phase shift characterizes the respiratory oscillator, and called as a phase response curve. Description of phase-resetting characteristics have focused on only one phase of the respiratory oscillator, i.e., the onset or the offset of phrenic nerve activity (e.g. Paydarfar *et al.*[2]). We monitored expiratory as well as inspiratory activities to examine the transient behavior of this oscillator in every phase.

The objective of the present study was to test whether a three phase model of the respiratory oscillator can predict respiratory activities and phase-resetting characteristics following a brief perturbation using an electrical stimulation of the superior laryngeal nerve. We constructed a neuronal network model based upon recent neurophysiological findings. Computer simulation produced a stable limit cycle oscillation and predicted the experimentally-derived response characteristics. The results provide a conceptual basis for understanding control of three phase motor outputs following perturbation.

We recorded phrenic, thyroarytenoid, and triangularis sterni motor neuronal activities in decerebrate, vagotomized and paralyzed cats. Like St. John's group[4], we define Stage I expiration as the phase with thyroarytenoid activity and Stage II expiration as the phase with triangularis sterni activity. These motor activities may reflect brainstem neuronal activities underlying a central pattern generator. We used an electrical stimulation of superior laryngeal nerve (100 Hz, 0.2 ms pulse duration, 50 ms train duration) as perturbation. Stimuli were delivered at every 2% of the respiratory cycle interposing 9 unperturbed breaths between stimuli. Our analyses focused on the durations of motor activities. Each duration was normalized by its control value. (The pertinence of the normalization procedure was questioned by Dr. S.M. Yamashiro because neural events occurred on an actual time basis. However, breathing variability was very low so the normalization procedure did not disguise the relationship of the stimulus pulse train to the respiratory cycle between trials.)

Control of Breathing and Its Modeling Perspective, Edited by
Y. Honda *et al.*, Plenum Press, New York, 1992

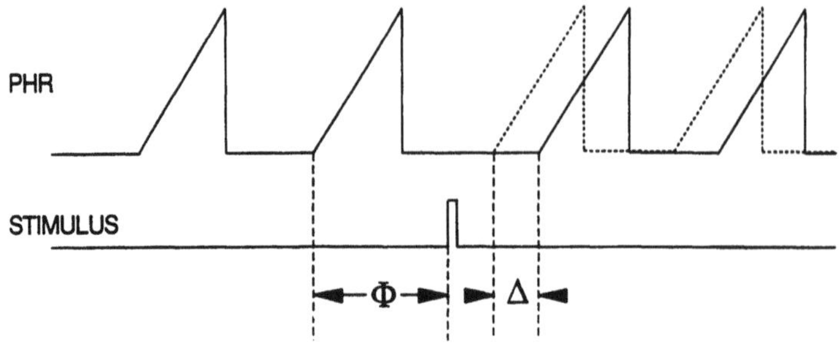

Figure 1. Schematic representation of old phase and phase shift. Old phase (Φ) is defined as the time from the onset of phrenic (PHR) activity to the onset of the stimulus. Phase shift (Δ) is defined as the difference in time between the onset of phrenic activity following the stimulus (solid line) and that which is predicted from unperturbed breaths (dotted line).

Perturbation during inspiration inhibited phrenic activity either reversibly or irreversibly depending on the strength and timing of the stimulus. Reversible inhibition was associated with transient activation of thyroarytenoid activity followed by a reactivation of phrenic activity. Premature inspiratory termination was followed by either shortening or lengthening of the duration of Stage I and consistently by shortening of the duration of Stage II expiration. Perturbation during Stage I expiration augmented thyroarytenoid activity and prolonged the duration of Stage I but did not affect the duration of succeeding Stage II expiration until the transition between stages. Perturbation during Stage II expiration evoked thyroarytenoid activity and inhibited triangularis sterni activity transiently. Following transient inhibition, triangularis sterni was re-activated, and the duration of Stage II expiration prolonged.

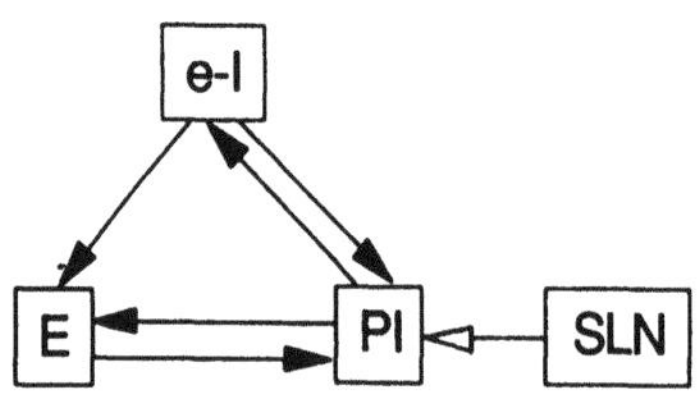

Figure 2. The schematic representation of our neuronal network model.

To describe an underlying mechanism for the results, we constructed a neuronal network model with three components based upon Richter's model[3] as shown in Figure 2. We assumed 1) reciprocal inhibitory connections between post-inspiratory and early-inspiratory neuron groups, and between post-inspiratory and expiratory neuron groups; 2) an inhibitory connection from early-inspiratory neuron group to expiratory neuron group; and 3) tonic excitatory input on all three neuronal groups, which is not shown in the schema. We did not incorporate other neuronal properties such as accommodation or self-excitation. Superior laryngeal nerve stimulation was assumed to excite post-inspiratory neuron's membrane potential. The model was translated into derivative equations widely used in this field (e.g. Botros & Bruce[1]), and these equations were solved numerically.

Computer simulation produced a stable limit cycle oscillation. Although these neuronal discharges overlap in the simulation, if we assume arbitrary thresholds for each neuronal activity, we can define inspiration and Stage I and II expiration as the duration of each neuronal discharge. Figure 3 compares the simulated responses of membrane trajectories and experimentally-derived responses when the stimulus was delivered during early inspiration. The model predicted the shortening of the durations of both post-inspiratory and expiratory neuronal discharges following premature inspiratory termination. The model prediction agrees with the experimental result. Figure 4 compares the simulated responses of membrane trajectories and experimentally-derived responses when the stimulus was delivered during Stage II expiration. The model predicted activation of post-inspiratory neuron group, inhibition of expiratory neuron group and its subsequent re-excitation. Again, the model prediction is in accordance with the experimental result.

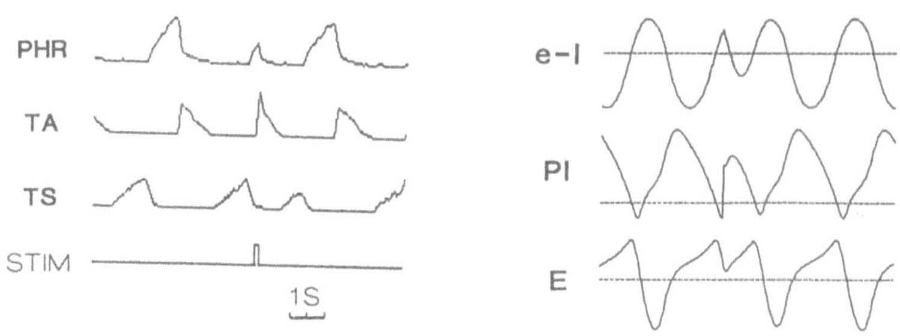

Figure 3. The comparison between the simulated responses of membrane trajectories and experimentally-derived responses when the stimulus was delivered during early inspiration.

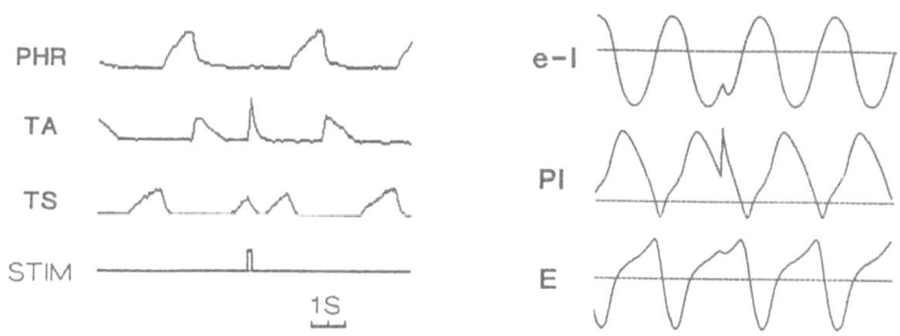

Figure 4. The comparison between the simulated responses of membrane trajectories and experimentally-derived responses when the stimulus was delivered during Stage II expiration.

Figure 5 shows the comparison between a simulated phase response curve and experimental results. The general shape of the simulated phase response curve fits well with experimental data. The magnitude of resetting or phase shift in both simulated and experimental response curves was greatest in early inspiration, gradually decreased toward late inspiration where the respiratory rhythm was affected least by perturbation. However, detailed phase-dependent characteristics during expiration cannot be reproduced by this simple model. For perturbations during Stage I expiration, phase shift was least phase-dependent in experimental response curves, whereas it has a stronger phase-dependency during Stage I than Stage II expiration in the simulated response curve.

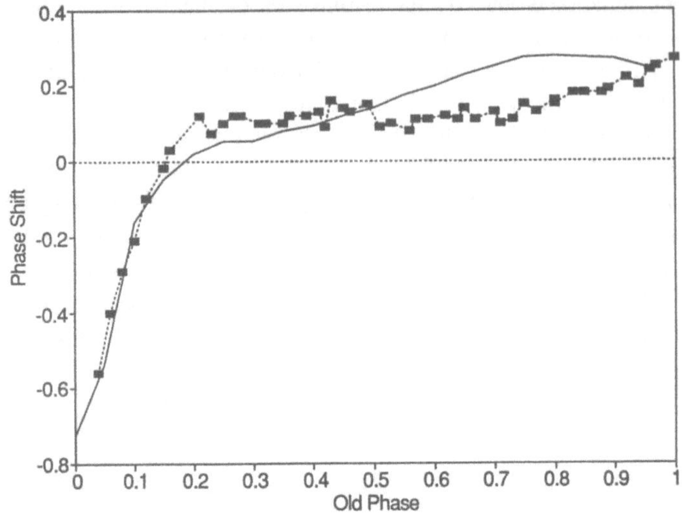

Figure 5. The comparison between a simulated phase response curve and experimental results.

In summary, we recorded three motor activities each representing a separate stage of respiratory cycle. Responses of these motor activity were related closely as indicated in case of premature inspiratory termination, but controlled differently as indicated in case of perturbation during Stage II expiration. Phase response curves showed a characteristic slope for each respiratory stage. These results suggest a three phase configuration of motor expression. Simulation results show the feasibility of a three phase neuronal network model to predict experimentally-derived transient responses of motor activities, however phase-dependency is not fully predicted with this simple model. Further, this analysis does not exclude the pacemaker neuronal mechanism of the respiratory rhythmogenesis since both neuronal network models and pacemaker models generate a limit cycle oscillation, and mathematical expression for these models may be the same.

ACKNOWLEDGEMENTS

We gratefully acknowledge the support of PHS, HL25830.

REFERENCES

1. Botros, S.M. and E.N. Bruce, Neural network implementation of the three-phase model of respiratory rhythm generation. *Biol. Cybern.* 63: 143-153 (1990).
2. Paydarfar, D., F.L. Eldridge, and J.P. Kiley, Resetting of mammalian respiratory rhythm: existence of a phase-singularity. *Am. J. Physiol.* 250: R721-R727 (1986).
3. Richter, D.W., D. Ballantyne, and J.E. Remmers, How is the respiratory rhythm generated? A model. *NIPS* 1: 109-112 (1986).
4. St John, W.M. and D. Zhou, Differing control of neural activities during various portions of expiration in the cat. *J. Physiol.* 418: 180-204 (1989).

LOWER LUMBAR AND SACRAL PROJECTIONS OF CAUDAL MEDULLARY EXPIRATORY NEURONS IN THE CAT

Sei-Ichi Sasaki[1], Hiroyuki Uchino[2], and Yoshio Uchino[1]

[1]Department of Physiology
[2]Department of Anesthesiology
Tokyo Medical College
6-1-1 Shinjuku, Shinjuku-ku
Tokyo 160, Japan

INTRODUCTION

Respiratory rhythmogenesis is organized in the brain stem and respiratory rhythm is then transmitted to the spinal respiratory motoneurons. Expiratory (E) neurons in the caudal nucleus retroambigualis are bulbospinal neurons[1] and have axonal branches in the thoracic and the upper lumbar spinal cord[2,3,4]. E neurons exert synaptic effects to the internal intercostal and the abdominal motoneurons monosynaptically and/or via interneurons[3,5,6]. Recently, the autoradiographic tracing method have revealed a projection from the region of the nucleus retroambigualis to the nucleus of Onuf in the sacral spinal cord[7]. In the present experiments, we have examined the lower lumbar (L6-L7) and sacral (S1-S3) spinal projections of physiological identified caudal medullary expiratory neurons.

MATERIALS AND METHODS

The experiments were performed on adult cats, anesthetized with sodium pentobarbital (Nembutal, initial dose:35-40 mg/kg, i.p.). Anesthesia was subsequently maintained by supplemental doses of sodium pentobarbital throughout the experiments (4-7 mg/kg/h). The femoral artery and the cephalic vein were cannulated to monitor blood pressure and to administer drugs, respectively. Arterial blood pressure was maintained at 100-130 mmHg by the intravenous administration of pressor agents, if needed.

Control of Breathing and Its Modeling Perspective, Edited by
Y. Honda *et al.*, Plenum Press, New York, 1992

Rectal temperature was maintained at 37°-$38^{o}C$. Phrenic nerves were dissected free, ligated and cut distally. The spinal cord was exposed by a laminectomy. The brain stem was exposed by a caudal craniotomy. Dorsal roots from L1 to S3 were cut at the root entry to expose the lateral surface of the spinal cord. Animals were paralyzed by a intravenous administration of pancuronium bromide (Mioblock. Sankyo, Organon) and kept on artificial ventilation. Glass micropipettes filled with 2 M NaCl solution saturated with Fast Green FCF dye (1-2 M Ω in resistance) were used for extracellular recordings of single E neurons. The spinal projection of E neurons was tested by monopolar stimulations through a small Ag-AgCl ball electrode. Microstimulation was performed with a movable glass insulated tungsten stimulating microelectrode (exposed tip:5-10 µm in diameter and 10-20 µm in length) with cathodal current at a maximal intensity of 100 µA (0.15 msec duration).

RESULTS

Unit spikes were recorded extracellularly from a total of 37 expiratory (E) neurons in the region between the obex and the rostral first cervical segment. Neurons were classified as E neurons on the basis of the relationship of their expiratory discharge pattern to that of the phrenic nerve (Fig. 1A). The stimulating ball electrode placed on the surface of the ventrolateral funiculus was shifted in a caudal direction along the spinal cord until antidromic spikes could no longer be elicited. When the stimulus voltage increased sharply and antidromic latency became almost constant, electrical stimulation was stopped, suggesting that the descending axon ended around that spinal level. To determine the exact spinal level of caudal end of descending axons, we mapped effective sites for antidromic activations within the spinal cord at the suspected terminal area with a movable glass insulated tungsten stimulating microelectrode. As a result, both local stimulation and surface stimulation yielded the same result. Thus the most caudal end of E neurons were determined following stimulation of the ventrolateral funiculus in most cases. Figure 1B represents the antidromic activation following stimulation of the caudal end of the descending axon (at the rostral L7 segment in this neuron). Antidromic spikes were confirmed with collision tests (Fig. 1C). Many E neurons (68%) extended their descending axons in the lower lumbar or the sacral segments (S1-S3), 19% of E neurons extended their descending axons in the upper lumbar segments (L2-L3). Figure 1D shows the relation between antidromic latencies and the distance from the recording sites. Antidromic latencies in the descending stem axons increased proportionally to the distance along the spinal cord in the cervical (C) and the thoracic (T) segments, and increased remarkably in the lumbar (L) and the sacral (S) segments.

This accumulative slowing in the lumbar and sacral segments was also observed in many other E neurons.

DISCUSSION

The present results show that many E neurons descend to the lower lumbar or the sacral spinal segments. Previous physiological studies have shown the axon collateral projections of E neurons in the upper lumbar segments[3,4]

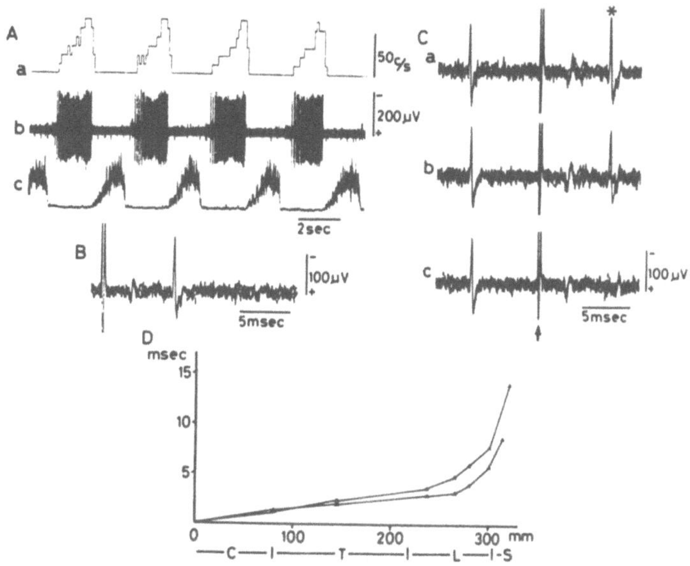

Fig. 1. Identification of expiratory (E) neurons and antidromic responses. A:Discharge pattern of an E neuron. The frequency histogram of spike discharge (a), the extracellular spike activity (b), the rectified and integrated phrenic nerve activity (c). B:Antidromic responses. C:Collision test. Arrows indicate stimulus artifacts. Stars indicate the activated spikes. D:Latency vs. distance in two E neurons. The ordinate indicates latencies and the abscissa the spinal level.

where abdominal motoneurons are located[8] and the lumbar region from the L5 to L7 segments[9]. These axon collateral projections may be involved in the expiratory abdominal motoneuronal activities[3,10,11] and the respiratory modulation of motor activities in the lower lumbar segments.

The nucleus of Onuf contains motoneurons innervating the pelvic floor muscles[12,13]. Thus, further studies are necessary to elucidate the precise locations of collateral arborizations of E neurons in the sacral segments in order to examine the co-ordination of an increase in the intra-abdominal pressure and the activities of the pelvic floor motoneurons by E neurons.

REFERENCES

1. H. Arita, N. Kogo, and N. Koshiya, Morphological and physiological properties of caudal medullary expiratory neurons of the cat, Brain Res. 401:258-266(1987).
2. E.G. Merrill, and J. Lipski, Inputs to intercostal motoneurons from ventrolateral medullary respiratory neurons in the cat, J. Neurophysiol. 57:1837-1853(1987).
3. A.D. Miller, K. Ezure, and I. Suzuki, Control of abdominal muscles by brain stem respiratory neurons in the cat, J. Neurophysiol. 54:155-167(1985).
4. S.-I. Sasaki, H. Uchino, M. Imagawa, T. Miyake, and Y. Uchino, Lumbar and sacral spinal projection of medullary expiratory neurons in the cat, Jap. J. Physiol. 40:Suppl. S49(1990).
5. M.I. Cohen, J.L. Feldman, and D.Sommer, Caudal medullary expiratory neurone and internal intercostal nerve discharges in the cat:effects of lung inflation, J. Physiol. 368:147-178(1985).
6. P.A. Kirkwood, and T.A. Sears, Monosynaptic excitation of thoracic motoneurones from lateral respiratory neurones in the medulla of the cat, J. Physiol. 234:87-89P(1973).
7. G. Holstege, and J. Tan, Supraspinal control of motoneurons innervating the striated muscles of the pelvic floor including urethral and anal sphincters in the cat, Brain. 110:1323-1344(1987).
8. A.D. Miller, Localization of motoneurons innervating individual abdominal muscles of the cat, J. Comp. Neurol. 256:600-606(1987).
9. S.-I. Sasaki, H. Uchino, M. Imagawa, T. Miyake, and Y. Uchino, Lower lumbar branching of caudal medullary expiratory neurons in the cat, Brain Res. 553:159-162(1991).
10. B. Bishop, Reflex control of abdominal muscles during positive-pressure breathing, J. Appl. Physiol. 19:224-232(1964).
11. S.-I. Sasaki, K. Yokogushi, and M. Aoki, Synaptic events in cat abdominal motoneurons during respiration, Neurosci. Res. Suppl. 9:S90(1989).
12. B. Onuf, On the arrangement and function of the cell groups of the sacral region of the spinal cord in man, Arch. Neurol. Psychopathol. 3:387-412(1900).
13. T. Ueyama, N. Mizuno, S. Nomura, A. Konishi, K. Itoh, and H. Arakawa, Central distribution of afferent and efferent components of the pudendal nerve in cat, J. Comp. Neurol. 222:38-46(1984).

REFLEX CONTROL OF RESPIRATORY ACTIVITY (Short Overview)

Giuseppe Sant'Ambrogio

Department of Physiology and Biophysics
The University of Texas Medical Branch at Galveston
Galveston, Texas 77550-2774, U.S.A.

All the several respiratory related activities can be modified by environmental stimuli, changes in energy requirements as well as by various behavioral functions like vocalization, emotional expressions, olfaction etc... These changes involve striated muscles acting on the respiratory bellows and the upper airway, tracheobronchial smooth muscles and mucus glands, blood vessels supplying various districts of the respiratory apparatus, etc...

Some of these responses are elicited through the stimulation of specialized sensors connected to the integrating centers of the nervous system which process the incoming information and convert it into the "required" command to the effector organs. This type of control, based on some kind of positive or negative feed-back from the periphery, is the basis for the reflex mechanisms. The respiratory system is critically dependent for its optimal performance on a great variety of reflexes and sensory feed-back circuits that, working together, adjust its functions to the continuously changing demands for gas exchange as well as for nonmetabolic behavioral activities.

Respiratory reflexes can be divided in two categories: regulatory reflexes that involve changes in breathing pattern as, for example, in exercise or exposure to an hypoxic environment and defense reflexes aimed at protecting the respiratory tract from potentially harmful agents. Vagally mediated reflexes have been found to be of the foremost importance for both regulatory and defensive reflexes.

The search for a specific receptor type as the sole contributor to a specific reflex response has not always been fruitful. In fact, in most situations, especially in physiological circumstances, the concomitant involvement of other afferent endings present in nearby structures, is very probable, as its is the subsequent recruitment of yet other sensors being stimulated by the initial primary reflex response. This may be the case of reflex responses to cold air inhalation which results in a concomitant stimulation of upper airway cold receptors and inhibition of upper airway mechanoreceptors. Also, inhalation of odoriferous substances may elicit responses due to stimulation of both irritant and olfactory receptors.

Another general consideration about reflex responses relates to the possibility of changing their strength irrespective of the magnitude of the stimulus. In fact, although a specific stimulus could be preferentially transduced by a specific receptor type (pressure receptor, cold receptor...), the response may be significantly modified by the conditions existing in the microenvironment of the ending (ionic composition, osmolality, temperature, etc...). Thus the resulting reflex response could be either

Control of Breathing and Its Modeling Perspective, Edited by
Y. Honda *et al.*, Plenum Press, New York, 1992

strengthened or diminished even when the 'proper' stimulus is kept constant.

As it is the case for other physiological functions reflex mechanisms play an important role in the maintenance of respiratory homeostasis. The following examples, based on actual experimental evidence, will serve to illustrate this point.

The role of vagal afferents distributed to the tracheobronchial tree can be demonstrated by indiscriminately blocking the corresponding endings with aerosols of local anesthetics. A selective block of slowly adapting stretch receptors can be obtained in certain species by inhalation of sulphur dioxide-air mixtures (200 p.p.m.). Both procedures change rate and depth of breathing in a manner closely resembling that observed after cervical vagotomy, suggesting that interruption of the afferent activity from the airway plays a major role in the respiratory changes. After vagotomy the economy of respiration deteriorates, as indicated by the increased work of breathing[1,2] and the depressed ventilatory response to carbon dioxide[3].

In healthy subjects a thorough anesthetization of the larynx led to consistent decreases in peak inspiratory flow, as well as forced inspiratory flows at 25, 50, and 75% vital capacity[4]. These results suggest that information from laryngeal receptors play a role in the maintenance of upper airway patency in awake humans thus helping the economy of respiration.

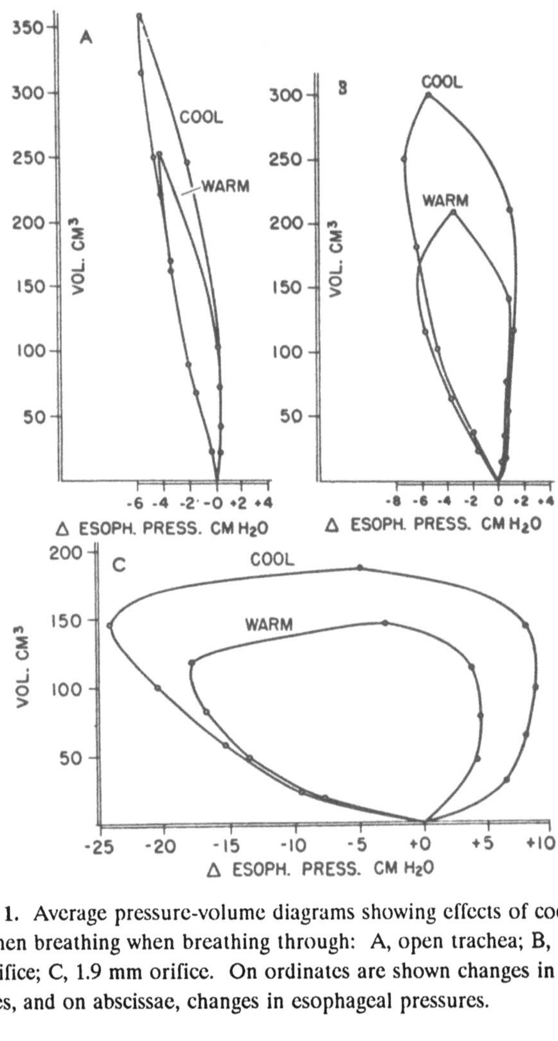

Figure 1. Average pressure-volume diagrams showing effects of cooling vagi when breathing when breathing through: A, open trachea; B, 3.2 mm orifice; C, 1.9 mm orifice. On ordinates are shown changes in lung volumes, and on abscissae, changes in esophageal pressures.

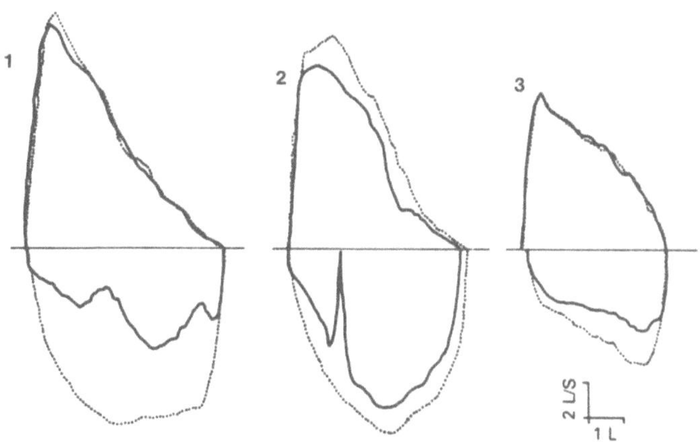

Figure 2. Effect of laryngeal airway anesthesia on a maximal forced inspiratory and expiratory maneuver in 3 subjects. For each subject, the flow volume loop just prior to anesthesia (dotted line) is superimposed on the first loop performed after airway anesthesia (solid line). The loops are aligned on the volume axis at TLC (total lung capacity).

Vagal afferents originating from the tracheobronchial tree are also involved in maintaining the patency of the larynx during hypoxia. Bartlett and coworkers[5,6] found that either vagotomy or blockade of airway stretch receptors induced a marked increase in expiratory laryngeal resistance due to a decreased expiratory activity of the posterior arytenoid muscles. This mechanism would also contribute to a reduction of the work of breathing.

With heart-lung transplantation the donor heart and lungs are implanted in the recipient's chest. The right atrium and the ascending aorta of the recipient are sutured to the donor's heart, while the trachea is sutured just above the carina. The donor's heart is not reinnervated beyond the atrial suture line and this accounts for the high resting heart rate observed in these patients who also have a delayed heart rate response to exercise[7]. All neural connections with the donor's lungs are also interrupted at the level of the tracheal suture line. The absence of a cough response to mechanical stimulation below the suture line indicates the absence of vagal reflex function after heart-lung transplantation. Also coughing in response to ultrasonically nebulized water was considerably impaired in heart-lungs transplanted patients as compared to normal subjects[8]. Evidence suggests that also muco-ciliary clearance is compromised. These observations suggest that lung denervation leads to a considerable impairment of protective mechanisms with grave consequences for respiratory economy[8].

Experimental evidence demonstrates that with aging perception threshold, reflex latencies and reaction time deteriorates. Similar deterioration applies to protective laryngeal reflexes[9]. By using ammonia-air mixtures at different concentrations as an irritant Pondoppidan and Beecher[9] determined the thresholds of a protective airway reflex in more than 100 subjects of various age. The median threshold was found to increase sixfold from the second to the eight and ninth decades of life. These results are consistent with the higher incidence of aspiration pneumonia in elderly patients. In fact the loss of airway protective reflexes as age progresses permits aspiration of foreign material, such as food, regurgitated gastric content, or purulent material from draining sinuses[9].

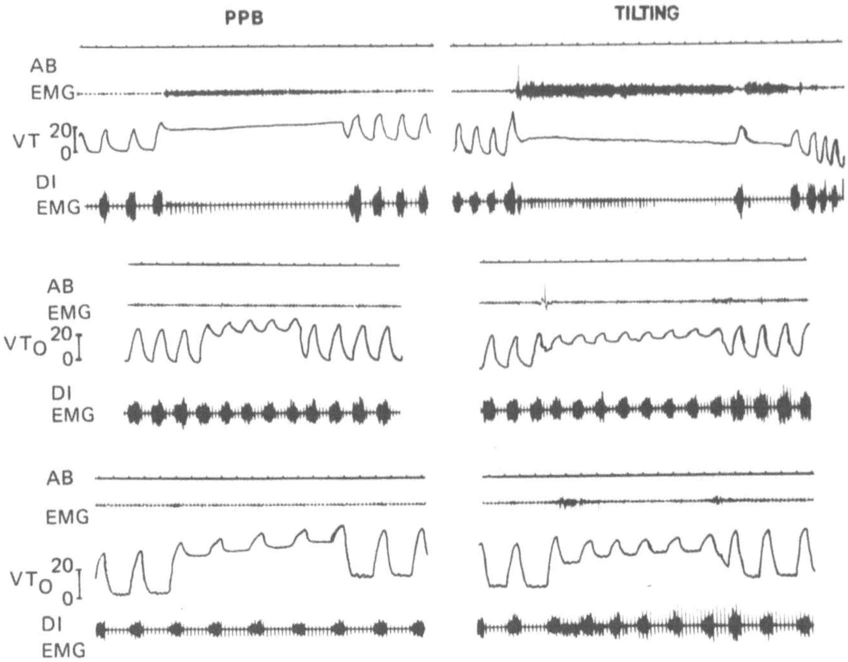

Figure 3 Anesthetized rabbit. Effect of positive pressure breathing (P.P.B., tracings on the left) and head-up tilting (tracings on the right) on the activity of the external oblique m. (ABD, emg), the diaphragm (DIA, emg) and the tidal volume (V_T) in control condition (top tracings), after SO_2 inhalation (middle tracings) and post-vagotomy (bottom tracings). Time marker = 1 sec.

The abdomen and thorax are mechanically interdependent and, during postural movements, there are marked changes in volume between these two compartments mostly due to gravitational forces acting on the abdominal content. Powerful reflexes elicited by increases in lung volume have been found to activate abdominal muscles[10]. The afferent pathway of these reflexes is mostly represented by vagal fibers originating from slowly adapting pulmonary stretch receptors[11]. In the absence of this reflex mechanism, as after vagal section or a selective block of slowly adapting stretch receptors, head-up tilting of experimental animals was found to cause a considerably greater increase in end-expiratory volume than in control conditions[11]. This vagally mediated reflex contributes to a homeostatic mechanism capable of minimizing changes in end-expiratory volume introduced by the gravitational influences of posture.

Another excellent example of respiratory homeostasis substantially affected by vagally mediated mechanisms is that provided by a comparison of the changes in relevant respiratory variables during induction and removal of an experimental pneumothorax[12]. As shown in fig. 4 changes in arterial blood gases (decrease in PO_2 and increase in PCO_2) become significantly greater after vagotomy, indicating a greater impairment of respiratory function due to differences in lung mechanics and pattern of breathing. In fact, the results show that vagal integrity resulted in smaller increases in resistance and absence of decrease in compliance during induction of pneumothorax when ventilation was significantly reduced[12].

In summary we may conclude that, although central neural mechanisms are capable by themselves to maintain a rhythmical breathing pattern even in the absence of any sensory feed-back, the economy of respiration is substantially

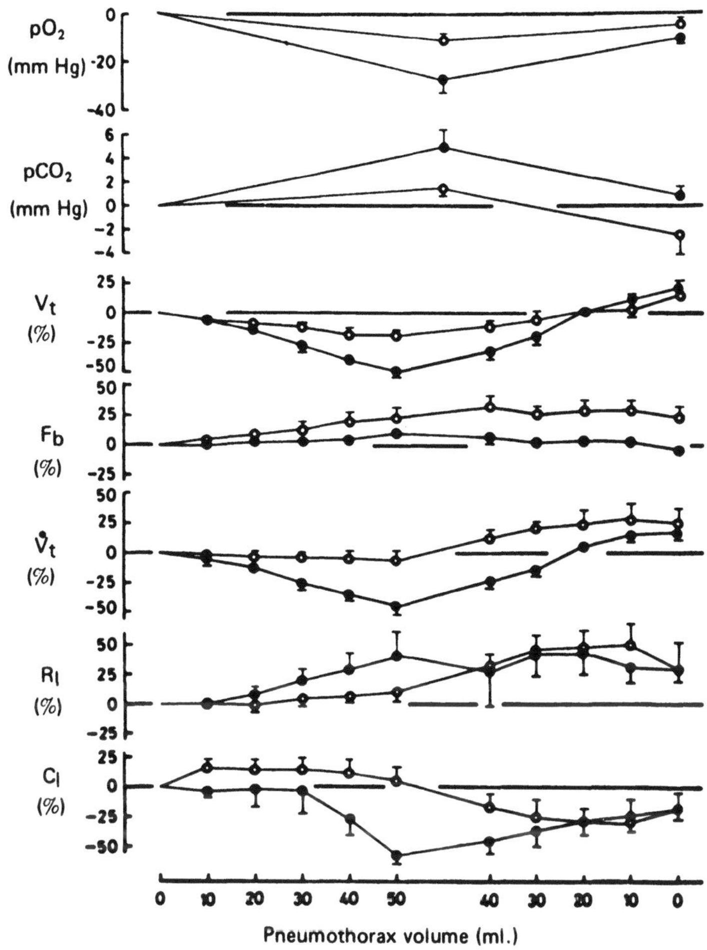

Figure 4. Sequential plot of mean changes in variables during induction and removal of pneumothorax. means of 8 experiments on rabbits with both vagus nerves intact. means of 6 experiments on rabbits with both vagus nerves cut. Vertical lines give S.E. of the means. Values are expressed as changes from pre-pneumothorax controls, and are percentage changes except for blood gases gas tensions which are absolute changes. Abscissa: volume of pneumothorax. Ordinates: arterial oxygen tension (Po_2); arterial carbon dioxide tension (Pco_2); tidal volume (V_T); breathing frequency (Fb); minute volume (Vt); total lung resistance (R_l); and lung compliance (C_l). Control values for blood gas tensions were for oxygen 64.8 ± 4.9 and 67.2 ± 5.8 mmHg, and for carbon dioxide 34.7 ± 1.4 and 32.8 ± 4.0 mm Hg for the vagus intact and vagotomized groups respectively.

dependent on a variety of reflexes that optimize its function according to the ever changing metabolic and behavioral demands.

Acknowledgments. The research work of Dr. G. Sant'Ambrogio is supported by NIH grant HL-20122.

References

1. F.W. Zechman, Jr., J. Salzano and F.G. Hall, Effect of cooling the cervical vagi on the work of breathing, *J.Appl.Physiol.* 12:301 (1958).
2. T.P.K. Lim, U.C. Luft and F. S. Grodins, Effects of cervical vagotomy on pulmonary ventilation and mechanics, *J.Appl.Physiol.* 13: 317 (1958).
3. A. Guz, M.I.M. Noble, J.G. Widdicombe, D. Trenchard and W.W. Mushin, The effect of bilateral block of vagus and glossopharyngeal nerves on the ventilatory response to CO2 of conscious man. *Respir.Physiol.* 1:206 (1966).
4. S. T. Kuna, G. E. Woodson, and G. Sant'Ambrogio, Effect of laryngeal anesthesia on pulmonary function testing in normal subjects, *Am.Rev.Respir.Dis.* 137: 656 (1988).
5. D. Bartlett, Jr., Effects of vagal afferents on laryngeal responses to hypercapnia and hypoxia, *Respir.Physiol.* 42: 189 (1980).
6. D. Bartlett, Jr., S. L. Knuth and K. V. Knuth, Effects of pulmonary stretch receptor blockade on laryngeal responses to hypercapnia and hypoxia, *Respir.Physiol.* 45: 67 (1981).
7. N. R. Banner, A. Guz, R. Heaton, J.A. Innes, K. Murphy and M. Yacoub, Ventilatory and circulatory responses at the onset of exercise in man following heart or heart-lung transplantation, *J.Physiol.Lond.* 399:437 (1988).
8. T. Higenbottam,B. A. Otulana, and J. Wallwork,. The Physiology of heart-lung transplantation in humans, *NIPS* 5: 71 (1990).
9. H. Pontoppidan and H. K. Beecher, Progressive loss of protective reflexes in the airway with the advance of age, *J.A.M.A.* 174: 77 (1960).
10. B. Bishop, Reflex control of abdominal muscles during positive pressure breathing, *J.Appl.Physiol.* 19: 224 (1964).
11. A. Davies, F.B. Sant'Ambrogio and G. Sant'Ambrogio, Control of postural changes of end expiratory volume (FRC) by airways slowly adapting mechanoreceptors, *Respir.Physiol.* 41: 211 (1980).
12. H. Sellick and J. G. Widdicombe, The activity of lung irritant receptors during pneumothorax, hyperpnea and pulmonary vascular congestion, *J.Physiol.Lond.* 203: 359 (1969).

ACID INFUSION ELICITS THROMBOXANE-MEDIATED CARDIO-RESPIRATORY EFFECTS

P. Scheid[1], H. Shams[1], W. Karla[1], J.A. Orr[2]

[1]Institut für Physiologie
Ruhr-Universität Bochum
3400 Bochum, FRG
and
[2]Department of Physiology and Cell Biology
University of Kansas
Lawrence, KS 66045, USA

INTRODUCTION

Involvement of peripheral chemoreceptors in the control of respiration is still an open issue. While arterial chemoreceptors are well described, controversy exists about the existence of such receptors in the venous system. During the search in the cat for CO_2-sensitive receptors in the larger venous vessels or the right heart, we have intravenously infused 0.25 M HCl, as has been performed in many studies before us. We have, however, observed rather unexpected cardio-respiratory responses which we describe in the following brief review. It compiles the results from a number of studies that we conducted in the cat. For more detailed information about the methodology of each study, we refer the reader to the published papers.

CARDIO-RESPIRATORY EFFECTS OF ACID INFUSION ARE UNRELATED TO pH

The effects that we observed during intravenous acid infusion consisted of a significant increase in pulmonary artery pressure, with a systolic peak pressure of above 50 mmHg, combined with an increase in respiratory frequency and a decrease in tidal volume[1]. These responses at an infusion rate of only 0.2 $mmol \cdot min^{-1}$ could be so vigorous that the infusion had to be stopped or else the cat would die. We had expected at least an increase in tidal volume, and we wondered whether the acid might act on the venous vessel wall to liberate mediators responsible for the effects.

We, therefore, installed an arteriovenous bypass in our cats, allowing a blood flow of 20 $ml \cdot min^{-1}$ from the femoral artery to the femoral vein. Infusion (0.25 M HCl, 0.2 $mmol \cdot min^{-1}$) into this bypass elicited the same rapid-shallow breathing and pulmonary

Control of Breathing and Its Modeling Perspective, Edited by
Y. Honda *et al.*, Plenum Press, New York, 1992

hypertension, which even persisted when we neutralized the infused acid by infusing base (0.25 M NaOH, 0.2 mmol·min^{-1}) into the bypass, some centimeters downstream from the acid infusion port. In this condition, the blood re-entering the body had normal pH. The obvious control, infusion of the externally mixed acid and base, remained without effects. Infusion of base alone showed no effect either.

An interesting observation in this, as in all experiments to follow, was that the effects were transient. Neither did they survive infusion times of more than about 20 min; nor could they be elicited when the acid/base was infused a second time.

We reasoned that some mediator might be released from blood cells, notably from platelets, at the site where the acid is emerging from the infusion catheter. Although the acid infusion would lead to an only 0.3 unit drop in the mixed blood in the bypass, the pH can be expected to be much lower in the direct vicinity of the tip of the acid infusion catheter. A candidate for mediators released by acid could be prostanoids, and we specifically determined the levels of some prostanoids in the blood.

INVOLVEMENT OF THROMBOXANE A_2

To avoid extensive surgery and possible side-effects from using roller pumps to create the bypass flow, we infused acid and base (at stoichiometrically balanced rates, as above) directly via the femoral veins. Aside from recording the cardio-respiratory effects, we sampled blood for determination of plasma levels of prostanoids[2]. We, thus, measured thromboxane(Tx)B_2 and 6-keto-prostaglandin(PG)$F_{1\alpha}$, as the stable metabolites of TxA_2 and PGI_2. The control levels of both metabolites were below the detection threshold (10 pg·ml^{-1}), indicating that our experimental procedures themselves did not produce significant levels of these vasoactive mediators.

During acid infusion, however, a strong correlation was observed between the cardiorespiratory effects and the level of TxB_2 in the plasma, which could go up to above 700 pg·ml^{-1}. This was also true for the attenuation during repeated acid infusion. The correlation for 6-keto-$PGF_{1\alpha}$ was not significant.

Further evidence for the involvement of TxA_2 as a mediator of these effects came from the use of the TX synthase blocker, Dazmegrel, and of the TxA_2 mimetic drug, U46,619[2]. After Dazmegrel, both the cardiorespiratory effects and the increase of the plasma TxB_2 levels were eliminated. But even then, infusion of U46,619 elicited the cardiorespiratory effects. This stimulation could be evoked repetitively. It was interesting to note that U46,619 was effective even when a second or third acid infusion remained without effect, suggesting that it was not the receptor that was blocked by the acid infusion. Indeed the TxB_2 levels remained virtually unelevated during the second and third acid infusion.

INVOLVEMENT OF THE VAGUS NERVE

At this stage, we wondered whether these cardiorespiratory effects, pulmonary hypertension with rapid-shallow breathing, might be reflex in nature, possibly involving the vagus nerve. We, therefore, repeated acid/base infusion in cats after reversibly blocking the vagus nerves of both body sides, by cooling them to 1°C, suggested to block all fiber activity[3]. Cooling the vagus nerves themselves showed the expected slowing and deepening of respiration with little effects on pulmonary blood pressure.

When acid and base were infused with the vagus nerves blocked, the hemodynamic response remained apparently unaltered from the situation with intact vagi. However, no respiratory effects were elicited by acid infusion with blocked vagi. Even infusing the TxA_2 mimetic, U46,619, did not affect respiration, despite its marked hypertensive effect in the pulmonary circulation. These effects were reversible: When the vagus nerve was re-warmed, U46,619 elicited the cardiorespiratory effects as before.

TYPES OF VAGAL FIBERS INVOLVED

We now wanted to know which type(s) of vagal afferent fibers from the lung might be responsible for the TxA_2-mediated respiratory effects. We recorded functional single-unit activity in fine strands of the vagus nerve in the cat and tested them for their sensitivity to U46,619[4]. We identified the three known pulmonary afferent fiber types, slowly-adapting (SAR) and rapidly-adapting stretch receptors (RAR or irritant receptors) and C-fibers, by known criteria: Their discharge pattern in the normal respiratory cycle (RAR and SAR fire in phase with breathing); their response to lung inflation (SAR, and some C-fibers maintain increased firing rate); their response to a brief injection of phenyl-biguanide (PBG; C-fibers increase their firing rate).

We found that SAR and RAR showed only a slight sensitivity to U46,619, and this was often linked to some increase in airway pressure. However, C-fibers were strongly excited by U46,619, independent of effects on airway pressure. This might suggest a direct effect of TxA_2 on the C-fiber endings, and these would be much more important than secondary effects, via constrictive actions on smooth muscles.

RESPIRATORY EFFECTS DURING ENDOTOXIN INFUSION

The significance of these effects is not very clear; it is, in particular, uncertain whether the effects described might be of relevance in the control of breathing in health and disease. We cannot conceive of an acid stimulus under physiologic conditions that might be strong enough to release TxA_2, comparable to our experimental situation.

However, in the adult respiratory distress syndrome (ARDS), rapid-shallow breathing often accompanies pulmonary hypertension. Since endotoxin infusion has been studied in many laboratories as a model of ARDS[5], we investigated whether TxA_2 release was involved in the effects observed during endotoxin infusion (J.A. Orr, H. Shams, W. Karla, B.A. Peskar and P. Scheid, unpublished).

We infused endotoxin (E. coli, strain B055, 1.6 $mg{\cdot}kg^{-1}$) intravenously over a period of 1 min. Within a few minutes following the infusion, we observed rapid-shallow breathing and pulmonary hypertension, following a marked decline in arterial blood pressure.

These cardio-respiratory effects could be totally abolished when the animals were pretreated with indomethacin, which is known to block the cyclooxygenase. This would indicate that prostanoids are involved in the endotoxin effects, but would not allow to point at TxA_2 as the only mediator.

However, also after specifically blocking the TxA_2 receptor by Daltroban, endotoxin infusion remained without any cardio-respiratory effect, and this would indicate that the effects observed by us were indeed mediated by TxA_2.

It should, however, be noted that the responses are transient in nature, occurring within minutes, much like those to acid infusion. Other effects have apparently been observed after longer periods, and persisting for a longer time; however, the cardio-respiratory responses that we review here, did not persist for more than about 1 hr. Its, physiological or pathophysiological significance needs, thus, to be further developed.

REFERENCES

1. J.A. Orr, H. Shams, M.R. Fedde, and P. Scheid, Cardiorespiratory changes during HCl infusion unrelated to decreases in circulating blood pH, *J. Appl. Physiol.* 62:2362 (1987).
2. H. Shams, B.A. Peskar, and P. Scheid, Acid infusion elicits thromboxane A_2-mediated effects on respiration and pulmonary hemodynamics in the cat, *Respir. Physiol.* 71:169 (1988).
3. H. Shams, and P. Scheid, Effects of thromboxane on respiration and pulmonary circulation in the cat: role of the vagus nerve, *J. Appl. Physiol.* 68:2042 (1990).
4. W. Karla, H. Shams, J.A. Orr, and P. Scheid, Effects of the thromboxane A_2 mimetic, U46,619 on pulmonary vagal afferent fibers in the cat, *Respir. Physiol.* 87:383 (1992).
5. K.L. Brigham, and B. Meyrick, Endotoxin and lung injury, *Am. Rev. Respir. Dis.* 133:913 (1986).

RESPIRATORY EFFECTS OF "RENZHONG XUE" IN RABBITS

Lei Liu, Gang Song, Heng Zhang*, Jian—xin Gao*,
Wei—yang Lu, and Min Zhang

Department of Physiology and *Central Laboratory
Shandong Medical University, Jinan, 250012, China

"Renzhong xue" (also called "Suigou xue")is an acupuncture locus used in respiratory and cardiovascular rescue since ancient time. According to Chinese historical records *Shiji*, Bian Que (350—? B. C.) successfully rescued Prince Guo from hemorrhage shock by acupuncturing the "Renzhong xue". The "Renzhong xue" is located at the middle of the sulcus nasolabial. Corresponding region in animals is considered to be the location of animal "Renzhong", although it is still debated whether the principle of *xue* (means locus or point) identification is similarly applicable to animals. Figure 1 shows the location of "Renzhong xue" in human and rabbit. Modern studies have shown that acupuncturing this locus did affect the blood pressure and respiration in patient with shock. In the present study we further investigated the effects of "Renzhong xue" with special reference on its effects on respiration.

Sixty rabbits, weighing 2. 0—3. 0 Kg, were used in this study. Animals were anesthetized with urethane (0. 8 — 1. 2g/Kg) and artificially ventilated

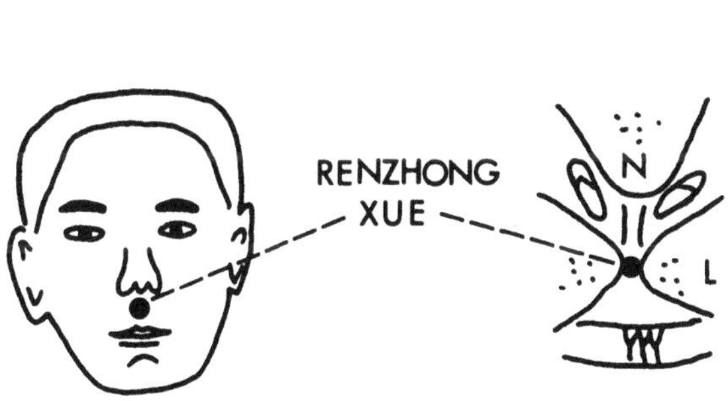

Fig. 1. Location of the "Renzhong xue" in human(left) and rabbit(right). N, nose; L, lip.

when necessary. The phrenic nerve discharge was recorded as an indicator of central respiratory outflow. Bipolar stainless steel electrodes (interpolar distance 5 mm) were inserted into the "Renzhong xue " area, as shown in figure 1, for the delivery of electrical stimulation. Blood pressure was monitored and maintained at 100—120 mmHg.

Effects on Respiration in Intact Rabbits: In intact rabbits, electrical stimulation of "Renzhong xue" caused two type of responses in respiration, i. e. facilitation and inhibition, depending on the stimulation intensity. Facilitation response was observed when the intensity was low, usually at 3—8 volts. As shown in figure 2, both the amplitude and duration of phrenic inspiratory discharge increased during stimulation. This kind of facilitation increased with the increase of stimulation intensity until a critical level (12 volts in most cases) was reached. Intensity higher than this critical level caused inhibition of inspiration. The phrenic discharge ceased immediately upon the start of the stimulation and remained inhibited for a certain time (inhibition time). Figure 3. The inhibition time increased with the increase of intensity. Intensity higher than 20 volts could hold the inspiration inhibited as long as the stimulation continuing.

Inspiration Starting Effects: In animals with expiratory apnea, stimulation of the "Renzhong xue" could elicit inspiration. The expiratory apnea was produced by infusing the carotid sinus with sodium citrus solution. A single injection of 0.

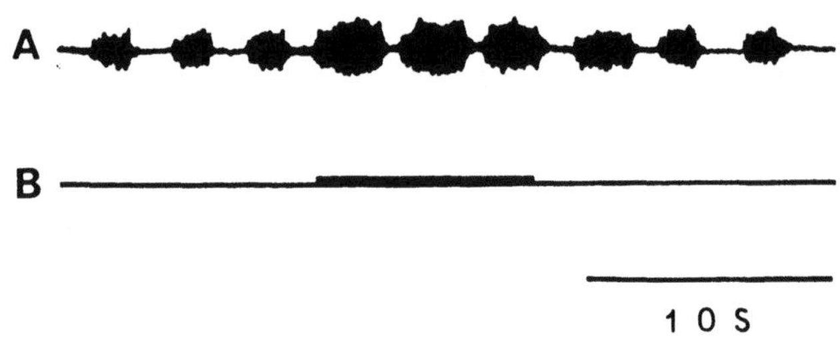

Fig. 2. Low intensity stimulation (6 volts, 100Hz) of the "Renzhong xue" facilitated inspiration in intact rabbits. The duration and amplitude of phrenic discharge increased during stimulation. A, phrenic discharge; B, stimulation marker.

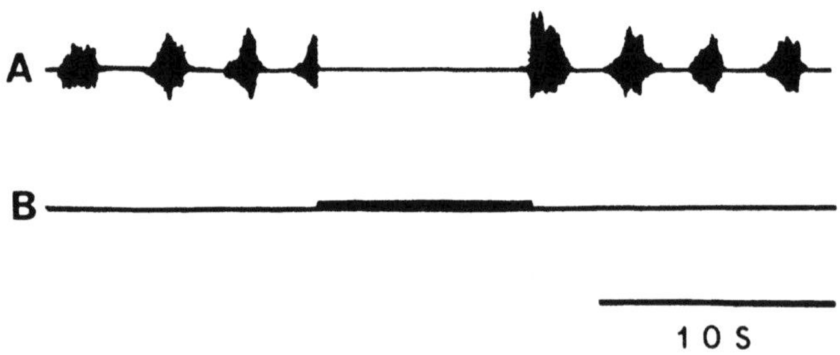

Fig. 3. High intensity stimulation (16 vollts, 100 Hz) of the "Renzhong xue" inhibited inspiration. The phrenic nerve discharge ceased completely during the stimulation. A, phrenic discharge; B, stimulation marker.

4 ml of 5% citrus solution into the carotid artery could induce expiratory apnea lasting 15—25 seconds[1]. An average value of 21.3±1.56 (mean±SE) seconds was observed in this study. During the apnea stimulation of the "Renzhong xue" elicited phrenic inspiratory discharge immediately and the effectiveness depended on the intensity and stimulation duration. Long train stimulation could induce a short period of rhythmic discharge. As can be seen in figure 4, the inspiration elicited was different from that of the normal respiration.

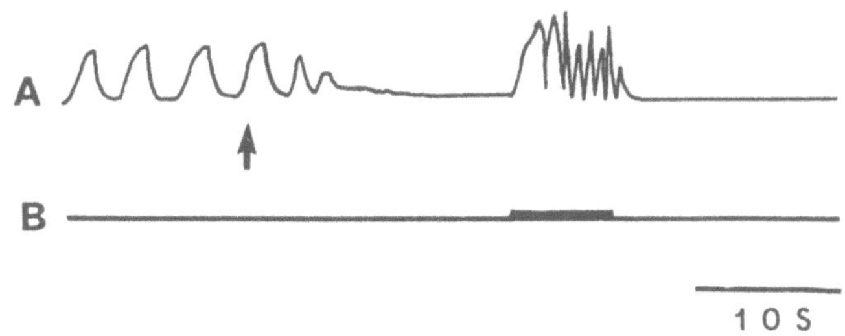

Fig. 4. Injection of citrus sodium solution into the carotid artery (arrow) caused cessation of inspiration. Stimulation of the "Renzhong xue" elicited a short period of rhythmic phrenic discharge. Note that the elicited inspiration is different from that of normal. A, integrated phrenic discharge; B, stimulation marker. Stimulation intensity was 1 mA at 100 Hz.

Respiratory Rescuing Effect in Animals with Central Respiratory Failure: The central respiratory failure was produced by destructing the dorso—medial area of the nucleus facialis (dmNF) or focally blocking this region with lidocaine. As we have reported, destruction or blocking of the dmNF in rabbits caused cessation or depression of inspiration[2]. In the present study complete cessation of inspiration was produced in 21 rabbits after blocking the dmNF with lidocaine (2μ). The animal was artificial ventilated and the blood pressure was kept at 100—120mmHg. Spontaneous respiration could gradully apprear after 40—50 minutes. Electrical stimulation of the "Renzhong xue" could elicit inspiration in 16 rabbits (80%) with complete apnea. A short train of pulses (20—40 pulses, 100Hz, 1mA) could elicit phrenic discharge. Sometimes a short period of rhythmic respiration (3—8 cycles) could be elicited. In 8 animals (39%) normal respiration appeared after repeated stimulation. Figure 5.

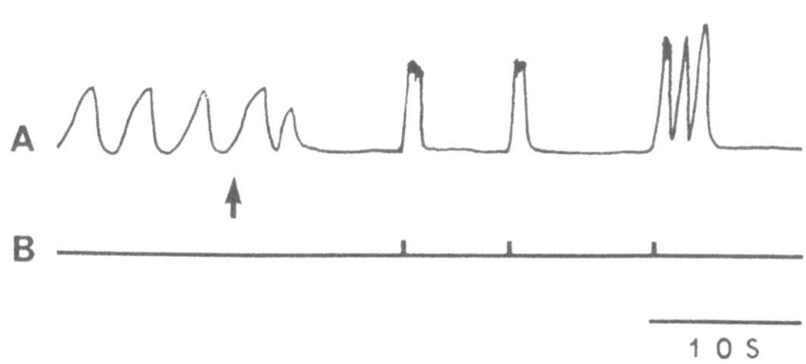

Fig. 5. Blocking of the dorso—medial area of nucleus facialis with lidocaine (2 ul, arrow) caused

cessation of inspiration. Short train stimulation of the "Renzhong xue" elicited gasp—like inspiration. In this case 3 successive inspiration could also be elicited. A, integrated phrenic discharge; B, stimulation marker.

Control Experiments: Stimulation of other points located in the trigeminal innervating region such as "Suliao xue " (at the tip of nose) and a point 2 cm beside the "Renzhong xue" had little effect on respiration and seldom elicited inspiration in apnea animals.

Results of the present study showed that stimulation of the "Renzhong xue" region produced respiratory responses in intact anesthetized rabbits and activated inspiration in apnea animals and animals with central respiratory failure . The "Renzhong xue" region is innervated by the ophalmic branches of the trigeminal nerves. Bilateral anesthetizing or cutting of this nerve blocked the respiratory effect. Recent studies in this laboratory showed that all those results could be mimicked by stimulating the ophalmic branches of the trigeminal nerves and the terminal nucleus of trigemini[3]. Thus it seems that the trigeminal afferent inputs are involved in the respiratory control. But the afferents from the "Renzhong xue " region are much more effective than afferents from other region innervated by the trigeminal nerve, since stimulation of the control loci in facial area only had weak effects; thus a different pattern of central projection specific to "Renzhong xue " region might exist . It seems paradoxical that strong stimulation activated inspiration in apnea animals but had inhibitory effect in intact animals . We propose that the inhibition is caused by the recruitment of fibers responsible for the "diving reflex[4]" by the strong stimulation, hence the inhibition of the dominant or "normal" respiratory rhythm generator. In apnea animals whose dominant rhythm generator was already interfered, the afferent inputs might activate additional or redundant networks possible to generate inspiration and the respiratory rhythm . We noticed that the inspiration and the respiratory rhythm elicited were different from that under the normal condition, showing the activation of other rhythm generators. Based on those results we suggest that afferents from the "Renzhong xue" region exert specific influence on the central respiratory rhythm generators. The physiological importance of this phenomenon remains to be studied.

REFERENCES

1. Gao, J. X. , G. Song, L. Liu. The inspiratory on—switch response of dorso—medial area of nucleus facialis . Acta Academiae Medicinae Shandong. 28: (3)22—25(1990).
2. Liu, L. , G. Song, W. Y. Lu. Studies on the inspiratory generating effect of the dorso—medial area of nucleus facialis. Respir. Physiol. 75: 65—74(1989).
3. Zhang, H. , M. Zhang, L. Liu. Comparison of effect in the change of respiration by stimulation three trigeminal subnucleus. Acta Physiol. Sinica. 41: 602—607(1989).
4. Elsner, R. , B. Gooden. Diving and Asphyxia. Monographs of the Physiological Society NO. 40. Cambridge, U. K. : Cambridge Univ. Press(1983).

EFFECTS OF HYPERCAPNIA ON NASAL REFLEX RESPONSES ELICITED BY NASAL INSUFFLATION OF ISOFLURANE IN LIGHTLY-ANESTHETIZED HUMANS

Takashi Nishino

Department of Anesthesiology, National Cancer Center Hospital
Tokyo 104, Japan

INTRODUCTION

Stimulation of nasal mucosa elicits nasal reflexes which are defensive in nature. In a previous study [1] we showed that in anesthetized humans nasal irritation with high concentrations of commonly-used volatile anesthetics causes changes in respiratory pattern characterized by prolongation of expirtatory time. Although it has been reported in several studies [2-5] that an increase in chemical ventilatory drive can attenuate various reflex responses to airway stimulation, whether CO_2 exerts a similar inhibitory effect on nasal reflexes has not been examined. In the present study we studied changes in respiration and circulation during nasal insufflation of isoflurane, a mildly-pungent volatile anesthetic, at a resting level of $P_{ET}CO_2$ and at an elevated level of $P_{ET}CO_2$ in lightly-anesthetized humans.

METHODS

The studies were performed on 13 spontaneously-breathing female patients whose ages ranged from 26-49 yr. All were scheduled for elective mastectomy or hysterectomy under general anesthesia. The protocol was approved by the Ethics Committee of our institution, and each patient gave informed consent. Anesthesia was induced with flunitrazepam and was maintained with pentazocine and nitrous oxide. The trachea was intubated with a cuffed endotrachcal tubc, which was connected to an experimental apparatus incorporated into a semi-closed anesthetic circuit. Ventilatory airflow was measured through a pneumotachograph and tidal volume (V_T) was obtained by electrical integration of the inspired flow. Tracheal pressure (Ptr) was measured with a pressure transducer. End-tidal CO_2 partial pressure ($P_{ET}CO_2$) was monitored with an infrared CO_2 analyzer. Systemic arterial blood pressure (BP), via radial artery catheter, and heart rate (HR) were continuously measured with a pressure transducer and heat rate counter connected to an electrocardiogram, respectively. A soft nasal tube that fits snugly in the patient's nostril was inserted 5 mm into the nostril and nasal insufflation of oxygen at a constant flow rate (3 l/min) was started via this nasal tube. When all the respiratory variables were stable, a sudden administration of 5% isoflurane in oxygen was performed through the nasal tube using pecalibrated vaporizer incorporated in the insufflation system. Nasal insufflation of isoflurane was continued for at least 60 s and was performed at two different levels of $P_{ET}CO_2$: $P_{ET}CO_2$ at resting breathing (resting $P_{ET}CO_2$) and hypercapnia ($P_{ET}CO_2$ 15 mmHg higher than resting $P_{ET}CO_2$). Control values for inspiratory time (T_I), expiratory time (T_E), V_T, BP, and HR were obtained by averaging the values for 20 s immediately preceding nasal insufflation of isoflurane.

Control of Breathing and Its Modeling Perspective, Edited by
Y. Honda *et al.*, Plenum Press, New York, 1992

Changes in respiration and circulation after nasal insufflation of isoflurane were analyzed in a breath-by-breath and in a beat-by-beat fashion , respectively.
In 8 of 13 patients, nasal insufflation of isoflurane was repeated after the nostrils were sprayed with 5-6 puffs of 8% lidocaine via an aerosol spray set. In the other 5 patients, in addition to nasal insufflation of isoflurane, insufflation of isoflurane into the pharynx through the mouth was performed. Statistical analysis was performed using two-way ANOVA and Dunnett's test. A paired t test was also performed where appropriate.

RESULTS

Figure 1 shows examples of reflex responses to nasal irritation.. Immediately after the start of nasal insufflation of isoflurane at resting breathing (A), respiratory frequency decreased due to prolongation of T_E. The prolongation of T_E gradually returned to the preinsufflation level during the continued insufflation of isoflurane. During the insufflation of isoflurane, there was a considerable rise in BP with a very small increase in HR. Compared with the respiratory responses, the onset of circulatory responses was considerably delayed. When the nasal irritation was performed at hypercapnia (B), there were almost no change in respiration, BP, and HR.

Table1 summarizes maximal responses of respiratory and circulatory variables to nasal insufflation of isoflurane obtained at two different levels of $P_{ET}CO_2$. Compared with the control values before nasal irritation, there were significant decreases in V_I and f immediately after the start of nasal irritation at both levels of $P_{ET}CO_2$, although the responses at the resting $P_{ET}CO_2$ were much more pronounced than those during hypercapnia. The decrease in V_I during nasal irritation was due to a decrease in f, and the decrease in f was solely due to prolongation of T_E. There was no significant difference between the values of BP at the resting $P_{ET}CO_2$ and at hypercapnia before nasal stimulation. Although BP increased significantly in response to nasal stimulation at both the resting $P_{ET}CO_2$ and hypercapnia, the increase in BP was more pronounced at the resting $P_{ET}CO_2$ than at hypercapnia. Similarly, the rise in HR at the resting $P_{ET}CO_2$ was more pronounced than the rise in HR at hypercapnia.
In the 8 patients whose nostrils had been sprayed with lidocaine, nasal insufflation of

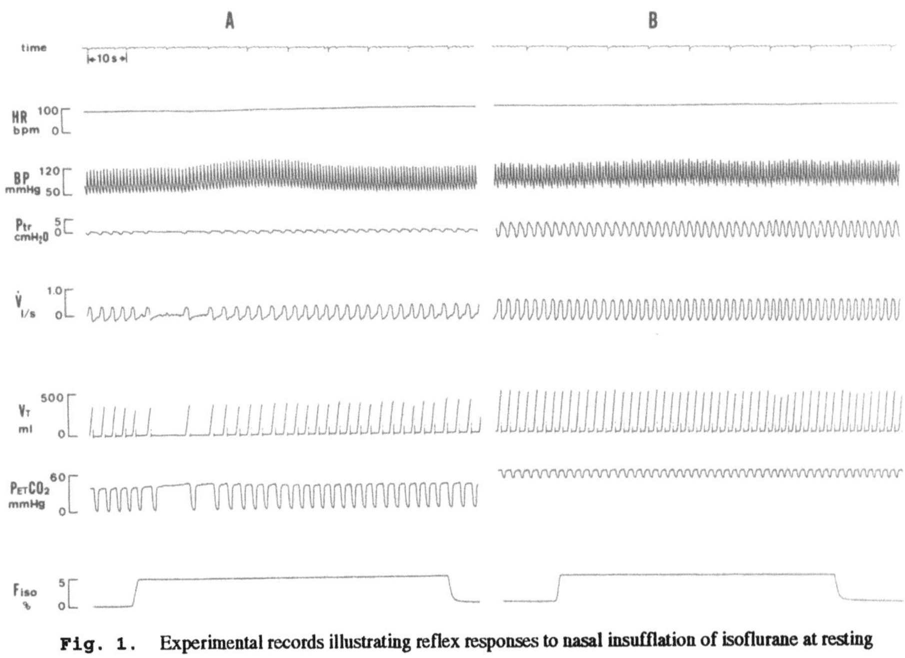

Fig. 1. Experimental records illustrating reflex responses to nasal insufflation of isoflurane at resting $P_{ET}CO_2$ (A) and hypercapnia(B).

Table 1. Maximal responses to nasal insufflation of isoflurane

	Resting P_{ETCO_2}		Hypercapnia	
	control	test	control	test
f (bpm)	17.3 ±1.0	10.5 ±1.2**	20.5 ±1.6	18.4 ±1.7*
V_T (ml)	359 ±20	291 ±37**	610 ±39	597 ±41
V_I (l/min)	6.1 ±0.4	3.4 ±0.4	12.2 ±1.1	10.9 ±1.1*
T_I (s)	1.3 ±0 .1	1.2 ±0.1	1.2 ±0.1	1.3 ±0.
T_E (s)	2.3 ±0.2	4.7 ±0.7	2.0 ±0.3	2.4 ±0.3*
Mean BP (mmHg)	94 ±2	100 ±2**	94 ±2	97 ±2*
HR (bpm)	68 ±3	73 ±3**	73 ±3	75 ±3
P_{ETCO_2} (mmHg)	43.1 ±1.4		57.8 ±0.7	

Values are mean ±SE (n =13) *$p<0.05$; **$p<0.01$, compared with control values

isoflurane did not cause any change in respiration and circulation. Also, in 5 patients who received insufflation of isoflurane into the pharynx, neither changes in breathing pattern nor changes in HR and BP were observed.

DISCUSSION

Nasal reflexes, like many other reflexes, consist of a nervous receptor, afferent pathway, central synapses, motor pathway, and effector organ. Although stimulation of afferent nerve endings or receptors is the natural starting point of reflex responses, the structure of afferent end-organs in the nasal mucosa, the patterns of discharge in their fibers and their responses to activation have not ben fully analyzed. Also, the correlation between the receptor activities and reflex responses to nasal stimulation has not been established. However, it is generally believed that the trigeminal nerves carry most of the fibers from nasal receptors responsible for nasal reflexes[6], and there is evidence that both myelinated and non-myelinated fibers in the trigeminal pathways are activated by odors and irritants[7].

In this study we confirmed our previous observation that nasal insufflation of isoflurane elicits respiratory reflex responses[1]. The major finding of this study is that both respiratory and circulatory responses were attenuated during hypercapnia. Since isoflurane is a mildly-pungent anesthetic agent, there is no doubt that the observed reflex responses were started with direct stimulation of nasal receptors. Our observations that the reflex responses were never observed after aerosol spray of lidocaine into the nostrils and during oral insufflation of isoflurane are in agreement with this notion. The respiratory reflex responses were characterized by a consistent prolongation of T_E with inconsistent changes in V_T and T_I.

The finding that the effect of nasal irritation on T_E was much less pronounced at hypercapnia than that at resting P_{ETCO_2} indicates that CO_2 exerts an inhibitory influence on respiratory responses to nasal stimulation. A similar inhibitory effect of CO_2 on reflex responses to airway irritation in humans has been reported in our previous study[2] in which respiratory reflexes were elicited by stimulation of tracheal mucosa with water. The inhibitory effect of CO_2 on airway defensive reflexes is not unique to human responses to airway stimulation. For instance, it has been shown in cats[3] that hypercapnia increases ventilation but decreases the degree and duration of laryngospasm induced by stimulation of superior laryngeal nerve. Thus, the attenuation of reflex activities during hypercapnia may be a common feature of airway defensive reflexes. Although it is possible that inhibitory effect of hypercapnia on nasal reflexes may develop at the receptor site, whether changes in CO_2 directly affect the activities of these fibers is unknown. The results of animal studies show that changes in alveolar P_{CO_2} has

little or no effect on irritant and c-fiber receptors in the vagus[8,9].
Assuming that myelinated and non-myelinated fibers in the trigeminus correspond to irritant and c-fiber receptors in the vagus, respectively, it is unlikely that attenuation of nasal reflexes occurs at the receptor site. It is more likely that the site of action of CO_2 is the central nervous system. It is conceivable that an increased respiratory activity due to hypercapnia counteracts the respiratory inhibition produced by nasal mucosa stimulation. It is also possible that CO_2 may exert some inhibitory effect on central nervous system responsible for nasal reflexes. In this connection, there is evidence in the literature that in the majority of neurons in the mammalian central nervous system, a rise in $PaCO_2$ is accompanied by a hyperpolarization, indicating a reduction in the excitability of these neurons[10]. Whatever the mechanisms of CO_2 inhibitory influence on nasal reflexes may be, attenuation of respiratory inhibition may be beneficial to maintenance of adequate ventilation in face with severe depression of respiration elicited by nasal stimulation.
Nasal insufflation of isoflurane caused significant increases in BP and HR. However, the onset of circulatory responses was much delayed, compared with the respiratory response. Also, the circulatory responses, particularly HR response, were less consistent in contrast to the consistent appearance of respiratory responses of nasal stimulation. These observations indicate that the responses of respiration and circulation do not necessarily parallel each other, presumably due to different thresholds for respiratory and circulatory responses.

In conclusion, we studied changes in respiration and circulation of anesthetized humans in response to stimulation of the nasal mucosa at two different levels of $P_{ET}CO_2$. At resting breathing nasal irritation characteristically caused prolongation of T_E with small increases in BP and HR. At an elevated level of $P_{ET}CO_2$ the changes in respiration, BP, and HR were significantly reduced. These results indicate that hypercapnia exerts an inhibitory influence on nasal reflexes.

REFERENCES

1. T. Nishino, et al.: Respiratory, laryngeal and tracheal responses to nasal insufflation of volatile anesthetics in anesthetized humans. Anesthesiology 75: 441-444,1991
2. T. Nishino et al.: Inhibitory effects of CO_2 on airway defensive reflexes in enflurane-anesthetized humans. J. Appl. Physiol.66: 2642-2646, 1989
3. T. Nishino, T. Yonezawa, and Y. Honda: Modification of laryngospasm in response to changes in $PaCO_2$ and PaO_2 in the cat. Anesthesiology 55: 286-291, 1981
4. T. Nishino and Y. Honda : Effects of $PaCO_2$ and PaO_2 on the threshold for the inspiratory-augmenting reflex in cats. J. Appl. Physiol. 53; 1152-1157, 1982
5. S.A. Coleman et al.: Inhalation induction of anaesthesia with isoflurane: Effect of added carbon dioxide. Br. J. Anaesth. 67: 257-261, 1991
6. R. Eccles: Neurological and pharmacological considerations. In: The Nose; Upper Airway Physiology and the Atmospheric Environment. Edited by D.F. Proctor, and I.B. Andersen. Elsevier Biomedical, Amsterdam, 1982, pp191-214
7. V.I. But, and V.I. Klimova-Cherkasova : Afferentation from upper respiratory tract. Bull. Exp. Biol. Med. 64: 13-16, 1967
8. H.M. Coleridge, J.C.G. Coleridge and R.B. Banzett : Effect of CO_2 on afferent vagal endings in the canine lung. Respir. Physiol. 34: 135-141, 1978
9. H. Sellick and J. G. Widdicombe: The activity of lung irritant receptors during pneumothorax, hyperpnea and pulmonary vascular congestion. J. Physiol. London 203: 359-381, 1969
10. R.A, Mitchell and D.A. Herbert: The effect of carbon dioxide on the membrane potential of medullary neurons. Brain Res. 75: 345-349, 1974

ELECTROPHYSIOLOGICAL AND ANATOMICAL STUDIES OF THE SECOND ORDER NEURONS IN THE REFLEX PATHWAY FROM PULMONARY RAPIDLY ADAPTING RECEPTORS IN THE CAT

Kazuhisa Ezure[1], Kazuyoshi Otake[2], Janusz Lipski[3], and Richard B. Wong She[3]

[1]Department of Neurobiology, Tokyo Metropolitan Institute for Neuroscience, Tokyo 183, Japan; [2]Department of Anatomy, Tokyo Medical and Dental University, Tokyo 113, Japan; [3]Department of Physiology, University of Auckland, Auckland, New Zealand

INTRODUCTION

Pulmonary rapidly adapting receptors (RARs)[7], also known as lung irritant receptors, elicit various respiratory reflexes including increase of inspiratory activity and shortening of expiration, increase of respiratory frequency (tachypnea), augmented breaths (sighs), coughing, mucus secretion and bronchoconstriction[2, 10, 11]. Despite the large number of reports on the properties of RARs, there have been relatively few studies on the central pathways through which the respiratory reflexes evoked by the activation of these receptors are mediated. The projections and central terminations of afferent fibers originating from RARs have been traced to the nucleus of the solitary tract (NTS) with antidromic mapping[4], spike-triggered averaging of field potentials[8], and intraaxonal labeling with HRP[6]. The presence of second order neurons was suggested in the medial and caudal portion of the NTS, especially in the commissural subnucleus (COM) caudal to the obex in both the cat[8] and the rat[1].

This study was conducted: (1) to locate and characterize the second-order neurons, denoted as RAR-cells (pulmonary rapidly adapting receptor-activated cells), in the pathway originating from RARs[9], and (2) to determine the efferent projections of RAR-cells[5].

RESULTS

Identification of RAR-cells

Experiments were conducted on cats which were initially anesthetized with Nembutal, paralyzed and artificially ventilated after a bilateral pneumothorax. Extra- and intracellular recordings were made from neurons located in the NTS between 1 mm rostral

Control of Breathing and Its Modeling Perspective, Edited by
Y. Honda *et al.*, Plenum Press, New York, 1992

and 2 mm caudal to the obex. For identification of RAR-cells, electrical stimulation of the vagus nerves and the four following 'physiological' stimuli (known to activate RARs[2, 10, 11]) were used: (1) Collapse of the lungs to atmospheric pressure. (2) Increases in tidal volume. (3) Maintained lung inflations. (4) Administration of ammonia vapor. Since none of these stimuli in isolation is specific for the activation of RARs, and it was often impossible to perform all the tests without losing a unit, a unit was designated as a RAR-cell if it was activated by electrical stimulation of myelinated afferents of the vagus nerve(s) (below 2 times the threshold for the activation of afferents from pulmonary stretch receptors) and by at least two of the above described 'physiological' stimuli.

A total of 121 RAR-cells were found extracellularly in the NTS area, with the majority (85%) located in the COM (Fig. 1). Under control ventilatory conditions, around half of these RAR-cells were either silent, or exhibited irregular activity not clearly related to the ventilator (Fig. 1A), their RAR-related activity becoming evident only after the 'physiological' stimuli were applied. Other RAR-cells showed 'spontaneous' ventilation-related activity, occurring primarily during ventilator-induced deflations or inflations. In some of the RAR-cells reversal of this ventilator-related modulation, from firing predominantly during lung deflations to lung inflations, and *vice versa*, could be induced by changing the lung volume or by the inhalation of ammonia vapor. Modulation of firing in synchrony with the central respiratory rhythm and/or cardiac pulses was also observed in the activity of 14% and 18% of the RAR-cells, respectively.

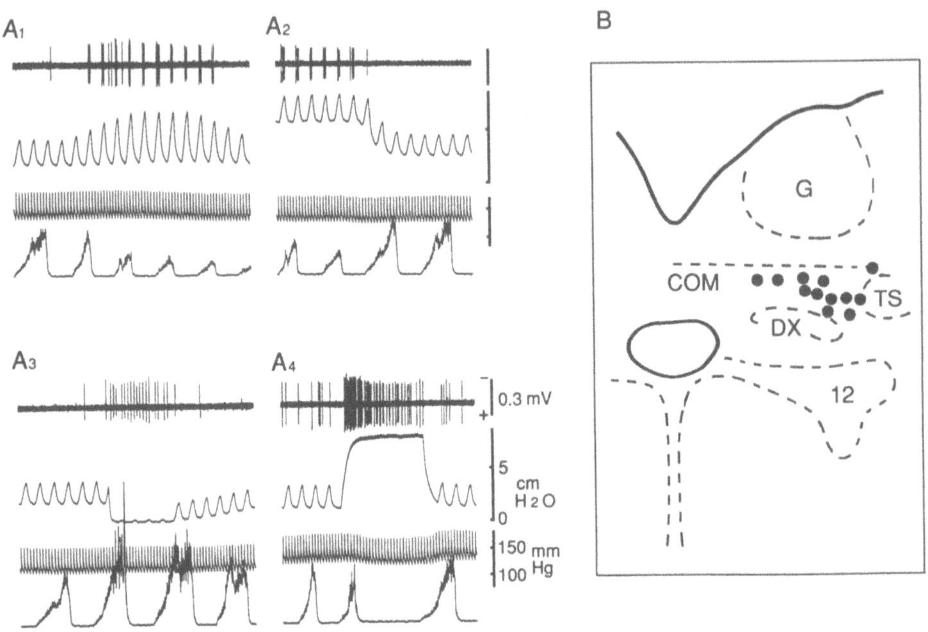

Figure 1. Identification and location of RAR-cells. A1-4: RAR-cell that exhibited weak irregular firing under control ventilatory conditions. The unit was orthodromically activated from both ipsi- and contralateral vagus nerves (not illustrated) and responded to increased tidal volume (A1), increased positive end-expiratory pressure (A2), collapse of the lungs to atmospheric pressure (A3), and maintained lung inflation (A4). Traces from top: unit activity, tracheal pressure, arterial blood pressure, and integrated (τ = 10 ms) phrenic nerve activity. B: Location of RAR-cells recorded caudal to the obex are reconstructed in the transverse plane (1mm caudal to the obex) based on deposits of Fast Green. DX, dorsal motor nucleus of the vagus; G, gracile nucleus; 12, hypoglossal nucleus; TS, solitary tract.

The majority (87%) of the RAR-cells recorded within the COM responded orthodromically to electrical stimulation of both ipsi- and contralateral vagal stimulation. The remaining RAR-cells, located in both the COM or the caudal portion of the medial subnucleus of the NTS rostral to the obex, responded to the ipsilateral vagus only. The latencies of the activated spikes were short (generally < 7 ms) and were stable at the stimulus intensity above 1.5 times threshold, suggesting that the excitation was monosynaptic. This monosynaptic nature was supported by intracellular recordings made from 22 COM neurons (including two identified RAR-cells) that received EPSPs of fixed latencies and fast rise times, from both ipsi- and contralateral vagus nerves.

In some animals, bilateral lesions in the COM were made by suction. The lesions abolished the reflex responses induced by ammonia inhalation or hyperinflation of the lungs, but not the Hering-Breuer reflex, confirming that a pathway through the COM is involved in the reflex arc originating from RARs.

Efferent projections of RAR-cells

A total of 80 RAR-cells were tested for axonal projections to respiration-related areas in the brainstem: the dorsolateral rostral pons; the dorsal respiratory group (DRG); the ventral respiratory group (VRG); or the Bötzinger complex (BOT), and/or the spinal cord. Twenty-two of the 47 (47%) RAR-cells tested for ipsilateral pontine projection could be antidromically activated, and in 8 cases evidence for axonal arborization was obtained. Estimated conduction velocities were less than 1 m/s, suggesting that projecting axons were unmyelinated. One of the 11 RAR-cells tested for DRG projection, and one of the 10 RAR-cells tested for VRG projection, were antidromically activated. No RAR-cells were activated from the BOT (n = 8) or from the C_3 - C_4 segments of the spinal cord (n = 11).

As the next step in tracing the efferent pathway originating from RAR-cells, biocytin was injected, as an anterograde tracer, into the COM, 0.8-1.0 mm lateral to the midline and 1-2 mm caudal to the obex, in 3 animals. After a survival period of 2-3 days, the animals were deeply anesthetized and transcardially perfused with a fixative. Labeled

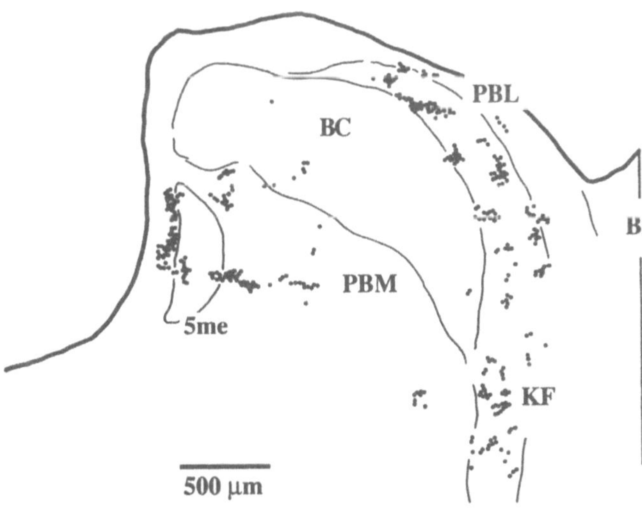

Figure 2. Terminal boutons (dots) in the ipsilateral dorsolateral pons found after injection of biocytin (0.3 μl) in the COM. Several consecutive sections, spanning 200 μm, are superimposed. BC, brachium conjunctivum; BP, brachium pontis; KF: Kölliker-Fuse nucleus; PBL, lateral parabrachial nucleus; PBM, medial parabrachial nucleus; 5me, mesencephalic trigeminal tract.

axons and boutons were examined in the brainstem from the rostral pons to the caudal medulla, and found mainly in the following regions: (1) The dorsolateral rostral pons around the brachium conjunctivum. Terminal boutons were observed in the lateral and medial parabrachial nuclei, Kölliker-Fuse nucleus, and around the mesencephalic trigeminal tract (Fig. 2). This area corresponds to the 'pontine respiratory group' also known as the 'pneumotaxic center'; (2) The pontine area dorsolateral to the superior olivary nucleus, the region known to contain the A5 noradrenergic cell group; (3) Near the ventral surface, below the facial nucleus. This area overlaps with the 'retrotrapezoid nucleus'[3]; (4) Respiration-related areas of the medulla, including the DRG, VRG and the BOT; (5) The dorsal motor nucleus of the vagus. This projection was particularly dense; (6) The area around the solitary tract. These projections were bilateral with generally ipsilateral predominance.

CONCLUSIONS

Second order relay neurons from RARs (RAR-cells) were identified in the medial and caudal portion of the NTS, particularly in the COM. The response characteristics of these RAR-cells largely correspond to known properties of RARs. Both antidromic stimulation experiments and biocytin injection to the COM have revealed prominent efferent projections to the dorsolateral pons, the area corresponding to the 'pontine respiratory group'. Thus the pathway to the dorsolateral pons via the COM is likely to play a role in mediating respiratory responses from RARs. Other efferent pathways found may represent additional projections from RAR-cells, or projections from as yet unidentified neurons which relay information from different afferents terminating in the COM, including afferents involved in autonomic control.

REFERENCES

1. A.C. Bonham and McCrimmon, D.R., Neurones in a discrete region of the nucleus tractus solitarius are required for the Breuer-Hering reflex in rat, *J. Physiol. (Lond.)*, 427: 261-280 (1990).
2. H.M. Coleridge and J.C.G. Coleridge, Reflexes evoked from tracheobronchial tree and lungs. In N.S. Cherniack and J.G. Widdicombe (Eds.), *Handbook of Physiology. Section 3, The Respiratory System Vol. II. Control of Breathing*, American Physiological Society, Washington, DC, pp 395-429 (1986).
3. C.A. Connelly, H.H. Ellenberger and J.L. Feldman, Respiratory activity in retrotrapezoid nucleus in cat. Am. J. Physiol. 258 (Lung Cell. Mol. Physiol. 2): L33-L44 (1990).
4. R.O. Davies and L. Kubin, Projection of pulmonary rapidly adapting receptors to the medulla of the cat: an antidromic mapping study, *J. Physiol. (Lond.)*, 373: 63-86 (1986).
5. K. Ezure, K. Otake, J. Lipski and R.B. Wong She, Efferent projections of pulmonary rapidly adapting receptor relay neurons in cats, *Brain Res.*, 564: 268-278 (1991).
6. M. Kalia and D. Richter, Rapidly adapting pulmonary receptor afferents: I. arborization in the nucleus of the tractus solitarius, *J. Comp. Neurol.*, 274: 560-573 (1988).
7. G.C. Knowlton and M.G. Larrabee, A unitary analysis of pulmonary volume receptors, *Am. J. Physiol.*, 147: 100-114 (1946).
8. L. Kubin and R.O. Davies, Sites of termination and relay of pulmonary rapidly adapting receptors as studied by spike-triggered averaging, *Brain Research*, 443: 215-221 (1988).
9. J. Lipski, K. Ezure and R.B. Wong She, Identification of neurons receiving input from pulmonary rapidly adapting receptors in the cat, *J. Physiol. (Lond.)*, 443: 55-77 (1991).
10. G. Sant'Ambrogio, Information arising from the tracheobronchial tree of mammals, *Physiol. Rev.*, 62: 531-569 (1982).
11. J.G. Widdicombe, Pulmonary and respiratory tract receptors, *J. Exp. Biol.*, 100: 41-57 (1982).

LARYNGEAL AFFERENT ACTIVITY AND ITS ROLE IN THE CONTROL OF BREATHING

Franca B. Sant'Ambrogio, James W. Anderson, and Giuseppe Sant'Ambrogio

Department of Physiology and Biophysics
The University of Texas Medical Branch at Galveston
Galveston, Texas 77550-2774, U.S.A.

The afferent activity emerging from the larynx travels through the superior laryngeal nerve and the recurrent laryngeal nerve. Recording whole nerve activity from the internal branch of the superior laryngeal nerve, the pathway which carries most of the laryngeal afferent information, reveals a clear respiratory-related activity. This respiratory modulation is related to changes in transmural pressure, activity of laryngeal muscles, pulling on the larynx due to the respiratory movements of the trachea and changes in temperature.

When single unit activity from the superior laryngeal nerve (SLN) is studied we can recognize three types of respiratory modulated receptors:

a) cold receptors - affected specifically by changes in laryngeal temperature and not by changes in transmural pressure or any other mechanical deformation of the receptor field;

b) pressure receptors - affected mainly or only by transmural pressure;

c) "drive" receptors - affected mainly by the activity of intrinsic and extrinsic laryngeal muscles, and the pull transmitted by the passive movements of the trachea (tracheal tug) due to the activity of the chest wall respiratory muscles. The vast majority of these endings are also affected to some extent by changes in laryngeal transmural pressure[1].

Besides these respiratory-modulated receptors, we can record activity from endings which are either silent or randomly active. These afferents are readily recruited by several mechanical and chemical nociceptive stimuli[2]. We refer to these endings as laryngeal irritant receptors.

Although each type of receptors is preferentially affected by one particular stimulus, the same stimulus can affect more than one type of ending. A case in point is provided by cooling of the laryngeal lumen which not only stimulates cold receptors[3] but also inhibits both pressure and "drive" receptors[4]. Moreover, the same stimulus can not only affect more than one type of receptor, but it can also do so through different mechanisms. For instance, distilled water stimulates pressure and "drive" receptors due to its hypo-osmolality, and irritant receptors due to the lack of chloride ions[5]. Indeed, iso-osmolal solutions lacking chloride ions, such as iso-osmolal dextrose or sodium gluconate, stimulate irritant receptors but do not affect pressure or "drive" receptors[5].

All this considered, reflex responses should be expected to depend, not only on the direct stimulation of a given receptor type, but also on the concurrent

Control of Breathing and Its Modeling Perspective, Edited by
Y. Honda *et al.*, Plenum Press, New York, 1992

influence of other types of receptors affected directly or indirectly by the same stimulus. Moreover, different stimuli impinging on the same receptor type can affect the overall response.

Receptors stimulated by negative pressure are thought to be involved in the maintenance of laryngeal patency; in fact, laryngeal collapsing pressure reflexly increases the contraction of the posterior crico-arytenoid muscle, a laryngeal abductor[6].

We wanted to study the interaction of pressure and water stimulation on the activity of laryngeal slowly adapting mechanoreceptors[7]. Experiments were performed in dogs anesthetized, breathing spontaneously through a tracheostomy and with the upper airway functionally isolated. Single unit action potentials were recorded from the peripheral cut end of the SLN. Upper airway occlusion (laryngeal collapsing pressure) was performed after administration of either saline or H_2O aerosols into the upper airway. Eleven receptors were studied. For similar changes in esophageal pressure, the rate of discharge of the endings was 39% higher after H_2O aerosol as compared to after saline aerosol (fig. 1).

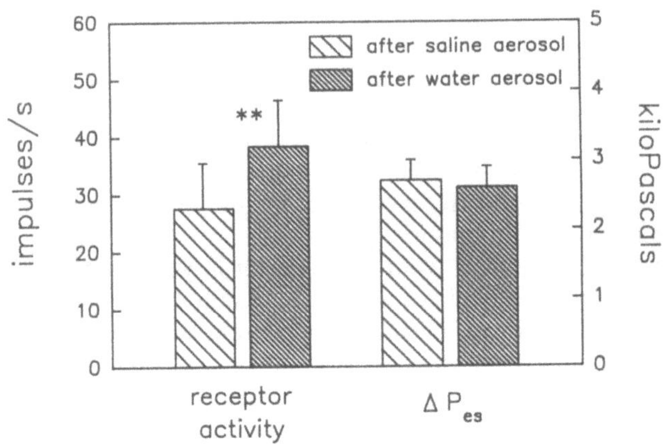

Figure 1. Effect of water aerosols on the response to negative pressure of laryngeal respiratory modulated receptors (mean ± SE; n = 11). The stimulation of the receptors by upper airway occlusions of a similar magnitude (ΔP_{es}) is greater after H_2O aerosols than after saline aerosols. ** = $P < 0.01$.

In preliminary experiments we also showed the effect of the interaction of stimuli on the reflex response of the posterior cricoarytenoid muscle. We found that when the laryngeal innervation is intact, water aerosols challenges enhance the response of the PCA to laryngeal negative pressure.

In humans, low chloride solutions have a tussigenic effect[8] which is attenuated by furosemide[9]. Irritant receptors are thought to be involved in the cough response. We wanted to ascertain if the stimulation of laryngeal irritant receptors by low-chloride solutions was affected by furosemide. Also this study was conducted on dogs, anesthetized, breathing spontaneously through a tracheostomy, and with the upper airway functionally isolated[10]. Single unit action potentials were recorded from the peripheral cut end of the SLN. Ten receptors were studied. The larynx was challenged with instillation of 4 ml solution of iso-osmolar dextrose before and after upper airway aerosolization of furosemide (a chloride ion transport inhibitor) for 8 min. After furosemide, the mean rate of discharge of the irritant receptors during

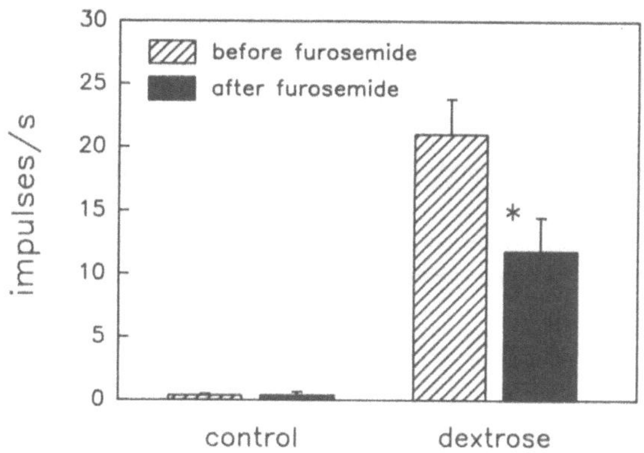

Figure 2. Effect of furosemide aerosols on the response of laryngeal irritant receptors to iso-osmolar solutions of dextrose (mean ± SE; n = 10). Furosemide diminishes the stimulation of the receptors by dextrose, but does not affect their control activity. * = $P < 0.05$.

the first 10 seconds following the administration of the challenge, was 56% of that before furosemide (fig. 2).

In summary, one must be aware that most reflex responses are probably not attributable to a single type of receptor since the same stimulus can affect more than one type of receptor and more than one stimulus can impinge on the same type of ending modifying its reflex effects. In particular, we may infer that either changes in airway surface liquid composition or pharmacological interventions which interfere with ion transport in the laryngeal mucosa modify the activity of laryngeal receptors and may provide a tool for "tuning" the reflex responses attributable to these endings.

Acknowledgments. Supported by NIH Grant HL-20122.

References

1. G. Sant'Ambrogio, O.P. Mathew, J.T. Fisher, and F.B. Sant'Ambrogio, Laryngeal receptors responding to transmural pressure, airflow and local muscle activity, *Respir.Physiol.* 54:317 (1983).
2. G. Sant'Ambrogio, J.W. Anderson, F.B. Sant'Ambrogio, and O.P. Mathew, Response of laryngeal receptors to water solutions of different osmolality and ionic composition, *Respir.Med.* 85:57 (1991).
3. G. Sant'Ambrogio, O.P. Mathew, and F.B. Sant'Ambrogio, Laryngeal cold receptors, *Respir.Physiol.* 59:35 (1985).
4. G. Sant'Ambrogio, F. Brambilla-Sant'Ambrogio, and O.P. Mathew, Effect of cold air on laryngeal mechanoreceptors in the dog, *Respir.Physiol.* 64:45 (1986).
5. J.W. Anderson, F.B. Sant'Ambrogio, O.P. Mathew, and G. Sant'Ambrogio, Water-responsive laryngeal receptors in the dog are not specialized endings, *Respir.Physiol.* 79:33 (1990).
6. F.B. Sant'Ambrogio, O.P. Mathew, W.D. Clark, and G. Sant'Ambrogio, Laryngeal influences on breathing pattern and posterior cricoarytenoid muscle activity, *J.Appl.Physiol.* 58:1298 (1985).
7. J.W. Anderson, F.B. Sant'Ambrogio, and G. Sant'Ambrogio, Changes in osmolality modify the pressure response of laryngeal receptors, *FASEB J.* 5:A1119 (1991).

8. W.L. Eschenbacker, H.A. Boushey, and D. Sheppard, Alteration in osmolarity of inhaled aerosols cause bronchoconstriction and cough, but absence of a permeant anion causes cough alone, *Am.Rev.Respir.Dis.* 129:211 (1984).
9. P.G. Ventresca, G.M. Nichol, P.J. Barnes, and K.F. Chung, Inhaled furosemide inhibits cough induced by low chloride content solutions but not by capsaicin, *Am.Rev.Respir.Dis.* 142:143 (1990).
10. F.B. Sant'Ambrogio, G. Sant'Ambrogio, and J.W. Anderson, Furosemide modifies the response of laryngeal "irritant" receptors to low-chloride solutions, *FASEB J.* 5:A1119 (1991).

SHORT OVERVIEW: CAROTID BODY ACTIVITIES IN MAN

Yoshiyuki Honda

Department of Physiology
School of Medicine
Chiba University
Chiba, 260 Japan

INTRODUCTION

In Japan during the late 1940s and 1950s, a considerable number of patients with bronchial asthma had therapeutic carotid body resection and some of them still alive in relatively asymptomatic condition. The operations were mainly done in the Department of Surgery of the Chiba University, where Dr. K. Nakayama, Professor of Surgery, developed a procedure for removing only the carotid chemoreceptors, while preserving the baroreceptor function. On the basis of the studies on these patients with carotid body resection, respiratory and circulatory activities of human carotid bodies were assessed.[1]

Response to Progressive Hypoxia and Transient Hypoxia

Figure 1 illustrates the ventilatory response to progressive hypoxia in the patients who had undergone bilateral carotid body resection.[2] As reflected in the unaltered profile of airway Pco_2, either augmentation nor depression of ventilation occurred with development of hypoxia. The right side of this figure indicates the hypoxic ventilatory response in terms of $\Delta\dot{V}_{40}$, the increment of the pulmonary ventilation from an end-tidal Po_2 (P_{ETo_2}) of 100 to a P_{ETo_2} of 40 mmHg while an end-tidal Pco_2 (P_{ETco_2}) was maintained at the resting level. Three groups of patients (C) and uni- and bilateral carotid body-resection (UR and BR) were compared. The UR group exhibited less response than the C group, and the BR group exhibited hardly any response; in the latter case, $\Delta\dot{V}_{40}$ did not differ significantly from zero.

Although patients with bilateral carotid body resection did not respond to progressive hypoxia, a significant elevation of ventilation was transiently seen until the hypoxic influence arrived at the central nervous system (CNS).[2] This is demonstrated in Fig. 2 where a vital capacity volume of a gas mixture of 15 % CO_2 in N_2 (severe hypoxic

Control of Breathing and Its Modeling Perspective, Edited by
Y. Honda *et al.*, Plenum Press, New York, 1992

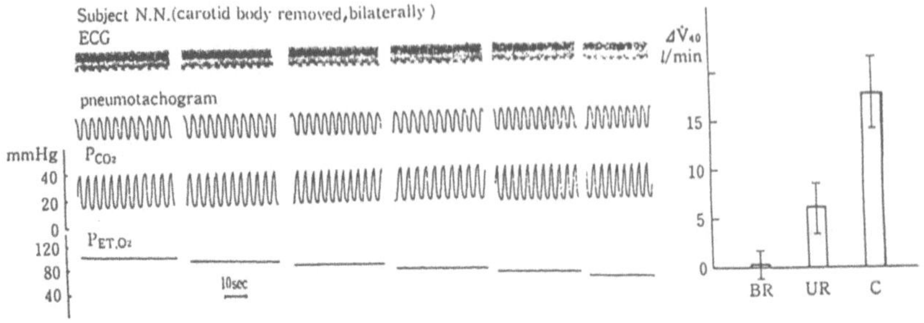

Figure 1. Ventilatory response to progressive hypoxia in subjects with carotid body resection.
P_{ETO_2} was progressively decreased at a rate of 10 mmHg/min from 100 to 40 mmHg. Despite of advancing hypoxia, no change in ventilation was seen and this was clearly reflected by an unaltered profile of airway P_{CO_2}. The right side of the figure indicates the hypoxic ventilatory response in terms of $\Delta\dot{V}_{40}$(mean±SE), i.e., the increment in ventilation as P_{ETO_2} dropped from 100 to 40 mmHg, while keeping P_{ETCO_2} at the resting level. BR, UR and C are the patients with bi- and unilateral carotid body resection and control, respectively. $\Delta\dot{V}_{40}$ of the BR group is statistically not different from zero, whereas that of the UR group is between the BR and C groups.[1] Reproduced by permission from J. Appl. Physiol.

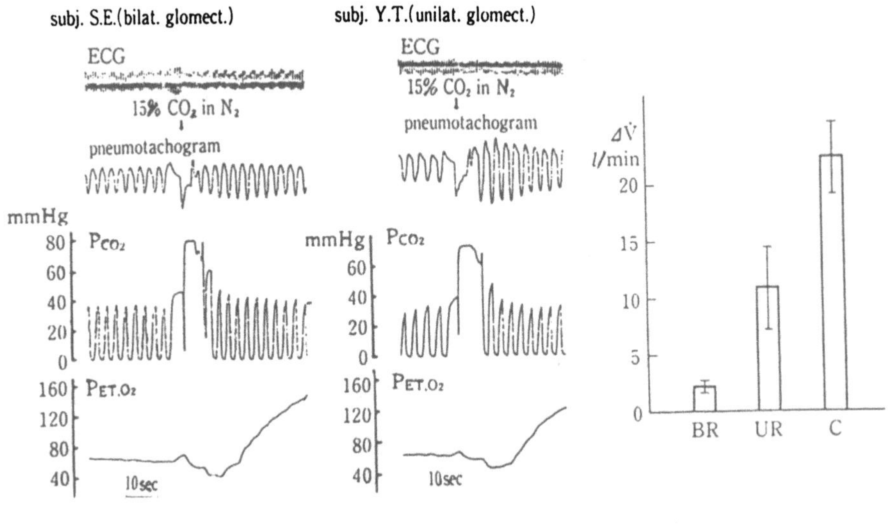

Figure 2. Ventilatory response to a single vital capacity breath test (Single VC breath test) in BR and UR patients.
This response is not affected by possible hypoxic ventilatory depression which is elicited when the CNS is exposed to continuous hypoxemia. The right side of the figure illustrates the response to single VC breath test, $\Delta\dot{V}$, of the three groups. In contrast to the ventilatory response to progressive hypoxia, the $\Delta\dot{V}_{40}$ shown in Fig. 1, the BR group now shows the $\Delta\dot{V}$ to be significantly greater than zero.[1] Reproduced by permission from J. Appl. Physiol.

hypercapnia) was inhaled in one breath following maximal expiration, i.e. a single VC breath test.[3] Bouverot et al.[4] also detected a significant response in two dogs by a similar technique. Swanson et al.[5] demonstrated in 5 patients with bilateral carotid body resection that hypoxic air inhalation under hypercapnic background exhibited a weak but still significant ventilatory response. They assumed that the aortic body is responsible for this ventilatory chemosensitivity. Judging from the magnitude of ventilatory augmentation by single VC breath test, as well as by the drop in ventilation following two breaths of O_2 with spontaneous respiration (withdrawal test), we estimated this residual chemosensitivity to be 5-10 % of the full hypoxic response of the control patients group.[2]

Hypoxic Ventilatory Depression

The fact that the weak hypoxic ventilatory response detected by the single VC breath test disappeared during progressive hypoxia as described above led us to assume that this was caused by an inhibitory influence of the CNS as it became exposed to hypoxia. This hypoxic ventilatory depression was more directly demonstrated by comparing the CO_2-ventilation responses in the control and carotid body-resected patient groups.[6] The response curves of the two groups shifted in opposite directions with hypoxia, i.e., leftward in the control and rightward in the carotid body-resected group. Therefore, the ventilation of the latter group at a given P_{ETCO_2} of 40 mmHg was significantly depressed compared to that of the control group. A significant rightward shift of the CO_2-ventilation response curve by hypoxia has also been reported after carotid endarterectomy in humans by Wade et al.[7]

No Significant Restoration in Peripheral Chemosensitivity

Following carotid body denervation, Smith and Mills[8] and Bisgard et al.[7] demonstrated an 85 % and 30 % recovery of hypoxic chemosensitivity at 215-260 days in 7 cats and 22 months in 8 ponies, respectively. They further found that this restoration disappeared after aortic nerve section,

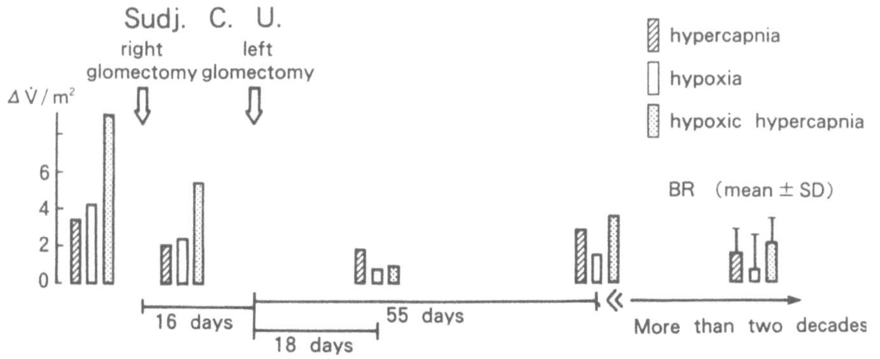

Figure 3. Chronological change in peripheral chemosensitivity detected by single VC breath test before and after carotid body resection. For details, see explanation in text.

suggesting an augmented gain of the aortic body chemoreflex some period after carotid body denervation.

Figure 3 illustrates the profile of chronological peripheral chemosensitivity detected by the single VC breath test following carotid body resection. Peripheral chemosensitivity was evaluated by the responses elicited from hypercapnic (15 % CO_2 in O_2), hypoxic (5 % CO_2 in N_2) and hypercapnic-hypoxic gas inhalation (15 % CO_2 in N_2). In one subject C. U. these responses were detected before and after uni- and bilateral carotid body resection. After bilateral resection, the responses were evaluated at 18th and 55th days. These magnitude were compared with those of patients operated more than two decades ago, and found no significant augmentation even after 20 yrs.

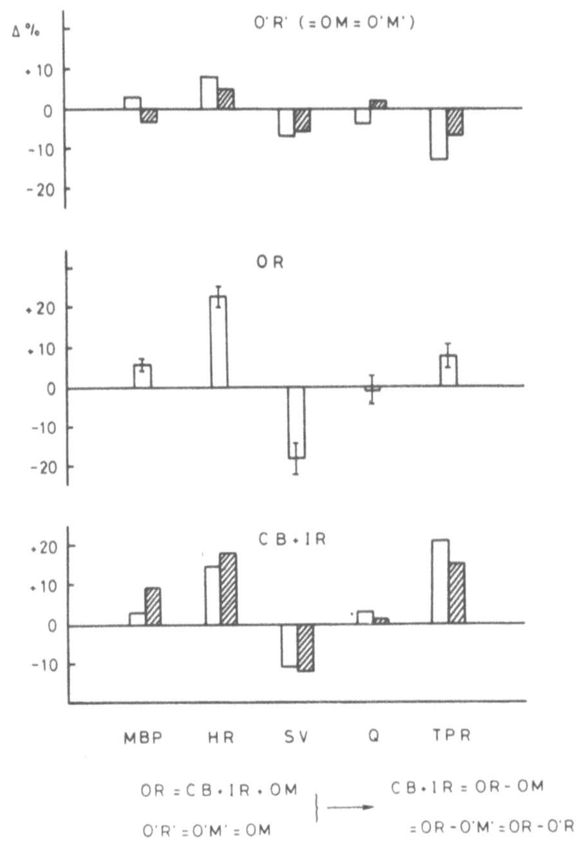

Figure 4. Estimation of the combined effect of carotid body stimulation and pulmonary inflation reflex.
MBP: mean blood pressure HR: heart rate SV: stroke volume
$\dot{Q}$: cardiac output TPR: total peripheral resistance
For details, see explanation in text. Reproduced by permission from J. Appl. Physiol.

Cardiovascular Activities in Carotid Body-Resected Humans

Daly[10] mentioned in his review that the primary effects of carotid body activity are bradycardia, negative left ventricular inotropic response, decreased cardiac output ($\dot{Q}$) and increased systemic and pulmonary vascular resistance. These changes are most clearly demonstrated in diving mammals. However, considerable species difference is known to exist.[10,11,12] Moreover, the above primary effects are substantially reversed by the pulmonary inflation reflex,[13,14] and modified more or less by additional factors such as catecholamines, autonomic nervous tone, Pao_2 and $Paco_2$ levels, anesthesia, and local vascular and myocardiac conditions.

In humans, no comprehensive studies of the influences of the carotid body on cardiovascular parameters are available. Investigations on carotid body-resected humans, however, should provide important clues for unravelling this problem, though such studies are, as yet, few, and only fragmentary informations are available from the literatures.[7,15,16,17,18]

The uppermost and middle panels of Fig. 4 represent the quantitative responses of mean blood pressure (MBP), heart rate (HR), stroke volume (SV), cardiac output ($\dot{Q}$) and systemic total peripheral resistance (TPR) to sustained moderate hypoxia (Ca 80 % Sao_2) lasting for 20 min in two patients with bilateral carotid body resected and 8 normal control subjects, respectively.

In carotid body-resected patients, HR increased and stroke volume (SV) and systemic total peripheral resistance (TPR) decreased while no marked changes were seen in mean blood pressure (MBP) and $\dot{Q}$. These changes are in contrast with normal subjects who exhibited a marked elevation in HR, significant augmentation in both MBP and TPR and substantial depression in SV.

The overall circulatory response (OR) to hypoxia in normal subjects may be determined by the sum of carotid body stimulation (CB), the reflex effect of pulmonary inflation (IR) and other modifying influence (OM) as described before,

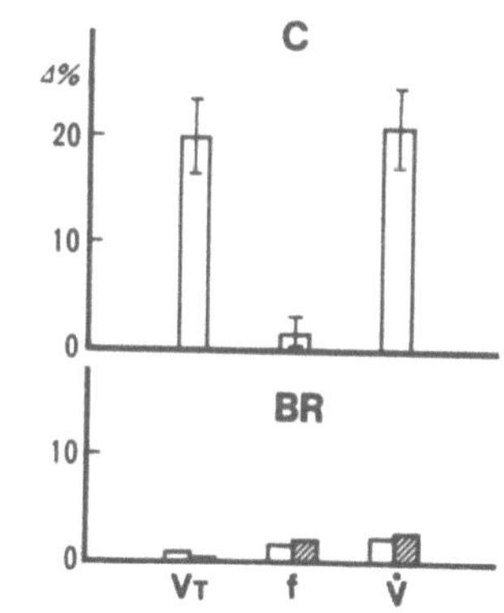

Figure 5. Percentile increments in tidal volume (V_T), respiratory frequency (f) and minute ventilation ($\dot{V}$) in response to sustained mild hypoxia (ca. 80 % Sao_2) in 8 control and 2 BR subjects.
No appreciable increment in f, $\dot{V}$ and particularly in V_T was seen in BR patients.

i.e., OR=CB+IR+OM. Now, as represented by Fig. 5 no significant augmentation in tidal volume (V_T) compared to the control subjects was seen in the patients with bilateral carotid body resection. Accordingly, due to lack of both carotid body and hypoxic hyperventilation, the overall response in the carotid body-resected patients (O'R') may be solely determined by other modifying influences on the circulation (O'M'), namely O'R'=O'M'. If we postulate in relative magnitude that OM equals O'M', then CB+IR can be calculated as OR-O'R'. The upper panel of Fig. 6 shows the result of this calculation in two patients with bilateral carotid body resection.

When we take into consideration Daly's statement[10] that the primary effects of carotid body stimulation are the elevation of MBP, $\dot{Q}$ and TPR and the depression of HR and SV as indicated by the filled arrows in Fig. 6, comparison of the upper against lower panel in this figure leads us to assume that MBP, SV and TPR are predominantly determined by carotid body stimulation whereas HR is effectively reversed from the depressed state by the carotid body stimulation to augmentation by pulmonary inflation reflex.

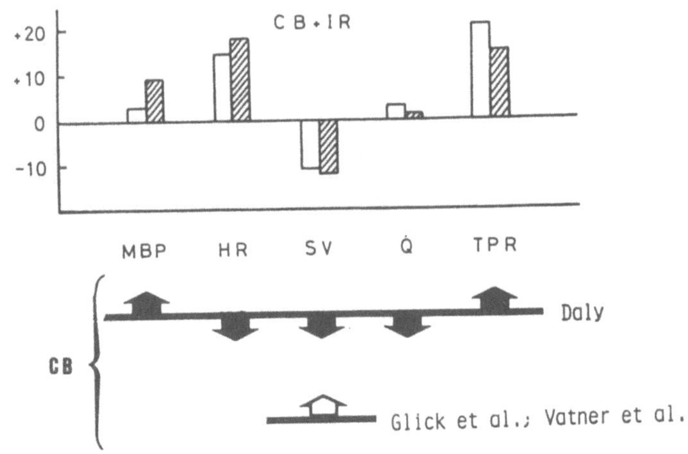

Figure 6. Estimated combined effects of carotid body stimulation (CB) and pulmonary inflation reflex (IR) on the circulatory parameters were compared against the known primary influences of carotid body stimulation. For details, see explanation in text.

Contrary to Daly's view Glick et al.[19] and Vatner and Rutherford[20] claimed enhanced left ventricular dp/dt by carotid body stimulation in anesthetized and awake dogs, respectively. If this is indeed the case, diminished SV by combined CB and IR may be mainly determined by IR. Thus, the main cardiac changes, i.e., HR and SV, are predominantly controlled by IR, whereas vascular changes, i.e., MBP and TPR, are mainly governed by carotid body stimulation. Hilton and Marshall[21] found in cats that vasoconstrictor effects on renal and mesenteric vessels by carotid body stimulation were not influenced by the pulmonary inflation reflex. The predominant role of the CB on vascular change found in our analysis in man appear to be in line with their findings.

Although we postulated that effects of CB, IR and OM are simply additive, evidence for on interaction between CB and OM has been reported by Hilton and Marshall.[21] They showed that carotid body stimulation elicited augmented sympathetic activity and adrenaline release. If this is the case, the magnitude of the CB's effect may have been underestimated in our analysis. However, Cherniack et al.[22] recently demonstrated marked sympathetic activity by hypoxic stimulation in anesthetized and peripheraly chemodenervated cats. Therefore, substantial amount of sympathetic contribution to circulatory activity in hypoxia seems to be expected without carotid body activation.

Summary

1. Specific features of human carotid body activities in respiration and circulation are presented.
2. From the cardiovascular activities in the carotid body-resected humans, influence of carotid body stimulation and pulmonary inflation reflex are analyzed.

REFERENCES

1. Y. Honda, Brief Review: Respiratory and circulatory activities in carotid body-resected humans, J. Appl. Physiol. 73:1 (1992).
2. Y. Honda, S. Watanabe, I. Hashizume, Y. Satomura, N. Hata, Y. Sakakibara, and J.W. Severinghaus, Hypoxic chemosensitivity in asthmatic patients two decades after carotid body resection, J. Appl. Physiol.: Respirat. Environ. Exercise Physiol. 46:632 (1979).
3. R.A. Gabel, R.S. Kronenberg, and J.W. Severinghaus, Vital capacity breaths of 5 % or 15 % CO_2 in N_2 or O_2 to test carotid chemosensitivity, Respir. Physiol. 17:195 (1973).
4. P. Bouverot, V. Candas, and J.P. Liberi, J.P., Role of the arterial chemoreceptors in ventilatory adaptation to hypoxia of awake dogs and rabbits, Respir. Physiol. 17:209 (1973).
5. G.D. Swanson, B.J. Whipp, R.D. Kaufman, K.A. Aqleh, B. Winter, and J. W. Bellville, Effect of hypercapnia on hypoxic ventilatory drive in carotid body-resected man, J. Appl. Physiol. Respirat. Environ. Exercise Physiol. 45:971 (1978).
6. Y. Honda, and I. Hashizume, Evidence for hypoxic depression of CO_2-ventilation response in carotid body-resected humans, J. Appl. Physiol. 70:590 (1991).
7. J.G. Wade, C.P.Jr. Larson, R.F. Hickey, W.K. Ehrenfeld, and J.W. Severinghaus, Effect of endarterectomy on carotid chemoreceptor and baroreceptor function in man, N. Engl.J. Med. 282:823 (1970).
8. U. Smith, and E. Mills, Respiration of reflex ventilatory response to hypoxia after removal of carotid bodies in the cat, Neuroscience 5:573 (1970).
9. G.E. Bisgard, H.V. Forster, and J.P. Klein, Recovery of peripheral chemoreceptor function after denervation in ponies, J. Appl. Physiol.: Respirat. Environ. Exercise Physiol. 49:964 (1980).
10. M. De B. Daly, Interaction between respiration and circulation, in: "Handbook of Physiology. The Respiratory System. Control of Breathing," A.P. Fishman, N.S. Cherniack, J.G. Widdicombe and S. R. Geiger, eds., Am. Physiol. Soc., Bethesda, sect. 3, vol. II, pt. 2, p.529 (1986).
11. J.C.G. Coleridge, and H.M. Coleridge, Chemoreflex regulation of the heart, in: "Handbook of Physiology. The Cardiovascular System.

The Heart," R.M. Berne, N. Sperelakins and T. Geiger, eds., Am. Physiol. Soc., Bethesda, sect. 2, vol. I, p.653 (1979).
12. R.S. Fitzgerald, and S. Lahiri, Reflex responses to chemoreceptor stimulation, in: "Handbook of Physiology. The Respiratory System. Control of Breathing," A.P. Fishman, N.S. Cherniack, J.G. Widdicombe and S.R. Geiger, eds., Am. Physiol. Soc., Bethesda, sect. 3, vol. II, pt. 1, p.313 (1986).
13. M. De B. Daly, and M.J. Scott, The cardiovascular effects of hypoxia in the dog with special reference to the contribution of the carotid body chemoreceptors, J. Physiol. 173:201 (1964).
14. J.E. Angell-James, and D. De B. Daly, Cardiovascular responses in apnoeic asphyxia: role of arterial chemoreceptors and the modifications of their effects by a pulmonary vagal inflation reflex, J. Physiol. 201:87 (1969).
15. P. Hilton, and J.B. Wood, The effects of bilateral removal of the carotid bodies and denervation of the carotid sinuses in two human subjects, J. Physiol. 181:365 (1965).
16. R. Lugliani, B.J. Whipp, C. Seard, and K. Wasserman, Effect of bilateral carotid body-resection on ventilatory control at rest and during exercise in man, N. Engl. J. Med. 285:1105 (1971).
17. R. Lugliani, B.J. Whipp, and K. Wasserman, A role for the carotid body in cardiovascular control in man. Chest 63:744 (1973).
18. P.M. Gross, B.J. Whipp, J.T. Davidson, S.N. Koyal, and K. Wasserman, Role of the carotid bodies in the heart rate response to breath holding in man, J. Appl. Physiol. 41:336 (1976).
19. G. Glick, A.S. Wechsler, and S.E. Epstein, with R.M. Lewis, and R.D. McGill, Reflex cardiovascular depression produced by stimulation of pulmonary stretch receptors in the dog, J. Clin. Invest. 48: 467 (1969).
20. S.F. Vatner, and J.D. Rutherford, Interaction of carotid chemoreceptor and pulmonary inflation reflexes in circulatory regulation in conscious dogs, Fed. Proc. 40:2188 (1981).
21. S.M. Hilton, and J.M. Marshall, The pattern of cardiovascular response to carotid chemoreceptor stimulation in the cat, J. Physiol. 326:495 (1982).
22. N.S. Cherniack, J. Mitra, N. Prabhakar, D. Lust, and P. Eransberger, Effect of CNS hypoxia on respiratory and sympathetic activity, in: "1991 Oxford Conference: 5th Meeting on Control of Breathing and Its Modelling Perspective (Abstract)," p.19 (1991).

DYNAMICAL ANALYSIS OF THE VENTILATORY RESPONSE TO CHANGES IN $PETO_2$: SEPARATION OF CENTRAL AND PERIPHERAL EFFECTS

A. Berkenbosch,[1] J. DeGoede,[1] C.N. Olievier[1] and D.S.Ward[2]

[1]Department of Physiology, University of Leiden, 2300 RC Leiden, The Netherlands
[2]Department of Anesthesiology, University of California, Los Angeles, California 90024, U.S.A.

INTRODUCTION

In adult humans and animals as well as in newborns the ventilatory response to an isocapnic stepwise change in end-tidal PO_2 ($PETO_2$) shows an initial fast increase frequently followed by a slow decline.[1,2,3,4,5] The ventilatory increase is generally ascribed to an increased drive from the peripheral chemoreceptors. The mechanism of the subsequent hypoxic ventilatory depression is not well understood.[6] To gain insight into the mechanism of this depression it is necessary to first separate and quantify stimulatory and depressant effects. An attractive technique to do this is the dynamic end-tidal forcing (DEF) technique, since it is non-invasive and can therefore be applied to human beings. However, for the dynamical analysis of the experimental data a mathematical model is needed, the validation of which is not easy a task.

In this study we propose a mathematical model for the ventilatory response to isocapnic changes in $PETO_2$ of anesthetized cats. From experiments using the artificial brain stem perfusion (ABP) technique, it is known that the stimulation of ventilation stems from the peripheral chemoreceptors, while the depression is of central origin.[7] We compare the estimated central and peripheral contributions to the ventilation due to a step change in $PETO_2$ from the DEF technique with those obtained from the ABP technique in which these contributions also can be measured in isolation. It will be shown that it is possible to estimate central and peripheral effects of hypoxia satisfactorily using the DEF technique together with the mathematical model in anesthetized cats.

MATHEMATICAL MODEL

From experiments using the ABP technique it was shown that the ventilatory response following changes in arterial PO_2 of the blood perfusing the brain stem can

Control of Breathing and Its Modeling Perspective, Edited by
Y. Honda *et al.*, Plenum Press, New York, 1992

be described by a single component,[8] while the ventilatory response of the peripheral chemoreflex loop to changes in $PETO_2$ has to be described by a fast and a slow component.[9] We therefore propose a three-component model (M) for the dynamics of the ventilatory response to isocapnic induction or relief of hypoxia, *viz.*

$$\tau_{p1}\, d\dot{V}_{p1}/dt + \dot{V}_{p1} = G_{p1} \exp[-D\ PETO_2(t-T_p)] \tag{1}$$

$$\tau_{p2}\, d\dot{V}_{p2}/dt + \dot{V}_{p2} = G_{p2} \exp[-D\ PETO_2(t-T_p)] \tag{2}$$

$$\tau_c\, d\dot{V}_c/dt + \dot{V}_c = -\ G_c \exp[-D\ PETO_2(t-T_c)] \tag{3}$$

$$\dot{V}_E = \dot{V}_{p1} + \dot{V}_{p2} + \dot{V}_c + C \tag{4}$$

In equations (1) and (2) G_{p1} and G_{p2} are the hypoxic sensitivities of the fast and slow peripheral component with time constants τ_{p1}, τ_{p2} and shape parameter D, while T_p is the transport delay time of the O_2 challenge from lungs to the sites of the peripheral chemoreceptors. Similarly for the central component (eq. 3) G_c is the hypoxic sensitivity, τ_c the time constant and T_c the transport delay time. The total ventilation ($\dot{V}_E$) is given by eq. (4) in which C denotes the ventilation during hyperoxia. In the steady state M reduces to

$$\dot{V}_E = G \exp(-D\ PETO_2) + C \tag{5}$$

with

$$G = G_{p1} + G_{p2} - G_c \tag{6}$$

In previous studies[10] we have found that in the steady state the peripheral and central hypoxic response as well as the overall response are adequately described by exponential expressions with the same shape parameter D.

Dynamical analysis

From each set of experimental data the parameters of M were estimated using a least squares parameter estimation technique. For a step change in $PETO_2$ it is not possible to estimate both the hypoxic sensitivities and the parameter D. We therefore set the parameter D to a fixed value, which was assessed by fitting the steady-state ventilation of each cat as a function of $PETO_2$ (eq. 5). We search for optimal time delays between 0 and 9 seconds with increments of 1 second for the peripheral components and between 0 and 100 seconds with increments of 5 seconds for the central component.

METHODS

Experiments were performed on 7 with α-chloralose-urethan anesthetized cats. Since both the ABP technique[11] and the DEF technique[12] have been previously published we restrict ourselves to brief descriptions.

In the DEF technique the $PETO_2$ is forced to follow a specific pattern in time while the $PETCO_2$ is kept constant. The ventilatory response following a prescribed change in $PETO_2$ is assessed on a breath-by-breath basis.

With the ABP technique the ponto-medullary region is perfused artificially with blood from a femoral artery, which after equilibration in a gas exchanger with a gas of known composition, is pumped into the medulla oblongata and pons. The central chemosensitive structures and respiratory integrating centres are therefore perfused with blood of which the O_2 and CO_2 tensions (called central arterial O_2 tension, Pa^cO_2, and central arterial CO_2 tension, Pa^cCO_2) can be controlled independently from the gas tensions in the blood of the systemic circulation which perfuses the carotid and aortic bodies (called Pa^pO_2 and Pa^pCO_2 respectively). In this way peripheral effects of O_2 can be isolated from the central effects.

Experimental designs

Each cat was subjected to both DEF and ABP experiments. In a DEF experiment each run started with a period of steady-state ventilation of about 2 minutes during which the $PETO_2$ and $PETCO_2$ were held constant. The $PETO_2$ was then stepwise changed from hyperoxia (50 kPa) to hypoxia (6.0 - 8.0 kPa) or the reverse for about 7 minutes. The $PETCO_2$ was kept constant at a few tenths of a kPa above the resting $PETCO_2$ during hyperoxia by giving the animal about 2 per cent CO_2 to inhale. In each cat the $PETCO_2$ was kept constant during the DEF and ABP experiments at the same value. The $PETCO_2$ varied between cats from 4.5 to 5.7 kPa (mean 5.0 kPa). In each cat 2 to 7 runs (mean 4) into hypoxia and out of hypoxia were performed. Besides the DEF runs steady-state ventilation was determined in each cat at 5 different $PETO_2$ values in the range of 50 to 6 kPa.

After the DEF runs, which were measured in non-artificially perfused (intact) cats, artificial perfusion of the brain stem was initiated with hyperoxic blood (Pa^cO_2 50 kPa).

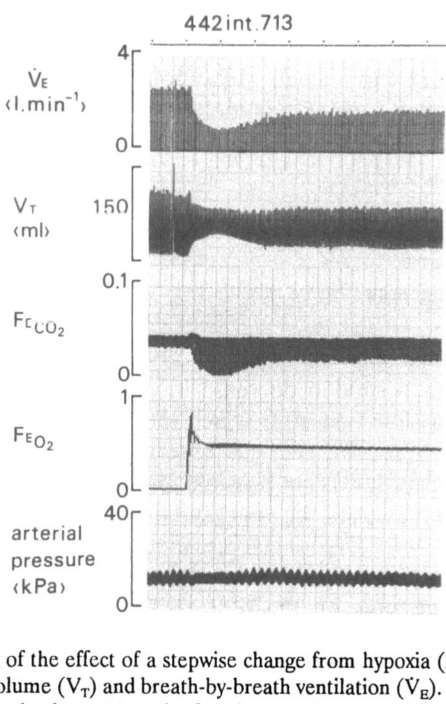

Figure 1. Actual recording of the effect of a stepwise change from hypoxia ($PETO_2$ 6.0 kPa) to hyperoxia ($PETO_2$ 50 kPa) on tidal volume (V_T) and breath-by-breath ventilation ($\dot{V}_E$). FEO_2 and $FECO_2$ denote the fractions O_2 and CO_2 in tracheal gas. Note the fast decrease in ventilation followed by a slower one and the relief of the hypoxic depression shown by a slow increase in ventilation. The upper trace indicates one minute time intervals.

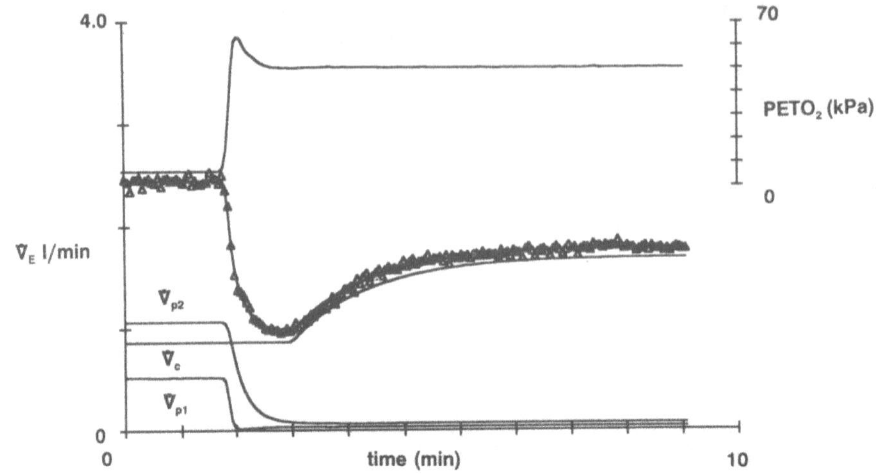

Figure 2. Average response and model fit for a step from hypoxia into hyperoxia. The estimated parameters are: G_c -1.46, G_{p1} 0.82 and G_{p2} 1.58 $l.min^{-1}$; τ_c 70, τ_{p1} 0.3, τ_{p2} 16 s; T_c 65, T_p 2 s; C 1.72 $l.min^{-1}$.

The Pa^cCO_2 was constant but varied between cats from 4.8 to 6.2 kPa. The Pa^pCO_2 in each cat was the same as the $PETCO_2$ during the DEF runs in the intact cat. Usually 15 minutes after the start of the perfusion of the brain stem the ventilation hadreached a steady state and measurements could be started. Stepwise changes in PO_2 in the systemic circulation only, were made by changing the $PETO_2$ in the same way as in the intact cats. Besides 2 to 5 peripheral responses into and out of hypoxia at constant Pa^cO_2 also changes in the Pa^cO_2 at constant Pa^pO_2 were performed to assess the magnitude of the central depressant effect of hypoxia. To this end steady-state ventilation was measured at 4 to 5 levels of Pa^cO_2 in the range of 50 to 5 kPa and was fitted as function of the Pa^cO_2 using equations (3) and (4), setting $PETO_2$ equal to Pa^cO_2. The stimulating effect of hypoxia on the peripheral chemoreceptors was assessed using equations (1), (2) and (4).

RESULTS

DEF experiments in intact cats

An actual recording of a stepwise increase in $PETO_2$ from hypoxia to hyperoxia is shown in fig.1. There was a sudden decrease in tidal volume and breath-by-breath ventilation, followed by a slow one after which the relief of the depressant effect of hypoxia became visible. Since we did not observe an overshoot for the stepwise decrease in $PETO_2$ in all cats, it was not possible to estimate the ventilatory depression from these curves and we only used the ventilatory response of the step out of hypoxia for further analyses.

In order to reduce the noise we interpolated the breath-by-breath data at 3 seconds intervals, time aligned the runs on the time of the step transitions and ensemble averaged the runs of each cat. Figure 2 shows the model fit to the average response of a cat for a step out of hypoxia. The dots are the breath-by-breath ventilation data. The curve through the data is the least squares model fit. The total ventilation is broken up into a fast ($\dot{V}_{p1}$) and a slow ($\dot{V}_{p2}$) peripheral stimulating component and a slow central depressant one ($\dot{V}_c$).

ABP experiments

In each cat we determined the steady-state ventilation as function of Pa^cO_2 while the other blood gas tensions were kept constant. Fig. 3 illustrates a fit to the data obtained from such an experiment using equations (3) and (4). The central O_2 sensitivity obtained in this way was compared with the value of G_c from the DEF experiments in intact cats.

In the DEF runs during ABP we observed no slow peripheral component in 3 of the 7 cats, in concordance with the DEF runs in the intact cats (see Table 1). During ABP an undershoot in the ventilatory response was never found from which it follows that the undershoot in the intact cats is due to the effect of hypoxia in the brain stem.

Comparison of techniques

There is good correspondence between the O_2 sensitivities obtained from the DEF and those obtained from the ABP technique with the exception of one cat which did not show an undershoot in the intact ventilatory response, although a depression of ventilation by brain stem hypoxia was observed during ABP. However, in this cat the depressant effect was small compared to the stimulatory one. The results of all experiments are summarized in table 1.

Since the estimated values for the parameter D varied between cats the correspondence between the peripheral and central contribution to the change in ventilation due to hypoxia can not directly be seen from the values of the sensitivities. We therefore calculated the change in ventilation due to the peripheral chemoreflex ($\Delta\dot{V}_{p1} + \Delta\dot{V}_{p2}$) and the central depressant effect $\Delta\dot{V}_c$ from the estimated parameters of the intact responses and the $PETO_2$ used in the DEF experiments and compared them with the corresponding measured values during ABP experiments at the same oxygen tension. The results are shown in fig. 4. The differences between the corresponding ventilatory components are small and did not reach the level of significance.

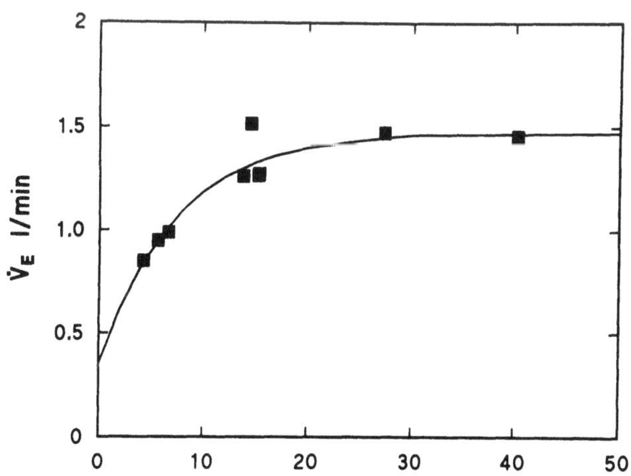

Figure 3. Ventilation as function of the arterial PO_2 of the blood perfusing the brain stem. The drawn curve is the fit using equations (3) and (4) with the value for D found from the intact responses.

DISCUSSION

To gain insight into the mechanisms which play a role in the ventilatory response to hypoxia it is necessary to quantitatively estimate the peripheral and central effects of hypoxia on ventilation. To do this invasive techniques such as sectioning, cold blocking of neural pathways and artificial perfusion of the brain stem or the carotid bodies have been used in experiments on animals. A non-invasive test to estimate the contribution of the peripheral chemoreceptors is the dynamical analysis of the ventilation after a few breaths of oxygen or nitrogen.[13] However, the slowness of the peripheral chemoreflex loop makes the use of this method questionable for quantitative purposes.[9,14] An attractive non-invasive method to separate central and peripheral effects of hypoxia is the DEF technique used together with a mathematical model which describes the dynamics of ventilation. Combination of this method with the ABP technique gives the possibility to test the performance of the dynamical analysis.

Table 1. O_2 sensitivities for DEF and ABP experiments

	Intact			ABP		
Cat No.	G_c l/min	G_{p1} l/min	G_{p2} l/min	G_c l/min	G_{p1} l/min	G_{p2} l/min
1	1.27	2.54	1.26	1.08	3.17	1.75
2	1.37	2.32	0	1.13	2.56	0
3	1.46	0.82	1.58	1.45	0.92	1.88
4	1.33	2.48	0	1.50	1.92	0
5	1.22	1.57	1.23	1.17	1.14	1.25
6	0.97	1.44	0	0.79	0.88	0
7	0	3.23	1.56	0.68	2.67	1.46
mean	1.09	2.06	0.80	1.11	1.89	0.91

In a previous study we proposed a two-component model to extract peripheral and central effects from the dynamics of the ventilatory response to changes in O_2 tension.[7] However, using the ABP technique we have recently shown that the ventilatory response of the peripheral chemoreflex loop to changes in $PETO_2$ often had to be described by two components, a fast one with a time constant of a few seconds and a slow one with a time constant of about 50 s.[9] For the dynamics of the peripheral chemoreflex we assume that both components have the same transport delay time. The dynamics of the central effect of hypoxia was found to be adequately modelled by a first order differential equation with a time delay.[8] We therefore propose the model described by equations (1) to (4) to analyze the ventilatory response to stepwise increase in $PETO_2$ of intact cats.

Simulation studies have revealed that the amount of noise normally present on the breath-by-breath data of individual runs is too large to estimate the parameters meaningfully. To reduce the noise we ensemble averaged the runs of each cat and estimated the parameters from the average ventilation and averaged input $PETO_2$. This procedure is not easy to justify since one of the assumptions is that the parameters are time invariant, which is probably not entirely true. However, when we estimated the parameters of each run of the peripheral ventilatory response and compared the means of the sensitivities of the individual runs with the estimated sensitivities of the ensemble averaged data of each cat, the agreement was very good. This suggests that the values of the sensitivities obtained from the ensemble averaged data of each intact cat are rather reliable. The finding that values of the peripheral time delay (mean 3 s) and the central time delay (mean 60 s) are quite different greatly facilitate the possibility to separate the central and peripheral ventilatory component in the dynamical analysis. If the time delays are of about the same value, simulation studies suggest that the parameters of the model can not be estimated with any precision.

Unfortunately not all anesthetized cats showed a clear overshoot in ventilation following a step decrease in $PETO_2$ so that it is not possible to estimate the depressant effects from these curves. From the ABP experiments it could be demonstrated that the depressant effect of hypoxia was indeed present. The results from the dynamical analysis of the peripheral ventilatory response suggest that the manifestation of central depression is obscured by the occurrence of a slow stimulatory peripheral component. In cats with no ventilatory overshoot there was nearly always an undershoot present so that the central effects of oxygen could nevertheless be assessed.

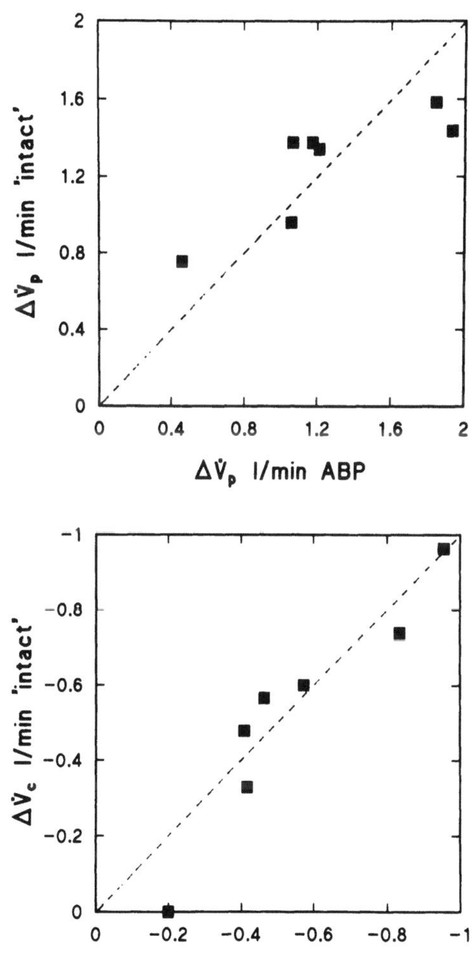

Figure 4. Scatter diagrams of peripheral and central contribution to the ventilation obtained from steps from hypoxia into hyperoxia in intact cats plotted against values from ABP experiments in the same cat. The dotted lines are the lines of equality.

We have shown that in the anesthetized cat the DEF technique together with our three-component model is able to correctly estimate the central and peripheral contribution to the ventilation following a step change in $PETO_2$ from hypoxia to hyperoxia. These results justify further efforts to apply the DEF technique especially in set-ups where the use of non-invasive techniques is a prerequisite to gain more insight into the mechanisms underlying the ventilatory response to hypoxia.

ACKNOWLEDGEMENTS

We are indebted to L. Philips for his skilful preparation of the animals. This research was supported by the Foundation for Medical Research (GMW) grant 900-519-064.

REFERENCES

1. S. Kagawa, M.J. Stafford, T.B. Waggener, and J.W. Severinghaus, No effect of naloxone on hypoxia-induced ventilatory depression in adults, *J. Appl. Physiol.* 52: 1030 (1982).
2. E.E. Lawson, and W.A. Long, Central origin of biphasic breathing pattern during hypoxia in newborns *J. Appl. Physiol.* 55: 483 (1983).
3. P.A. Easton, L.J. Slykerman, and N.R. Anthonisen, Ventilatory response to sustained hypoxia in normal adults, *J. Appl. Physiol.* 61: 906 (1986).
4. M. Vizek, C.K. Pickett, and J.V. Weil, Biphasic ventilatory response of adult cats to sustained hypoxia has central origin, *J. Appl. Physiol.* 63: 1658 (1987).
5. A. Suzuki, M. Nishimura, H. Yamamoto, K. Miyamoto, F. Kishi, and Y. Kawakami, No effect of brain blood flow on ventilatory depression during sustained hypoxia, *J. Appl. Physiol.* 66: 1674 (1989).
6. A. Berkenbosch, and J. DeGoede, Effects of brain hypoxia on ventilation, *Eur. Respir. J.* 1: 184 (1988).
7. J. DeGoede, N. VanDerHoeven, A. Berkenbosch, C.N. Olievier, and J.H.G.M. VanBeek, Ventilatory response to sudden isocapnic changes in end-tidal O_2 in cats, in: Modelling and Control of Breathing. B.J. Whipp and D.M. Wiberg, ed., Elsevier, New York, (1983).
8. D.S. Ward, A. Berkenbosch, J. DeGoede, and C.N. Olievier, Dynamics of the ventilatory response to central hypoxia in cats, *J. Appl. Physiol.* 68: 1107 (1990).
9. A. Berkenbosch, J. DeGoede, D.S. Ward, C.N. Olievier, and J. VanHartevelt, Dynamic response of the peripheral chemoreflex loop to changes in end-tidal O_2, *J. Appl. Physiol.* 71: 1123 (1991).
10. A. Berkenbosch and J. DeGoede, Actions and interactions of CO_2 and O_2 on central and peripheral chemoreceptive structures, in: Neurobiology of the Control of Breathing. C. von Euler and H. Lagercrantz, ed., Raven Press, New York, (1986).
11. A. Berkenbosch, J. Heeringa, C.N. Olievier, and E.W. Kruyt, Artificial perfusion of the pontomedullary region of cats. A method for separation of central an peripheral effects of chemical stimulation of ventilation, *Respir. Physiol.* 37: 347 (1979).
12. J. DeGoede, A. Berkenbosch, D.S. Ward, J.W. Bellville, and C.N. Olievier, 1985, Comparison between chemoreflex gains obtained with two different methods in cats, *J. Appl. Physiol.* 59: 170 (1985).
13. P. Dejours, Chemoreflexes in breathing, *Physiol. Rev.* 42: 335 (1962).
14. W.N. Gardner, The pattern of breathing following step changes of alveolar partial pressures of carbon dioxide and oxygen in man, *J. Physiol.* 300: 55 (1980).

SINGLE BREATH TESTS OF THE VENTILATORY RESPONSE TO CARBON DIOXIDE: EXPERIMENTAL AND MODELLING RESULTS

S.S.D. Fernando and K.B. Saunders

Department of Medicine, St. George's Hospital Medical School
Cranmer Terrace, LONDON SW17 ORE U.K.

INTRODUCTION

Single breath tests have several obvious advantages. They are easy for the subject to do. They may be undetectable by the subject, if tidal volume inhalations of low inspired CO_2 concentrations ($FICO_2$) are given. The short duration of the response makes baseline change less troublesome and minimises confounding changes in cerebral blood flow.

On the other hand, multiple repetitions are needed, which is tedious for the subjects. Automated data handling is essential in order to apply some form of ensemble averaging, for the response to a single small stimulus cannot be effectively distinguished from the normal background noise. The need for such averaging may be decreased by using high $FICO_2$ or large inspirations, but the stimulus is then obvious to the subject and may induce an additional cortical response. High $FICO_2$ concentrations may induce coughing, which completely invalidates the procedure.

This paper tests the effect of moderate levels of $FICO_2$ (5-8%) given in single voluntary deep inspirations from functional residual capacity to total lung capacity, that is inspiratory capacity breaths, in normal subjects.

We wish to find out, first, whether the smallest stimuli would produce a detectable ventilatory response, that is whether the system was acting above or below any CO_2 threshold level. Second, we asked whether, for these brief small stimuli, the ventilatory response was in accordance with that expected from a standard two-compartment controller model (1). To this end we compared experimental results with simulation from a published model of the cardiopulmonary system (2).

METHODS

Protocols

Five normal subjects attended the laboratory on three days. They rested for 20 min. before performing the procedures. On each day they breathed for 6 periods on the respiratory circuit, interspersed by rest periods of 10-15 min. Each experimental period contained 5 voluntary deep inspirations of test gas, 5,6 or 8% CO_2, with 2-3 min between each test. The order of administration of the different concentrations was randomised.

Control of Breathing and Its Modeling Perspective, Edited by
Y. Honda *et al.*, Plenum Press, New York, 1992

Thus at the end of the three days there was a total of 30 breaths at each concentration for each subject.

Statistical Analysis

This was decided after consultation with Dr. Martin Bland (Dept. of Public Health, St. George's Hospital Medical School) and is illustrated in Fig. 1. First (Fig 1, line 1), the records for each individual were lined up according to breath number with 10 control breaths before the test inspiration, and a desired number of breaths (7 in the example illustrated) after the test breath. These were then ensemble-averaged both vertically with regard to the desired variable (e.g. end-tidal PCO_2, $PetCO_2$) and horizontally for each breath number to give a real-time estimation of the measurement (Fig. 1, line 2). The means from these data were taken as the basic experimental result (Fig. 1, line 3) and the mean and standard deviation of the 10 control points taken as the control value (Fig. 1 line 4). Finally each post-test value was tested against control with regard to the 95% confidence interval (Fig. 1, line 5) and also by Student's t test. Results were essentially identical by either method.

For group results, as presented in the paper, we used Student's paired t test for the control and each post-test breath for each subject.

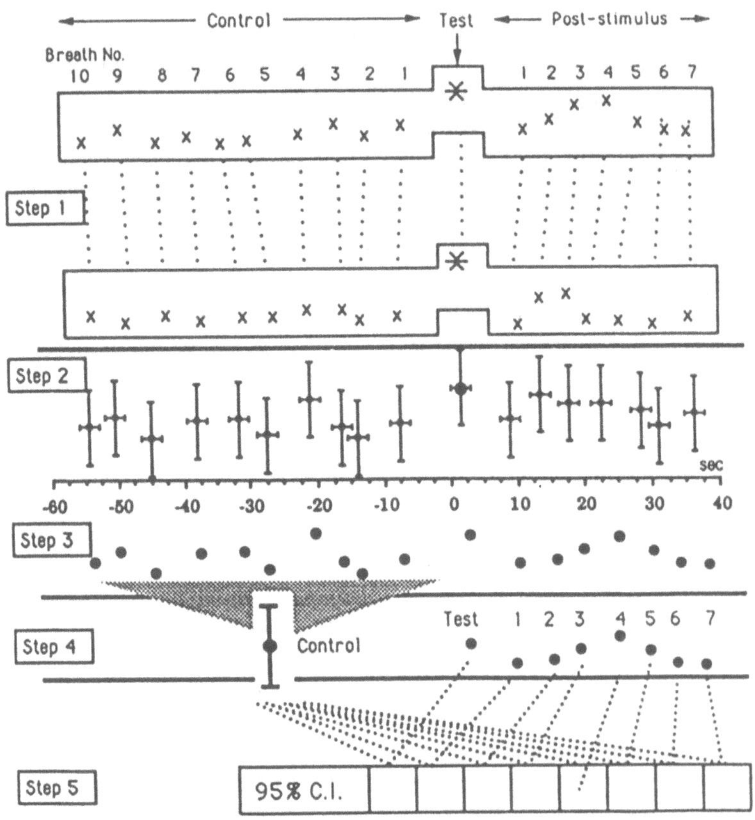

Fig. 1 Breath-by-breath averaging and statistical analysis. See text.

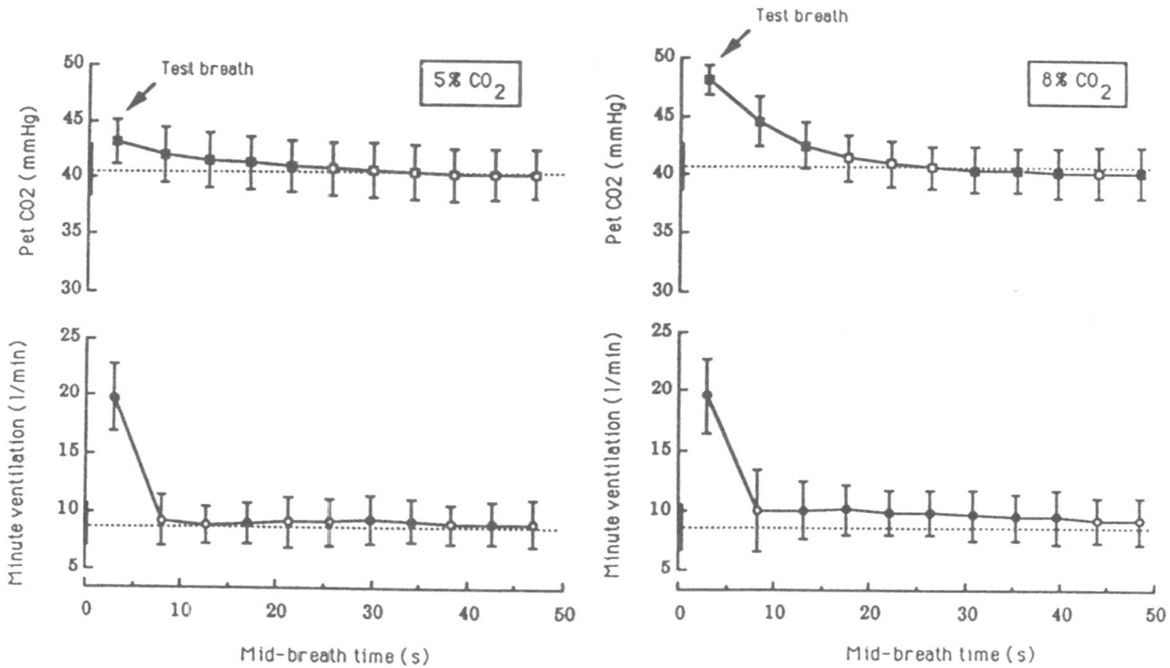

Fig. 2 Group experimental results for inhalations of 5% and 8% CO_2. Control value is shown at time zero. The next breath is the test inhalation. Filled symbols denote values significantly different from control ($P < 0.05$).

Modelling

We used an adaptation of the Grodins model (3) which differs chiefly by the addition of an expansible lung compartment to simulate tidal breathing (2), and of shunt and dead space components. In this application the model was driven by a controller with two first order components: "peripheral" - lag 8s, time constant 15s and gain 0.7 $l.min^{-1}.mmHg^{-1}$, and "central" - lag 12s, time constant 180s and gain 1.4 $l.min^{-1}.mmHg^{-1}$. This is the normal controller configuration (1). The simulations were also performed with the central controller only (peripheral controller deleted and all gain ($0.7 + 1.4 = 2.1$) at the central controller; and in open-loop conditions, when after the test breath, ventilation returns to and remains at the control value. For each set of simulations we used for the test breath an inspiratory volume and duration determined from the mean of the experimental results.

RESULTS

Experimental

Results for 5% and 8% breaths are shown in Fig. 2 for variables $PetCO_2$ and minute ventilation ($\dot{V}$). Filled points denote values significantly different from control ($P < 0.05$). Control $PetCO_2$ was between 40 and 41 mmHg for all three inspired gas fractions. Mean rises of 2.2, 3.8 and 7.6 mmHg were seen in the test breaths of 5, 6 and 8% respectively. Significant increases in ventilation were not seen in the first post-test breath, but were seen in breaths 3, 6, 7 and 9 (5%) 2-7 (6%) and 2-8 (8%). Peak ventilatory response was 0.6 (5%), 1.6 (6%) and 2.0 (8%) $l.min^{-1}.mmHg^{-1}$.

These small ventilatory responses were sufficient to produce undershoots in $PetCO_2$ at breaths 8 and 9 (40-45 sec) for 6%, and at breaths 6-8 and 10 (30-50 sec) for 8%.

Modelling

Results for an 8% CO_2 inhalation are shown in Fig. 3. With open loop conditions ventilation returns immediately to control levels, and there is no $PetCO_2$ undershoot, indeed $PetCO_2$ remains slightly above control values for the next minute. With the standard central plus peripheral controller there is a peak ventilatory response of 2 $l.min^{-1}$ at about 25s and a $PetCO_2$ undershoot of about 2.5mmHg at 30s. If the central controller alone is active the ventilatory response and undershoot are diminished in amplitude and prolonged. Similar results, but of lower amplitude, were obtained for 5 and 6% inhalations.

Comparison of experimental and model results.

We decided to compare the responses according to the following features

- Height of CO_2 impulse
- Height of peak ventilation response
- Time of peak ventilatory response
- Depth of $PetCO_2$ undershoot
- Time of trough $PetCO_2$ undershoot

The model predicted the change in $PetCO_2$ produced by the test breath with acceptable accuracy (Fig. 4). The height of the ventilatory response is better represented by the central plus peripheral controller configuration, the depth of the $PetCO_2$ trough by the central controller alone. Timing of these features was not discriminatory (Fig. 5).

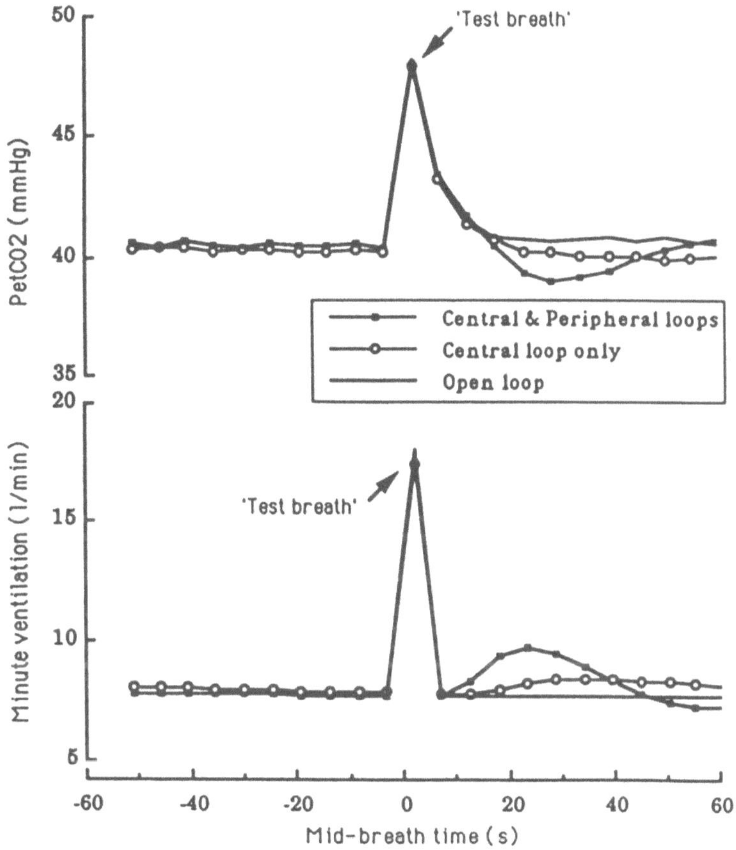

Fig. 3 Model results from single inhalation of 8% CO_2 under three controller configurations.

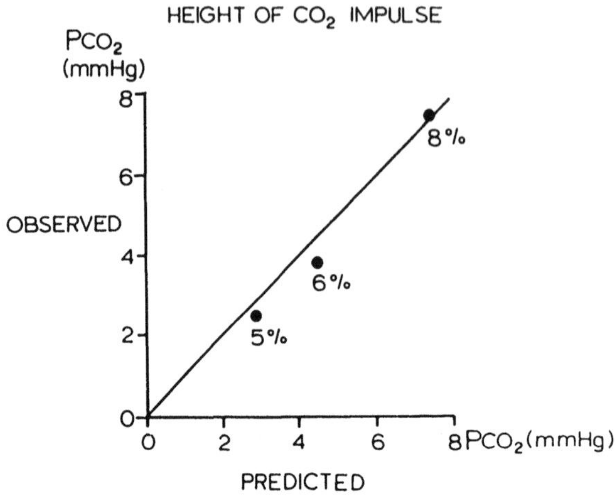

Fig. 4 Experimental (observed) versus model (predicted) height of CO_2 impulse, that is the change in $PetCO_2$ from control to the test breath.

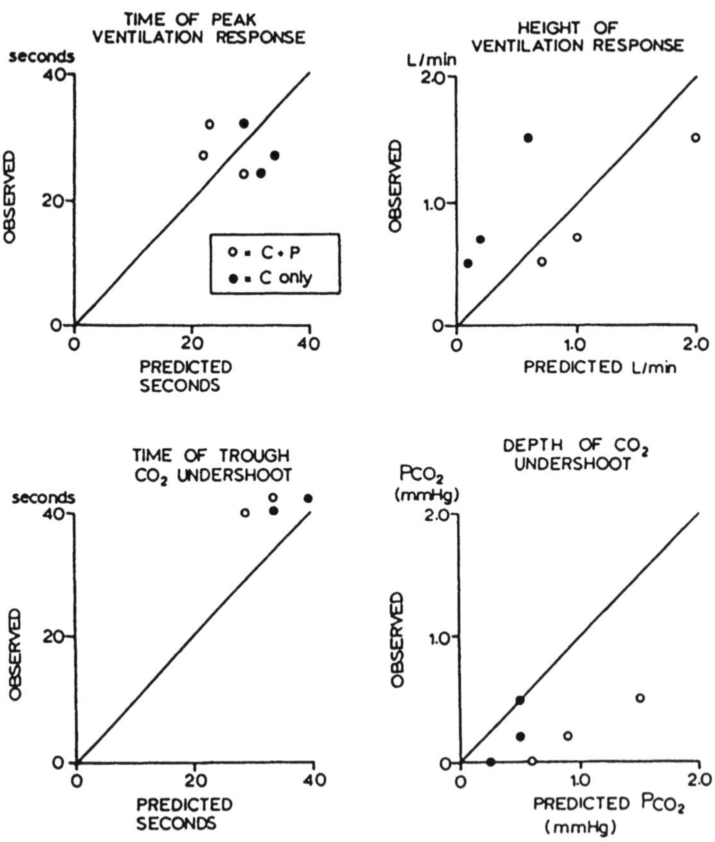

Fig. 5 Model versus experimental results for the main features of the response to inhalation of 5, 6 and 8% CO_2, under two controller configurations: central controller only (●) and central plus peripheral controller (O). An undershoot in $PetCO_2$ was not observed experimentally for 5% CO_2.

DISCUSSION

In 1958 Dejours and colleagues (4) gave single tidal volume inhalations of 7% CO_2 in air to a normal subject, averaged the results from seven experiments, and showed that ventilation increased after a delay of 5-10s. Similar results were obtained in anaesthetised dogs (5) where two tidal volumes of 6% CO_2 again increased ventilation after 5-10s. After chemodenervation, a smaller response occurred with a lag of 15s. It was concluded that the early response in the dog, and the normal response in man, were mediated by peripheral chemoreceptors.

In order to simplify the procedure and reduce the need for averaging, subsequent investigators have generally used test breaths larger than tidal volume, higher CO_2 concentrations, or both. For example Gabel et al. (6) gave vital capcity breaths of 15% CO_2, and Sorensen and Cruz (7) tested vital capacity breaths of 15 and 20% CO_2. More recently, McClean and colleagues developed a single-breath test using a tidal volume inhalation of 13% CO_2 (8). The resulting ventilatory response is large enough not to require averaging, but such concentrations are invariable tasted and provoke coughing in some subjects.

This is we believe the first occasion when automatic data handling techniques and ensemble-averaging have been used to define the results of single breath tests. We were able to detect changes in ventilation down to about 0.2 $l.min^{-1}$ and changes in $PetCO_2$ down to about 0.2 mmHg.

We were particularly interested in the possibility that very small, brief stimuli might cause no ventilatory response. The presence and position of any ventilatory threshold to CO_2 in normal conscious euoxic man is a point of some controvercy. At present, experiments using re-breathing techniques (9) and mechanical hyperventilation (10) suggest a threshold at about 40mmHg. CO_2 unloading in patients with renal failure (11) suggests a significantly lower threshold. When small transient CO_2 stimuli were given to subjects during post-hyperventilation hypocapnia (12) one subject behaved in the first and two in the second way. Certainly in this study with control $PetCO_2$ very close to 40mmHg, a response to the small 5% $FICO_2$ stimulus could routinely be found.

It is customary in transient response techniques to obtain an estimate of overall gain by dividing the peak ventilation by the peak of the $PetCO_2$ input. The dynamics of the controlled system ensure that the value of "gain" thus obtained will be less than the true steady-state value. This can be easily seen from the relative rates of rise of $PetCO_2$ and $\dot{V}$ in response to a step input of $FICO_2$ (13). $PetCO_2$ rises rapidly, $\dot{V}$ much more slowly, with 5-20 min required to reach steady-state, depending on the value of $FICO_2$. Thus if the input is prematurely terminated, as in a transient response technique, $\dot{V}$ will not have risen to a level appropriate to the simultaneous value $PetCO_2$, and the value obtained for $\dot{V}/PetCO_2$ will be low. For example a model simulation of McCleans' (8) results (14) suggests that that technique will give a gain of about 40% of the true steady-state gain. The gain obtained from such techniques will be largely but not entirely due to the peripheral chemoreceptor. In our results the demonstration of a significant increase in ventilation at the second post-test breath strongly suggests the involvement of the peripheral chemoreceptor, but (Figs 3, 5) the central chemoreceptor must clearly also be involved. (Indeed, if the central chemoreceptor is deleted in the model simulation, the ventilatory response to a single inhalation of CO_2 oscillates indefinitely (14).)

In conclusion the technique and analysis described allowed detection of small changes in $\dot{V}$ and $PetCO_2$, about 0.2 $l.min^{-1}$ and 0.2mmHg respectively. Responses to single inhalations of CO_2 are in keeping with the standard 2-compartment model (1), with peak ventilation response at about 30s. The ventilation responses favour the hypothesis that both peripheral and central chemoreceptors are active, but undershoots in $PetCO_2$ are less than would be expected.

ACKNOWLEDGEMENTS

We thank Mrs. Rita Perry for preparing the manuscript.

REFERENCES

1. J.W. Bellville,B.J. Whipp, R.D. Kaufman, G.D. Swanson, K.A. Aqleh and D.M. Wiberg. Central and peripheral chemoreflex loop gain in normal and carotid body-resected subjects. J.Appl.Physiol. 46:843-853 (1979)
2. K.B. Saunders, H.N. Bali and E.R. Carson. A breathing model of the respiratory system: The controlled system. J.Theor.Biol. 84: 135-161 (1980a)
3. F.S. Grodins, J. Buell and A.J. Bart. Mathematical analysis and digital simulation of the respiratory control system. J.Appl.Physiol. 22: 260-276 (1967)
4. P. Dejours, Y. Labrousse, J. Raynaud and R. Flandrois. Etude du stimulus gaz carbonique de la ventilation chez l'homme. J.Physiol, Paris 50: 239-43 (1958)
5. P. Bouverot, R. Flandrois and R. Grandpierre. A propos du mécanisme d'action chémoréflexe ou central du 'stimulus CO_2' de la ventilation. Compt.Rend.Acad.Sci. Paris 252: 790-792 (1961)
6. R.A. Gabel, R.S. Kronenberg and J.W. Severinghaus. Vital capacity breaths of 5% or 15% CO_2 in N_2 or O_2 to test carotid chemosensitivity. Respir.Physiol. 17: 195-208 (1973)
7. S.C. Serensen and J.C. Cruz. Ventilatory response to a single breath of CO_2 in O_2 normal man at sea level and high altitude. J.Appl.Physiol. 27: 186-190 (1969)
8. P.A. McClean, E.A. Phillipson, D. Martinez and N. Zamel. Single breath of CO_2 as a clinical test of the peripheral chemoreflex. J.Appl.Physiol. 64: 84-89 (1988)
9. J. Duffin and G.V. McAvoy. The peripheral-chemoreceptor threshold to carbon dioxide in man. J.Physiol. 406: 15-26 (1988)
10. R.J. Castele, A.F. Connors and M.D. Altose. Effects of changes in CO_2 partial pressure on the sensation of respiratory drive. J.Appl.Physiol. 59: 1747-1751 (1985)
11. W. De Backer, R.M. Heyrman, W. Wittesaele, J.P. Van Waeleghem, P.A. Vermiere, and M.E. De Broe. Ventilation and breathing patterns during haemodialysis induced CO_2 unloading. Am.Rev.Respir.Dis. 136: 406-410 (1987)
12. A.R.C. Cummin, V.S. Sidhu, R.J. Telford and K.B. Saunders. Ventilatory responsiveness to carbon dioxide below the normal control point in conscious normoxic humans. Eur.Resp.J. in press (1992)
13. W.J. Reynolds, H.T. Milhorn and G.H. Holloman. Transient ventilatory response to graded hypercapnia in man. J.Appl.Physiol. 33: 47-54 (1972)
14. M.S. Jacobi, C.P. Patil and K.B. Saunders. Estimates of chemoreceptor gain from single breaths or short pulses of inhaled CO_2 in man. J.Physiol. (Lond.) 407: 39P (1988)

VENTILATORY RESPONSES TO RESPIRATORY AND ACUTE METABOLIC ACIDOSIS IN MAN

I.D.Clement, D.A.Bascom, K.L.Dorrington, R.Painter, J.J.Pandit and P.A.Robbins

University Laboratory of Physiology
Parks Road
Oxford, OX1 3PT
U.K.

INTRODUCTION

The Lloyd-Cunningham model[1] of the chemical control of breathing in man describes the steady-state central and peripheral chemoreflex drives as being independent and additive. The model is supported by observations in the anaesthetised cat using the technique of independent perfusion of the central and peripheral chemoreceptors[2]. However there is also some evidence suggesting that there may be a mutual interaction between the central and peripheral chemoreflex sensitivities. Bellville et al.[3] fitted a two-compartment model to the ventilatory responses of human subjects to steps in PET_{CO_2} and found the slow compartment (central chemoreceptor) gain was greater in hypoxia than euoxia in 6 out of 7 human subjects. Robbins[4] found evidence for interaction between the hypoxic ventilatory response and the level of central P_{CO_2} in two out of three subjects.

The object of this study was to test for interaction between the central and peripheral chemoreflexes. If there is interaction then the response from one of the chemoreflexes will depend in some way on the stimulus to the other chemoreflex. The experiment we carried out made use of respiratory and metabolic acidoses to allow us to assess the peripheral chemoreflex sensitivity at two different levels of central chemoreceptor activity. Acute metabolic acidosis should selectively stimulate the peripheral chemoreceptors since H^+ does not easily cross the blood-brain barrier [5,6]. This is in contrast to CO_2 inhalation where the resultant acidosis affects both central and peripheral chemoreceptors. These results have been published elsewhere[7].

PROTOCOL

This experiment was undertaken by seven healthy young male subjects. The complete experiment on each subject consisted of 4 experimental periods. In 2 of these,

Control of Breathing and Its Modeling Perspective, Edited by
Y. Honda *et al.*, Plenum Press, New York, 1992

acute metabolic acidosis was generated by a 1-2 min bout of hard exercise on a bicycle ergometer. Measurements were made over the following 30 min recovery period. Ventilation was measured during euoxia (PET_{O_2} = 100 Torr) and during 6 isocapnic hypoxic steps (PET_{O_2} = 50 Torr) each lasting 1.5 min. In the other two experimental periods respiratory acidosis was generated by inhaled CO_2. PET_{CO_2} was elevated to 45 Torr in one period and 50 Torr in the other, while in each case PET_{O_2} was held at 100 Torr during euoxia and at 50 Torr during four 1.5 min steps of hypoxia.

Arterial pH (pHa) and Pa_{CO_2} were measured from blood sampled from an arterial catheter. The subject breathed on a mouthpiece with the nose occluded and ventilatory volumes and flows were measured using a turbine and a pneumotachograph. Gas composition at the mouth was sampled continuously by a mass spectrometer and changes in end-tidal gas composition were imposed using a computer-controlled dynamic end-tidal forcing system[8].

ANALYSIS OF RESULTS

Figure 1 is a schematic illustration of the methods used to analyse the results in each subject. First, looking at the data from the CO_2-inhalation experiments, regression lines were fitted to the euoxic and hypoxic data points and these lines were used to estimate the hypoxic response for a given value of Pa_{CO_2} (shown in Figure 1(i) as $\Delta\dot{V}_E^{CO_2}$). The particular value of Pa_{CO_2} used for each subject was the mean of all the data points in the plot for that subject. For the next stage of the analysis Pa_{CO_2} and pHa from the CO_2-inhalation experiment were plotted on a Siggaard-Andersen plot (Figure 1(ii)) to determine the change in pHa which would be effected by the step from euoxia to hypoxia at this constant Pa_{CO_2}. This change in pHa arises because the blood becomes slightly alkaline as the haemoglobin desaturates (Haldane effect). Finally the data from the exercise recovery experiments were

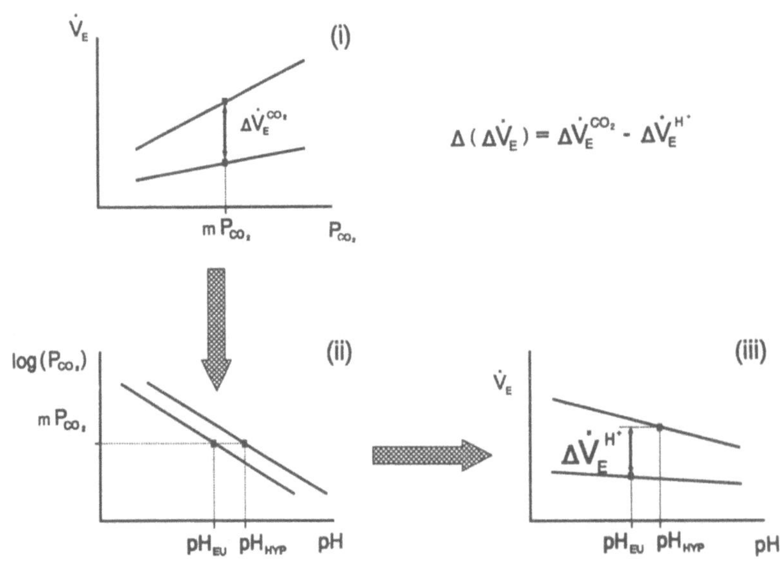

Figure 1. Stages of calculations performed in comparison of hypoxic responses between CO_2-inhalation and exercise recovery experiments. $\Delta\dot{V}_E^{CO_2}$, hypoxic response at mean Pa_{CO_2} (mPa_{CO_2}) for CO_2-inhalation experiment. $\Delta\dot{V}_E^{H+}$, hypoxic response for metabolic acidosis experiment. pHEU, pHHYP, euoxic and hypoxic pHa values at mean Pa_{CO_2} (mPa_{CO_2}) from CO_2-inhalation experiments. Plot (ii) is a Siggaard-Andersen plot.

plotted as ventilation against pHa (Figure 1(iii)). The euoxic and hypoxic regression lines together with the euoxic and hypoxic values for pHa were used to estimate the hypoxic response, shown in Figure 1(iii) as $\Delta\dot{V}_E^{H+}$. If the peripheral chemoreflex response is independent of central chemoreflex activity we would expect the response in each condition to be the same and hence the difference, $\Delta(\Delta\dot{V}_E)$, to be zero.

Implicit in this analysis are two assumptions regarding the actions of H^+ and CO_2 at the central and peripheral chemoreceptors. The first is that there is no molecular effect of CO_2 at the peripheral chemoreceptors and that CO_2 stimulates them only via its effect on H^+ [9]. The second assumption is that the central chemoreceptors respond to CO_2 in the arterial blood to a degree which is greater than that due entirely to the change in pHa.

RESULTS

Analysis of the results showed that the magnitude of the hypoxic stimulus was not always exactly the same in both protocols. To overcome this problem the ventilatory responses to hypoxia ($\Delta\dot{V}_E^{CO_2}$ and $\Delta\dot{V}_E^{H+}$) were scaled according to the size of the hypoxic stimulus using the Lloyd-Cunningham[1] hyperbolic relationship between hypoxia and P_{O_2}. The results of this analysis are shown for each subject in Table 1. The units for the ventilatory responses are l/min per arbitrary hypoxic unit. The results show that $\Delta(\Delta\dot{V}_E)$ is not significantly different from zero in any of the subjects.

DISCUSSION

The results of this study appear to support the hypothesis that the central and peripheral chemoreflex responses are independent of each other.

In comparing the results from the exercise recovery and metabolic acidosis protocols

Table 1. Ventilatory responses to hypoxia[1] (see text).

Subject	$\Delta\dot{V}_E^{CO_2}$	$\Delta\dot{V}_E^{H+}$	$\Delta(\Delta\dot{V}_E)$	SE [2]	DF [3]	*P*-value [4]
766	2.835	2.088	0.747	0.643	31	(NS)
724	3.651	4.070	-0.419	0.473	32	(NS)
786	1.146	1.781	-0.634	0.722	18	(NS)
787	3.768	3.200	0.568	0.653	25	(NS)
789	3.609	4.709	-1.101	0.642	27	(NS)
791	3.423	3.220	0.203	0.454	27	(NS)
785	4.674	3.795	0.879	0.637	35	(NS)
Mean (SD)	3.301 (1.095)	3.266 (1.049)	0.035 (0.761)			

[1] Ventilatory sensitivities expressed as L/min per unit hypoxia where hypoxia has been defined as $100/(P_{O_2} - C)$; C = 32.2 Torr[1].
[2] Standard error for $\Delta(\Delta\dot{V}_E)$.
[3] Degrees of freedom for $\Delta(\Delta\dot{V}_E)$.
[4] Probability that $\Delta(\Delta\dot{V}_E)$ does not differ from zero (NS, *P*-value > 0.05).

we assume that in the post-exercise period no factors resulting from the exercise, other than lactate, are affecting the ventilation. During exercise the arterial potassium concentration is elevated[10] and this is known to enhance hypoxic sensitivity in the cat[11]. However the arterial potassium concentration falls rapidly after the end of exercise[10] and since the first measurements were not made until 4 min after the end of exercise the arterial potassium concentration should be normal throughout the period of study. Hypoxic sensitivity is known to be enhanced during exercise by mechanisms which are not fully understood and the persistence of this enhanced sensitivity into the post-exercise period cannot be ruled out.

These results differ from those of Robbins[4] who used the differing speeds of response of the central and peripheral chemoreceptors to give the same peripheral chemoreceptor stimulus (an isocapnic step into hypoxia) in conditions of either eucapnia or residual hypercapnia in the central chemoreceptor. He found that in two out three subjects hypoxic sensitivity was significantly greater in conditions of central chemoreceptor hypercapnia. Why the results of this procedure appear to conflict with the results presented here remains unclear, but may reflect a breakdown of one or more of the assumptions made either in this study or in the Robbins[4] study.

REFERENCES

1. B.B. Lloyd, M.G.M. Jukes and D.J.C. Cunningham. The relation between alveolar oxygen pressure and the respiratory response to carbon dioxide in man. *Quart. J. Exp. Physiol.* 43:214 (1958).

2. J.H.G.M. Van Beek, A. Berkenbosch, J. De Goede and C.N. Olievier. Influence of peripheral O_2 tension on the ventilatory response to CO_2 in cats. *Respir. Physiol.* 51:379 (1983).

3. J.W. Bellville, B.J. Whipp, R.D. Kaufman, G.D. Swanson, K.A. Aqleh and D.M. Wiberg. Central and peripheral chemoreflex loop gain in normal and carotid body-resected subjects. *J. Appl. Physiol.* 46:843 (1979).

4. P.A. Robbins. Evidence for interaction between the contributions to ventilation from the central and peripheral chemoreceptors in man. *J. Physiol. (London).* 401:503 (1988).

5. R.L. Knill and J.L. Clement. Ventilatory responses to acute metabolic acidemia in humans awake, sedated, and anesthetized with halothane. *Anesthesiology.* 62:745 (1985).

6. R.A. Mitchell and M.M. Singer. Respiration and cerebrospinal fluid pH in metabolic acidosis and alkalosis. *J. Appl. Physiol.* 20:905 (1965).

7. I.D. Clement, D.A. Bascom, J. Conway, K.L. Dorrington, D.F. O'Connor, R. Painter, D.J. Paterson and P.A. Robbins. An assessment of central-peripheral ventilatory chemoreflex interaction in humans. *Respir. Physiol.* In press (1992).

8. M.G. Howson, S. Khamnei, M.E. McIntyre, D.F. O'Connor and P.A. Robbins. A rapid computer-controlled binary gas-mixing system for studies in respiratory control. *J. Physiol. (London).* 394:7P (1987).

9. T.F. Hornbein and A. Roos. Specificity of H ion concentration as a carotid chemoreceptor stimulus. *J. Appl. Physiol.* 18:580 (1963).

10. D.J. Paterson, P.A. Robbins and J. Conway. Changes in arterial plasma potassium and ventilation during exercise in man. *Respir. Physiol.* 78:323 (1989).

11. R.E. Burger, J.A. Estavillo, P. Kumar, P.C.G. Nye and D.J. Paterson. Effects of potassium, oxygen and carbon dioxide on the steady-state discharge of cat carotid body chemoreceptors. *J. Physiol. (London).* 401:519 (1988).

ROLE OF Cl^--HCO_3^- EXCHANGER AND ANION CHANNEL IN THE CAT CAROTID BODY FUNCTION

Rodrigo Iturriaga and Sukhamay Lahiri

Department of Physiology, School of Medicine
University of Pennsylvania. Philadelphia, PA 19104, USA

INTRODUCTION

Carotid body (CB) chemosensory responses to respiratory and metabolic acidosis are well demonstrated both in vivo and in vitro[1]. The responses are expected to originate from the chemoreceptor cells which usually should manifest parallel phenomena. The consensus model of the chemoreceptor unit is that the glomus cells are the presynaptic chemoreceptor cells and the sensory fibers are the postsynaptic elements which, we supposed, should reflect the events in the presynaptic glomus cells. However, Buckler et al.[2] and Wilding et al.[3] reported that the glomus cells possessed at least three ion-exchangers which regulated the intracellular pH (pH_i). If the chemosensory responses are coupled to the pH_i of the glomus cell and its pH_i is well regulated then there will be a lack of correspondence between the glomus cell pH_i and the sensory response to CO_2-H^+. Accordingly the role of the ion-exchangers would appear ambivalent.

Presence and absence of CO_2-HCO_3^- can influence the functions of the HCO_3^- dependent ion exchangers. In the presence of CO_2-HCO_3^- and normal $[Cl^-]_o$ there is a net efflux of HCO_3^- through the Na^+-independent Cl^-/HCO_3^- exchanger. The process could be reversed by removing Cl^- and suppressed by the removal of CO_2-HCO_3^- in the medium[4]. Consequently the pH_i would be alkaline. Additionally, Na^+/H^+ exchanger, if active, could continue to extrude H^+ and add to the alkalinity. The alkaline pH_i would diminish the baseline chemosensory activity, the peak and steady chemosensory responses to hypoxia. Furthermore, Cl^- channel present in the glomus cell membrane[5] could conduct HCO_3^{-2}, and its blockade would attenuate HCO_3^- efflux and make the cell alkaline. The chemosensory responses would de expected to parallel these pH_i changes. We tested these predictions by studying CB chemosensory responses in vitro.

METHODS

The CBs were excised from the cats anesthetized with sodium pentobarbitone (35 mg/kg, i.p) and perfused and superfused at 36.5 ± 0.5 °C with a modified Tyrode solution as previously described[6]. The composition of the Tyrode was (in mM) Na^+ 154; Cl^- 103; K^+ 4.7; Ca^{2+} 2.2; Mg^{2+} 1.1; glutamate 22.0; HEPES 5.0,

HCO_3^-, and glucose 5.0. The pH was adjusted to 7.39 with NaOH. To study the effects of the nominal absence of CO_2-HCO_3^- the CBs were perfused-superfused initially with Tyrode, free of CO_2-HCO_3^- (HCO_3^- was replaced by glutamate, pH = 7.39), and then with Tyrode containing HEPES-HCO_3^- and equilibrated with 5% CO_2. The frequency of chemosensory discharges was recorded from the whole carotid sinus nerve. The chemosensory responses to hypoxia (PO_2= 30-50 Torr) and to acid hypercapnia (PCO_2= 60 Torr at pH 7.15) were tested before and after an experimental maneuver. The chloride channel blocker 9-anthracene carboxylic acid (9-ANC) was dissolved in Tyrode and perfused for 30 min.

RESULTS

Effects of CO_2-HCO_3^- on the carotid chemosensory response to hypoxia

Figure 1 shows the chemosensory responses to hypoxia (PO_2 = 30 Torr) in the absence (Fig. 1A) and presence (Fig. 1B) of CO_2-HCO_3^- at constant external pH. Hypoxia stimulated the chemosensory discharge both in the absence and presence of CO_2-HCO_3^-. However, the presence of CO_2-HCO_3^- augmented the baseline chemosensory activity and the speed and peak responses to hypoxia. Associated with the rapid response in the presence of CO_2-HCO_3^-, there was adaptation.

Effects of chloride channel blocker, 9-ANC on the carotid chemosensory responses

Figure 2 shows representative effects of 9-ANC (2 mM) on carotid chemosensory responses to acidic hypercapnia (PCO_2 = 60 Torr and pH 7.15, PO_2 of 120) and to hypoxia (PO_2 = 50 Torr, PCO_2 of 30 Torr and pH 7.38). Application of 9-ANC decreased the baseline activity, delayed and eliminated the initial peak responses, and hence adaptation to both hypercapnia and hypoxia.

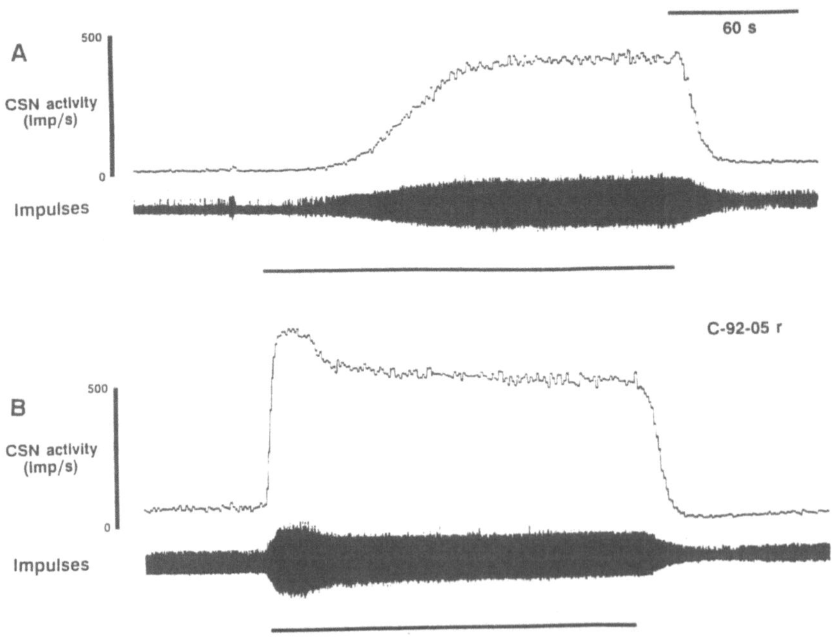

Figure 1. Effects of CO_2-HCO_3^- on the chemosensory response to hypoxia (PO_2 = 30 Torr at pH = 7.39). A, perfusion without CO_2-HCO_3^- and B, perfusion with CO_2-HCO_3^-. Before and after hypoxia perfusate PO_2 was 130 Torr.

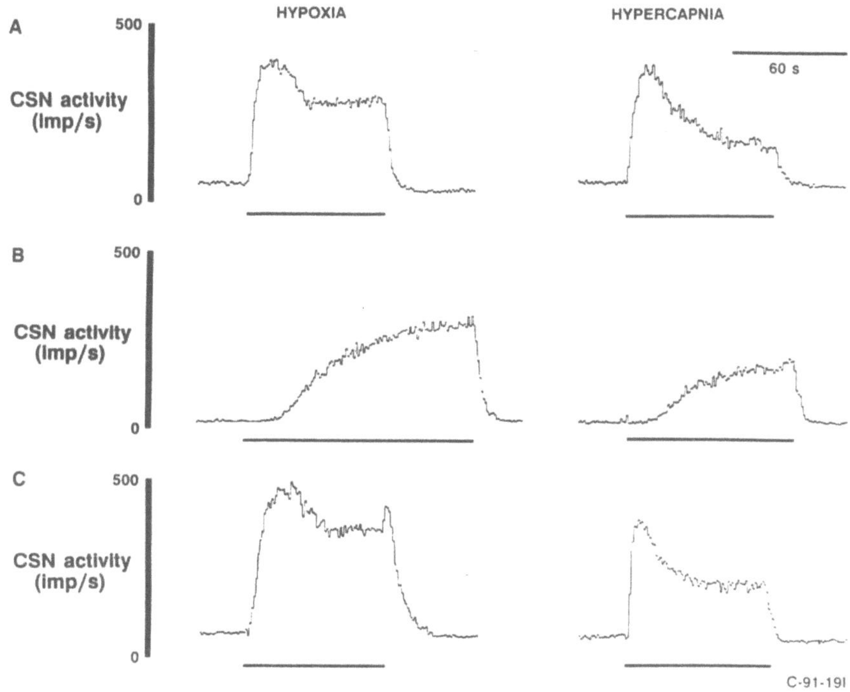

Figure 2. Effects of 9-ANC on the chemosensory response to hypoxia (left hand, PO_2 = 50 Torr, PCO_2 = 30 Torr and pH = 7.38) and to acidic hypercapnia (right hand, PCO_2 = 60 Torr at pH = 7.15, PO_2 = 120 Torr). A, control. B, during 9-ANC (2 mM) and C, recovery. Before and after hypoxia and hypercapnia perfusate PO_2 was 120 Torr, PCO_2 was 30 Torr and pH = 7.38.

However, the magnitude of the late responses were not significantly affected. The 9-ANC effects were readily reversible.

DISCUSSION

The augmenting effects of CO_2-HCO_3^- on the baseline chemosensory activity at a constant pH_o indicates its intracellular acidification effect, and is consistent with the observation on the glomus cells[2,3]. Exposure to CO_2-HCO_3^- decreased the pH_i not just because of rapid formation of carbonic acid, mediated by the enzyme carbonic anhydrase[1], but also because of the HCO_3^- extrusion through the Cl^--HCO_3^- exchanger mechanisms. In the nominal absence of CO_2-HCO_3^- these effects were diminished, and the pH_i was presumably alkaline which attenuated the chemosensory response to hypoxia and cyanide[7]. Shirahata and Fitzgerald[8], however, found that the hypoxic response was eliminated in the nominal absence of CO_2-HCO_3^-.

The initial alkalinity made the oxygen chemoreception less responsive, presumably by generating lesser amount of a second messenger, like Ca^{2+}. But O_2 chemoreception was not mediated by increasing the intracellular acidity. This conclusion is consistent with the observation that, whereas hypercapnia simultaneously increased glomeral acidity and chemosensory discharge, hypoxia increased chemosensory discharge without pH_i change[9]. The foregoing conclusion is also in conformity with the results from the Cl^- channel blockade which would made the chemoreceptor cell alkaline[2]. The decreased baseline chemosensory activity observed after 9-ANC application support this interpretation.

The mechanisms of adaptation or desensitization to a stimulus, hypercapnia or hypoxia is not known. It seems to be associated with the high speed of the stimulus application and the sensory response. Clearly, normal function of the Cl^- /HCO_3^- exchanger and Cl^--HCO_3^- channel seem to be necessary for a rapid movement of these ions, and pH_i and membrane voltage responses.

In summary, the results indicate that the Cl^--HCO_3^- exchanger and the Cl^- channel in the glomus cells contributed to the "resting" pH_i regulation, which affected the responses to hypoxia. Also, the rapidity of the response and the initial adaptation of the responses were dependent on the normal movement of these ions across the plasma membrane. However, the steady-state responses to hypercapnia indicated that the pH_i was not well regulated. The effects of hypoxia on the chemosensory discharges were modulated but not initiated by these ion movement and pH_i regulation.

ACKNOWLEDGMENT

Supported in part by NIH grant HL-43413.

REFERENCES

1 C. Eyzaguirre, R. S. Fitzgerald, S. Lahiri, and P. Zapata, Arterial Chemoreceptors, in: Handbook of Physiology. The Cardiovascular System, Peripheral Circulation and Organ Flow, American Physiol Soc, Williams and Wilkins, Baltimore, p. 557-621 (1983).

2 K. J. Buckler, R.D. Vaughan-Jones, C. Peers and P.C.G. Nye, Intracellular pH and its regulation in the isolated type I carotid body cells of the neonatal rat, J. Physiol. Lond., 436: 107-129 (1991).

3 T. J. Wilding, B. Cheng, and A, Roos, The relationship between extracellular pH (pH_o) and intracellular pH (pH_i) in adult rat carotid body glomus cells, Biophys. J., 59: 184a (1991).

4 A. Roos, and W.F. Boron, Intracellular pH, Physiol Rev., 61: 296-434 (1984).

5 A. Stea, and C. Nurse, Chloride channels in cultured glomus cells of the rat carotid body, Am. J. Physiol., 257: C174-C181 (1989).

6 R. Iturriaga, W.L. Rumsey, A. Mokashi, D. Spergel, D.F. Wilson, and S. Lahiri, In vitro perfused-superfused cat carotid body for physiological and pharmacological studies, J. Appl. Physiol., 70: 1393-1400 (1991).

7 R. Iturriaga, and S. Lahiri, Carotid body chemoreception in the absence and presence of CO_2-HCO_3^-, Brain Res., 568: 253-260 (1991).

8 M. Shirahata, and R.S. Fitzgerald, The presence of CO_2/HCO_3^- is essential for hypoxic chemotransduction in the in vivo perfused carotid body, Brain Res., 545: 297-300 (1991).

9 R. Iturriaga, W.L. Rumsey, S. Lahiri, D. Spergel, and D.F. Wilson, Intracellular pH and O_2 chemoreception in the cat carotid body, J. Appl. Physiol., 72: (1992). (In press).

A MATHEMATICAL MODEL OF THE VENTILATORY RESPONSE TO A PERIOD OF SUSTAINED ISOCAPNIC HYPOXIA IN HUMANS

Rosemary Painter, Saeed Khamnei and Peter Robbins

University Laboratory of Physiology
Parks Road
Oxford
OX1 3PT
U.K.

The human ventilatory response to a period of isocapnic hypoxia is biphasic, with a fast initial increase in ventilation after the step into hypoxia, followed by a slower decline towards its initial value. At the end of the hypoxic period there is another rapid change in ventilation but the decrease is not as great as the initial increase. We have developed a mathematical model to describe the response to hypoxia, taking into account the rapid changes in ventilation at the steps of P_{O2}, the gradual decline, and the asymmetry between the on and the off responses.

Khamnei and Robbins[1] have suggested that the decline in ventilation during the hypoxic period is brought about by a reduction in peripheral chemoreflex sensitivity. The purpose of this study was to develop a dynamic model based on this idea, and to see if the model can describe experimental data.

DEVELOPMENT OF MODEL

There were three main steps in the development of the model. The starting point was a steady-state model which ignored hypoxic ventilatory decline; the next step was to extend the model so that hypoxic ventilatory decline could be incorporated; the last step was to introduce the dynamics into the model.

The model was developed initially from the steady-state model of Lloyd et al.[2], where ventilation $\dot{V}$ is separated into peripheral and central components:

$$\dot{V} = \dot{V}_p + \dot{V}_c = A\,D\,(P_{CO2} - B) / (P_{O2} - C) + D\,(P_{CO2} - C) \qquad (1)$$

Two gains, AD and D, and two constants, B and C, are needed to characterise the response. For our model, the central component of ventilation was not considered since the P_{CO2} was constant throughout. The P_{O2} in the peripheral component was replaced by oxygen saturation S to avoid the hyperbolic relation between the peripheral component of ventilation and P_{O2}. The P_{CO2} term was incorporated into the gain term, resulting in an equation for the peripheral component of ventilation, $\dot{V}_p$:

$$\dot{V}_p = g_p\,(1 - S + k_p) \qquad (2)$$

with g_p the peripheral gain and k_p a constant used so that the peripheral component of ventilation does not have to be zero when the oxygen saturation is 100%.

Control of Breathing and Its Modeling Perspective, Edited by
Y. Honda *et al.*, Plenum Press, New York, 1992

Hypoxic ventilatory decline was then incorporated into the model by having an additional equation, equivalent to making the gain AD in Equation (1) no longer constant, but affected by hypoxia. The peripheral gain g_p was made a linear function of the oxygen saturation, with a gain and a constant:

$$g_p = g_h S + k_h \tag{3}$$

A dynamic model was then derived from Equations (2) and (3):

$$d\dot{V}_p/dt = (1/T_p) \{ g_p (1 - S(t-d_p) + k_p) - \dot{V}_p \} \tag{4}$$

and

$$dg_p/dt = (1/T_h) \{ g_h S(t-d_p) + k_h - g_p \} \tag{5}$$

These reduce to the steady-state equations when the first, changing terms are zero. The equations each incorporate a time constant, T_p and T_h, and there is a peripheral delay term, d_p, for the oxygen saturation. T_p is in essence the fast time constant associated with the rapid changes in ventilation at the steps, and T_h is the much slower time constant associated with the slow decline in ventilation during the hypoxic period. The dynamic model thus has six parameters in two equations.

DATA

Having developed the model, the next step was to see whether it could describe some real experimental data. The data used were from a study by Khamnei and Robbins[1], in which they looked at five subjects, with two protocols repeated six times for each subject. In each protocol, there was a 20 min period of isocapnic hypoxia, with the end-tidal P_{O2} kept at 50 Torr. In the first protocol the P_{CO2} was kept 2 Torr above the resting value, and in the second, it was kept 8 Torr above. The model was fitted to data from 3 min before the step into hypoxia, the hypoxic period, and 10 minutes after the step out of hypoxia (although there were only 2 min after the step available for two of the subjects in the lower P_{CO2} protocol).

FITTING THE MODEL

The model was fitted to each protocol using all 6 repeats together, to obtain the parameters for each subject and protocol. The equations were solved for each breath analytically and so gave a series of difference equations; these were then used in conjunction with a non-linear least squares fitting routine to obtain values for the parameters.

SENSITIVITY ANALYSES

Sensitivity analyses were carried out to check whether the parameters of the model were identifiable. This involved generating model data for each of the two protocols using the median parameter values from the five subjects. The value of each parameter was then fixed to a series of different values, and the model was fitted for the other parameters, and the sum of squares of errors obtained. Plots of the sums of squares of errors against parameter value were then drawn for each parameter. At the model value, the sum of squares is zero. The dependence of the model fit on the value of the parameter could be seen by looking at how flat or steep the line was on either side of the model value.

This technique was carried out for all the parameters and each protocol, and showed that for both protocols all the parameters were identifiable.

RESULTS

The results of fitting the model to the ventilatory responses for one representative subject are shown in Figure 1. These responses are from the lower P_{CO2} protocol, and each plot shows one of the six sets of data. The dots represent the measured breath by breath ventilations, and the lines are the model ventilations using the parameters which minimise the errors for all the data. The variability breath by breath and between runs can be seen. In order to assess how well the model fitted the data, the responses and the errors were averaged across all six sets of data for each subject and protocol. These could then be plotted as in Figure 2. This shows the averaged data and model responses for the subject whose individual lower P_{CO2} runs are given in Figure 1. The averaged data are represented by the dots, and the averaged model response by the lines, for each of the two protocols. Also shown are the 95% confidence limits for the mean deviations of the data from the model.

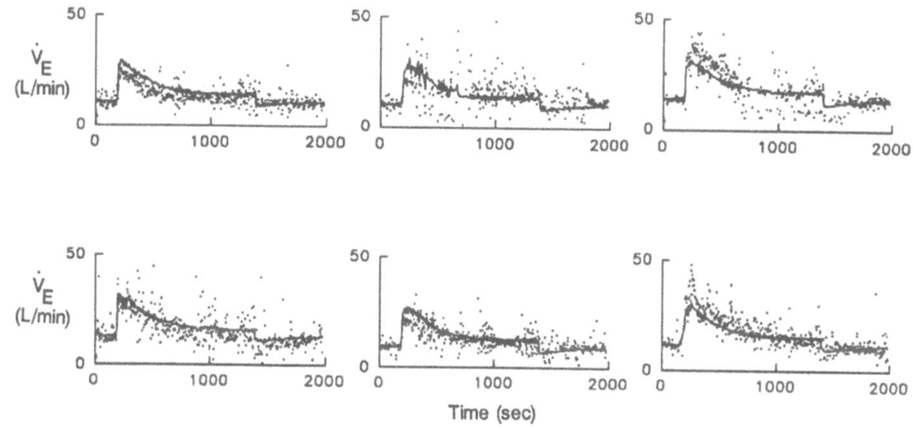

Figure 1. Ventilatory responses and model fit for the individual runs for one subject for the lower P_{CO2} protocol.

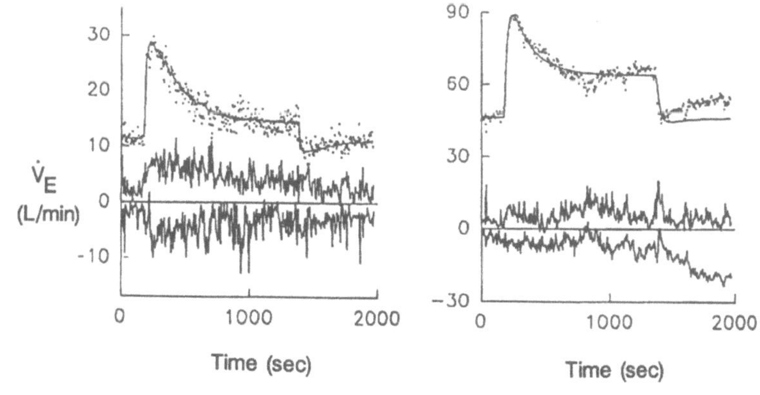

Figure 2. Ventilatory responses and model fit averaged over the 6 runs for one subject, for the two different protocols. The lower lines show the 95% confidence limits of the mean error (positive for model minus data).

Averaging the data shows that the fit is generally good, even though for individual runs it may not be. In the lower P_{CO_2} protocol, the model fits well at the rapid increase, the subsequent decline and the off transient, and gives an asymmetric response at the steps of P_{O_2}. It can be seen to fit well by looking at the confidence limits, which rarely cross the zero line. In the higher P_{CO_2} protocol, the model fits well too, although there are some problems after the step back to normoxia.

CONCLUSIONS

In conclusion, the plots show that the model can describe the experimental data well, including the different sizes of the on and off transients. Ours is the first model to do this and is in marked contrast to other dynamic models that have been fitted to experimental data, which fail to generate the asymmetry. The earlier models were based on the physiological idea that the ventilatory decline arose as a component which was independent of the peripheral chemoreflex, whereas our model is based on the physiological idea that the ventilatory decline arises as an adaptation of the peripheral chemoreflex sensitivity during hypoxia.

ACKNOWLEDGEMENTS

This work was supported by the Wellcome Trust. RP was a Schorstein Medical Research Fellow.

REFERENCES

1. S. Khamnei and P.A. Robbins, Hypoxic depression of ventilation in humans: alternative models for the chemoreflexes, *Respir. Physiol.* 81:117-134 (1990).
2. B.B. Lloyd, M.G.M. Jukes and D.J.C. Cunningham, The relation between alveolar oxygen pressure and the respiratory response to carbon dioxide in man, *Q. J. Exp. Physiol.* 43:214-226 (1958).

CARBON DIOXIDE — ESSENTIAL INGREDIENT FOR *IN VIVO* CAROTID BODY CHEMOTRANSDUCTION

Machiko Shirahata[1], Tohru Ide[2], and Robert S. Fitzgerald[3]

Departments of Environmental Health Sciences[1,2,3], Anesthesiology/ Critical Care Medicine[1], Medicine[3], and Physiology[3], The Johns Hopkins Medical Institutions, 615 N. Wolfe Street, Baltimore, MD 21205, USA

INTRODUCTION

The carotid body responds to the changes in arterial Po_2, Pco_2, and pH. Although different mechanisms have been suggested for the chemotransduction of hypoxia and hypercapnia in the carotid body, it is also true that the response of the carotid body to one stimuli is not totally independent of other stimuli. Several studies have shown that lowering arterial Pco_2 can dramatically reduce the carotid body response to hypoxia[2,3,4,5]. In addition, we have recently demonstrated that, whereas selective perfusion of the carotid body with hypoxic perfusate containing CO_2/HCO_3^- increased carotid chemoreceptor neural activity, perfusion with CO_2/HCO_3^--free hypoxic perfusate did not. This suggested a crucial role for CO_2/HCO_3^- in hypoxic chemotransduction[7]. Pertinent to these studies a recent report showed that intracellular pH of the type I cell from the rat carotid body was very alkalotic in the CO_2/HCO_3^--free media[1]. Based on these findings we hypothesized that CO_2/HCO_3^- plays a fundamental role for the carotid body chemotransduction, possibly as an essential ingredient in the regulatory mechanisms of intracellular pH (pH*i*) in the chemosensitive unit. Decrease in CO_2/HCO_3^- would increase pH*i* of the chemosensitive unit. Alkalinization of the unit would turn off the process of chemotransduction. We tested this hypothesis using a technique of *in vivo* selective perfusion of the carotid body.

METHODS

Twelve cats were anesthetized, and canulas were inserted in the trachea, femoral artery and vein. The cats were artificially ventilated with oxygen-enriched air. Glucose and saline were continuously administered. The carotid body area was prepared for selective and intermittent perfusion as described previously[6,8]. In short, a loop catheter was inserted in the common carotid artery. The lingual artery was catheterized. Snares were placed around the common carotid artery (proximal of the loop catheter) and around the external carotid artery. Other arterial branches except for the carotid body artery were tied. To perfuse the carotid body selectively blood flow was stopped by occluding the snares and cell-free solutions were perfused at a rate of 5-8 ml/min. The perfusion pressure was adjusted to systemic arterial pressure. Except for the short perfusion period (90 sec), carotid body was supplied with its own natural blood. More than 30 min was allowed between the cell-free perfusions. Chemoreceptor neural activity was recorded from the whole carotid sinus nerve from which the baroreceptor component was eliminated by mechanical and thermal destruction of the baroreceptor nerve endings. Chemoreceptor neural activity, systemic arterial pressure, and common carotid arterial pressure were continuously monitored. Arterial blood was withdrawn periodically to measure arterial pH, Pco_2, and Po_2. Rectal temperature was kept at 38°C using a warm water blanket.

Experimental Protocol

In the first set of experiments, we examined the effect of the presence or absence of CO_2/HCO_3^- on the acidotic chemotransduction of the carotid body. The carotid body was perfused for 90 sec with one of the following Krebs Ringer bicarbonate solution (KRB), or HEPES buffered solution (HBS). Perfusion solutions were; KRB exposed to 5 % CO_2 in 95 % O_2 (composition in mM: NaCl 120, KCl 3.5, $CaCl_2$ 1.8, $MgCl_2$ 0.6, NaH_2PO_4 0.6, $NaHCO_3$ 19, glucose 20; pH=7.4), KRB exposed to 10 % CO_2 in 90 % O_2 (pH=7.1), HBS exposed to 100 % O_2 (HBS-; Composition was the same as KRB except $NaHCO_3$ was replaced by NaCl and 10 mM HEPES was added. pH=7.1).

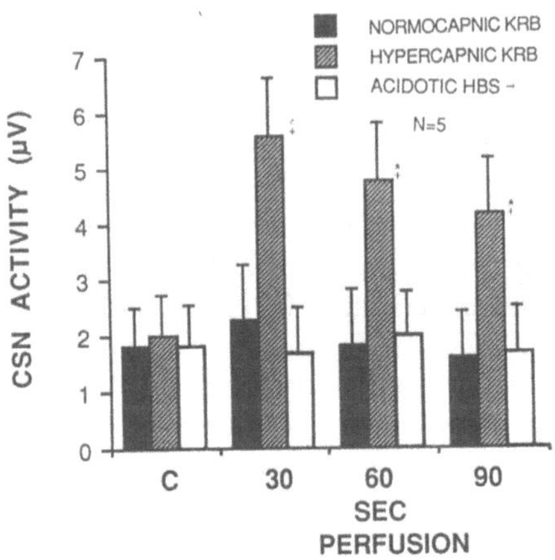

Figure 1. Carotid chemoreceptor neural responses to the selective perfusion of the carotid body with hyperoxic normocapnic Krebs Ringer bicarbonate solution (KRB), hyperoxic hypercapnic KRB, and hyperoxic acidotic HEPES buffered solution without CO_2/HCO_3^- (HBS-). CSN, carotid sinus nerve; C, pre-perfusion control; *, significantly different from own control; +, significantly different from other perfusions.

In the second set of experiments, we antagonized possible alkalinization of the chemosensitive unit during CO_2/HCO_3^--free perfusion by adding butyrate. The carotid body was perfused with either HBS exposed to 5 % CO_2 in 95 % N_2, HBS exposed to 100 % N_2, butyrate-containing HBS exposed to 100 % N_2, HBS exposed to 10 % CO_2 in 90 % O_2 (pH=7.1), HBS exposed to 100 % O_2 (pH=7.1), and butyrate-containing HBS exposed to 100 % O_2 (pH=7.1). Butyrate was added in a form of sodium butyrate (10 mM), and 10 mM NaCl was reduced in the solution containing butyrate.

Data Analysis

Data were reported as mean±SEM. They were analyzed with two or three way analysis of variance and Dancan's new multiple-range test.

RESULTS AND DISCUSSION

Figure 1 shows effects of extracellular acidosis on carotid chemoreceptor neural activity with or without CO_2/HCO_3^-. A selective perfusion of the carotid body with hyperoxic normocapnic KRB (pH 7.478±0.026, Pco_2 30±3 torr, Po_2 451±26 torr) did not change carotid chemoreceptor neural activity from pre-perfusion hyperoxic normocapnic level (pHa 7.384±0.007, $Paco_2$ 34±2 torr, Pao_2 328±47 torr). Hyperoxic hypercapnic KRB (pH 7.119±0.017, $Pco_2$59±2 torr, Po_2 453±21 torr) significantly increased neural activity. Hyperoxic acidotic HBS- (pH 7.087±0.025, Pco_2 1±0 torr, Po_2 510±20 torr) did not increase neural activity. The results suggest that extracellular acidity *per se* is ineffective in evoking increased neural activity.

Here, it is important to note that without CO_2/HCO_3^- the carotid body is not only unresponsive to acidosis but also insensitive to hypoxia as shown in Fig. 2, a and b (see also Ref. 7). The simplest explanation of the results would be that CO_2/HCO_3^- is necessary for chemotransduction of the carotid body. CO_2/HCO_3^- may play a crucial role

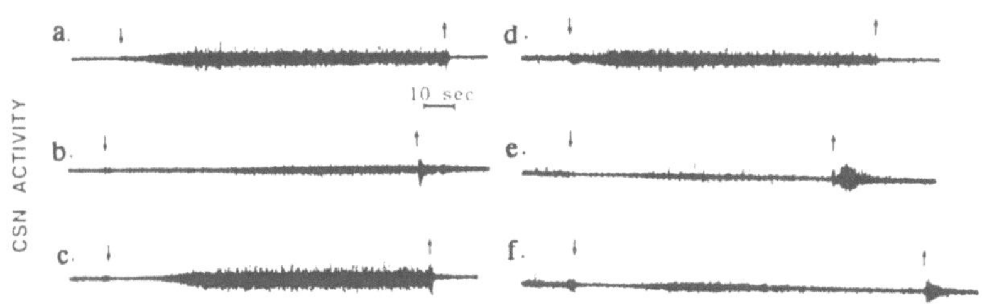

Figure 2. Responses of the carotid body to hypoxic HBS (trace a), hypoxic HBS- (b), hypoxic HBS-containing 10 mM sodium butyrate (c), hyperoxic hypercapnic HBS+ (d), hyperoxic acidotic HBS- (e), and hyperoxic acidotic HBS- containing butyrate (f). CSN, carotid sinus nerve; ↓, start of perfusion; ↑, end of perfusion.

for the regulation of pH*i* in the chemosensitive unit. During HBS- perfusion intracellular HCO_3^- would bind H^+ ions and exit cells as CO_2. Normally intracellular HCO_3^- and H^+ are in mM and nM range[9], respectively, and therefore, the process would leave the cell alkalotic. Perfusion time was only 90 sec, and readjustment of pH*i* to normal pH would not occur during this time period. This alkalinization of the chemosensitive unit may block the process of hypoxic and acidic chemotransduction. A recent report suggested this may be the case: Buckler *et al.*[1], found that the pH*i* of type I cells was 7.3 if the cells had been kept in the CO_2/HCO_3^--containing solution. However, if the cells were initially kept in the CO_2/HCO_3^--containing media and transferred to the CO_2/HCO_3^--free solution, pH*i* was 7.8.

If alkalosis in the chemosensitive unit would block hypoxic and acidotic chemotransduction, we could preserve the responses during HBS- perfusion by preventing alkalinization of the chemosensitive unit. Butyrate was used for this purpose. Like any other weak acid, a butyrate-containing solution has both butyrate ions and undissociated butyric acid. Undissociated butyric acid easily and rapidly enters the cells and dissociates to H^+ and butyrate ion. The process continues until the concentration of undissociated butyrate in the cells reaches the same level as that in the extracellular fluid[9]. This process certainly would have some antagonistic effect on the alkalinization of the chemosensitive unit during CO_2/HCO_3^--free perfusion.

The experimental results partly support the hypothesis. When the carotid body was perfused with hypoxic HBS- containing butyrate, chemoreceptor neural activity increased as much as during the perfusion with hypoxic HBS+ (Fig. 2, trace a-c). The results strongly suggest that certain amount of hydrogen ions in the chemosensitive unit is necessary for hypoxic chemotransduction. However, butyrate did not restore the carotid chemoreceptor response to CO_2/HCO_3^--free acidosis (Fig. 2, trace d-f): A selective perfusion of the carotid body with hyperoxic hypercapnic HBS+ increased chemoreceptor neural activity, but not with hyperoxic acidotic HBS-. Hyperoxic acidotic HBS- containing butyrate slightly increased neural activity, but not so much as during hypercapnic HBS+ perfusion. It is possible that intracellular acidosis is a necessary step for increase in chemoreceptor neural activity during extracellular acidosis, and that butyrate in HBS- solution cannot provide enough hydrogen ions to make cells in the chemosensitive unit acidotic.

In summary, we have demonstrated that the carotid chemoreceptor neural activity significantly increased when the carotid body was perfused with hypoxic or acidotic perfusate containing CO_2/HCO_3^-, but not with CO_2/HCO_3^--free perfusate. When butyrate was added in the CO_2/HCO_3^--free perfusate, hypoxic response was restored, but not acidotic response. The results suggest that certain levels of intracellular hydrogen ions is absolutely necessary for the carotid body to respond hypoxia. For acidotic chemotransduction increase in intracellular hydrogen ions seems to be a key step.

ACKNOWLEDGEMENTS

This work was supported by HL 10342 .

REFERENCES

1. BUCKLER, K. J., R. D. VAUGHAN-JONES, C. PEERS, AND P. C. G. NYE. Intracellular pH and its regulation in isolated type I carotid body cells of the neonatal rat. *J. Physiol.* 436:107-129, 1991.
2. EYZAGUIRRE, C., AND J. LEWIN. Chemoreceptor activity of the carotid body of the cat. *J. Physiol.* 159:222-237,1961
3. FITZGERALD, R. S., AND D. C. PARKS. Effect of hypoxia on carotid chemoreceptor response to carbon dioxide in cats. *Respir. Physiol. 12:218-229,1971.*
4. HORNBEIN, T. F., Z. J. GRIFFO, AND A. ROOS. Quantitation of chemoreceptor activity: Interrelation of hypoxia and hypercapnia. *J. Neurophysiol.* 24:561-568, 1961.
5. LAHIRI, S., AND R. G. DELANEY. Stimulus interaction in the responses of carotid body single afferent fibers. *Respir. Physiol.* 24:249-266, 1975.
6. SHIRAHATA, M., S. ANDRONIKOU, AND S. LAHIRI. Differential effects of oligomycin on carotid chemoreceptor responses to O_2 and CO_2 in the cat. *J. Appl. Physiol.* 63:2084-2092, 1987.
7. SHIRAHATA, M., AND R. S. FITZGERALD. The presence of CO_2/HCO_3^- is essential for hypoxic chemotransduction in the in vivo perfused carotid body. *Brain Research* 545:297-300, 1991.
8. SHIRAHATA, M., AND R. S. FITZGERALD. Dependency of hypoxic chemotransduction in cat carotid body on voltage-gated calcium channels. *J. Appl. Physiol.* 71:1062-1069, 1991.
9. THOMAS, R. C. Experimental displacement of intracellular pH and the mechanism of its subsequent recovery. *J. Physiol.* 354:3P-22P, 1984.

HYPOXIC DEPRESSION OF RESPIRATION

Neil S. Cherniack, Jyoti Mitra, and Nanduri R. Prabhakar

Departments of Medicine and Physiology and Biophysics
Case Western Reserve University School of Medicine
Cleveland, OH 44106, U.S.A.

Hypoxia is one of the more common and serious stresses challenging homeostasis. Apart from its effects on specialized O_2 sensing chemoreceptors, hypoxia has direct effects on all the tissues of the body. The tissues of the central nervous system (CNS) are probably the most vulnerable to the injurious actions of low levels of oxygen. Because of the importance of the CNS as an organizer of integrated responses, CNS reactions to subnormal oxygenation underlie the responses of the body as a whole to acute and chronic hypoxia. The response of the CNS to hypoxia is of obvious importance in unraveling the mechanisms that participate in altitude adaptation, diseases which produce global hypoxia (chronic lung disease of the adult and infant), and diseases which produce local cerebral hypoxia (cerebral ischemia and convulsive disorders).

Several explanations have been proposed for the depressive actions of CNS hypoxia on respiration, but none deal explicitly with the vasomotor excitation that also occurs as oxygen is reduced. Hypoxic reduction of respiratory activity has, for example, been attributed to increased cerebral blood flow or changes in energy supply which interfere with the function of respiratory neurons.[1] Metabolic or membrane differences in vasomotor and respiratory neurons at the cellular level could explain their different responses to CNS hypoxia.

An increasingly accepted idea is that moderately severe hypoxia affects the activity of neurotransmitters in the CNS, tipping the balance between inhibitory and excitatory influences on respiratory neurons.[1] These effects may be local and/or global. Hypoxia may act on local areas of the CNS to inhibit respiration but excite sympathetic activity. Sympathetic excitation might then occur as a result of a disinhibition (a release of inhibitory effects) on vasomotor neurons. Alternatively, an overabundance of excitatory neuromodulators (which directly stimulate vasomotor neurons and excite inhibitory respiratory interneurons) might also explain the CNS effects of hypoxia. Indeed, even with respect to breathing, a number of excitatory effects of central hypoxia, such as tachypnea, have been described during CNS hypoxia, particularly in unanesthetized animals.[2]

Hypoxic Effects on Respiration

In an early study of hypoxic depression, we showed a depressive effect of hypoxia on phrenic nerve activity in anesthetized, paralyzed and artificially ventilated dogs (with chemoreceptors intact).[3] These animals were exposed in separate trials to asphyxia (increasing Pco_2 and hypoxia); apneic oxygenation (hyperoxic hypercapnia); infusion of Tris buffer intravenously during asphyxia (hypoxia with isocapnia); and apneic oxygenation with Tris

Control of Breathing and Its Modeling Perspective, Edited by
Y. Honda *et al.*, Plenum Press, New York, 1992

buffer infusion (a control condition in which neither Pco_2 nor PO_2 changed). Phasic movements of the chest were eliminated so that pulmonary stretch receptors and muscle proprioreceptor effects were minimal. A clear depressive effect of hypoxia could be observed during both isocapnic and hypercapnic conditions. The level of hypoxia at which phrenic nerve activity decreased from its peak was higher during hypercapnia than during isocapnic hypoxia. Tris buffer infusion, while it maintained isocapnia, regularly produced an increase in arterial pH despite the occurrence of hypoxia. In light of more recent experiments by Neubauer et al.[4] showing that prevention of acidosis forestalled respiratory depression of cats exposed to progressive increases in carbon monoxide, it is possible that the Tris buffer by counteracting acidosis contributed to the greater resistance of the isocapnic animals to hypoxic depression.

Studies by our group indicate that hypoxic depression can differentially affect the early and late discharging fibers of the phrenic nerves[5] and the activity of various respiratory muscles. Van Lunteren et al.[6] have demonstrated a much more rapid decline in activity in the expiratory muscle, the triangularis sterni, than in the diaphragm in spontaneously breathing cats exposed to steady-state hypoxia. Interestingly, another group of investigators has observed a post-hypoxic rebound in expiratory muscle activity.[7] Studies by Carlo et al.[8] in this laboratory have demonstrated a less depressive effect on hypoglossal activity of hypoxia than on phrenic nerve activity. In addition, Hutt and coworkers[9] have reported that carbon monoxide induced hypoxemia in goats seems to affect the threshold level of Pco_2 at which hypoglossal activity begins less than the phrenic.

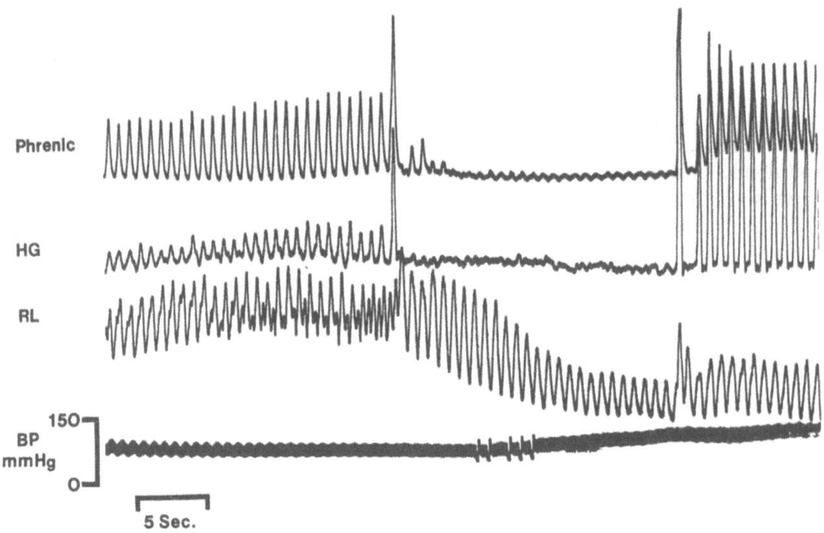

Figure 1. Differing behavior of hypoglossal (HG) and recurrent laryngeal (RL) activity during depression of phrenic nerve activity by CNS hypoxia.

We have examined the effects of hypoxia on urethane anesthetized rats. Hypoxia was produced by adding increasing amounts of N_2 to the inspired air. In rats with intact peripheral chemoreceptors respiratory depression evenually occurred accompanied or preceded slightly by an elevation of sympathetic discharge.

In experiments in which rapid and progressive hypoxia was produced in chemodenervated artificially ventilated rats, expiratory recurrent laryngeal activity was maintained even when hypoxia caused phrenic nerve and hypoglossal nerve apnea (see Figure 1). This observation suggests that central nervous system hypoxia may affect different respiratory motor outputs non-uniformly.

Despite the small size of the rat, in a few experiments we have been able to record from at least two sympathetic nerves simultaneously in peripherally chemo- and

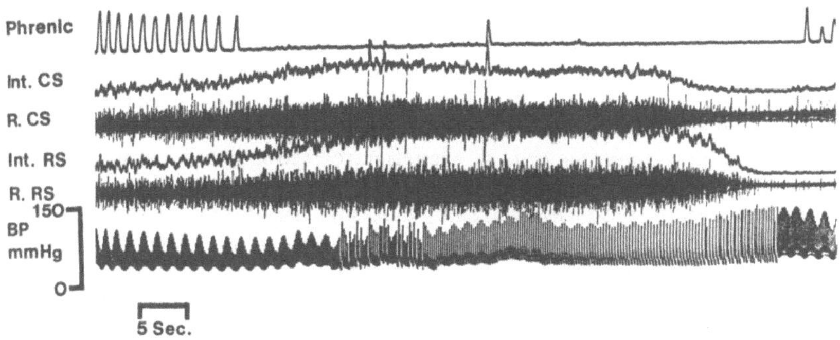

Figure 2. Rise in cervical (CS) and renal sympathetic (RS) activity during phrenic nerve depression by CNS hypoxia in a rat.

barodenervated animals. An example of one such experiment is shown in Figure 2. As can be seen from this figure, renal and cervical sympathetic activity appear to increase at the same time. In two other animals sphlanchnic nerve activity also increased at nearly the same moment as cervical sympathetic activity. While the response of all sympathetic branches to CNS hypoxia thus far tested seems similar in terms of timing of the response, further study is needed.

To determine the effects of baroreceptor input during systemic hypoxia, the responses to hypoxia were examined in paralyzed, anesthetized vagotomized cats in the following circumstances: with intact carotid chemo- and baroreceptors; after antimycin injection into the lingual artery which eliminated the response of the carotid body to hypoxia but left baroreceptor function intact; and then, after the carotid sinus nerves were cut, to remove both carotid body and baroreceptor function. As can be seen in Figure 3, systemic hypoxia in the presence of intact carotid chemo- and baroreceptors caused a transient initial rise in both respiratory and sympathetic activity. As hypoxia continued, respiratory activity (measured from the phrenic nerve) decreased but sympathetic activity rose once again. After elimination of the hypoxic response of the carotid body with antimycin, the initial increases in both respiratory and sympathetic activity were eliminated. The results were the same after carotid sinus nerve denervation. These results suggest that baroreceptor function remains even after the hypoxic response of the carotid body is abolished by antimycin and that baroreceptor activity may not significantly contribute to the changes in sympathetic activity occurring during systemic hypoxia.

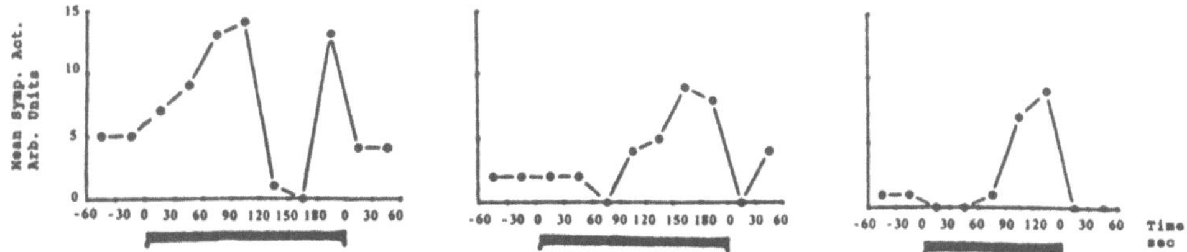

Figure 3. Effect of selective carotid body denervation on efferent phrenic and sympathetic nerve activities during hypoxia.

We could not eliminate either the respiratory depressive or the sympathetic excitatory effects of central hypoxia in the rat by decerebration although the animals seemed to endure hypoxia for longer periods of time before respiration or vasomotor activity changed. In another two peripherally chemodenervated animals we have observed that microinjection of lidocaine into the RVLM produced phrenic apnea. Although sympathetic excitation by hypoxia was not prevented, the level of excitation seemed less than in the control state and could have resulted from a spinal mechanism.

Sodium cyanide produces hypoxia by blocking tissue oxidation and by a direct effect on neurons.[10] Therefore, in another series of experiments we produced local hypoxia of the brain stem of anesthetized cats by injecting sodium cyanide unilaterally into the vertebral artery at the spino-medullary junction. Unlike inhalation of hypoxic gas mixtures, intravertebral NaCN injections have the advantage of avoiding the systemic effects of hypoxia and limiting the effects of hypoxia only to the brain. We also compared the effects of NaCN injection with the responses produced by local hypoxia caused by the infusion of saline equilibrated with 3% CO_2 in N_2.

Experiments were performed on anesthetized, tracheotomized, vagotomized, sinus denervated, paralyzed and artificially ventilated cats weighing 2-4 kg. The ventral surface of the medulla was exposed between the foramen caecum and the first cervical nerve using usual techniques. The right vertebral artery, before it joins the contralateral to form the basilar artery, was cannulated for subsequent injection of sodium cyanide or infusion of hypoxic saline. The basilar artery near the foramen caecum was occluded with a vascular clip.

Sodium cyanide solution was prepared in saline. Hypoxic saline was prepared by bubbling 3% CO_2 in N_2 gas in saline. Control oxygenated saline was prepared by bubbling 3% CO_2 in O_2 gas.

NaCN was given as a bolus in a volume of 0.1 ml over a period of 5 sec (1.2 ml/min). During hypoxic saline infusion, the vertebral artery was infused at a constant rate (1.2 ml/min) with warm (37°C) saline either bubbled with 3% CO_2 in O_2 (oxygenated) or 3% CO_2 in N_2 (hypoxic) gases for a period of one and a half minute. The amplitudes of the nerve signals were expressed in arbitrary units.

Three dose levels of sodium cyanide were used for the intravertebral injections (1, 10 and 20 μg) in 8 animals ventilated with air. Phrenic nerve amplitude decreased with increasing sodium cyanide dose. Occasionally respiratory frequency increased with depression of phrenic amplitude, but it was not a consistent finding. All animals became apneic after receiving the higher dose (20 μg) of sodium cyanide. The average times for the onset of phrenic depression and apnea were 4 to 8 seconds following injection. The duration of apnea varied considerably (20.0 ± 10.0 sec).

The depression of phrenic nerve activity produced by two different dose levels of sodium cyanide (5 and 10 μg) was also determined in the same group of animals breathing 7% CO_2 in O_2. At both doses tested the depression of phrenic activity was less when animals breathed 7% CO_2 ($P < 0.05$).

In animals breathing room air, sympathetic activity increased progressively with increasing doses of sodium cyanide (1, 10 and 20 μg)($P < 0.05$). The effect of sympathetic excitation following injection of two doses of sodium cyanide (5 and 10 μg) was compared while animals breathed either room air or 7% CO_2 in O_2. Unlike phrenic nerve activity, there was no statistical difference when animals breathed the two different gases.

Our above observations suggest that in peripherally chemodenervated animals the depressive effect of central hypoxia could be attenuated by high respiratory drive. However, higher respiratory drive was ineffective in altering sympathetic excitation.

In another group of five animals the effect of intravertebral injection of sodium cyanide on sympathetic nerve activity before and after transection of spinal cord at the cervical level (C_1) was tested. After spinal cord transection the effect of sodium cyanide on sympathetic nerve activity disappeared for all doses mentioned above. However, sympathetic nerve excitation could still be obtained in the same animals by a single large dose of sodium cyanide (600 μg) given intravenously (29.0 ± 8.0%, $P < 0.05$). This suggests that hypoxia can influence sympathetic nerve activity through spinal cord circuitry even after elimination of supraspinal inputs.

Like sympathetic activity, blood pressure in the main group of 8 cats increased progressively as the dose of intravertebral sodium cyanide increased from 1 to 10 to 20 μg (in room air) and was statistically significant ($P < 0.05$).

However, blood pressure elevation produced by two different doses of sodium cyanide (5 and 10 μg) was more during breathing of air than 7% CO_2 in O_2. At the higher dose of sodium cyanide (10 μg) this blood pressure increase became significant ($P < 0.05$).

Since sodium cyanide is a potent vasodilator, the responses observed so far could be, in part, due to CO_2 washout from the "central chemoreceptors". Hence we repeated the experiment in another group of 6 animals in which the following protocol was followed. The vertebral artery was first infused either with oxygenated or hypoxic saline at a rate of 1.2 ml/min. At this rate of infusion oxygenated saline caused no changes in phrenic amplitude, sympathetic basal activity or blood pressure. On the other hand, hypoxic saline infusion consistently depressed peak phrenic amplitude (before 30.4 ± 4.4, after 18.5 ± 4.0, $P < 0.05$) and elevated both sympathetic activity (before 87.0 ± 3.2, after 90.0 ± 2.0, $P < 0.05$) and blood pressure (before 121.4 ± 2.0 mmHg, after 130.0 ± 2.0 mmHg, $P < 0.05$). The responses were essentially similar to sodium cyanide injection. This confirms the idea that the responses to sodium cyanide were due to central hypoxia. Furthermore, infusion at a faster rate 3.6 ml/min) with saline bubbled with 100% O_2 (to washout CO_2), not only depressed phrenic amplitude (before 36.0 ± 4.0, after 28.0 ± 4.0, $P < 0.05$) but also depressed sympathetic activity (before 82.4 ± 3.0, after 79.2 ± 3.0, $P < 0.05$) and blood pressure (before 131.0 ± 8.0 mmHg, after 124.5 ± 6.0 mmHg, $P < 0.05$). These results show that compared to hypoxic responses, CO_2 washout has an opposite effect on sympathetic activity and blood pressure.

Since sodium cyanide given intravertebrally may reach brain stem sites, it is difficult to localize precisely the regions involved in eliciting the responses we observed. An additional set of experiments was designed to answer this question. In these experiments we micro-injected sodium cyanide in different parts of the medulla while recording from phrenic and cervical sympathetic nerves as well as monitoring systemic blood pressure.

The animal preparation was similar to that of the earlier series of experiments, except there was no vertebral cannulation. Sodium cyanide was pressure injected into the medulla using a micropipette (tip 50 μ). The volume (100 nl) and concentration of sodium cyanide (100 ng) were kept constant throughout the experiment. The depth (1 mm) of injection was also kept constant. The area of the medulla explored with microinjection ranged from 1 to 12 mm caudal to the foramen caecum and from 1 to 5 mm lateral to the midline. For initial analysis, the region of the ventral surface of the medulla was divided into "rostral" or R, "intermediate" or I, and "caudal" or C. The "R" region corresponds 1-4 mm, the "I" region 5-8 mm, and the "C" region 9-12 mm caudal to foramen caecum. Data were analyzed and expressed as in the previous experiment.

The main finding of this experiment is that sodium cyanide microinjections into the "R" area had no effect either on phrenic nerve activity or blood pressure. The response to sympathetic nerve was variable and statistically not significant. Microinjections of sodium cyanide into the "I" area between 3 and 5 mm lateral to the midline had strong depressive effect on phrenic nerve activity (before 26.0 ± 1.6, after 18.7 ± 2.6, $P < 0.05$) and excitatory effects on sympathetic (before 63.3 ± 3.4, after 68.4 ± 3.2, $P< 0.05$) and blood pressure (before 118.1 ± 4.1, after 137.3 ± 5.0 mmHg). On the other hand, sodium cyanide injections into the "C" area did not have any effect on phrenic nerve activity. However, sympathetic nerve activity and blood pressure gave variable responses. The results suggest that a large part of the central hypoxic response may originate near the "I" region of the medulla.

These results indicate (1) that hypoxia acting centrally can depress respiratory activity in spinal and cranial nerves; (2) that at about the same time, excitation occurs in several sympathetic nerves; and (3) that the ventral medulla may be an important site at which hypoxia causes its sympathetic and respiratory effects.

REFERENCES

1. J.A. Neubauer, J.E. Melton and N.H. Edelman. Modulation of respiration during brain hypoxia. *J. Appl. Physiol.* 68:441-451 (1990).
2. H.W. Davenport, G. Brewer, A.H. Chambers and S. Goldschmidt. The respiratory response to anoxemia of unanesthetized dogs with chronically denervated aortic and carotid chemoreceptors and their causes. *Am. J. Physiol.* 148:406-417 (1947).
3. N.S. Cherniack, N.H. Edelman and S. Lahiri. Hypoxia and hypercapnia as respiratory stimulants and depressants. *Respir. Physiol.* 11:113-126 (1970).
4. J.A. Neubauer, A. Simone and N.H. Edelman. Role of brain lactic acidosis in hypoxic depression of respiration. *J. Appl. Physiol.* 65:1324-31 (1988).
5. N.R. Prabhakar, J. Mitra, J.L. Overholt and N.S. Cherniack. Analysis of postinspiratory activity of phrenic motoneurons with chemical and vagal reflexes. *J. Appl. Physiol.* 61:1499-1509 (1986).

6. E. van Lunteren, R.J. Martin, M.A. Haxhiu and W.A. Carlo. Diaphragm, genioglossus and triangularis sterni responses to poikilocapnic hypoxia. *J. Appl. Physiol.* 67:2303-2310 (1989).
7. L.M. Oyer, J.E. Melton, J.A. Neubauer and N.H. Edelman. Enhancement of expiratory triangularis sterni nerve (TSN) activity following severe hypoxia. *FASEB J.* 3:A1159 (1989).
8. W.A. Carlo, J.M. DiFiore and R.J. Martin. Increased upper airway muscle activity during hypoxemia-induced respiratory depression in preterm infants. *Pediatr. Res.* 25:373A (1989).
9. D.A. Hutt, R.A. Parisi, T.V. Santiago and N.H. Edelman. Brain hypoxia preferentially stimulates genioglossal EMG response to CO_2. *J. Appl. Physiol.* 66:51-56 (1989).
10. P.G. Aitken and D.J. Braitman. The effects of cyanide on neural and synaptic function in hippocampal slices. *Neurotoxicology* 10:239-248 (1989).

IDENTIFICATION OF CLOSED-LOOP CHEMOREFLEX DYNAMICS USING PSEUDORANDOM STIMULI

Eugene N. Bruce, Mohammad Modarreszadeh[+], and Kenneth Kump

Center for Biomedical Engineering
University of Kentucky
Lexington, KY 40506

INTRODUCTION

A brief disturbance to respiratory chemical feedback loops may initiate an interval of oscillatory fluctuations in ventilation[1]. The propensity for such oscillations to occur often is quantified[1,2] in terms of chemical feedback "loop gain". A more direct and more complete characterization of the stability of ventilation can be obtained, however, by measuring the time course of the ventilatory response to a standard, brief disturbance[3,4]. With appropriate parameterization of such responses, one can obtain indices which are highly correlated with loop gain, or can calculate loop gain itself. To characterize and compare the dynamics of central and peripheral chemoreflex loops of man to disturbances, we have assessed the closed-loop ventilatory responses to single-breath inhalations of hypercapnic and hypoxic gases respectively. Such assessments in awake man are hampered by the typically large ventilatory variability relative to the small ventilatory responses elicited by such brief chemical stimuli. To overcome this limitation we have used pseudorandom chemical stimulation instead of single-breath inhalations. The responses were parameterized using the prediction-error method of transfer function estimation[5].

METHODS

To assess central chemoreflex responses to single-breath inhalation of a hypercapnic gas, we presented breath-by-breath pseudorandom changes in F_ICO_2 while maintaining $F_IO_2 > 0.94$. Eight normal, awake subjects sat in a comfortable chair, breathing through a facemask and non-rebreathing valve. Airflow was measured with a Fleisch pneumotachometer and integrated digitally to obtain tidal volume. Airway CO_2 was measured with an infrared analyzer. After 5 minutes of breathing 100% O_2, F_ICO_2 was varied pseudorandomly between 0.0 and 0.05 according to a 63-breath pseudorandom binary sequence (PRBS). F_ICO_2 was changed by a computer during expiration by electronically switching a large-bore, 2-way solenoid valve which connected

[+]Dept. of Biomedical Engineering, Case Western Reserve University, Cleveland, OH 44106 the inspiratory inlet of the non-rebreathing valve to one of two sources of inspired gas. Between ten and sixteen 63-breath sequences were presented to each subject. Details of these methods are given elsewhere[6].

In another group of 5 awake normal subjects we assessed the closed-loop ventilatory responses to single-breath inhalation of an hypoxic gas by switching the inspired oxygen level between 12% and 21% while maintaining the average F_ICO_2 at 0.02. In addition to the above measurements, airway O_2 fraction was measured with a rapid analyzer. Two tests were performed on each subject. In the first test we used a generalized minimum variance, digital adaptive controller to adjust F_ICO_2 to oppose changes in $P_{ET}CO_2$ on a breath by breath basis[7]. In this way during pseudorandom hypoxic stimulation $P_{ET}CO_2$ was held very close to its mean level before initiating pseudorandom hypoxia. In the second test F_ICO_2 was held constant at 0.02 and $P_{ET}CO_2$ was allowed to vary in response to ventilatory changes.

In all studies airflow and airway CO_2 and O_2 were sampled by a computer. Respiratory rate, tidal volume, minute ventilation, F_ICO_2, F_IO_2, $F_{ET}CO_2$, and $F_{ET}O_2$ were calculated for each breath. Estimates of the responses of each variable to a single-breath inhalation of the stimulus gas (i.e., 1% CO_2 on a background of >94% O_2 plus 2.5% CO_2, or 1% O_2 on a background of 16.5% O_2 plus 2% CO_2) were computed using the cross-correlation method[8]. In addition, a discrete-time second-order transfer function estimate between the input signal (i.e., F_ICO_2 or F_IO_2) and ventilation was calculated using the prediction-error method[5]. Then the equivalent continuous-time transfer function was determined using the impulse invariance method, from which the natural frequency and damping factor of the system were calculated. The unit impulse response of the continuous-time, second-order model was considered to be the estimated ventilatory response to a single breath inhalation containing 1% of the stimulating gas.

RESULTS

Responses of $F_{ET}CO_2$ and ventilation to a single-breath inhalation of 1% CO_2 in hyperoxia, calculated using the cross-correlation method, have been presented previously[6]. When the ventilatory responses were fit using the prediction-error method as described above, a second-order model was adequate in all cases. Typically the calculated impulse responses of these models overlay the ventilatory responses based on cross-correlation but they were much less noisy. Figure 1 shows an example. In four subjects the transfer function relating the closed-loop ventilatory output to the input F_ICO_2 had complex conjugate poles with damping factors ranging from 0.33 to 0.59 and natural frequencies between 0.03 and 0.06 Hz. The impulse responses of these systems exhibited weak, highly-damped oscillations (e.g., figure 1). The transfer functions for the other four subjects had two negative real poles. The latter impulse responses of ventilation exhibited an initial rise which reached a peak at about 30-40 seconds, then a slow decline without undershoot which reached the baseline at about 80-100 seconds.

During pseudorandom hypoxic stimulation we attempted to prevent any change in the CO_2 partial pressure at the peripheral chemoreceptors by regulating $P_{ET}CO_2$ using an adaptive control scheme. The effectiveness of regulation was assessed by calculating the cross-correlation function between F_IO_2 and $P_{ET}CO_2$. $P_{ET}CO_2$ did show a small response that was correlated with the F_IO_2 stimulus. After an initial decrease that averaged about 1 Torr, $P_{ET}CO_2$ returned to baseline over 4-5 breaths. However, in the absence of regulation the initial decrease was 2-3 times as large and the return to baseline required 10 or more breaths.

For each subject a second-order model relating the input, F_IO_2, to the output, ventilation, was fit to the data obtained during pseudorandom hypoxic stimulation. The ventilatory response to a decrease of F_IO_2 of 0.01 for a single

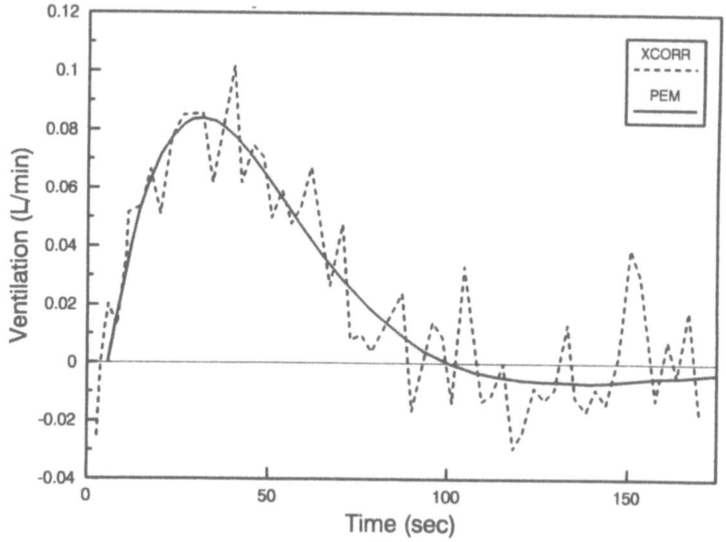

Figure 1. Ventilatory impulse response to CO_2 in hyperoxia for one subject calculated using cross-correlation method (dashed) and prediction-error method (solid).

inhalation depended on whether $F_{ET}CO_2$ was regulated or allowed to vary. When $F_{ET}CO_2$ was regulated the ventilatory impulse responses peaked at 20-30 seconds, then returned to baseline by 60-90 seconds. In 4 of these cases the ventilatory impulses responses were nearly critically damped or overdamped (figure 2). Two of the 5 subjects showed small undershoots in their impulse responses (damping factors of 0.68 and 0.46). When $F_{ET}CO_2$ was allowed to change as a consequence of the ventilatory changes, 4 of the impulse responses became underdamped while the frequency of oscillation (i.e., the magnitude of the imaginary part of the second-order poles) increased in each case (figure 2). While both effects should increase the amount of undershoot in the ventilatory impulse response, the difference between the regulated and unregulated $P_{ET}CO_2$ cases was usually small (figure 3).

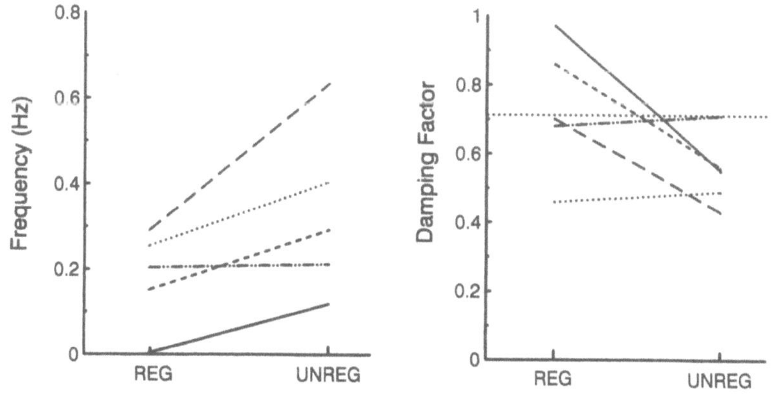

Figure 2. Parameters from the second-order fits to the pseudorandom hypoxia data when $P_{ET}CO_2$ was regulated (REG) and allowed to vary (UNREG). Damping factor (right) and frequency of oscillation (left), which is the magnitude of the imaginary part of the second-order poles. Light dotted line at right indicates critical damping factor of 0.707.

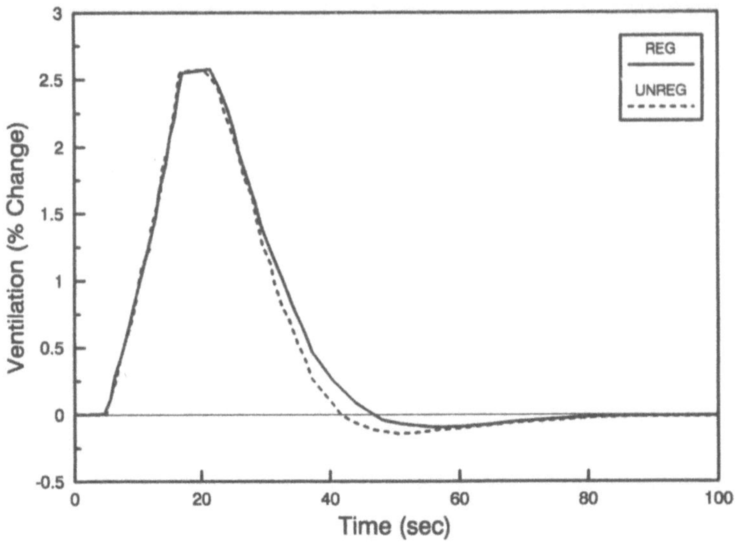

Figure 3. Ventilatory impulse responses for one subject calculated using prediction-error method. Solid: $P_{ET}CO_2$ regulated; dashed: $P_{ET}CO_2$ unregulated.

DISCUSSION

Periodic breathing is thought to arise in many cases because of unstable operation of respiratory chemical feedback loops[1,2,9,10]. Within this framework the stability of ventilation (or equivalently, the propensity for periodic breathing to develop) can be determined by assessing the degree to which a brief disturbance to chemical feedback loops elicits oscillatory fluctuations in ventilation[3,4]. If a subject has a "loop gain" of one or greater, then such a disturbance should lead to sustained oscillations, whereas a loop gain less than approximately 0.71 should be associated with damped oscillations in ventilation consequent to a brief disturbance. Furthermore it is usually concluded that the central chemoreflex pathway does not participate substantially in such responses to disturbances since it is "too slow" unless peripheral chemosensitivity is low[1,2,8,11]. In the present studies we have directly measured the ventilatory responses of individual subjects to respiratory chemical disturbances, rather than inferring such responses from estimates of loop gain. We found that the central pathway can contribute to such responses, even producing damped oscillations in some subjects. We also found that the tendency for damped oscillations to occur in response to a hypoxic disturbance was increased when $F_{ET}CO_2$ was allowed to vary in response to the ventilatory changes (as would occur naturally). Previously we have shown both in anesthetized rats and in simulation studies of human respiratory control that these estimated single-breath responses derived from use of pseudorandom stimulation are good approximations to averaged responses to true single-breath stimuli[12].

Although it is desirable to measure these responses while the subjects are breathing room air, the experimental techniques did not permit that situation. Because F_ICO_2 cannot be lower than zero, the mean F_ICO_2 during PRBS CO_2 stimulation (i.e., the mean of the two levels used) was 2.5%. Likewise, the mean F_IO_2 during PRBS hypoxia was 16.5%. Furthermore, to regulate $P_{ET}CO_2$ it was necessary to be able to adjust F_ICO_2 both above and below its mean level. Thus, pseudorandom hypoxia was performed on a background of 2% CO_2. The elevated mean F_ICO_2 may have decreased slightly the tendency for ventilatory oscillations to occur[10] whereas the lowered mean F_IO_2 during PRBS hypoxia may have increased this tendency[1,2].

As discussed in detail previously[6], we have assumed that hyperoxia effectively prevents any peripheral chemoreceptor responsiveness to CO_2 stimulation. Thus we conclude that the central chemoreflex pathway does respond to small, transient disturbances in pulmonary CO_2 exchange, such as those that occur on a breath to breath basis due to ventilatory variability during wakefulness[13,14]. This pathway might contribute as much as 35-52% of the ventilatory response to small disturbances in normal subjects on room air[6]. Our results imply that in some subjects such disturbances could elicit highly damped ventilatory oscillations via central chemoreflex pathways. Comparison of figures 1 and 3 suggests that these oscillations would have lower frequencies than those elicited via peripheral chemoreflex mechanisms. It remains to be determined whether similar conclusions would apply during any stage of sleep and whether a small ventilatory or gas exchange disturbance in sleep (such as a partial airway obstruction) might evoke a transient ventilatory oscillation via central chemoreflexes.

The regulation of $P_{ET}CO_2$ during pseudorandom hypoxia was not perfect but one can infer that any contribution to the ventilatory response from CO_2 changes was much smaller when $P_{ET}CO_2$ was regulated. Inferring therefore that the ventilatory response when $P_{ET}CO_2$ was regulated was essentially due to hypoxia alone, we conclude that a single breath of hypoxic inspirate has ventilatory consequences that last for at least 30 seconds in awake humans.

In natural situations $P_{ET}CO_2$ varies consequent to the ventilatory response to hypoxia. Thus the initial hyperventilation should change both $P_{ET}O_2$ and $P_{ET}CO_2$ in directions to diminish stimulation of peripheral chemoreceptors. Whereas the effect on $P_{ET}O_2$ compensates for a decrease in F_IO_2, the effect on $P_{ET}CO_2$ is simply to lower its stimulation of chemoreceptors. We would expect then an increasing tendency for ventilation to undershoot its baseline. We observed this effect when $P_{ET}CO_2$ was not regulated. In waking subjects, however, this difference in response might well be obscured by the normal ventilatory variability. The question remains whether in quiet sleep it would be more apparent, and whether it then may be possible to observe ventilatory oscillations in quiet sleep even when loop gain is less than one.

Although we observed the abovementioned effect on the ventilatory response when $P_{ET}CO_2$ was regulated, the effect was small when expressed in terms of the second-order fits to the data. On the other hand, visual inspection of the responses calculated using the cross-correlation method suggested that all of their dynamic features were not always captured by a second-order model. In particular, the model behavior during the decline to baseline was sensitive to the orders of the system and noise models used in the prediction-error method. We chose to use the simplest system model even if the stopping criterion (i.e., AIC or FPE) was slightly larger than that of higher order models. The effect of regulating $P_{ET}CO_2$ may be greater than these simple models indicate.

Simulation studies which have addressed the conditions which would lead to periodic breathing sometimes have considered that hypoxia is a factor only to the extent that it alters the mean peripheral sensitivity to CO_2[1,10]. Our results, however, suggest that the dynamic interaction between hypoxic and hypercapnic stimuli increase the tendency for periodic breathing to a greater degree than is accounted for by the mean change in peripheral chemosensitivity alone. This suggestion could be tested experimentally using PRBS CO_2 stimulation with and without regulation of $P_{ET}O_2$. Our results also may explain some of the discrepancies between loop gain estimates and ventilatory responses to CO_2 disturbances reported by Carley, *et al.*[3]. We would argue, in fact, that the latter are more useful tests of the propensity of individual subjects to develop periodic breathing. Furthermore it is no more difficult to obtain these transient ventilatory responses, using pseudorandom stimulation and the prediction-error method to parameterize the responses, than to obtain the necessary parameters for calculating loop gain.

REFERENCES

1. Carley, D. W., and D. C. Shannon, A minimal mathematical model of human periodic breathing. *J. Appl. Physiol.* 65: 1400(1988).
2. Khoo, M. C. K., R.E. Kronauer, K. Strohl, and A. S. Slutsky, Factors inducing periodic breathing in humans: a general model. *J. Appl. Physiol.* 53:644(1982).
3. Carley, D. W., and D. C. Shannon, Relative stability of human respiration during hypoxia. *J. Appl. Physiol.* 65: 1389(1988).
4. Fleming, P. J., A. L Concalves, M. R. Levine, and S. Woolard, The developemnt of stability of respiration in human infants: changes in ventilatory responses to spontaneous sihgs. *J. Physiol. Lond.* 347:1(1984).
5. Ljung, L., 1987, "System Identification: Theory for the User", Prentice-Hall, Englewood Cliffs, NJ.
6. Modarreszadeh, M., and E. N. Bruce, Long-lasting ventilatory response of humans to a single breath of hypercapnia in hyperoxia. *J. Appl. Physiol.* 72:242(1992).
7. Modarreszadeh, M., Systems Analysis of Breath to Breath Ventilatory Variations in Man: Role of CO_2 Feedback. Ph. D. Dissertation: Case Western Reserve University, Cleveland, OH (1990).
8. Sohrab, S., and S. M. Yamashiro, Pseudorandom testing of ventilatory response to inspired carbon dioxide in man. *J. Appl. Physiol.* 49:1000(1980).
9. Brusil, P. J., T. B. Wagener, R. E. Kronauer, and P. Gulesian, Methods for identifying respiratory oscillations disclose altitude effects. *J. Appl. Physiol.* 48:545(1980).
10. ElHefnawy, A., G. M. Saidel, E. N. Bruce, and N. S. Cherniack, Stability analysis of CO_2 control of ventilation. *J. Appl. Physiol.* 69:498(1990).
11. Khoo, M. C. K., and V. Z. Marmarelis, Estimation of peripheral chemoreflex gain from spontaneous sigh responses. *J. Appl. Physiol.* 68:393(1990).
12. Dhawale, P., and E. N. Bruce, Ventilatory response to hypoxia: Estimation using pseudorandom technique. *FASEB J.* 4:A1106(1990).
13. Modarreszadeh, M., E. N. Bruce, and B. Gothe, Nonrandom variability in respiratory cycle parameters of man during quiet sleep. *J. Appl. Physiol.* 69:630(1990).
14. Khatib, M. F., Y. Oku, and E. N.Bruce, Contribution of chemical feedback loops to breath-to-breath variability of tidal volume. *Respir. Physiol.* 83:115(1991).

LACK OF VENTILATORY REACTION TO ENDOGENOUS AND EXOGENOUS METABOLIC BRAIN-STEM ACIDOSIS

H. Kiwull-Schöne and P. Kiwull

Department of Physiology
Ruhr-University
D-4630 Bochum, Germany

INTRODUCTION

It is generally assumed that metabolic acid-base disturbances in the extracellular body fluids are compensated by adjustment of the arterial P_{CO_2} through pulmonary ventilation. Since the pH of the extracellular fluid (ECF) is one of the most accurately controlled variables, Winterstein[1] postulated the H^+-ions dissociated either from carbonic acid or from non-volatile acids as being the unique stimulus for the chemical control of breathing (Reaction Theory). Loeschcke[2] adapted this theory to the central chemosensitive system and conducted a considerable amount of experiments supporting the ECF-pH at the ventral surface of the medulla oblongata to be the essential chemical signal to drive pulmonary ventilation. Indeed, ventilation in cats was highly sensitive against ECF-pH changes at the ventral medullary surface, if induced by CO_2-inhalation[2, 3]. Unexpectedly, we did not observe any ventilatory reaction to even more pronounced ECF-pH changes if caused by post-hypoxic lactacidosis after carotid chemodenervation[3]. Endogenous accumulation of organic acid thus failed to drive ventilation, unlike exogenous infusion of strong anorganic acids[4, 5, 6]. Therefore, in the present study, the ventilatory reaction to equimolar plasma lactic acid concentrations was compared, either endogenously developed during a limited period of hypoxia or exogenously induced by intravenous infusion.

METHODS

The experiments were performed in 15 rabbits (2.9 ±0.2 kg), anaesthetized by an initial dose of 43.3 ±1.8 mg/kg pentobarbital sodium i.v., followed by continuous infusion of 6.3 ±0.35 mg/kg/h. All animals were peripherally chemodenervated and vagotomized. They were either spontaneously breathing (Group A, N=9) or paralysed by Alloferin® and artificially ventilated (Group B, N=6). By adding oxygen to the inspired air, the arterial P_{O_2} was kept above 20 kPa. During spontaneous breathing, tidal volume (V_T) and inspiratory/expiratory durations (T_I, T_E) were continuously measured by pneumotachography. In the artificially ventilated animals, the integrated compound potential of one desheathed phrenic nerve (IPNA) was taken as a measure of central nervous respiratory

Control of Breathing and Its Modeling Perspective, Edited by
Y. Honda *et al.*, Plenum Press, New York, 1992

output. Concomitantly, the ECF-pH at the exposed ventral surface of the medulla oblongata was continuously recorded by a balanced pH/reference-electrode combination[3]. In both groups, the end-tidal P_{CO_2} (P_{ETCO_2}) was estimated by infrared absorption. From arterial blood samples, the oxygen pressure (Pa_{O_2}) was measured by electrode, the CO_2 pressure (Pa_{CO_2}) indirectly as pH after equilibration with CO_2, yielding also concentrations of standard bicarbonate (HCO^-_{3st}) and the CO_2-buffering capacity ($\beta co_2 = \Delta \log Pa_{CO_2}/\Delta pHa$). Lactic acid concentrations (Lac^-) were determined enzymatically. Differences of group means ±SEM were regarded as being significant at the level $P_D \leq 0.05$ of a paired t-test.

RESULTS

Endogenous lactacidosis: The 9 animals of Group A initially inhaled O_2-enriched air until a steady state was reached. Accumulation of lactic acid was achieved by a 30 min period of inspiratory hypoxia at arterial P_{O_2}-levels between 9.7 and 3.2 kPa. Going back to inspiratory hyperoxia, the post-hypoxic steady state was compared with the pre-hypoxic control condition. Table 1 shows a significant rise in plasma lactic acid concentration, accompanied by a fall in standard HCO_3^-. The considerable fall in pH is entirely caused by metabolic acidosis, since the Pa_{CO_2} does not change significantly. Accordingly, there are significant changes neither in pulmonary ventilation ($\dot{V}$) nor in tidal volume (V_T).

Exogenous lactacidosis: In the 6 mechanically ventilated animals of Group B, accumulation of lactic acid was achieved by 10 min intravenous infusion of 1mmol/min lactic acid. Table 1 shows plasma lactic acid concentrations and acid-base conditions in the order of magnitude as for Group A. 60 min after infusion stop, steady state changes neither in Pa_{CO_2}, nor in tidal (IPNA) or minute phrenic activity (IPNA·f) could be shown, although the metabolic acidosis was still persisting in the arterial blood (Table) and in the brain-stem extracellular fluid (see below), ECF-pH being changed from the control value of 7.365 ±0.032 to 7.304 ±0.035. There are, however, transient rises in IPNA·f by up to 35% (Fig. 1A) and in arterial P_{CO_2} (by up to 0.8 kPa).

In order to differentiate between the CO_2-induced change in ECF-pH and the acid-induced portion, the arterial P_{CO_2} was varied in the range between 2.7 and 5.3 kPa by the pumping rate of the respirator. ECF-pH changed on an average by 0.065 ±0.012 units per kPa, corresponding to an in vivo CO_2-buffering capacity $\Delta Pa_{CO_2}/\Delta pH_{ECF} = -1.74 \pm 0.20$,

Table 1. Respiratory data and arterial acid-base conditions during metabolic acidosis of different origin.

		Group A: Endogenous Lactacidosis		*Group B*: Exogenous Lactacidosis	
		control	after hypoxia	control	after infusion
$\dot{V}$	[ml·min⁻¹·kg⁻¹]	370 ±28	390 ±20		
V_T	[ml·kg⁻¹]	13.4 ±1.1	13.0 ±0.8		
IPNA·f	[units·min⁻¹]			684 ±128	636 ±115
IPNA	[units]			16.3 ±3.0	17.0 ±2.9
P_{ETCO_2}	[kPa]	3.88 ±0.13	3.73 ±0.11	2.87 ±0.24	2.57 ±0.22*
Pa_{CO_2}	[kPa]	4.60 ±0.15	4.47 ±0.12	3.12 ±0.15	3.24 ±0.47
pHa		7.392±0.021	7.236±0.029*	7.441±0.032	7.327±0.027*
HCO^-_{3st}	[mM]	21.5 ±1.1	15.1 ±1.0*	19.2 ±1.2	14.9 ±1.1*
Lac^-	[mM]	6.2 ±1.2	10.9 ±1.0*	4.6 ±1.0	10.0 ±1.4*

Mean values ±SEM in carotid chemodenervated vagotomized rabbits. Group A (N=9): spontaneously breathing. Group B (N=6): artificially ventilated.
* Significant changes by acidosis

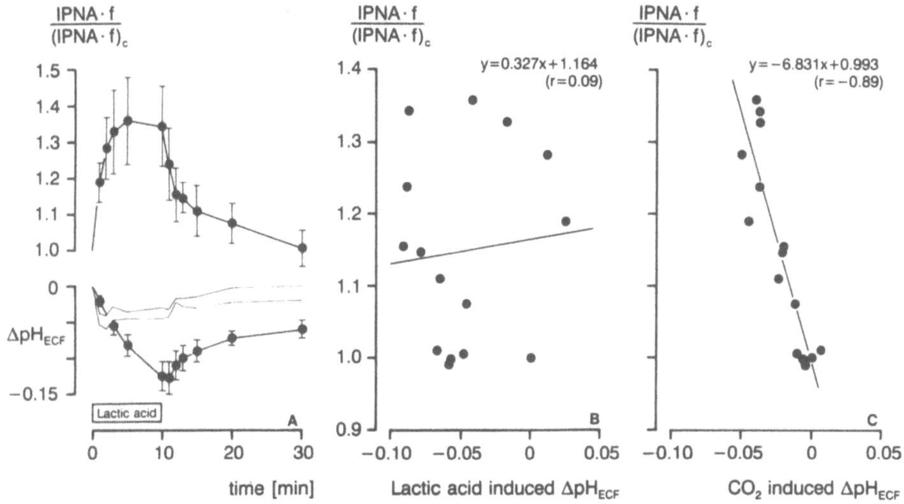

Figure 1. Respiratory effects of intravenous lactic acid infusion in carotid chemodenervated vagotomized rabbits. Group B (N=6), artificially ventilated. Mean values ±SEM of phrenic minute activity (relative changes) and of medullary surface ECF-pH (absolute changes) in response to 10 min acid infusion (A). Thin lines indicate the range of estimated CO_2-induced portion of pH-change. Same data of phrenic minute activity (for clarity without SEM) plotted either against the lactic acid-induced (B) or CO_2-induced (C) change in ECF-pH.

rather similar to the arterial β_{CO_2} in vitro of -1.73 ± 0.07. Based on this relationship, the CO_2-induced change of ECF-pH in response to acid infusion can be separated (Fig. 1A, C) and subtracted (Fig. 1B) from the total pH-change. If the changes in phrenic minute activity are related to ECF-pH changes of either origin, there is rather no correlation in case of lactic acid-induced changes (Fig. 1B), but a considerable high correlation in case of CO_2-induced changes (Fig. 1C). Thus, both the immediate deflection of the brain-stem surface-pH and the prompt but transient respiratory drive observed in response to acute acid infusion appear to be predominantly caused by a transient release of CO_2 from HCO_3^-.

DISCUSSION

In spontaneously breathing rabbits, a distinct endogenous metabolic acidosis caused by hypoxia-induced accumulation of lactic acid was not compensated by hyperventilation and reduction of Pa_{CO_2}. Exogenous metabolic acidosis as a consequence of intravenous lactic acid infusion in artificially ventilated rabbits led only to a transient respiratory drive, which could be explained by secondary effects of the transient release of CO_2 from HCO_3^-, being quickly exchanged across the blood-brain barrier.

Due to the discrepancy between this finding and the high sensitivity of pulmonary ventilation against CO_2-induced changes in ECF-pH[2,3], investigators were led to replace the unique significance of brain-stem ECF-pH for central chemoreception (Reaction Theory) by a possible role of the intra/extracellular H^+-gradient[3,7]. During hypercapnia, a decreasing gradient as a tentative respiratory drive is to be expected, due to a more effective buffering of CO_2 in the intracellular than in the extracellular compartment[7,8]. Furthermore, intracellular lactic acid during hypoxia would increase the gradient in agreement with ventilatory depression[9]. However, the lacking respiratory effect of exogenous metabolic acidosis we found is not compatible with a unifying transmembrane H^+-gradient theory, since there is no reason to believe that extracellular metabolic acidosis is less effectively buffered inside the cell than respiratory acidosis, hence comparable gradients should elicit comparable respiratory responses.

At first sight, our finding appears to be completely at variance with those of other investigators[4, 5, 6], using infusions of strong mineral acids. However, some of the reported considerable respiratory drives in response to acid infusion may be explained by secondary effects of the transient release of CO_2 (see above). Furthermore, there is a general problem to establish true "isocapnic" conditions when controlling only endtidal Pco_2[4,5]. In our experiments, although arterial isocapnia was reached as soon as 10 min after end of lactic acid infusion, the endtidal Pco_2 progressively decreased.

Possibly, the lack of ventilatory reaction to exogenous lactacidosis in our experiments is mainly a quantitative problem, since the metabolic brain-stem acidosis did not exceed 0.068 $\pm$0,014 pH-units within one hour. Thereby, a small but not significant rise in IPNA was typically accompanied by a reduction in respiratory rate. Correspondingly, rather strong metabolic acidification by low HCO_3^- of isocapnic mock cerebrospinal fluid (0.25 to 0.73 pH-units) has been used in medullary superfusion experiments to increase tidal volume by about 10% and 15% per 0.1 pH-units in spontaneously breathing rats and cats, respectively[2, 10]. This means only about 5% of the ventilatory sensitivity against changes in ECF-pH by inhaled CO_2[3].

The striking differences observed in sensitivity of the responses to metabolic and respiratory brain-stem acidosis as well as some characteristics in breathing pattern may point to different sites of action and/or underlying mechanisms, although the present data do not allow farther reaching conclusions. The general question arises, whether under physiological conditions, the respiratory compensation of a systemic metabolic acidosis can be performed by one and the same central chemosensitive mechanism being to such a degree protected against severe local metabolic acid-base changes and rather insensitive to them.

ACKNOWLEDGEMENTS

Professor Dr. Dr. h.c. Hans H. Loeschcke (1912-1986) in memory of his 80th birthday

REFERENCES

1. H. Winterstein, Die Regulierung der Atmung durch das Blut, *Pflügers Arch.* 138: 167 (1911)
2. H.H. Loeschcke, Central chemosensitivity and the reaction theory, *J. Physiol.(Lond.)* 322:1 (1982)
3. H. Kiwull-Schöne and P. Kiwull, Hypoxic modulation of central chemosensitivity, *in*: "Central Neurone Environment and the Control Systems of Breathing and Circulation", M.E. Schläfke, H.P. Koepchen, and W.R. See, eds., Springer, Berlin-Heidelberg-New York (1983)
4. L.J. Teppema, P.W.J.A. Barts, H.Th. Folgering and J.A.M. Evers, Effects of respiratory and (isocapnic) metabolic arterial acid-base disturbances on medullary extracellular fluid pH and ventilation in cats, *Respir. Physiol.* 53:379 (1983)
5. F.L. Eldridge, J.P. Kiley and D.E. Millhorn, Respiratory responses to medullary hydrogen ion changes in cats: Different effects of respiratory and metabolic acidoses, *J. Physiol.(Lond.)* 358:285 (1985)
6. H. Shams, Differential effects of CO_2 and H^+ as central stimuli of respiration in the cat, *J. Appl. Physiol.* 58:357 (1985)
7. F.D. Xu, M.J. Spellman Jr., M. Sato, J.E. Baumgartner, S.F. Ciricillo and J.W. Severinghaus, Anomalous hypoxic acidification of medullary ventral surface, *J. Appl. Physiol.* 71:2211 (1991)
8. K.E. Jensen, C. Thomsen and O. Henriksen, In vivo measurement of intracellular pH in human brain during different tensions of carbon dioxide in arterial blood. A ^{31}P-NMR study, *Acta Physiol. Scand.* 134:295 (1988)
9. J.A. Neubauer, A. Simone and N.H. Edelman, Role of brain lactic acidosis in hypoxic depression of respiration, *J. Appl. Physiol.* 65:1324 (1988)
10. H. Tojima, T. Kuriyama and Y. Fukuda, Differential respiratory effects of HCO_3^- and CO_2 applied on ventral medullary surface of rats, *J. Appl. Physiol.* 70:2217 (1991)

SYNCHRONIZATION OF RESPIRATORY RHYTHM WITH MECHANICAL VENTILATION BY RESPIRATORY CO_2 OSCILLATIONS IN VAGOTOMIZED DOGS

[1]Eiji Takahashi, Kazunori Tejima, and Isao Tateishi

Division of Biomedical Systems Engineering
Faculty of Engineering, Hokkaido University
Sapporo, 060 Japan

INTRODUCTION

The medullary respiratory rhythm generator of animals on a fixed rate mechanical ventilation tends to synchronize to the rate of mechanical ventilation. Since vagotomy abolishes this synchronization, phasic vagal afferent discharges arising from pulmonary stretch receptors are believed to carry the information of phasic lung movements (caused by mechanical ventilation).[1] Another source of phasic afferent input to the respiratory center is the respiratory oscillations of arterial Pco_2 (CO_2 oscillations), which can be sensed by the peripheral chemoreceptor. Therefore, it may be natural to postulate that, in the absence of vagal afferent input, the central respiratory rhythm could entrain to the phasic chemical afferent input arising from respiratory CO_2 oscillations. However, the effect of CO_2 oscillations seems to depend upon their magnitude.[2] In the present study, we determined a level of CO_2 oscillations which is large enough to induce the synchronization in anesthetized, paralyzed and vagotomized dogs. The possible role of the entrainment by CO_2 oscillations in the respiratory control is also discussed.

METHODS

We used 10 urethane-chloralose anesthetized, paralyzed and vagotomized dogs. The dogs were tracheostomized and mechanically ventilated at a fixed rate. Arterial catheterization was conducted to measure arterial blood pressure (P231D, Gould Statham) and for arterial blood sampling. The right superior laryngeal nerve (SLN) was separated just above the larynx, cut, and placed on a unipolar platinum electrode for electrical stimulation. The C_5 branch of the right phrenic nerve was separated at the level of the neck, cut, and placed on a bipolar platinum electrode. The neural discharge was amplified, full-rectified, and moving time averaged to give a phrenic neurogram. Following a total heparinization, a catheter-tip ion-sensitive field effect

---

[1] Present address: Department of Physiology, Yamagata University School of Medicine, Yamagata, 990-23 Japan.

Control of Breathing and Its Modeling Perspective, Edited by
Y. Honda *et al.*, Plenum Press, New York, 1992

transistor pH electrode (PH1035, Kuraray)[3] was inserted into the descending aorta to measure respiratory oscillations of arterial pH, which was subsequently converted to CO_2 oscillations.[4] In 5 dogs, veno-venous bypass was conducted using a membrane lung (Capiox-II 08, Terumo) to alter CO_2 flow to the lung. Arterial blood gases were analyzed by standard electrodes (ABL-30, Radiometer). Alveolar gas was sampled from the tracheostomy tube and analyzed (1H21A, San-ei) to calculate CO_2 output from the lung ($\dot{V}co_2$). Bilateral thoracotomy was conducted in all the dogs to minimize fluctuations of arterial blood pressure caused by phasic lung inflations.

Firstly, the rate of mechanical ventilation was adjusted so that it matched the dog's spontaneous respiratory rate as judged by the phrenic neurogram. Tidal volume was regulated to maintain isocapnia. Phase difference between the onset of the burst of phrenic neurogram (determined by the dog's respiratory center) and the respiratory oscillations of arterial pH

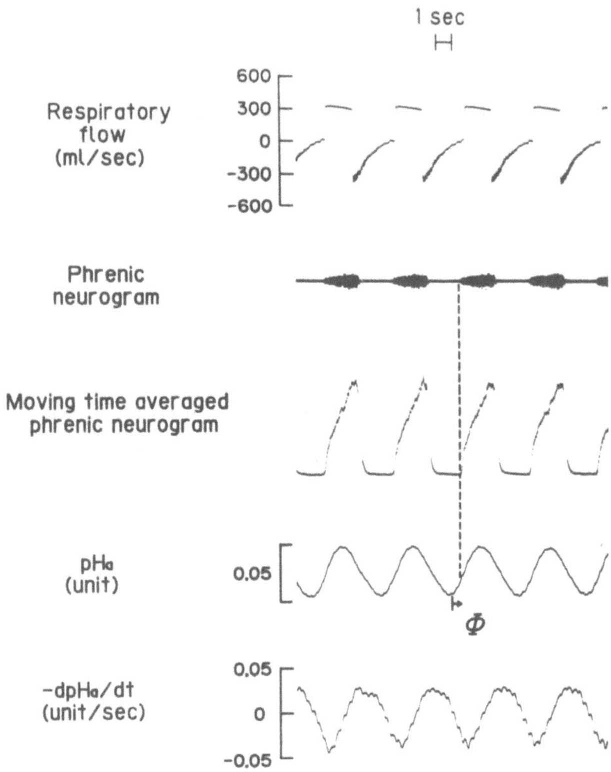

Figure 1. Tracings from one dog demonstrating (from top to bottom) respiratory flow (inspiration upward), raw phrenic neurogram, moving time average of the phrenic neurogram, intraarterial pH (alkaline upward), and the rate of change of arterial pH. Phase (Φ) was defined as the time difference from the trough of respiratory pH oscillations to onset of inspiratory activity divided by total respiratory duration.

(determined by the mechanical ventilation) was calculated from 5 successive breaths, and designated as the initial phase, Φ_0 (Fig.1). Then this phase relationship was temporarily disturbed by a brief (*ca.* 200 msec) electrical stimulation of SLN. When the phase relationship reached a new steady state, the phase relationship was again calculated from 5 successive breaths. Deviation of this phase from the initial phase was reported as $\Delta\Phi$. The data were discarded if, following electrical stimulation of SLN, the constant phase relationship was not established. Above procedures were repeated after increasing $\dot{V}co_2$ by venous CO_2 loading. Oxygen was added to the inspired gas in the hyperoxic CO_2 loading experiment.

RESULTS

In the normoxic normocapnic control experiment where venous CO_2 loading was not conducted (n=37), the amplitude and maximum rate of rise of CO_2 oscillations calculated from measured respiratory pH oscillations were 2.4±1.1 mmHg and 1.6±0.6 mmHg/sec, respectively. The initial phase (Φ_0) did not show any grouping but was randomly scattered within the phase plane (Table 1). Also, the phase did not return to the prestimulation level following a brief electrical stimulation of SLN (Table 1). These results together imply that there was no interaction between the respiratory center and the phasic chemical afferent input arising from CO_2 oscillations at resting $\dot{V}co_2$.

In the normoxic venous CO_2 loading experiment (n=25), $\dot{V}co_2$ was increased on average by a factor of 2.5. The amplitude and the maximum rate of rise of CO_2 oscillations were considerably increased to 8.1±3.2 mmHg and 4.3±1.4 mmHg/sec, respectively. As shown in Table 1, the steady state phase shift following brief electrical stimulation of SLN was almost null ($P<0.001$) implying that the respiratory rhythm can resume the initial phase following temporal disturbance (synchronization). The mean phase angle for the initial phase was 0.70 deg (Table 1), which implies that the phase of the onset of spontaneous respiratory activity was locked to the peak of respiratory CO_2 oscillations (see Fig.1 for the definition of phase angle).

Oxygen was added to the airway to functionally block the carotid body. In the hyperoxic (Pao_2=337±135 mmHg) venous CO_2 loading experiment (n=24), $\dot{V}co_2$ was increased by a factor of 2.7, which resulted in considerable increases in the amplitude and maximum rate of rise of CO_2 oscillations (7.2±3.6 mmHg and 4.9±1.0 mmHg/sec, respectively). Despite the significantly augmented respiratory CO_2 oscillations by venous CO_2 loading, $\Delta\Phi$ was randomly distributed within the phase plane and synchronization was not demonstrated (Table 1).

The blood transport delay from the lung to the intraarterial pH electrode was 3.1±0.6 sec, which may be comparable to the blood transport delay from the lung to the carotid body.

Table 1. Circular analysis for the initial phase (Φ_0) and phase shift ($\Delta\Phi$).

	Control	Normoxic CO_2 loading	Hyperoxic CO_2 loading
Φ_0 (deg)	125 ± 106	0.698 ± 25.9*	31.1 ± 68.7*
$\Delta\Phi$ (deg)	102 ± 106	1.91 ± 26.0*	354 ± 91.3

Values are represented in the mean angle ± circular standard deviation.
*: significantly different from a uniform distribution ($P<0.05$) as analyzed by the Reyleigh's test.

DISCUSSION

We have demonstrated in anesthetized and paralyzed dogs that the central respiratory rhythm generated in the brain stem respiratory center could synchronize to the rate of mechanical ventilation in the absence of phasic vagal afferent input. Prerequisites for the synchronization were an increase in $\dot{V}co_2$ and functioning carotid bodies. Since venous CO_2 loading considerably increased the magnitude of respiratory CO_2 oscillations, we conclude that the central respiratory rhythm was entrained to the phasic chemical afferent input arising from respiratory CO_2 oscillations which was sensed by the carotid body.

One of the important findings of the present study was that the effect of CO_2 oscillations on the respiratory center output appeared to depend on their magnitude, since it was necessary to increase $\dot{V}co_2$ 2.5 fold so that CO_2 oscillations affect the respiratory control. The present result is consistent with our previous finding that sudden elimination of CO_2 oscillations without change in the mean $Paco_2$ had no effect on the respiratory center output at resting $\dot{V}co_2$.[2] Therefore, we conclude that the respiratory CO_2 oscillations may be involved in the control of respiration when they are considerably increased by, for example, muscular exercise.

Then how would respiratory CO_2 oscillations affect the respiratory control during exercise? In the previous study,[3] we have demonstrated that respiratory CO_2 oscillations may be important in the control of timing of respiration in vagotomized dogs. Namely, significant prolongation of expiratory duration commonly seen in vagotomized dogs can be, at least in part, accounted for by the large alkaline swing of respiratory pH oscillations caused by augmented tidal volume due to the loss of volume feedback by vagotomy. Probably, respiratory CO_2 oscillations could affect the timing of the respiration (i.e., regulation of expiratory duration *or* inspiratory onset) in vagi intact animals in the same line. In the present study, we have demonstrated that the onset of inspiration can be locked to the trough of respiratory pH oscillations (peak of CO_2 oscillations) when the oscillations were sufficiently large. At this phase, the afferent neural discharge of the carotid body would be maximal and the medullary respiratory neurones would receive maximal facilitatory chemical input, which would then trigger the next inspiration.

At least the following two points should be considered with regard to the physiological role of the synchronization. Firstly, this phase locking mechanism may stabilize the respiratory control during exercise. It is empirically known that breath-to-breath fluctuations in respiratory rate are less during exercise compared to resting state. Furthermore, Petersen *et al.*[5] found in human subjects that coupling of the lung-to-ear transit time of blood and respiratory phase became tighter with increasing exercise intensity. These together with the present results suggest that respiratory CO_2 oscillations may be involved in the control of respiration, particularly of the timing, when $\dot{V}co_2$ is increased. Secondly, if the phase locking is tight enough, the respiratory rate would increase in concert with increase in cardiac output, since increase in cardiac output during muscular exercise shortens the lung to carotid body blood transit time. This mechanism may operate in favor of matching ventilation with circulation. However, the effectiveness of these reflex effects is a complex function of the amplitude and rate of change of CO_2 oscillations, and the lung to carotid body blood transport delay.

REFERENCES

1. G.A. Petrillo and L. Glass, A theory for phase locking of respiration in cats to a mechanical ventilator, *Am. J. Physiol.* 246(Regulatory Integrative Comp. Physiol. 15):R311-R320 (1984).
2. E. Takahashi, I. Tateishi, K. Yamamoto, and T. Mikami, Effect of withdrawal of respiratory CO_2 oscillations on respiratory control at rest, *J. Appl. Physiol.* 70:1601-1606 (1991).
3. E. Takahashi, A.S. Menon, H. Kato, A.S. Slutsky and E.A. Phillipson, Control of expiratory duration by arterial CO_2 oscillations in vagotomized dogs, *Respir. Physiol.* 79:45-56 (1990).
4. E. Takahashi and K.A. Ashe, Role of carbon dioxide oscillation in the control of respiration in the anesthetized dog, *Jpn. J. Physiol.* 39:267-281 (1989).
5. E.S. Petersen, B.J. Whipp, D.B. Drysdale, and D.J.C. Cunningham, Carotid arterial blood gas oscillations and the phase of the respiratory cycle during exercise in man: testing a model, *in*: "The Regulation of Respiration During Sleep and Anesthesia," R.S. Fitzgerald, H. Gautier, and S. Lahiri, ed., Plenum, New York, pp. 335-342 (1978).

FACTORS INFLUENCING DAY TO DAY VARIABILITY OF HUMAN VENTILATORY SENSITIVITY TO CARBON DIOXIDE

Edward S. G. Semple and Alison K. McConnell

Department of Human Sciences
Loughborough University of Technology
Loughborough
Leicestershire
LE11 3TU
UK

INTRODUCTION

It is well accepted that the ventilatory response to inhaled carbon dioxide displays large inter- and intra-subject variability. However, the published literature contains very little quantitative information to support this belief, and certainly nothing that addresses the issue specifically. In his 1967 paper, Read[5] examined the reproducibility of his rebreathing technique and obtained a coefficient of variation of about 16% (range 7%-20%). More recently, Berkenbosch and coworkers[1] compared the rebreathing and steady-state methods for measuring CO_2 sensitivity, and in doing so, made repeated rebreathing measurements during a single day. The coefficient of variation was about 15% (range 2%-34%). These, and other studies, indicate that for the rebreathing technique, the within day coefficient of variation is about 16%.

To our knowledge, the issue of variation in CO_2 sensitivity from *day to day* has not been addressed. Moreover, there is no information about the variability of measurements made using steady state techniques, a method which recent evidence suggests is more reliable and accurate than the rebreathing technique[1,2]. Furthermore, the influence of exercise upon the variability of CO_2 sensitivity, and of factors such as athletic training remains unexplored. There remains considerable controversy about directional changes in CO_2 sensitivity elicited by exercise. In a previous study [3] we compared the CO_2 sensitivity of endurance trained and non endurance trained subjects. All endurance trained subjects exhibited a rise in CO_2 sensitivity from rest to exercise. In contrast, the response of the sedentary group was variable. In a subsequent study[4], we examined the reproducibility of this response in the endurance trained, by measuring the day to day variability of CO_2 sensitivity in a single subject on 10 occasions. For this subject, CO_2 sensitivity always increased from rest to exercise, and the coefficient of variation was 21.1% at rest, and 15.3% during exercise.

These studies raised a number of issues; were the relatively high coefficients of variation for our single endurance trained subject a reflection of our steady state technique, of his endurance training, or simply unique to this subjects? Do non-athletic subjects who show small reductions in CO_2 sensitivity during exercise display the same homogeneity of response from day to day as our single endurance trained subject? Do other types of athletic training influence CO_2 sensitivity? Finally, and most importantly, in view of the large coefficient of variation of this measurement, what is the value of measuring CO_2 sensitivity on a single occasion?

Control of Breathing and Its Modeling Perspective, Edited by
Y. Honda *et al.*, Plenum Press, New York, 1992

The present study addressed these issues by examining the characteristics of the CO_2 response curve over 5 days in 3 differing groups of subjects. The subjects were either: endurance trained, sprint trained, or sedentary.

METHODS

Subjects

Some physical characteristics of the subjects are given in table 1. Each group contained 2 men and 2 women. The only significant difference between the groups was in maximal oxygen uptake, which was significantly different between the sedentary and sprint trained groups ($P<0.05$) and between the sedentary and endurance trained groups ($P<0.01$), but not between the sprint trained and endurance trained groups.

Table 1. Some physical characteristics of the subjects (mean ± SD); **$P<0.01$; *$P<0.05$ (students *t* -test); $\dot{V}O_2max$ (maximal oxygen uptake).

Group	Age (years)	Height (cm)	Weight (kg)	$\dot{V}O_2max$ ($ml.min^{-1}.kg^{-1}$)
Endurance	23.2±2.6	174.7±13.1	66.0±13.7	63.5±9.1**
Sprint	20.7±3.0	170.6± 8.2	67.2±11.3	55.6±8.8*
Sedentary	27.0±9.2	166.7± 4.0	61.9± 6.5	38.9±6.7

Apparatus

Carbon dioxide sensitivity was measured at rest and during the steady state of moderate cycle ergometer exercise at 30% of maximal oxygen uptake. The ventilatory response to CO_2 was measured using an open circuit apparatus by a steady state method. Air was blown through a wide bore tube (inspiratory line) at a rate sufficient to exceed the subject's peak inspiratory flow. Subjects respired via an ultrasonic flowmeter attached to a side arm on the inspiratory line. Subjects were interfaced to the flowmeter using a mouth-piece which was used in the usual way with a nose clip. Pure CO_2 was bled into the inspiratory line at its proximal end to obtain constant inspired concentrations of between 0% and 4% CO_2 in air.

The signal from the flowmeter was analysed by computer (Amstrad PC1512). Real-time breath-by-breath analysis was carried out and all breaths occurring within the measurement period were averaged. The flowmeter was calibrated volumetrically before and after each experiment using a 1 litre syringe.

Carbon dioxide concentration was measured continuously at the mouth by mass spectrometer (Airspec MGA 2000) and recorded on a chart recorder. The mass spectrometer was calibrated before and after each experiment using standard gases analysed previously by the Lloyd-Haldane technique.

Protocol

Room air and 3 concentrations of CO_2 (2%, 3% and 4% CO_2) were used to establish a relationship between each subject's minute ventilation ($\dot{V}_E$) and mean alveolar P_{CO2} ($P_{A,CO2}$). Each concentration of CO_2, was administered for 14 minutes in total. For the first 8 minutes the subject rested on the cycle ergometer breathing the gas mixture from the mouthpiece; measurements were made for the final 2 minutes. The subject then exercised for 6 minutes breathing exactly the same concentration of gases. Measurements were made during the final minute of exercise. The order in which the CO_2 concentrations were administered was randomised. The same procedures were used for the 'air point' measurements.

Subjects were fasted, and the athletic subjects continued to train normally in the evenings; measurements were made at the same time of day on five consecutive days.

Analysis

The mean alveolar partial pressure of CO_2 ($P_{A,CO2}$) was obtained breath-by-breath from the chart recorded CO_2 concentrations using the graphical technique, as described by Ward and Whipp[6]. All breaths occurring during the sampling period were averaged.

Steady state values for $\dot{V}_E$ and $P_{A,CO2}$ measured at each level of CO_2 inhalation were used to define equation 1 by least squares linear regression; where parameter S is the gradient of the relationship (CO_2 sensitivity) and parameter B is the point at which the extrapolated relationship intersects the CO_2 axis.

Equation 1: $$\dot{V}_E = S\ (P_{A,CO2} - B)$$

Statistical Analysis

Variability was quantified in terms of a coefficient of variation for the 5 daily measurements. Since the ratio was normally distributed, a mean coefficient of variation was derived for each group; differences between groups were tested using a *t*-test. The significance level was set at 5%.

RESULTS

Endurance trained subjects always showed an increase in CO_2 sensitivity (a rise in parameter S) from rest to exercise. The response of the control group to exercise was variable, with some subjects showing small increases in S, and some small decreases. However, for a given subject, the direction of the change was consistent from day to day (with the exception of a single subject whose response changed direction on one day). The sprint trained group displayed variability both in terms of the direction of the change in S during exercise, and also in the direction of the change from day to day.

For parameters S and B, the endurance trained group exhibited the largest coefficient of variation both at rest and during exercise (see figure 1). The control group exhibited the smallest coefficient of variation for all parameters, with those of the sprint trained group falling roughly between the endurance and sedentary groups. None of these differences was statistically significant. In contrast to the variability of parameters S and B, *normocapnic* $\dot{V}_E$

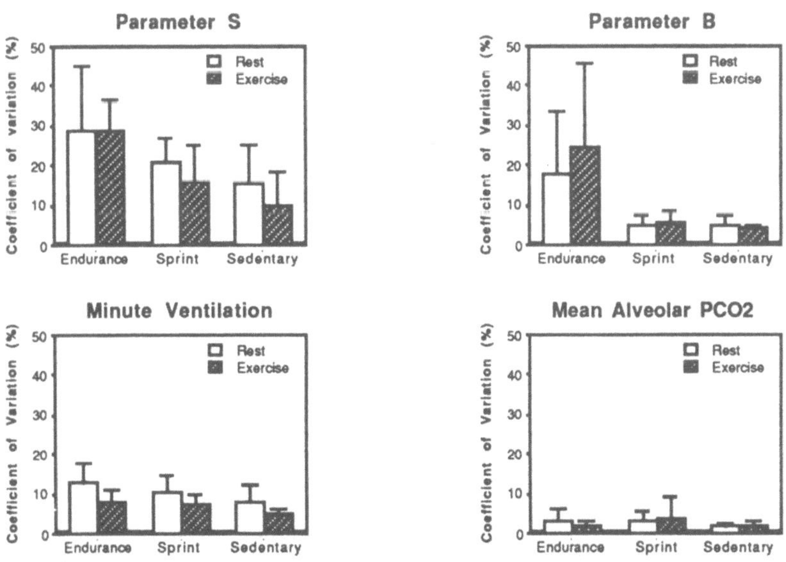

Figure 1. Mean coefficients of variation (± 2SE) for each of the groups studied.

and $P_{A,CO2}$ remained relatively unchanged from day to day in all subjects. For $\dot{V}_E$, the variability was greatest for the endurance trained group and smallest for the sedentary group, with the sprinters falling between the two. These differences were not statistically significant. There was no difference in the coefficients of variation of men and women.

DISCUSSION

One of the purposes of this study was to quantify the day to day variability of CO_2 sensitivity using a steady state inhalation technique. The data from the present study suggest that the relatively large coefficient of variation measured in our previous study[4] was a reflection of that subjects endurance training, and not of the steady state method used to derive the CO_2 response. The coefficient of variation for parameter S measured on our sedentary subjects is virtually identical to that measured by previous investigators using the rebreathing technique[1,5]. The fact that the present study quantified *day to day* variability and previous studies have quantified *within day* variability stills begs a question. However, it is reasonable to suggest that the within day variability of steady state measurements may be less than for the rebreathing technique.

The other questions we wished to answer related to the influence of exercise, and of athletic training. So far as the exercise induced changes in S are concerned, the data suggest that, on the whole, subjects were consistent in the direction of this change irrespective of the starting point. In other words, an endurance trained subject always showed an increase in S during exercise, even if his/her resting CO_2 sensitivity on, say, visit 2 was the same as the value it was during exercise on visit 1. However, the sprint trained group did not show this homogeneity; their response was varied both from subject to subject, and from day to day; we are presently unable to speculate upon a mechanism for this difference.

With regard to the influence of athletic training upon the variability of the CO_2 response curve, the data suggest that athletic training increases this variability, and/or that individuals who are successful athletes are in some way genetically predisposed to a greater variability. The fact that the endurance trained group showed the greatest variability, the sedentary the lowest, and the sprint trained lay between the two, implies that the mechanism may be linked to endurance performance. However, the absence of any statistical significance to the observed differences makes these data equivocal. We attribute the lack of significance to the small size of our experimental groups (n=4).

The final question that we wished to answer was perhaps the most important. Is a *single* measurement of CO_2 sensitivity truly representative of that individuals' sensitivity to CO_2? The large coefficients of variation suggest that measurements made on a single occasion are of limited value, particularly for endurance trained subjects. However, one can view the question from a different perspective. We compared the mean values for CO_2 sensitivity derived for the endurance trained group of the present study, with those of a previous study in which we measured CO_2 sensitivity in a group of 10 endurance trained subjects on a single occasion[3]. There was no significant difference between the absolute values for CO_2 sensitivity of the two studies either at rest or during exercise. Thus, measurements made on a single occasion are probably representative of the CO_2 sensitivity for the group, but not for the individual.

REFERENCES

1. A. Berkenbosch, J.G. Bovill, A. Dahan, J. DeGoede and I.C.W. Olievier. The ventilatory CO_2 sensitivities from Read's rebreathing method and the steady-state method are not equal in man, *J. Physiol.* 411:367-377 (1989).
2. M.S. Jacobi, C.P. Patil and K.B. Saunders. Transient, steady-state and rebreathing responses to carbon dioxide in man, at rest and during light exercise, *J. Physiol.* 411:85-96 (1987).
3. A.K. McConnell and E.S.G. Semple. Ventilatory sensitivity to CO_2 at rest and during exercise in endurance trained and non endurance trained humans, *J. Physiol.* 412:39P(1989a).
4. A.K. McConnell and E.S.G. Semple. Day to day variability of human ventilatory sensitivity to CO_2, *J. Physiol.* 417:122P (1989b).
5. Read, D.J.C. A clinical method for assessing the ventilatory response to carbon dioxide. *Aust. Ann. Med.* 16:20-32 (1967).
6. S.A. Ward and B.J. Whipp. Ventilatory control during exercise with increased dead space, *J. Appl. Physiol.* 48:225-231 (1980).

RESPIRATORY-ASSOCIATED FIRING OF MIDBRAIN AND THALAMIC NEURONS: POSSIBLE RELATION TO SENSATION OF DYSPNEA

Frederic L. Eldridge, Zibin Chen, and Paul G. Wagner

Departments of Medicine and Physiology
University of North Carolina
Chapel Hill, North Carolina 27599

INTRODUCTION

Although suprapontine brain is not necessary for normal breathing patterns and responses to chemical and other stimuli, it is known that neurons in the mesencephalon and thalamus do under some conditions develop a rhythmic pattern of firing that has the timing of the respiratory cycle[1,2]. We became interested in the question because of observations we had made in unanesthetized, decerebrate, vagotomized cats. Some mesencephalic neurons, which fired irregularly or tonically when Pco_2 was low and phrenic activity small, developed a respiratory-linked rhythm when respiratory activity had been increased by electrical stimulation of a carotid sinus nerve. The present studies were designed to investigate systematically the characteristics and source of the respiratory-associated neuronal rhythm[3,4].

METHODS

Single unit activities of neurons in the midbrain and thalamus were measured extracellularly with stainless steel electrodes in 21 supracollicularly and in 16 suprathalamically decerebrated adult cats. The decerebrations were performed under ether anesthesia which was then stopped. Vagus nerves were cut bilaterally. Both carotid sinus nerves (CSN) were cut and one was placed on a bipolar electrode for stimulation. The animals were paralyzed and ventilated with a volume-cycled ventilator. Integrated phrenic nerve activity was used to assess the level of respiratory activity, or "drive."

After preparations had been completed, end-tidal Pco_2 was set (27 mm Hg) to produce a low level of rhythmic phrenic activity. We then searched the mesencephalon or thalamus for stable spontaneously firing neurons. Having found such a neuron, its activity was recorded continuously during variations in the level of respiratory stimulation, or "drive," which was changed in two ways: 1) the CSN was stimulated continuously (0.5 ms, 25 Hz) at varied voltages to obtain graded increases of phrenic activity; 2) Pco_2 was increased or decreased by changing the ventilator settings. In addition to searching for midbrain and thalamic neurons that developed a respiratory-associated rhythm when respiratory drive was increased, we tested the effect of spinal cord transection at C_{1-2} and the effect of a short-lasting anesthetic agent (Saffan) on the rhythmic firing of the neurons.

RESULTS

Respiratory-associated Rhythm

The purpose of the studies was to find mesencephalic and thalamic neurons, which were silent, irregular or tonically firing at low respiratory drive levels, that developed increased firing with a respiratory rhythm when the level of respiratory drive, reflected by phrenic activity, was high. An example is shown in Fig. 1a for a mesencephalic neuron that fired irregularly when drive was low (Pco_2 = 27.5 mm Hg), but that developed regular respiratory-associated bursts of firing in late inspiration and post-inspiration when drive was increased (Fig. 1b, Pco_2 = 32.4 mm Hg).

The testing of 354 mesencephalic neurons yielded 135 that developed these respiratory-associated rhythmic changes of firing when respiratory drive was increased. The responding neurons were located in the central tegmental field (FTC) of the mesencephalon 1 to 3 mm anterior to stereotaxic zero and 1 to 3 mm lateral to midline. Although at the lowest respiratory drive levels, there was no rhythmicity to the neuronal firing or indeed no firing at all, increased firing with the respiratory rhythm appeared

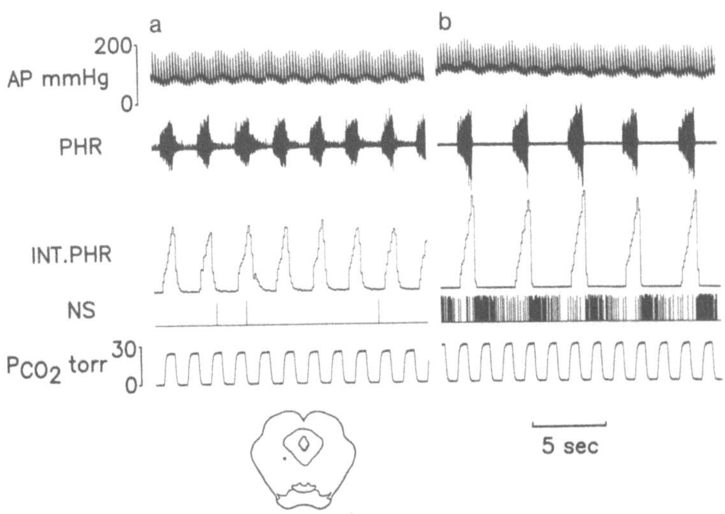

Figure 1. A: recording of arterial pressure (AP), phrenic nerve and integrated phrenic activity, and spikes from a mesencephalic neuron (location indicated) at end-tidal PCO_2 of 27.5 mm Hg. Neural activity is irregular and minimal. B: recording during increased respiratory drive after increase of PCO_2 to 32 mm Hg. Neuron has developed regular respiratory-associated rhythmic firing. (from reference 3)

when respiratory drive had increased above an apparent threshold of 40 to 50% of maximum phrenic activity, the latter determined separately. Above this threshold level there was a clear graded relation between the increased firing and the amplitude of respiratory activity. Figure 2 shows a neuron: that was silent (panel A) when P_{CO_2} was 28.8 mm Hg and phrenic activity 36% of maximum; that developed a small amount of rhythmic activity (panel B) when P_{CO_2} was 32.9 mm Hg and phrenic activity 56% of maximum; and that had markedly increased rhythmic neuronal firing when phrenic activity had increased to 91% of maximum during CSN stimulation although P_{CO_2} remained 32.9 mm Hg (panel C).

Similar rhythmic firing occurred in 76 of 545 thalamic neurons tested. The firing was always tonic and irregular at low respiratory drive levels, as in Fig. 3A where phrenic activity was only 38% of maximum. However,

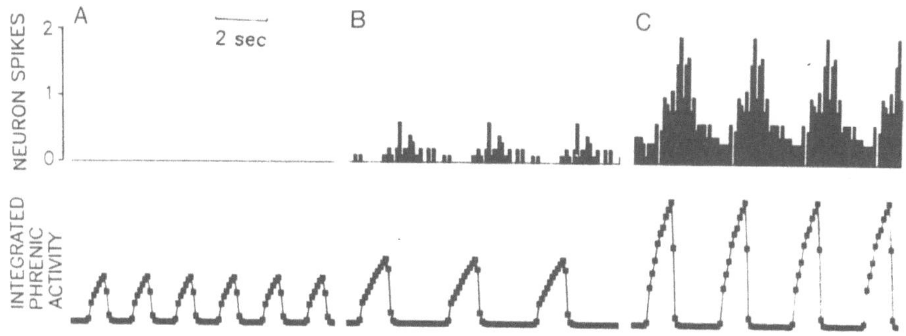

Figure 2. Histograms of mesencephalic neuron's firing rate (spikes per 0.1 s) and integrated phrenic activity averaged for 6-10 breaths, showing time relations between the two and the effect of increasing respiratory drive. **A**: No firing at low drive (36% of max. phrenic, P_{CO_2} = 28.8 mm Hg). **B**: small amount of rhythmic firing (drive = 56%, P_{CO_2} = 32.9 mm Hg). **C**: large increase of rhythmic firing during CSN stimulation (drive = 91%, P_{CO_2} = 32.9 mm Hg). In this and in Figs.3, 4 and 5, the repeats of rhythmic events in each panel are replottings of the same data.

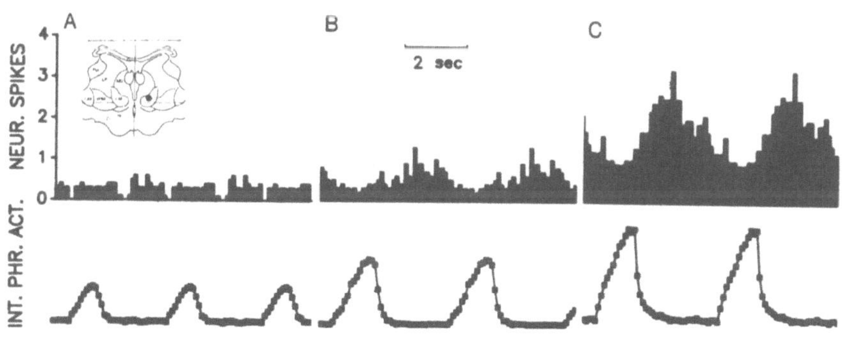

Figure 3. Histograms of thalamic neuron's firing rate (spikes/0.1 s) and integrated phrenic activity averaged over 8-10 breaths; effect of increasing respiratory drive. **A**: irregular firing at low drive (38% of max.phrenic activity, P_{CO_2} = 28 mm Hg). **B**: small amount of rhythmic firing (drive = 65%, P_{CO_2} = 33 mm Hg). **C**: large increase of rhythmic firing at high drive during CSN stimulation (91%, P_{CO_2} = 33 mm Hg).

when respiratory drive was raised to above a threshold of 60% of maximum phrenic activity (65% in Fig. 3B), the respiratory-associated rhythm appeared. Figure 3C, obtained when phrenic activity was 91% of maximum, shows that the graded response occurs in thalamic as well as mesencephalic neurons.

Ten pairs of simultaneous recordings of mesencephalic and thalamic neurons were made in five cats. Figure 4 (A,B,C,D) shows four of these neuron pairs at high levels of phrenic activity (>70% of maximum). The similarities of the firings in the different neurons are apparent but there are differences: the mesencephalic neurons begin their increased firing sooner after the onset of inspiration than do the thalamic neurons; and the thalamic neurons have a higher threshold level for firing (60-70% of maximum phrenic activity) than do the mesencephalic (40-50% of maximum). Cross-correlation histograms of relations in the ten simultaneously recorded pairs of mesencephalic and thalamic neurons revealed no sharp peaks or valleys but only a broad symmetrical peak with its center near zero. The findings show that the respiratory-associated relationship between the neurons is not due to mono- or pauci-synaptic connections; they are consistent with the existence of a common source that drives both.

Several additional procedures were performed. The effect of transection of the spinal cord at C_{1-2} on the respiratory-associated rhythm in mesencephalic neurons was studied in three cats which had continuous recordings of hypoglossal nerve as well as phrenic nerve activity. We first found neurons that developed the rhythmic activity associated with respiration, as

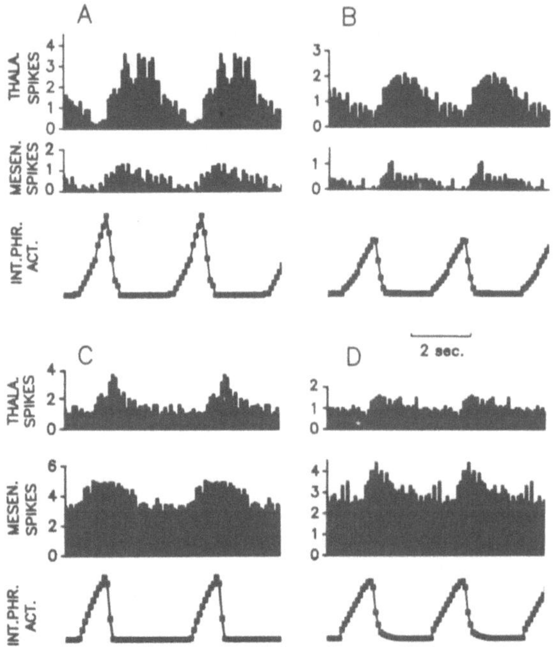

Figure 4. Histograms of firing rates (spikes per 0.1 s)of four different simultaneously recorded thalamic and mesencephalic neuron pairs (A,B,C,D) and integrated phrenic activity at high levels of respiratory drive (>70 % of maximum). Both neurons in each pair show the respiratory-associated rhythm but onsets and peaks of rhythmic firing occur earlier in mesencephalic than in thalamic neurons.

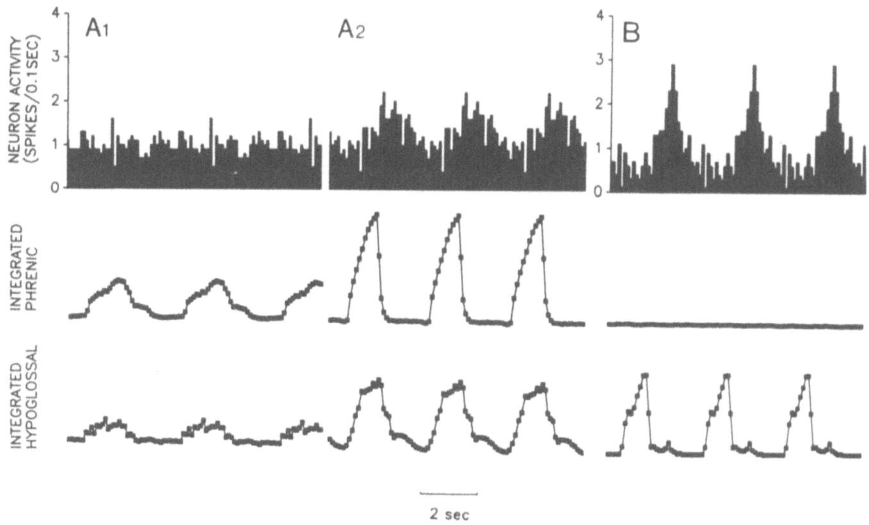

Figure 5. Lack of effect of transection of spinal cord on midbrain neuronal respiratory rhythm. A1:tonic firing (spikes per 0.1 s) of mesencephalic neuron at low respiratory drive (P_{CO_2} = 22 mm Hg) with recordings of phrenic and hypoglossal nerve activities. A2: increase of respiratory-associated rhythmic firing when respiratory drive increased by CSN stimulation. B: same cat after transection of cord at $C_{1\text{-}2}$, with loss of phrenic activity but still showing hypoglossal activity and respiratory-associated rhythm. Each panel represents the averages of 6-10 breaths. (from reference 3)

shown in simultaneous phrenic and hypoglossal activities. Figure $5A_1$ shows a tonically firing neuron at low respiratory drive (31% of maximum) along with phrenic and hypoglossal activities when the spinal cord was intact. When drive had been increased by CSN stimulation to 80% of maximum (Fig. $5A_2$), typical respiratory-associated rhythmic activity appeared in the neuron. The cutting of the spinal cord at $C_{1\text{-}2}$ led, as expected, to loss of rhythmic phrenic activity but the medulla continued to generate rhythmic respiration as evidenced by hypoglossal nerve activity (Fig. 5B). A mesencephalic neuron also continued, at the high level of respiratory drive, to exhibit the respiratory-associated rhythm (Fig. 5B). This experiment rules out afferent input from chest wall or muscles in the generation of the rhythm (although see below for effects of input from chest) and show that a corollary discharge from medulla is responsible.

The purpose of the second procedure was to use a very low dose of a very short-lasting anesthetic agent (Saffan) to eliminate the respiratory-associated activity in mesencephalic neurons present when drive was high, and to determine if this had any effect on respiratory period or its components, inspiratory duration (T_I) and expiratory duration (T_E). Saffan infusion led to elimination of the rhythmic component and most of the firing of the neurons, which returned after cessation of the infusion. Despite loss of rhythmic mesencephalic activity, no significant change in the rhythmic components of phrenic activity occurred. These findings indicate that the rhythmic firing of the neurons was not contributing significantly to the generation or modulation of respiratory rhythm.

Ventilator-related Effects

In addition to the rhythmic activity that was associated with neural respiration, we found neurons in both mesencephalon and thalamus that fired with each pumping cycle produced by the ventilator. Figure 6 is an example of a thalamic neuron with this rhythm in a preparation that had no regular respiratory rhythm but only a few phrenic bursts; the close relation between firings of the neuron and expansions of the chest, indexed by the fall of airway Pco_2. is apparent. These ventilator-related bursts disappeared when the ventilator was turned off and reappeared when it was turned back on. Forty thalamic neurons with this type of activity were found in 13 cats studied. Similar ventilator-related activity was found in a number of mesencephalic neurons (55 neurons in 12 cats).

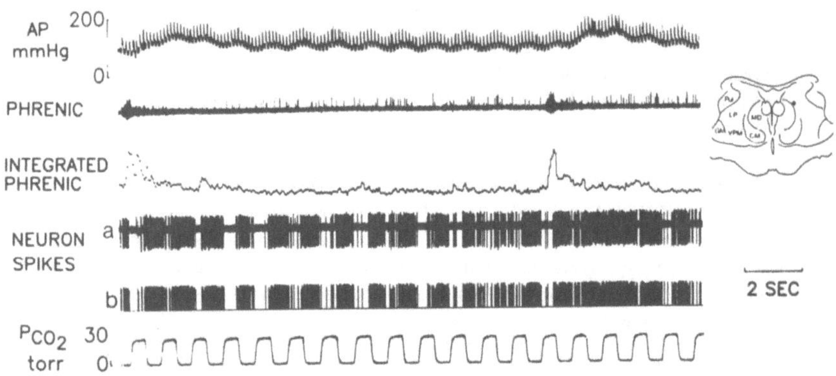

Figure 6. Example of rhythmic firing, related to expansion of chest, of thalamic neuron in vagotomized cat which has minimal spontaneous rhythmic respiratory activity (Pco_2 = 25 mm Hg). AP, arterial pressure. Neuron spikes: a, raw recording; b, standard pulses from window discriminator.

In both thalamus and mesencephalon, some of the same neurons that exhibited the ventilator-related rhythm also developed the respiratory-associated rhythm when respiratory drive was high. An example is found in Fig. 7 where a mesencephalic neuron that showed only the central respiratory rhythm when the ventilator was turned off (Fig. 7B) added the ventilator-related component when the ventilator was turned back on (Fig. 7A). No ventilator-related rhythm was found after the spinal cord had been cut at C_{1-2}.

DISCUSSION

We studied neurons in the reticular formation (central tegmental field) of the mesencephalon and in thalamic nuclei related to the thalamocortical system involving ascending reticular influences[5]. Neurons that were silent or irregularly tonic at low neural respiratory drives often developed (25% of mesencephalic, 14% of thalamic neurons tested) a clear respiratory-associated increase of firing in late inspiration and postinspiration at high respir-

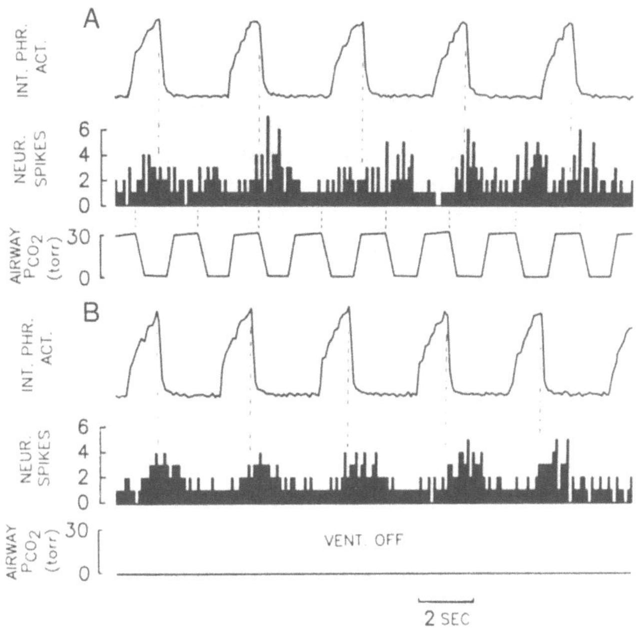

Figure 7. Example (panel A) of mesencephalic neuron in vagotomized cat that shows both respiratory-associated rhythmic increases of firing during and following phrenic activity and those related to expansion of chest by ventilator, indexed by fall of airway P_{CO_2}. When ventilator has been turned off (panel B), only the respiratory-associated rhythm remains. (from reference 3).

tory drive levels, when the latter exceeded a threshold level (40-50% of maximum phrenic activity for mesencephalic, 60-70% for thalamic). Above this level the magnitude of firing increased as the drive level increased.

The source of rhythmic activity cannot be via the vagi which were cut, or the carotid bodies which were denervated. Our experiments in which the spinal cord was transected just below the medulla rule out afferent input from chest wall and respiratory muscles. Finally, the lack of effect on respiratory rhythm of an anesthetic that abolished mesencephalic rhythmic activity suggests that the neuronal rhythmicity is not importantly involved in respiratory rhythm or pattern generation. The findings are consistent with an interpretation that the medullary neurons that drive respiration are also the source of the signals that drive the respiratory-associated rhythms in mesencephalic and thalamic neurons[3,4] The signals thus represent a corollary discharge, neural activity that is sent in parallel from medullary neurons to respiratory motoneurons and to the suprapontine neurons.

The experiments show that in addition to stimulation of these suprapontine neurons by central inspiratory activity, there is also stimulation associated with expansion of the chest during artificial ventilation. Because vagi and CSN had been eliminated in the preparation, the signals must have originated in chest wall or respiratory muscular receptors and been transmitted via the spinal cord. It is important to note that some of the neurons that responded to chest wall input also responded to the central neural respiratory drive signals (Fig. 7). Both the mesencephalic and the thalamic neurons thus must provide some integration of the two inputs.

The findings indicate that the respiratory-associated rhythmic firing of the neurons we have studied is not primarily involved in the generation or modulation of the motor functions of the respiratory oscillator. Rather, it is a part of an afferent sensory mechanism conveying information to cortex from medullary neurons (as a corollary discharge) about magnitude of central neural respiratory drive, as well as spinally transmitted information from receptors in chest wall and respiratory muscle.

Other studies support our suggestion that the respiratory-associated and ventilator/chest wall-related information can reach the cortex, since respiratory-associated cortical EEG activity has been demonstrated in awake cats[6] and short-latency somatosensory cortical potentials can be evoked in awake humans by sudden added inspiratory loads[7].

Sensory information about magnitude of medullary respiratory drive and events in the chest wall and respiratory muscles could have several roles. One would be the generation of a conscious sensation, which may be the unpleasant sense of "air hunger" or dyspnea. In this regard, our studies are consistent with the psychophysiologic studies of "air hunger" in ventilated human quadriplegics[8] in whom the sensation cannot be due to afferent input involving respiratory muscular contraction, or chest wall or upper airway receptors. The second potential role would be that leading to arousals, such as occur in periodic breathing, *e.g.,* obstructive sleep apnea. Such arousals have been shown to be related not to levels of chemical stimuli *per se* but rather to the level of respiratory drive[9].

ACKNOWLEDGEMENT

This work was supported by NIH MERIT Award Grant HL-17689.

REFERENCES

1. Vibert, J.F., Caille, D., Bertrand, F., Gromysz, H., and Hugelin, A., Ascending projection from the respiratory centre to mesencephalon and diencephalon. *Neuroscience Letters* 11:29 (1979).
2.. Orem, J., and Netick, A., Characteristics of midbrain respiratory neurons in sleep and wakefulness in the cat, *Brain Research* 244:231 (1982).
3. Chen, Z., Eldridge, F.L., and Wagner, P.G., Respiratory-associated rhythmic firing of midbrain neurons in cats: Relation to level of respiratory drive, *J. Physiol. (London)* 437:305 (1991).
4. Chen, Z., Eldridge, F.L., and Wagner, P.G., Respiratory associated rhythmic firing of thalamic neurons: Relation to level of respiratory drive, *FASEB J.* 5:A666 (1991).
5. Hobson, J.A., and Steriade, M., Neuronal basis of behavioral control. In *Handbook of Physoiology, section I, The Nervous System, Vol. IV,* American Physiol. Soc., Bethesda, MD, USA, pp. 701-803 (1986).
6. Kumagai, H., Sakai, F., Sakuma, A., and Hukuhara, T., Relationship between activity of respiratory center and EEG. *Prog. Brain Res.* 21:98 (1966).
7. Davenport, P.W., Friedman, W.A., Thompson, F.J., and Franzen, O., Respiratory-related cortical potentials evoked by inspiratory occlusion in humans. *J. Appl. Physiol.* 60:1843 (1986).
8. Banzett, R.B., Lansing, R.W., Reid, M.B., Adams, L., and Brown, L. "Air hunger" arising from increased Pco2 in mechanically ventilated quadriplegics. *Respir. Physiol.* 76:53 (1989).
9. Gleeson, K., Zwillich, C.W., Ventilation, arousal from sleep, and adenosine stimulation., *Clinical Res.* 38:790A (1990).

WHERE IS THE CORTICAL REPRESENTATION OF THE DIAPHRAGM IN MAN?

Kevin Murphy, David Maskill, Anne Mier and Abraham Guz

Department of Medicine
Charing Cross and Westminster Medical School
London W6 8RF

INTRODUCTION

The use of transcranial stimulation, initially using electrical techniques[1] and subsequently using magnetic techniques[2] have demonstrated a rapid oligosynaptic pathway from the cortex to the diaphragm. These studies do not however establish the exact area of the cortex that is associated with activation of the diaphragm. The early work of Foerster[3] identified a site near the vertex which when stimulated produced hiccough but no detail is available. Use has been made of an improved magnetic coil design (figure-of-eight) to 'Map' the unilateral cortical representation of the inspiratory muscles; essentially the diaphragm. The work of Rösler et al[4] had showed that a figure-of-eight coil was better at exciting specific areas of the motor cortex; essentially showing an extended peak current along the intersection of the two coils. He had also shown the importance of the direction of the principal stimulating current vector.

METHODS

Electromyogram Recording

Electrodes were placed in the 7th or 8th intercostal space 3 cm lateral to the anterior costal margin in the anterior axillary line on both sides of the chest. This site has been shown as optimal for recording the diaphragm in man[5] . The **emg** signals were filtered (10-1000Hz) and the compound muscle action potentials induced by the stimulation captured on a computer averaging system (1401, Cambridge Electronic Design, England).

Magnetic Stimulator

A Magstim 200 (Magstim Co. Ltd. UK) magnetic stimulator using a figure of eight coil which could produce a peak output of approximately 2.1 Tesla was used to perform the stimulations. The machine was triggered by the value of the integrated **emg** activity exceeding a fixed level, to remove the dependence of the response on the level of inspiratory drive'[1]. This was equivalent to a tidal volume of 1 - 2 l. To increase the size of the response the subjects increased inspiratory drive by inspiring through an inspiratory load (20cm$H_2$0).

Mapping Protocol

Studies were performed on 5 male subjects. To enable accurate placement of the stimulating coil, each subject wore a rubber swimming cap upon which was draw a mid sagittal line and a transverse line between the ears. Previous studies had shown that the most likely 'focus' was at a position 3 cm lateral and 2 cm anterior to the vertex, a grid consisting of points 2cm apart around this point was drawn on the right side of the head. With the stimulator set at 90% power, two stimuli were given at each point on the grid and then the coil was rotated 45 degrees and so on, giving a total of 16 stimuli at each point. Each pair of responses were averaged and the peak to peak amplitude of any compound muscle action potential measured. This arrangement is shown in Figure 1.

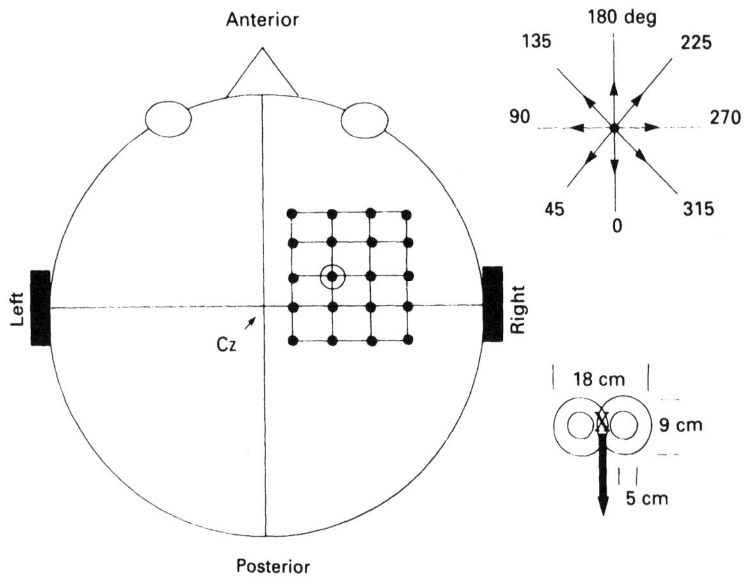

Figure 1. A view on the top of a subjects head showing the line from the inion to the nasion passing through the line from the left to right ear. Cz marks the intersection of these lines; the vertex. The stimulation grid (2 cm apart); the angles at which the stimuli were applied is shown with the arrow representing the direction of current flow. The dimension of the coil are also given.

RESULTS

Subjects tolerated the study well and did not complain of any untoward effects. Figure 2 shows the results, from one typical subject, of varying the angle of the stimulating current vector whilst remaining fixed over the optimal spot in the right cortex. The recordings are from the left side of the chest. The heavy dependence of the response with angle can clearly be seen with a maximum at about 315 degrees. This dependence was seen in all subjects and also at other non optimal positions on the grid. The optimal angle ranged from 315 to 270 degrees.

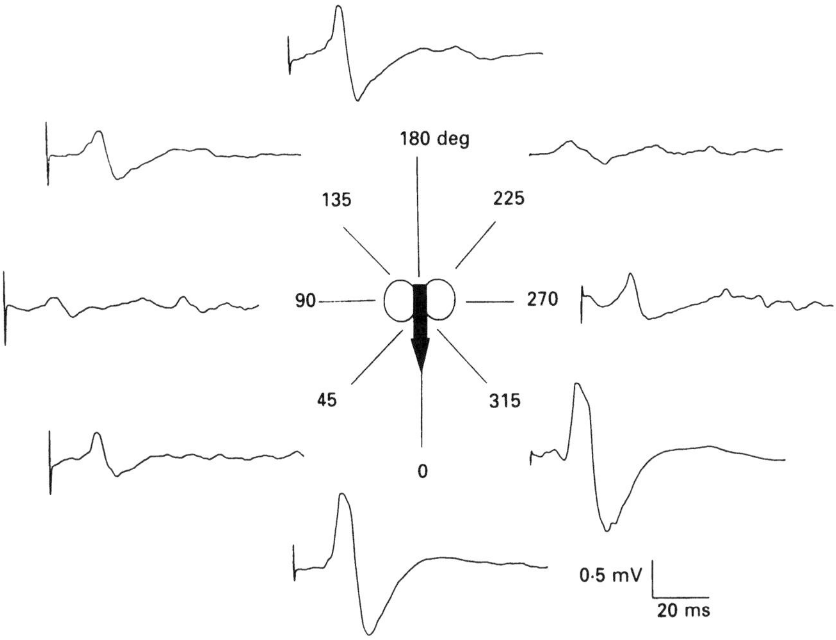

Figure 2. Recordings obtained from the left 7th intercostal space in response to magnetic stimulation over the right cortex. Each trace is the average of the two responses obtained at each of the 8 angles studied. The variation in the response with angle is clearly seen.

Figure 3 shows a representation of the stimulation grid with filled circles whose area are related to the maximum amplitude of the compound muscle action potential at that site; the arrows point in the direction of the coil at that time. The stimulation was performed over the right cortex and the results show the response on the left and right hand side of the chest. This response appears to be almost entirely contralateral, but with a small significant ipsilateral response still seen. The mean result from all subjects showed a stimulation focus at approximately 3 to 4 cm lateral to the vertex about 2-3 cm anterior to the auricular plane.

In a continuation of this protocol, which is fully reported by Maskill et al[6], ultrasonic imaging techniques were used to visualise the movement of the diaphragm in response to 'focal' transcranial magnetic stimulation. It was shown that movement of the contralateral diaphragm could be seen whilst the ipsilateral side did not move. Placing the stimulation coil in a central position where both sides of the diaphragm would be expected to move, did indeed produce movement of both leaves of the diaphragm.

DISCUSSION

This technique has been able to define a site of cortical representation of the inspiratory muscles in man unilaterally and when this site is stimulated produces a predominately contralateral response.

Only the right cortex has been mapped formally, because, to repeat this study on the other side at the same time would require what was felt was an excessive number of stimuli. The resolution of the assessment of the optimal angle could also have been improved by using greater numbers of stimuli. The advantages of the figure of eight coil

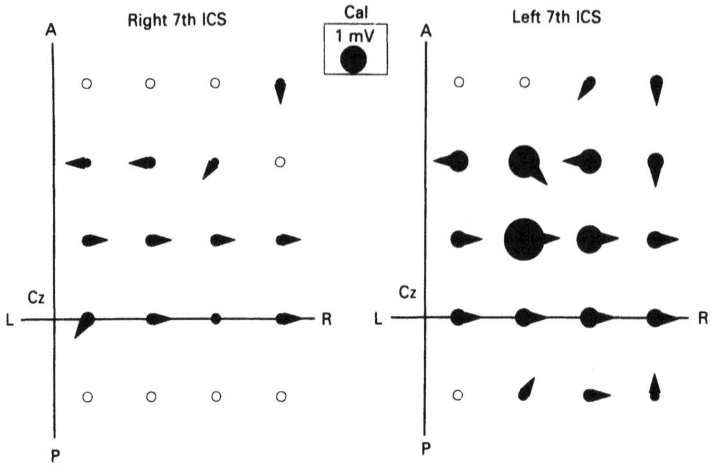

Figure 3. The stimulation grid showing the anterior posterior line; the inter-aural line. The area of the filled circles is proportional to the amplitude of the compound muscle action potential response obtained as a result of stimulation over the right cortex. The arrows indicate the direction of the current vector at that maximum point. Open circles indicate that no response was seen at any stimulation angle. The 'focus' can be seen clearly and the non zero response on the ipsilateral side of the body can also be seen.

can clearly be seen as the orientation of the main stimulating current vector can be simply altered. The mean site found is in agreement with that previously indicated[3] and also agrees with the observations of Colebatch et al[7] who used positron emission tomography to highlight those areas of the cortex that are involved with volitional inspiration; they showed activation in both the left and right motor cortex. These result raise the question of where the integration of the left and right diaphragm movement occurs during a voluntary breath.

ACKNOWLEDGEMENTS

The Figures are from Maskill[6] with permission of the Journal of Physiology.

REFERENCES

1. S.C. Gandevia and J.C. Rothwell, Activation of the human diaphragm from the motor cortex, *J. Physiol.* 384:109-118 (1987).
2. K. Murphy, Anne Mier, L. Adams and A. Guz, Putative cerebral cortical involvement in the ventilatory response to inhaled CO_2 in man, *J. Physiol.* 420:1-18(1990).
3. O. Foerster, Motorische felder und bahen *in*: Handbook der neurologie, O. Bumke and O. Foerster ed., Springer, Berlin (1936).
4. K.M. Rösler, C.W. Hess, R. Heckmann and H.P. Ludin, Significance of shape and size of the stimulating coil in magnetic stimulation of the human motor cortex, *Neurosci. Lett.*, 100:347-352(1989).
5. R. Lansing and J. Saville, Chest surface recording of diaphragm potentials in man. *Electroencephalogr. Clin. Neurosci.*, 72:59-68(1989).
6. D. Maskill, K. Murphy, A. Mier, M. Owen and A. Guz, Motor cortical representation of the diaphragm in man. *J. Physiol.* 443:105-121(1991).
7. J.G. Colebatch, L. Adams, K. Murphy, A.J. Martin, A.A. Lammertsma, H.J. Touchon-Danguy, J.C. Clark, K.J. Friston and A. Guz, Regional cerebral blood flow during volitional breathing in man. *J. Physiol.* 443:91-103(1991).

DYSPNEA AND RESPIRATORY COMPENSATION RESPONSE TO RESISTIVE LOADING IN ELDERLY HUMANS

Yasushi Akiyama, Masaharu Nishimura, Shuichi Kobayashi, Makoto Yamamoto, Kenji Miyamoto, and Yoshikazu Kawakami

First Department of Medicine, Hokkaido University School of Medicine
Sapporo 060, Japan

INTRODUCTION

Compensatory mechanisms in response to respiratory mechanical loading involve various steps in the control of breathing.[1] These include changes in the intrinsic properties of the respiratory muscles, reflexes from mechanoreceptors in the thorax, lungs and airways, and the chemoreceptor reflexes. In addition, conscious appreciation of the load in the higher brains is implicated as a behavioral component of respiratory compensation to loading. Tack and coworkers reported, using the magnitude-estimation method, that the respiratory sensation of resistive loads as well as elastic ones is diminished in elderly healthy subjects.[2, 3] However, effects of aging on the load compensation reflex and the sensation of dyspnea have not been simultaneously studied. To examine this, we studied, in healthy volunteers, the ventilatory and occlusion pressure responses to progressive hypercapnia with and without inspiratory flow-resistive loading while the intensity of dyspnea was simultaneously assessed by visual analogue scaling.

METHODS

Twenty-eight healthy male volunteers, from whom informed consent had been obtained, served as subjects. Among them, 14 subjects (61 to 79 yr of age) were classified into the older group, and the other 14 (19 to 48 yr of age) into the control group. Subjects in the older group were members of local senior-citizen societies organized for social purposes. All subjects were naive to measurements of respiratory indices and were not aware of the physiologic purpose of this study. Six subjects in the older group and 7 in the control group were current smokers.

For all experiments, the subject was in a supine position and breathed spontaneously through a mouthpiece connected to a J valve. A dual-control system[4] was used to regulate arterial PO_2 and PCO_2 simultaneously and independently. Briefly, the system utilized end-tidal PO_2 ($P_{ET}O_2$) and PCO_2 ($P_{ET}CO_2$) as guides to control arterial blood gases by automatically changing inspiratory gas composition. Minute ventilation ($\dot{V}_E$) was measured every 15 s by electrical integration of the flow signals obtained from a hot-wire respiratory flow meter. Respiratory gases were continuously monitored by a mass spectrometer. A valve in the inspiratory limb of the circuit was automatically closed during expiration once every 15 s. The occlusion pressure 0.1 s after the initiation of inspiration ($P_{0.1}$) was measured at the inspiratory line near the mouthpiece by a differential pressure transducer and the valve was opened automatically soon after each measurement of $P_{0.1}$. The signals

Control of Breathing and Its Modeling Perspective, Edited by
Y. Honda *et al.*, Plenum Press, New York, 1992

of $\dot{V}_E$, $P_{ET}O_2$, $P_{ET}CO_2$, and $P_{0.1}$ were continuously monitored and recorded on a multichannel recorder and at the same time stored at 15-s intervals in an on-line signal processing computer for later analysis. For the assessment of dyspnea, subjects were instructed to concentrate on the sensation of difficulty in breathing, and asked to continuously quantify that sensation by visual analogue scaling. The scale was 10 cm in height, and its lower and upper ends were designated "not at all" and "maximum imaginable" dyspnea, respectively. Subjects indicated the intensity of dyspnea by changing the height of a red-colored column with a finger-controlled potentiometer. Subjects were not given a further definition for the term "difficulty in breathing." Dyspnea intensity was continuously recorded on the recorder. Data for dyspnea were obtained every 15 s from the chart and were analyzed with other indices.

Subjects were first trained to get accustomed to the experimental setup and how to rate dyspnea intensity by visual analogue scaling. After a 30-min interval, the experimental protocol was started as follows. The subject breathed room air spontaneously through the mouthpiece for about 3 min. Then the subject breathed the initial gas, which contained 3% CO_2, 24% O_2 and the balance N_2. $P_{ET}O_2$ was maintained automatically at 160 Torr throughout each run. After respiration had become stable for at least 3 min, $P_{ET}CO_2$ was gradually increased to 60 Torr for about 6 min (hyperoxic progressive hypercapnia). If the subject rated the intensity of dyspnea sensation as "maximum imaginable," the experiment was discontinued whatever the level of $P_{ET}CO_2$. With an interval of 30 min, the above protocol was repeated with a wire-mesh inspiratory flow-resistive load (17 cm H_2O/L/s) in the inspiratory circuit.

Ventilatory and $P_{0.1}$ responses to hyperoxic progressive hypercapnia were assessed by the slopes of response lines ($\Delta\dot{V}_E/\Delta P_{ET}CO_2$ and $\Delta P_{0.1}/\Delta P_{ET}CO_2$, respectively) using the least-squares method. $\Delta\dot{V}_E/\Delta P_{ET}CO_2$ was standardized by body surface area (BSA). Data were analyzed by paired and unpaired *t* tests. P values of less than 0.05 were accepted as significant.

RESULTS

The mean age was 29±SE3 yr for the control group and 69±1 yr for the older group. Body height, body weight and body surface area were not different between the two groups. Forced vital capacity was smaller in the older group than in the control group (2.99±0.10 and 4.13±0.17 L, respectively, $p < 0.001$), although this was not different when expressed as the percentage of predicted values (93±3 and 101±3%, respectively). Forced expiratory volume in 1 s was smaller in the older group than in the control group (2.34±0.10 and 3.61±0.19, respectively, $p < 0.001$), but the percentage of predicted values was not different between the two groups (103±5 and 95±2%, respectively).

Table 1. Effects of resistive loading on ventilatory and occlusion pressure responses to hyperoxic progressive hypercapnia.[1]

	Control Group			Older Group		
	Unloaded	Loaded	p value[2]	Unloaded	Loaded	p value
$\Delta\dot{V}_E/\Delta P_{ET}CO_2$/BSA, L/min/Torr/m^2	1.32±0.18	1.27±0.19	NS	1.25±0.15	0.93±0.11	< 0.01
$\Delta P_{0.1}/\Delta P_{ET}CO_2$, cm H_2O/Torr	0.31±0.06	0.58±0.12	< 0.01	0.29±0.05	0.27±0.05	NS

Definition of abbreviation: NS = not significant.
[1] Values are means ± SE.
[2] P values are for intragroup comparisons between unloaded and loaded experiments.

There were no differences between the two groups in the ventilatory or $P_{0.1}$ responses to hyperoxic progressive hypercapnia without loading (Table 1). In the control group, the $P_{0.1}$ response to progressive hypercapnia increased with loading, and thus the ventilatory response was maintained (Table 1). In contrast, the ventilatory response to progressive hypercapnia decreased with loading because the $P_{0.1}$ response did not increase in the older group (Table 1). In the 28 subjects as a whole, an inverse correlation was found between age and the change in $P_{0.1}$ responses to progressive hypercapnia between unloaded and loaded studies ($r = -0.53$, $p < 0.01$), i.e., aging attenuated the respiratory compensation response to inspiratory flow-resistive loading.

Table 2. Effects of loading on occlusion pressure and dyspnea intensity at a $P_{ET}CO_2$ of 55 Torr.[1]

	Control Group			Older Group		
	Unloaded	Loaded	p value[2]	Unloaded	Loaded	p value
$P_{0.1}$, cm H_2O	4.8±0.9	8.0±1.5	< 0.01	4.5±0.7	4.9±0.7	NS
Dyspnea intensity, cm	3.3±0.6	6.0±0.8	< 0.001	5.3±0.7	7.8±0.7	< 0.001

[1] Values are means ± SE.
[2] P values are for intragroup comparisons between unloaded and loaded experiments.

The intensity of dyspnea as well as the $P_{0.1}$ at a $P_{ET}CO_2$ of 55 Torr increased with loading in the control group (Table 2). In the older group, the intensity of dyspnea at a $P_{ET}CO_2$ of 55 Torr increased with loading, although there was no increase in $P_{0.1}$ (Table 2). At this level of $P_{ET}CO_2$, the intensity of dyspnea in the older group was greater without loading ($p< 0.05$) and tended to be greater with loading ($p = 0.1$) than in the control group. The $P_{0.1}$ was not different between the two groups without loading at this $P_{ET}CO_2$ level, and was greater in the control group than in the older group with loading, although this was not statistically significant ($p = 0.07$).

DISCUSSION

Respiratory muscle strength is known to decrease with age. However, this change does not seem to totally explain the lack of increase in the mouth occlusion pressure response to loading in the elderly because the measured occlusion pressure values were less than a twentieth of the maximal inspiratory pressures predicted by the equation of Black and Hyatt.[5] Other factors[1] may be involved in diminished load-compensation with age.

The chemoreceptor reflexes are implicated in the mechanisms that overcome respiratory mechanical loading.[1] Rubin and coworkers[6] reported that aging does not affect hypercapnic respiratory chemosensitivity without loading in healthy humans. Although several studies have indicated that aging decreases the ventilatory response to hypercapnia,[7, 8] their subjects were small in number or included obese persons. We have recently reported in an 8-10 yr-long[9] longitudinal study, as well as in a cross-sectional one[10] that aging does not attenuate the magnitude of human ventilatory responses to hypercapnia. The present results again confirmed our previous results in a total of 28 healthy males. The weaker compensatory response to resistive loads in the elderly in this study thus seems to be inexplicable by a blunted hypercapnic respiratory chemoreflex.

There might be a change in breathing pattern with loading via various mechanoreceptors in the thorax, lungs and airways. If the inspiratory time is prolonged, the relations of the occlusion pressure and the neural outputs to respiratory muscles will change. Thus, this could be another reason for the diminished occlusion pressure response.

Conscious subjects such as those in the present study perceive respiratory sensations during hypercapnic challenge with resistive loading. They are considered to adjust breathing after processing of the sensation in the suprapontine mechanism in the light of previous experience and anticipation. The intensity of the dyspnea sensation seems to play a very important role in load compensation. By the magnitude-estimation method, the sensation of respiratory loads in the higher brain was reported to diminish with age[2, 3] like other sensations such as tactile, visual and auditory sensations. However, each intensity of loads was presented to subjects only for three consecutive breaths with this method and the respiratory muscle power, especially aspects of endurance, does not seem to have been completely taken into account in those studies. Inspiratory muscle strength training diminished the sensation of respiratory muscle force in normal subjects and detraining increased the sensation to the previous level.[11] Therefore, the greater dyspnea intensity observed in the older group in the present study may have resulted from age-induced respiratory muscle weakness. Deviation of the breathing pattern with loading from the preload pattern also may have led to the increased dyspnea sensation with aging.[12]

Finally, we have to mention the possibility of secondary phenomena due to the stronger dyspnea in the elderly subjects. Acute respiratory stress releases endogenous opioids and these opioid peptides inhibit breathing in healthy humans.[13, 14] Such inhibition of respiratory outputs in the brainstem, in addition to factors in respiratory muscles, may have led to the attenuated load compensation.

In conclusion, aging attenuates the compensatory response to inspiratory flow-resistive loading during progressive hypercapnia. This finding does not seem to be explained by a blunted dyspnea sensation.

REFERENCES

1. N.S. Cherniack, and M.D. Altose, Respiratory responses to ventilatory loading, in: "Regulation of Breathing," Lung Biology in Health and Disease, vol. 17, part II, T.F. Hornbein, ed., Marcel Dekker, Inc., New York (1981)
2. M. Tack, M.D. Altose, and N.S. Cherniack, Effect of aging on respiratory sensations produced by elastic loads, *J. Appl. Physiol.* 50: 844 (1981).
3. M. Tack, M.D. Altose, and N.S. Cherniack, Effect of aging on the perception of resistive ventilatory loads, *Am. Rev. Respir. Dis.* 126: 463 (1982).
4. Y. Kawakami, T. Yoshikawa, Y. Asanuma, and M. Murao, A control system for arterial blood gases, *J. Appl. Physiol.* 50: 1362 (1981).
5. L.F. Black, and R.E. Hyatt, Maximal respiratory pressures: normal values and relationship to age and sex, *Am. Rev. Respir. Dis.* 99: 696 (1969).
6. S. Rubin, M. Tack, and N.S. Cherniack, Effect of aging on respiratory responses to CO_2 and inspiratory resistive loads, *J. Gerontol.* 37: 306 (1982).
7. R.S. Kronenberg, and C.W. Drage, Attenuation of the ventilatory and heart rate responses to hypoxia and hypercapnia with aging in normal men, *J. Clin. Invest.* 52: 1812 (1973).
8. D.D. Peterson, A.I. Pack, D.A. Silage, and A.P. Fishman, Effects of aging on ventilatory and occlusion pressure responses to hypoxia and hypercapnia, *Am. Rev. Respir. Dis.* 124: 387 (1981).
9. M. Nishimura, M. Yamamoto, A. Yoshioka, Y. Akiyama, F. Kishi, and Y. Kawakami, Longitudinal analyses of respiratory chemosensitivity in normal subjects, *Am. Rev. Respir. Dis.* 143: 1278 (1991).
10. Y. Kawakami, H. Yamamoto, T. Yoshikawa, and A. Shida, Age-related variation of respiratory chemosensitivity in monozygotic twins, *Am. Rev. Respir. Dis.* 132: 89 (1985).
11. S. Redline, S.B. Gottfried, and M.D. Altose, Effects of changes in inspiratory muscle strength on the sensation of respiratory force, *J. Appl. Physiol.* 70: 240 (1991).
12. M. Sakurai, Y. Kikuchi, Y. Chung, W. Hida, S. Okabe, S. Ebihara, T. Chonan, C. Shindoh, and T. Takishima, Effects of changes in breathing pattern on the sensation of dyspnea during loaded breathing (abstract), *Am. Rev. Respir. Dis.* 143, A594 (1991).
13. Y. Akiyama, M. Nishimura, A. Suzuki, M. Yamamoto, F. Kishi, and Y. Kawakami, Naloxone increases ventilatory response to hypercapnic hypoxia in healthy adult humans, *Am. Rev. Respir. Rev.* 142: 301 (1990).
14. Y. Akiyama, M. Nishimura, S. Kobayashi, A. Yoshioka, T. Hiraga, F. Kishi, and Y. Kawakami, Naloxone alters ventilatory response but not breathlessness during hypoxic hypercapnia with flow-resistive loading (abstract), *Am. Rev. Respir. Dis.* 141: A552 (1990).

REDUNDANCY STRUCTURES IN RESPIRATORY CONTROL

George D. Swanson

Department of Physical Education and
Pacific Wellness Institute
California State University
Chico, California

INTRODUCTION

The observation that arterial CO_2 tension is regulated (remains at the resting value) under a metabolic CO_2 load via exercise, but increases under an airway CO_2 load via inspiratory CO_2, has motivated numerous theories about the structure of the respiratory controller. This respiratory control system behavior appears to be most consistent with a feedforward/feedback control system structure.[1] The peripheral chemoreceptors (carotid body) and indirectly central brain chemoreceptors act as feedback mechanisms with respect to the regulation of arterial CO_2 tension. Feedforward mechanisms (signals related to metabolic CO_2 production during exercise) are more controversial. Part of this controversy relates to the traditional "search" for a "single" mechanism that will explain the exercise hyperpnea response.[2] However, in general, biological systems are characterized by redundancy, suggesting that many "signals" may be combined to form an appropriate feedforward stimulus.

The purpose of this paper is to explore alternative redundancy structures. Such structures are characterized by an inherent reliability such that if there should be a loss of a particular feedforward "signal" (e.g. as caused by denervation or disease), the resulting regulation error will be minimized. Our approach will be to consider the inherent reliability properties of the feedforward/feedback structure and then suggest alternative feedforward redundancy configurations. A *mutually inhibitory* configuration and a *efferent/afferent loop* configuration both can incorporate several feedforward signals and both provide for a zero regulation error in the steady state. However, only the *efferent/afferent loop* configuration yields appropriate transient state behavior.

FEEDFORWARD/FEEDBACK STRUCTURE

The respiratory feedforward/feedback controller structure can be visualized in terms of the fluid regulator shown in Fig. 1A. The inflow of fluid is analogous to metabolic CO_2 production. An increase in the fluid flow causes a increase in the pressure exerted on the concave surface attached to the left hand lever arm. This causes the lever arm to rise, pivoting on the stationary float attached to the right hand lever arm. The outflow valve, analogous to ventilation, opens, and if the system is calibrated correctly the increase

Control of Breathing and Its Modeling Perspective, Edited by
Y. Honda *et al.*, Plenum Press, New York, 1992

in the outflow will just equal the increase in the inflow. Thus the fluid level will remain constant, resulting in prefect regulation. In contrast, when CO_2 is added via the airway, the fluid is forced up the outflow spout. This flow causes the fluid level to increase in both reservoirs. This increase in fluid level is sensed by the float attached to the right hand lever, causing this lever arm to rise. This opens the outflow valve. The level of the fluid must rise until the value opening is sufficient so that the outflow now compensates for the added inflow. Thus a degraded regulation results.

This type of structure has the advantage that the feedback mechanism minimizes the error in regulation when the feedforward mechanism is not perfectly calibrated. Thus, the feedforward should yield a ventilation correlated with CO_2 production, but the correlation need not be perfect and the correlation slope need not be perfectly consistent with the resting value, since the feedback will correct for these errors.

Furthermore, this feedforward/feedback structure inherently yields a set-point accommodation, which tends to compensate for a change in feedforward calibration when signal loss occurs. That is, a loss of feedforward signal leads to a rise in the arterial PCO_2 set-point, which requires less of a feedforward signal for regulation.[3] The fluid regular shown in Fig. 1A, illustrates this concept. If the feedforward signal calibration is decreased, then the fluid level rises to a new steady state set-point. Because the column of fluid is higher, a smaller change in value opening will be needed to compensate for a

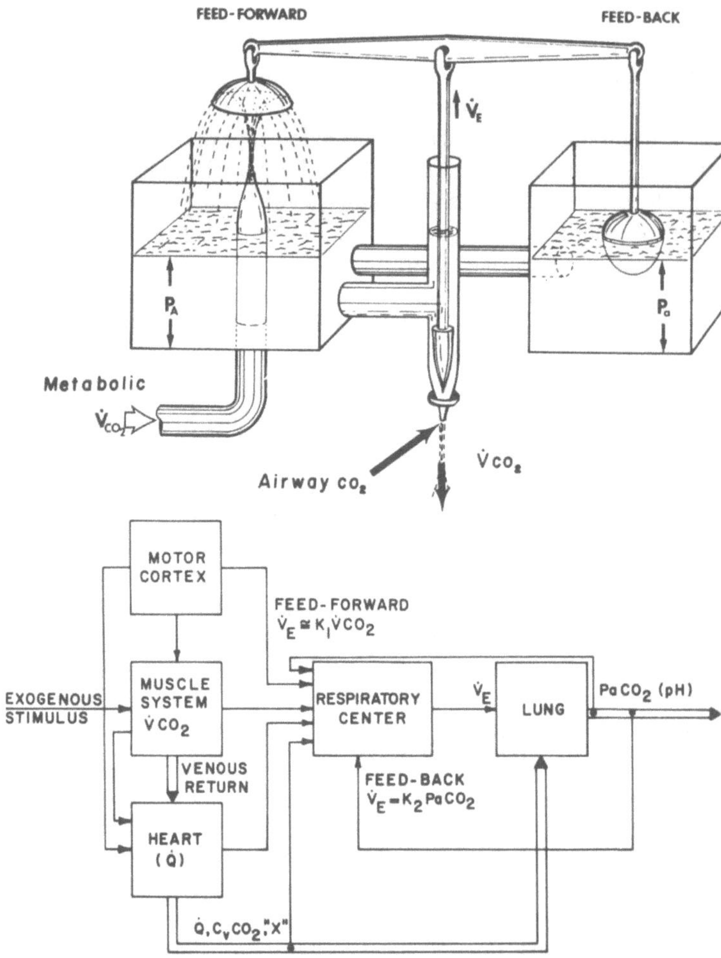

Fig. 1. Feedforward/feedback configuration for the respiratory Controller.

given increase in fluid inflow (analogous to metabolically produced CO_2). This compensation tends to minimize regulation error at the new set-point when CO_2 is loaded metabolically via exercise.

FEEDFORWARD REDUNDANCY STRUCTURES

There are, in fact, many candidates for a feedforward signal (See Fig. 1B). Since, in general, biological systems are characterized by redundancy, this suggests that many signals may be combined to form the feedforward stimulus. The simplest combination would be a so-called spatial summation; that is, at some point (perhaps the respiratory center in the brain) various signals correlated to CO_2 production are combined to yield a composite signal. The loss of a particular signal would lead to a regulation error but it would be minimized by feedback action.

An alternative configuration for providing feedforward redundancy is a *mutually inhibitory* configuration as illustrated in Fig. 2A for combining carotid sinus nerve signal (X_2) with perhaps a peripheral neural drive signal (X_1). With all pathways intact, the muscular CO_2 production X is transmitted to Y as the feedforward signal. However, if via surgical denervation or disease, the pathway is opened (dash denervation line), X is still transmitted to Y via X_1 so that the feedforward calibration is preserved. This scheme was used by Yamamoto to combine the blood gas oscillation signal with a neural signal from peripheral muscle CO_2 production.[4]

Alternatively, Swanson suggested an *efferent/afferent loop* configuration as a means for providing redundancy.[3] The carotid body (Fig. 2B) is a peripheral chemoreceptor that may be responsive to blood gas oscillations. The afferent branch of the carotid sinus nerve appears to transmit these oscillations to the brain respiratory center. Efferent nerve fibers to the carotid body inhibit afferent output and increased afferent activity increases efferent activity.[5,6] These observations are consistent with the efferent/afferent loop configuration shown in Fig. 2B.

If blood gas oscillations are a viable feedforward stimulus via the carotid body, then the redundancy configurations of Fig. 2 suggest that surgical denervation at the carotid body will not degrade steady-state CO_2 regulation during exercise. This concept is an important one because a group of surgically denervated patients exhibit a regulated response to exercise.[7]

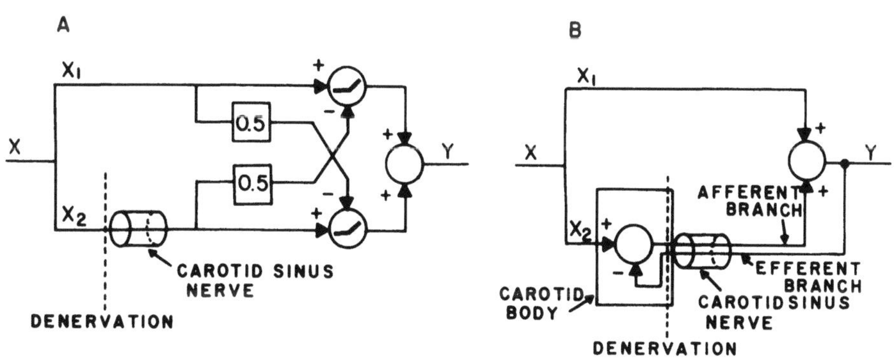

Fig. 2. Configuration for providing feedforward redundancy. A: *mututally inhibitory*; B: *efferent/afferent loop*.

Both the *mutually inhibitory* configuration and the *efferent/afferent loop* configuration can be generalized for multiple feedforward pathways. Fig. 3 illustrates the *mutually inhibiting* configuration for three feedforward inputs while the *efferent/afferent loop* is illustrated for four feedforward inputs. Note as long as at least one input remains (denervation of the other inputs including the feedback loop in the case of the efferent/afferent loop), the appropriate feedforward calibration remains intact. For a three input system, with all inputs intact, the composite output is given by $X_1/4+X_2/4+X_3/2$. For example, if X_2 input is denervated, the composite output is given by $X_1/2+X_3/2$. For a four input system, with all inputs intact, the composite output is given by

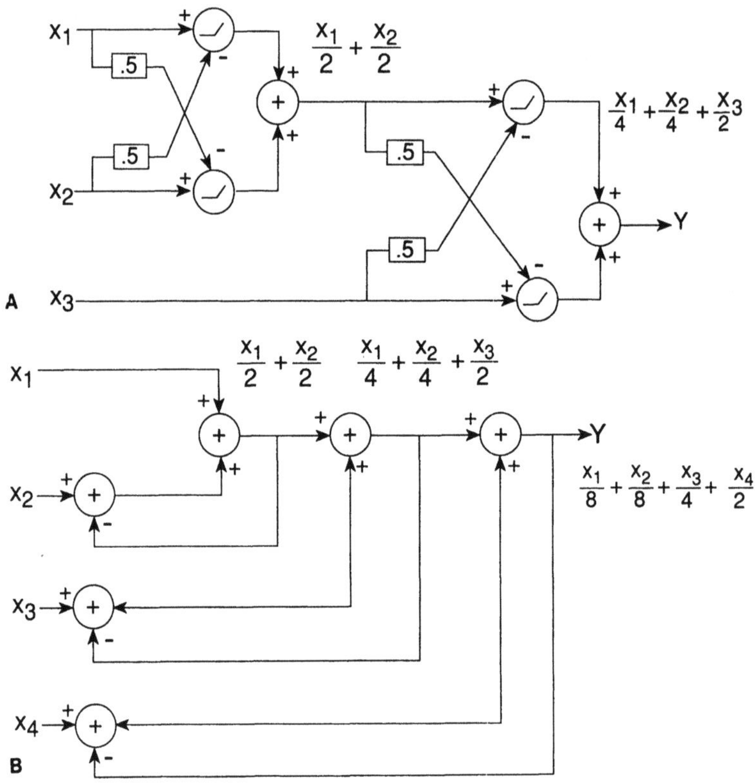

Fig. 3. Multiple input feedforward/feedback redundancy configurations.

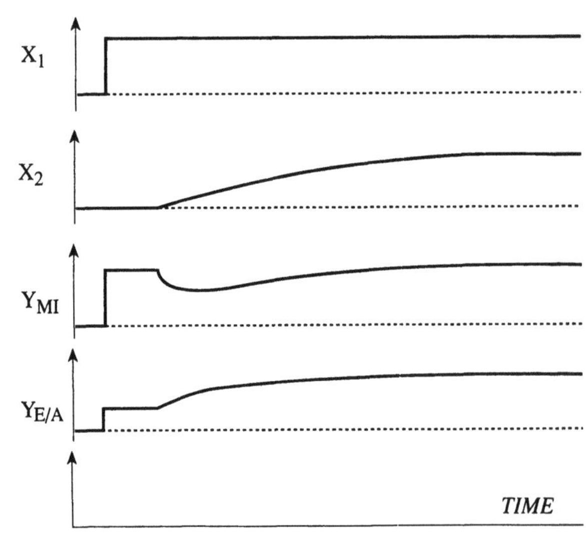

Fig. 4. Transient response of the *mutually inhibitory* (Y_{MI}) and the *efferent/afferent loop* ($Y_{E/A}$) configurations.

$X_1/8+X_2/8+X_3/4+X_4/2$. For example, if X_1 and X_3 are denervated, the composite input is given by $X_1/2+X_4/2$.

The transient response of the *mutally inhibitory* configuration and the *efferent/afferent loop* configuration is shown in Fig. 4. Note that the X_1 input is an abrupt increase (step input) while the X_2 input is a delayed exponential type function. The output from the *mutually inhibitory* configuration (Y_{MI}), as shown, reflects the full-on input of X_1 and transient "dip" as X_2 begins to rise. This response appears to be unrealistic. Alternatively, the response of the *efferent/afferent loop* configuration ($Y_{E/A}$) is such that the temporal aspects of the input transient are preserved.

DISCUSSION

Theories of respiratory controller structure suggest a controller with an intrinsic "exercise stimulus" (i.e., a signal that is correlated to metabolic CO_2 production and appropriately drives breathing). However, in spite of intensive experimental investigations beginning with Geppert and Zuntz in 1888, a mechanism that will explain all the data has not been identified. Perhaps, as Julius Comroe has suggested, "this means either that this seemingly simple, uncomplicated, problem is an exceedingly difficult one, or that respiratory physiologists have not been very perceptive, or both. One of the difficulties is that most physiologists have been—and still are—searching for a single measurable stimulus and mechanism that will explain all the data.[2]

An alternative theoretical point of view is that an "exercise stimulus" is not necessary. This concept can be imbedded in an optimal controller—a controller that operates so as to minimize a suitably defined cost that is associated with its behavior. From this perspective, arterial CO_2 tension is regulated under a metabolic CO_2 load because it is "cost effective" and rises under an airway CO_2 load because it is too "costly" in terms of work of breathing to maintain regulation.

In 1983, Poon proposed an optimal controller structure, which dictates ventilation so as to balance the "cost" of regulation with the "cost" of breathing.[8] This controller correctly predicts both an isocapnic (regulated arterial CO_2 tension) exercise response and an appropriate CO_2 response slope (increase in arterial CO_2) for resting man without the explicit need for an "exercise stimulus." However, the underlying structure of Poon's controller is a multiplicative one such that the response to an exercise-airway CO_2 interaction is multiplicative. As an alternative, an additive controller structure was proposed by Swanson and Robbins in 1986.[9] This additive controller structure also balances the "cost" of regulation with the "cost" of breathing.

These optimal controller configurations apparently eliminate the need (and the need for the search) for an explicit "exercise stimulus". However, every optimal controller configuration has a mathematically equivalent feedforward/feedback controller configuration.[10] The problem then is to decide whether the optimal controller or the feedforward/feedback controller is more appropriate.

Three experimental conditions suggest that a feedforward/feedback controller is appropriate.

1) The optimal controller model must have a "knowledge" of the actual ventilation achieved to calculate the "cost." This means that an experimental preparation would have to have an intact vagus feedback system in order to yield a viable exercise response. Experimental studies published in the literature are to the contrary.

2) The optimal controller model must have a "knowledge" of the arterial CO_2 tension to calculate the "cost." This means that a subject would have to have a nonzero CO_2 response slope in order to yield a viable exercise response. Subjects that have essentially a zero slope CO_2 response curve do exist and have essentially a normal exercise response.[11,12]

3) The optimal controller model cannot separate a metabolic CO_2 load from an airway CO_2 load via "slug" CO_2 breathing or added dead space.[13,14]. The optimal controller model predicts an isocapnic response to these airway CO_2 loads—contrary to the literature.

These considerations suggest that a feedforward/feedback controller configuration with feedforward redundancy is the more appropriate model for the respiratory control system. This model allows for denervation and/or disease effects that remove a feedforward input signal. That is, arterial CO_2 tension regulation remains even though one or more feedforward signals are removed. In fact, the motivation for developing the optimal controller configuration was to overcome the experimental observation that a potential feedforward input could be eliminated and the system would still be able to regulate.[1] Thus, it was argued that since a "single measurable stimulus and mechanism" could not be isolated, there must not be one. The concept of an optimal controller, which did not require a feedforward input, seemed attractive. However, as suggested, the optimal controller does not appear to be robust enough to mimic all experimental results.

We are, therefore, left with the conclusion that most appropriate model is the feedforward/feedback model with feedforward redundancy. This model is certainly consistent with biological systems in general where redundancy and, thus, reliability are inherent features.

Can we say that the "exercise hyperpnea dilemma" has now been solved.[14] Is there more to be considered?

Science would like to claim that ultimate understanding can be obtained. That is, we can discover the nature of the respiratory controller via clever experimental design and observation of stimulus-response data. Our understanding is complete when we generate a mathematical model, as we have just done, that mimics the observed physical data.

This notion that ''understanding is complete'' is based on the epistemological assumption that all knowledge is based on physical sense data. However, suppose we make the epistemological choice to include ''all evidence'', including self reports of pertinent subjective experience and collective intuition.[15]

From the beginning, breathing has been associated with a mysterious life force. For example, in western religious mythology we find, ''And the Lord God formed man of the dust of the ground and breathed into his nostrils the breath of life.'' The Hebrew writer of the book of Genesis speaks of neshemet ruach chayim. In the Hebrew, neshemet means ordinary breath of atmospheric air, and chayim means life, while ruach means spirit of life. It is the spirit of life concept that the eastern traditions have called prana, which comes from Sancrit meaning absolute energy.

Thus, the ''science of breath'' via Hindu tradition suggests that breathing not only delivers oxygen and removes carbon dioxide, it also fulfills the equally important function of delivering prana. Under exercise, more prana is needed, so a person breaths more, resulting in a regulated arterial CO_2 tension. Thus, prana may serve as an additional feedforward signal and if it is mystical in nature, we would indeed have trouble isolating it.

Yogi Ramacharaka puts it this way:

> ''The oxygen in the air is appropriated by the blood and is made use of by the circulatory system. The prana in the air is appropriated by the nervous system, and is used in its work. And as the oxygenated blood is carried to all parts of the system, building up and replenishing, so is the prana carried to all parts of the nervous system, adding strength and vitality. If we think of prana as being the active principle of what we call ''vitality,'' we will be able to form a much clearer idea of what an important part it plays in our lives. Just as in the oxygen in the blood used up by the wants of the system, so the supply of prana taken up by the nervous system is exhausted by our thinking, willing, acting, etc., and in consequence constant replenishing is necessary. Every thought, every act, every

effort of the will, every motion of the muscle, uses up a certain amount of what we call nerve force, which is really a form of prana. To move a muscle the brain sends out an impulse over the nerves, and the muscle contracts, and so much prana is expended. When it is remembered that the greater portion of prana acquired by man comes to him from the air inhaled, the importance of proper breathing is readily understood.''[16]

Shall we dismiss the notion of prana? Mythology often communicates collective wisdom and when pneumonosity results, the numbers *seven* and *four* are constilated. And so it is with respiratory control, the set-point of regulation is at an arterial CO_2 tension of 40 and an arterial blood pH of 7.4.

REFERENCES

1. G. D. Swanson, Overview of ventilating control during exercise, *Med. Sci. Sports* 11: 221-226 (1979).
2. J. H. Comroe, *Physiology of Respiration*, Year Book Medical, Chicago (1974).
3. G. D. Swanson, ''Respiratory control during exercise,'' in *Encyclopedia of Systems and Control*, M. Singh, Ed. Oxford: Pergamon, pp. 4045-4051 (1988).
4. W. S. Yamamoto, Computer simulation of ventilatory control by both neural and humoral CO_2 signals. *Am. J. Physiol.* 238: 28-35 (1980).
5. E. Neil, R. G. O'Regan, The effects of electrical stimulation of the distal and the cut sinus and aortic nerves on peripheral arterial chemoreception activity in the cat. *J. Physiol.* 215: 15-32 (1971).
6. E. Neil, R. G. O'Regan, Efferent and afferent impulse activity recorded from few-fibre preparations of outher intact sinus and aortic nerves. *J. Physiol.* 215: 33-47 (1971).
7. R. Lugliani, B. J. Whipp, C. Seard, K. Wasserman, Effects of bilateral carotid body resection on ventilatory control at rest and during exercise in man. *N. Eng. J. Med.*, 285: 1105-1111 (1971).
8. C. S. Poon, Optimal Control of ventilation in hypoxia, hypercapnia and exercise. In: B. J. Whipp, D. M. Wiberg (eds.) *Proc. Symp. Modelling and Control of Breathing*. Elsevier, Amsterdam, pp. 189-196 (1983).
9. G. D. Swanson and P. A. Robbins, Optimal respiratory controller structures, *IEEE Trans. Biomed Engr.* 33: 677-680 (1986).
10. G. D. Swanson, Reply to comments on "Optimal respiratory controller structures.'' *IEEE Trans. Biomed. Engr.* 35-395-397 (1988).
11. G. C. Moore, C. W. Zwillich, J. D. Battaglia, E. K. Cotton, and J. V. Weill, ''Respiratory failure associated with familial depression of ventilatory response of hypoxia and hypercapnia,'' *New England, J. Med.*, vol. 295, pp. 862-865 (1976).
12. S. A. Shea, L. P. Andres, R. B. Banzett, A. Guz and D. C. Shannon, The ventillatory response to exercise in the absence of CO_2 sensitivity, *Am. Rev. Respir, Dis.* 143: A593 (1991).
13. S. A. Ward and B. J. Whipp, ''Ventilatory control during exercise with increased external dead space,'' *J. Appl. Physiol.* vol. 48, pp. 225-231 (1980).
14. G. D. Swanson, The exercise hyperpnea dilemma, *Chest* 73: 270-272 (1978).
15. W. W. Harmon, ''A re-examination of the metaphysical foundations of modern science,'' *The Institute of Noctic Sciences*, Sausalito, California (1991).
16. Yogi Ramachavaka, ''A Science of Breath,'' *The Yogi Publication Society*, Chicago (1905).

CHEMICAL AND NON-CHEMICAL CONTRIBUTIONS TO BREATHING IN MAN, AWAKE AND ASLEEP

A.K. Datta[1,2], S.A. Shea[2] and A. Guz[2]

Departments of Medicine,
[1]United Medical & Dental Schools of Guy's and St. Thomas'
St. Thomas' Hospital
London, United Kingdom SE1 7EH
[2]Charing Cross & Westminster Medical School
London, United Kingdom W6 8RF

INTRODUCTION

Hypocapnia reproducibly elicits apnoea during anaesthesia in animals and man (Fredericq, 1901; Fink, 1961). However in conscious man during wakefulness there are very variable responses (Douglas and Haldane, 1909; Bainton and Mitchell, 1966). This variability in awake man may be due to behavioural influences upon breathing related to the subjects' expectations concerning the experiment - which would presumably disappear during sleep. We have therefore examined the effect of sleep onset on the respiratory rhythm during hypocapnia produced by passive mechanical hyperventilation (PMH) during quiet relaxed wakefulness and sleep.

METHODS

Ten healthy, post-prandial subjects (nine male, one female; age 21-63 years) were studied. Only one of the subjects was aware of the purpose of the study and had any prior expectations of the results. Subjects lay in the supine position on a bed in a room separate from the experimenters and monitoring equipment; they were observed remotely with a television camera. An airtight nose-mask (Respironics Inc) was attached to the subject's face with elastic straps and connected to a positive pressure volume-cycled ventilator (pneuPac Ltd., Luton) for passive mechanical hyperventilation. Electo-encephalogram (C3-A2 and C3-O1), electro-oculogram (F7-A2 and F8-A2), chest wall movements (inductance plethysmography, Respitrace), haemoglobin oxygen saturation (ear oximetry, Ohmeda Biox 3700) and $P_{et}CO_2$ (Beckman LB2), airflow (Fleisch pneumotachograph No 1) and pressure within the nasal mask were recorded on a chartwriter (Siemens Mingograph EEG 10, paper speed 15 mm s^{-1}).

Control of Breathing and Its Modeling Perspective, Edited by
Y. Honda *et al.*, Plenum Press, New York, 1992

Subjects were given the following typed instructions at the beginning of the study; no other instructions, verbal or written were made.

Instructions

1. It is important that you are as comfortable and relaxed as possible. You will be listening to a story through the headphones and will be asked some simple questions at the end of the study merely to ensure that you were concentrating throughout !

2. The mask over the nose is connected to an air pump which will provide all the air that you need. The settings of the air pump will change from time to time; if you wish to breathe more, then you can do so at any time and override the air pump.

Throughout the study subjects listened to a taped recitation by Sir John Gielgud of Waugh's 'Brideshead Revisited'. This was soporific; most subjects fell asleep within twenty minutes of lying down.

Test of endogenous breathing

Once steady hypocapnia with a stable breathing pattern was achieved, during wakefulness or sleep, the ventilator was abruptly disconnected during its inactiye expiratory phase; the subject's endogenous breathing and conscious state was then monitored. Apnoea was defined as zero airflow for longer than one sec. On resumption of the subject's endogenous breathing the ventilator was reconnected.

RESULTS

Data obtained when subjects were merely drowsy (EEG showing alpha rhythm interspersed with theta rhythm) were eliminated from the analysis. Stage I sleep was only accepted for analysis if there was a complete absence of alpha rhythm. Post-hyperventilation hypocapnic apnoea was more consistent and of longer duration during sleep than during wakefulness. In 168 trials the apnoea incidence was 44/82 (54%) during wakefulness (W), 43/49 (88%) during Stage I sleep (S1), 31/35 (89%) during Stage II sleep (S2), 1/1 during Stage III sleep (S3) and 1/1 during REM sleep. The subjects' mean(range) apnoea duration was 4.0(0-27) sec, 14.2(0-37.2) sec, 22.8(3-54.6) sec, 79.2 sec and 27 sec for W, S1, S2, S3 and REM sleep respectively. There was a significant difference in apnoea duration between wakefulness and both stage I and stage II non-REM sleep (paired t-test; $p=0.02$ for W vs S1 and W vs S2, $p=0.08$ for S1 vs S2).

After attachment of the transducers but before the nasal mask and ventilator was connected, a baseline measurement of breathing was made during relaxed wakefulness over three minutes; $P_{et}\,CO_2$ was monitored with a nasal probe. The subjects' mean(range) $P_{et}\,CO_2$ was 39.5(35.5-43.1) mmHg. During passive mechanical hyperventilation subjects' mean(range) $P_{et}\,CO_2$ before the trials was 31(23-42) mmHg, 31(19-41) mmHg, 34.3(29-39) mmHg, 26.7 mmHg and 32.5 mmHg for W, S1, S2, S3 and REM sleep respectively.

For the group there was a significant, albeit small, correlation between apnoea duration and pre-test $P_{et}\,CO_2$ only in stage II sleep ($r=-0.4$, $p=0.03$). In each subject, there were instances when resumption of breathing preceded, suceeded or were even unaccompanied by arousal from sleep. To determine whether prior arousal from sleep disrupts the relationship between apnoea duration and the degree of hypocapnia, an analysis was made of only those trials where the resumption of breathing preceded arousal or was unaccompanied by any change in sleep stage. In those cases, there was a significant correlation of apnoea duration against the pre-test $P_{et}CO_2$ (stage I sleep, $r= -0.96$, $p=0.048$, $n=12$, three subjects; stage II sleep, $r= -0.82$, $p=0.002$, $n=19$, two subjects).

The regression indicated that a pre-test P_{et} CO_2 less than a threshold of 41 mmHg could result in an apnoea during both stage I and II sleep; each mmHg below that threshold resulted in an apnoea of 3.1 sec in S1 and 3.7 sec in S2.

Prevention of hypoxia during tests by prior hyperoxia lengthened apnoea duration during both non-REM and REM sleep (two subjects). In one subject there was sufficient data for a paired analysis. During stage II sleep, mean $\pm$ s.d. apnoea duration was 13.6$\pm$10.4 sec, following hyperoxia mean $\pm$ s.d. apnoea duration was prolonged to 50.4$\pm$30.2 sec (paired t test, p=0.001).

After each study, subjects were systematically questioned about their subjective sensations during hypocapnic apnoeas and the resumption of breathing. all subjects reported an awareness of the ventilator and its disconnection when awake; four subjects said that they resumed breathing during the trials because of an uncomfortable feeling, the other six did not know why they resumed breathing. None of the subjects could recall those trials performed during sleep, even though they may have had transient arousal by EEG and EOG criteria.

DISCUSSION

Behavioural influences on breathing may alter the response to hypocapnia. To minimise these effects during wakefulness in this study, we studied naive subjects who were not given any prior expectations of the study; typed instructions only were given so that subjects could not be influenced by the experimenters; they were relaxed and lying on a bed in a quiet darkened room; subjects' attention was diverted by them listening to a taped story; hypocapnia was induced by passive mechanical hyperventilation with strict attention to criteria for passivity. Despite all these precautions, hypocapnia induced by passive mechanical ventilation results in inconsistent apnoea of only short duration. However even in the lightest of sleep, the same degree of hypocapnia results in a consistent apnoea of much longer duration. The deeper the sleep stage, the longer the apnoea. The results suggest that the maintenance of respiratory rhythm is critically dependent on arterial PCO_2 during sleep but during wakefulness other drives, including behavioural, supervene to a varying degree when the chemical drive is removed. Not all of these behavioural drives involve consciousness; following the study, subjects were unable to identify a common respiratory sensation preceding the resumption of breathing.

The duration of apnoea was not clearly related to the degree of pre-test hypocapnia during sleep. This is in contrast to the results of Skatrud and Dempsey (1983). In our study, subjects were made hypocapnic at different rates and so body stores of CO_2 and blood CO_2 were probably not at equilibrium during tests of endogenous breathging. Close analysis of the EEG in the present study has also showed for the first time that resumption of breathing in the same subject may be preceded by arousal, succeeded by arousal or unaccompanied by any change in sleep stage. This has a major influence in disrupting the relationship between the duration of apnoea and the pre-test P_{et} CO_2 ; in those tests where breathing preceded or was unaccompanied by arousal, regression analysis showed a clear relationship in both stage I and stage II sleep. Extrapolation of the linear regression reveals that the threshold P_{et} CO_2
during sleep below which apnoea occurs is only 41 mmHg, i.e. only 3 mmHg below the the normal resting P_{et} CO_2 during sleep (Gothe et al 1981 & 1982). Furthermore in the deeper sleep stage there was a steeper change in the duration of apnoea with each mmHg of hypocapnia below the threshold;this is consistent with a graded removal of a wakefulness related drive to breathe with deeper sleep.

The results of the present study confirm and extend those of Skatrud and Dempsey (1983). They found that during sleep but not wakefulness, PMH resulting in hypocapnia of 3-6 mmHg below non-REM sleep levels, produced consistent apnoea in two normal subjects. Thus apnoea occurs when chemical drives to breathe are reduced by inducing hypocapnia and when supervening behavioural drives related to wakefulness are removed by anaesthesia (Fink, 1961) or as in this study, by sleep.

ACKNOWLEDGMENTS

We thank Dr. K.McCrae for statistical advice and Dr. R. Horner for assistance with EEG analysis. Material from the Journal of Physiology is reproduced.

REFERENCES

Bainton, C.R. and Mitchell, R.A. (1966) Post hyperventilation apnoea in awake man. Journal of Applied Physiology 21: 411-415.

Douglas, C.J. and Haldane, J.S. (1909). The causes of periodic or cheyne-Stokes breathing. Journal of Physiology 33: 401-419.

Fink, B.R. (1961). Influence of cerebral activity in wakefulness on regulation of breathing. Journal of Applied Physiology 16: 15-20.

Fredericq, L. (1901) Sur la cause de l'apnee. Archives of Biology, 17: 561-576

Gothe, B., Altose, M.D., Goldman, M.D., and Cherniack, N.S. (1981) Effect of quiet sleep on resting and CO_2-stimulated breathing in humans. Journal of Applied Physiology 50: 724-730.

Gothe, B., Goldman, M.D., Cherniack, N.S. and Mantey, P. (1981) Effect of progressive hypoxia on breathing during sleep. American Review of Respiratory Disease, 126: 97-102.

Skatrud, J.B. and Dempsey, J.A. (1983) Interaction of sleep state and chemical stimuli in sustaining rhythmic ventilation. Journal of Applied Physiology, 55: 813-822.

DOES SUSTAINED HYPOXIA HAVE BIPHASIC EFFECTS ON THE SENSATION OF DYSPNEA?

Tatsuya Chonan, Wataru Hida, Shinichi Okabe, Yeontae Chung, Yoshihiro Kikuchi and Tamotsu Takishima

First Department of Internal Medicine, Tohoku University School of Medicine, Sendai 980 Japan

INTRODUCTION

The effect of sustained hypoxia on respiration is reported to be biphasic and comprised of an initial stimulatory phase mediated by peripheral chemoreceptors and a following depressive phase caused by central mechanisms (Neubauer, 1990). However, it is not clear how sustained hypoxia affects respiratory sensations. In this study we wished to ascertain whether sustained hypoxia has a biphasic effects on the sensation of dyspnea as well as on ventilation.

METHODS

Studies were carried out on 16 normal healthy males who ranged in age from 26 to 45 yr. All subjects had previously participated in hypoxic studies using the same circuit employed in this study, but most of the subjects had no knowledge of the hypothesis of the experiments.

The breathing apparatus consisted of a mouthpiece and a directional Hans-Rudolph low resistance valve. Expiratory airflow was measured with a Fleisch pneumotachograph (No. 3) and a differential pressure transducer (MP45 ± 5cmH_2O; Validyne, Northridge, CA) and integrated to obtain minute ventilation ($\dot{V}_E$). Mouth pressure was measured at the mouthpiece using a pressure transducer (MP45 ± 50cmH_2O; Validyne, Northridge, CA). End-tidal PO_2 and PCO_2 were continuously monitored at the directional valve with a mass spectrometer (WSMR-1400; Westron, Chiba, Japan). Arterial oxygen saturation (SaO_2) was measured with a pulse oximeter (Biox 3700; Ohmeda, Boulder, CO). The inspiratory side of the directional valve was connected to a three-way cock and a 150 liter reservoir bag which contained an initial gas mixture of 10 -11 % O_2 in N_2. Supplementary gas was continuously added to the reservoir from a gas blender (N3800; Bird, Palm Spring, CA) which mixes the air and N_2 at a variable ratio. In addition, pure CO_2 gas was independently added to the reservoir at a variably low rate, and a bypass circuit consisting of a CO_2 absorber and a variable fan was connected between the reservoir and the inspiratory line. Using this apparatus normocapnic hypoxia (PCO_2 = 40 ± 3 torr, SaO_2 = 80 ± 3%) was introduced, by changing the mixing ratio of O_2 and N_2 and CO_2 inflow, within 3 min and maintained for 20 min while the sensation of difficulty in breathing (ψ) was rated on visual analog scales at 1 minute intervals.

Subjects were divided into two groups. In the first group ψ was measured during free air breathing and while breathing through inspiratory resistance (15 cmH_2O L^{-1} s) at the same or twice the control level of minute ventilation ($\dot{V}_E$), before and after sustained hypoxia. Each breathing pattern during loading was voluntarily maintained for 2 min

Control of Breathing and Its Modeling Perspective, Edited by
Y. Honda *et al.*, Plenum Press, New York, 1992

under normocapnic normoxia using a visual targeting system reported previously (Chonan et al., 1990a). In the second group subjects sat on a bicycle ergometer throughout the experiment and performed mild exercise at 10 W for 3 min after 20 min of sustained hypoxia, while SaO_2 was maintained within 80 ± 3 % and ψ was rated at the end of the trial.

RESULTS

Figure 1 shows the average time course of changes in minute ventilation ($\dot{V}_E$) and the intensity of dyspnea (ψ) in eight subjects who participated in the first set of experiments. The level of $\dot{V}_E$ became maximum 1-7 min (average 2.8 min) after SaO_2 reached the target level of 80 ± 3 % and gradually decreased thereafter. Similarly, ψ showed an initial increase which was followed by a gradual attenuation while the steady level of isocapnic hypoxia was maintained.

Dyspea intensity during loaded breathing is shown in Figure 2. On the average, sensory magnitude increased linearly with the rise in peak mouth pressure as has been reported previously (Chonan et al., 1990b), and ψ at comparable levels of mouth pressure did not significantly differ before and after hypoxic exposure.

Table 1 shows the average data of eight subjects who performed mild exercise under sustained hypoxia. The value of SaO_2, $\dot{V}_E$ and ψ were analyzed in the control state, at peak ventilation in the initial hypoxic period, during the last minute of resting hypoxia and at the end of the exercise under hypoxia. Although the value of SaO_2 was carefully maintained throughout the experiment , both $\dot{V}_E$ and ψ increased initially and decreased in the last minute of the hypoxic period without exercise, which is basically the same as in the first set of experiments. However, the rise of sensory intensity was relatively smaller than that of $\dot{V}_E$ from rest to exercise; $\dot{V}_E$ during exercise exceeded the peak $\dot{V}_E$ from rest to exercise; $\dot{V}_E$ during resting hypoxic period, but ψ during exercise stayed below the level at peak ventilation in the resting hypoxic period.

Table 1 . Arterial oxygen saturation (SaO_2), minute ventilation ($\dot{V}_E$) and the intensity of dyspnea (ψ) in the control state, at peak ventilation during hypoxia (Peak), in the last minute of hypoxia without exercise (Late) and during exercise under hypoxia (Exercise) in 8 subjects.

	Control	Peak	Late	Exercise
SaO_2 (%)	96.8 ± 0.3	80.9 ± 0.7	79.0 ± 0.6	78.9 ± 0.5
$\dot{V}_E$ (L/min)	10.6 ± 1.2	19.7 ± 1.4	15.1 ± 1.6	24.0 ± 3.0
ψ (cm)	0.7 ± 0.3	3.9 ± 1.0	2.0 ± 0.7	2.3 ± 0.7

Data are means ± SE.

DISCUSSION

The results of this study indicate that sustained hypoxia has a biphasic, i.e. initially stimulatory and delayed depressant, effect on the sensation of dyspnea as well as on ventilation.

The initial stimulatory effect of hypoxia on dyspnea is probably, at least partly due to the increase in ventilation, because it has been suggested that a rise in respiratory motor output is a major factor which shapes the sense of effort and dyspnea (Killian et al., 1984; Chonan et al., 1990b). However, it is not clear whether there is an additional effect of hypoxia at a given respiratory motor output. Ward and Whipp(1989) have reported that

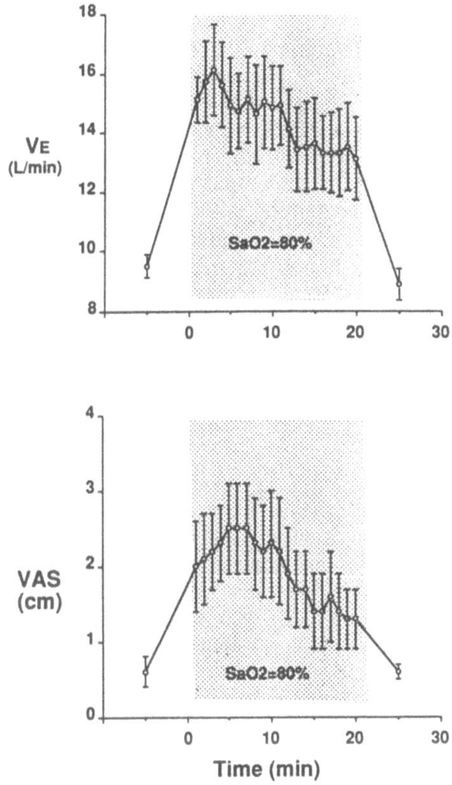

Figure 1. Average time course of changes in minute ventilation (VE) and the intensity of dyspnea (VAS) in 8 subjects. Data are shown as mean ± SE. Average values of both VE and VAS showed biphasic response, i.e. an initial increase followed by a decrease, although there were variations among subjects.

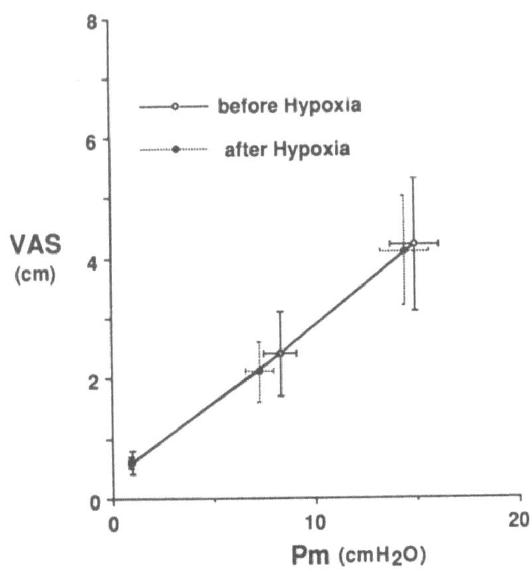

Figure 2. Intensity of dyspnea (VAS) plotted against peak inspiratory mouth pressure (Pm) during free breathing and while breathing through an inspiratory resistance (15cmH2O/L/s) at the same or twice the control level of ventilation, before and after hypoxic exposure. VAS at comparable levels of Pm did not significantly differ before and after hypoxia.

hypoxia auguments the sensation of difficulty in breathing at a given level of ventilation. The augmentative effect of hypercapnia on dyspnea has also been known (Chonan et al., 1990a). Therefore, it is possible that the initial increase in dyspnea was produced by two components, i.e., the rise in respiratory motor output and the afferent input from the peripheral chemoreceptors.

The sensory depression occurred during sustained hypoxia may not be due simply to the reduction in ventilation, because dyspnea was only mildly accentuated in spite of an apparent increase in ventilation during exercise. It is possible that the dyspnea sensing mechanism was supressed centrally by sustained hypoxia. This is inconsistent with the results of inspiratory loading trials done immediately after returning to normoxia, where the sensory intensity at comparable levels of respiratory motor output was unchanged as compared with the pre-hypoxic period. One possible explanation is that within the range expolored in this study the dyspnea sensing mechanism was suppressed only during sustained hypoxia, but it recovered shortly after returning to normoxia.

The inhibition of respiratory sensation by sustained hypoxia may be hazardous for the body in a sense that behavioral control of breathing is impaired (Cherniack, 1986, Chonan et at, 1990c), and the homeostasis of internal circumstance is endangered. In patients with severe hypoxemia, sustained hypoxia appears to reduce ventilation without a great accentuation of respiratory sensation and a worsening of the hypoxia.

REFERENCES

Cherniack, N.S., 1986, Potential role of optimization in alveolar hypoventilation and respiratory instability, in: "Neurobiology of the Control of Breathing, "C von. Euler and H. Lagercrants, eds., Raven Press, New York.

Chonan, T., Mulholland, M.B., Leitner, J., Altose, M.D., and Cherniack, N.S., 1990, Sensation of dyspnea during hypercapnia, exercise, and voluntary hyperventilation, J Appl Physiol. 68: 2100.

Chonan, T., Altose, M.D., and Cherniack, N.S., 1990, Effects of expiratory resistive loading on the sensationof dyspnea, J Appl Physiol. 69: 91.

Chonan, T., Mulholland, M.B., Altose, M.D., and Cherniack, N.S., 1990,Effects of changes in level and pattern of breathing of on the sensation of dyspnea, J Appl Physiol. 69: 1290.

Killian, K.J., Gandevia, S.C., Summers, E., and Campbell, E.J.M., 1984, Effect of increased lung volume on perception of breathlessness, effort and tension, J Appl Physiol. 57: 686.

Neubauer, J.A., Melton, J.E., and Edelman, N.H., 1990, Modulation of respiration during brain hypoxia, J Appl Physiol. 68: 441.

Ward, S.A., and Whipp, B.J., 1989, Effects of peripheral and central chemoreflex activation on the isopnoeic rating of breathig in exercising humans, J Physiol [London]. 411:27.

OVERVIEW: ROLE OF NEUROCHEMICALS AND HORMONES

Frederic L. Eldridge

Departments of Medicine and Physiology
University of North Carolina
Chapel Hill, North Carolina 27599

INTRODUCTION

Over the past decade there has been burgeoning interest in roles that various neurochemicals and hormones play in control of breathing. This of course reflects the more general attention the question has received in most areas of biology. Although chemical messengers of many types have been identified, a relatively small amount of information exists about the involement and actions of these agents on specific physiological responses in the area of respiratory control.

For control of breathing, there is the desire to understand mechanisms of actions of neurochemicals involved in several processes. One is the ways in which respiratory rhythm is generated and respiratory pattern formed. Another relates to those acute changes of breathing that result not from changes of Po2, Pco2 or pH acting upon the well-known chemoreceptors but from central neuronal mechanisms; an example of this type is the depression of breathing associated with hypoxia. A third and important need is to understand mechanisms by which various stimuli lead to prolonged changes of breathing; these changes can outlast the actual triggering stimulus by hours to days and also cannot be explained by traditional chemoreceptor mechanisms. Prolonged effects on respiratory neuronal function apparently underlie the effects on breathing. These long-lasting mechanisms may explain a number of physiological observations, such as "secondary acclimatization," the increase of breathing that accompanies prolonged exposure to altitude hypoxia and that can persist for days after return to sea level. They can be considered under the rubric of "neuronal plasticity," "long-term potentiation," "memory" or "learning." They may also be involved in the "optimization" of respiratory control, to be discussed later in this symposium.

In this presentation, I shall first consider some of the cellular and molecular mechanisms of neurochemical transmission and signal transduction and then present some examples of responses that typify neuronal receptor-transduction mechanisms that lead to long-lasting effects on respiration.

Control of Breathing and Its Modeling Perspective, Edited by
Y. Honda *et al.*, Plenum Press, New York, 1992

NEURONAL COMMUNICATION

Transmitters

Everyone will be aware that a neurochemical released from a presynaptic terminal binds to a specific recognition molecule (called a receptor) on the surface of a target cell, or on the presynaptic terminal itself, and initiates a series of events that ultimately leads to altered electrical activity and in some cases to a change in phenotype of the cell, i.e., activity-dependent regulation of gene expression. In the case hormonal neurochemicals the delivery is via circulating blood and the receptors are often intracellular. A list of some of the known neurotransmitters and modulators is given in Table 1 and an expanded list of neuroactive peptides in Table 2. Most of the latter have not been shown to play roles in respiratory control; it is not even clear that all of them act as neurotransmitters.

Table 1. Examples of neurotransmitters and modulators

Monoamines	Amino Acids	Peptides
Norepinephrine	Glutamic	Many
Dopamine	Aspartic	
Serotonin	GABA	Other
Histamine	Glycine	Adenosine
	Taurine	ATP
Acetylcholine		Progesterone

Table 2. Neuroactive peptides found in nervous tissue

Adrenocorticotropin	Luteinizing hormone
Angiotensin II	α-Melanocyte stim.
Bombesin	Motilin
Bradykinin	Neuokinin A
Calcitonin gene-related	Neuropeptide
Cholecystokinin	Neurophysin
Corticotropin-releasing	Neurotensin
β-Endorphin	Oxytocin
Leu-enkephalin	Prolactin
Met-enkaphalin	Secretin
Galanin	Sleep peptide
Gastrin	Somatostatin
Glucogon	Substance P
Growth hormone	Thyrotropin-releasing
Growth hormone-releasing	Vasoactive intestinal
Insulin	Vasopressin

Not long ago it was thought that a single neuron contained a single type of neurotransmitting chemical that acted on a single type of receptor on a post-synaptic neuron. It is now clear that the picture is more complicated. Thus, when one is considering what a given neurochemical does in a physiological system, it is important to understand several things about the mechanisms of neural communication. 1) the same presynaptic terminal can contain and release more than one neurotransmitter substance (co-localization, e.g., serotonin and substance P) and that they can act in different ways postsynaptically and presynaptically, so that the ultimate effect depends upon their differential release. A partial list of known co-localizations (Millhorn and Hokfelt, 1988) is given in Table 3.

Table 3. Examples of co-localizations of transmitters

Acetylcholine/Enkephalin	Serotonin/Substance P
Acetylcholine/Substance P	Serotonin/TRH
Acetylcholine/GABA	Serotonin/Enkephalin
	Serotonin/GABA
Dopamine/Cholecystokinin	
Dopamine/Neurotensin	GABA/Substance P
Dopamine/GABA	GABA/VIP
	GABA/Somatostatin
Noradrenaline/Enkephalin	GABA/Glycine
Noradrenaline/Neuropeptide Y	GABA/Histamine

2) some transmitters ("classical") are formed and regenerated with enzymatic action from available substrates in the nerve terminal; such transmitters are likely to be readily available for rapid and prolonged release. Peptide transmitters, on the other hand, are generated only in the neuron's soma and are transported relatively slowly to the site of release at the nerve terminal; peptides are thus not good candidates for mediation of sustained, rapid signals but may be more suited to modulating the effects of other messengers. 3) hormones are blood-borne and probably act mainly intracellularly, although there are recent studies that suggest a membrane effect in some cases (Orchinik et al., 1991).

Receptors

For a neurochemical to have an effect it must bind to one of many protein recognition molecules, a receptor. It is important to understand that the same transmitter can interact with several different types and subtypes of postsynaptic receptors, and even with receptors at the presynaptic terminal ("autoreceptors"). The cellular effects of the transmitter-receptor binding are thus "receptor-specific" rather than "transmitter-specific." In other words, there can be different and even opposing effects on a postsynaptic neuron of the same neurotransmitter, depending upon conditions and receptor type activated. Some classical transmitters such as serotonin, dopamine and glutamine have already been shown to have as many as 3 to 5 receptor subtypes and the number is growing.

There are three general kinds of receptors and mechanisms of action. **One type** mediates its effects directly by acting on ligand (transmitter)-gated ion channels (ionophores) which are characterized by an extracellular domain and an hydrophobic membrane-spanning region that functions, when activated, as an ion channel (Fig. 1). This type of receptor is well suited for rapid signalling, such as is involved in respiratory rhythmogenesis and the shaping of respiratory pattern. Examples are receptors for nicotinic acetylcholine and for excitatory (L-

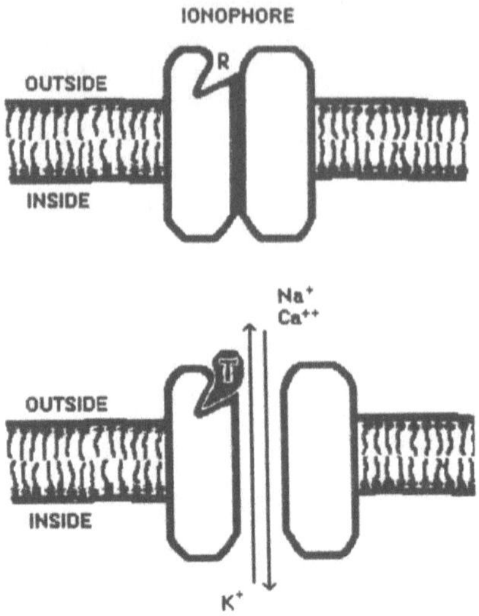

Figure 1. Top: Ligand-gated ion channel in closed state (R is receptor site). Bottom: binding of transmitter (T) to receptor causes channel to open.

glutamate) and inhibitory (GABA) amino acid transmitters. The **second** type of receptor mediates its effects on the neuron indirectly by linkage with a membrane-based guanine nucleotide-binding protein (G protein) that alters activity of the target cell by further linkage with intramembranous or cytoplasmic molecules (Fig. 2). Some examples are receptors for acetylcholine, catecholamines, serotonin, most peptides and probably adenosine. These second messenger pathways involve less rapid signalling, but can produce longer lasting effects involving phosphorylation of a membrane channel protein or, when the ultimate response element is on the gene, of regulation of gene expression.

A **third type** of receptor mechanism, exemplified by those for hydrophobic steroid hormones such as progesterone, has intracellular receptors which form an hormone-receptor complex which in turn binds to nuclear DNA. This then leads to changes

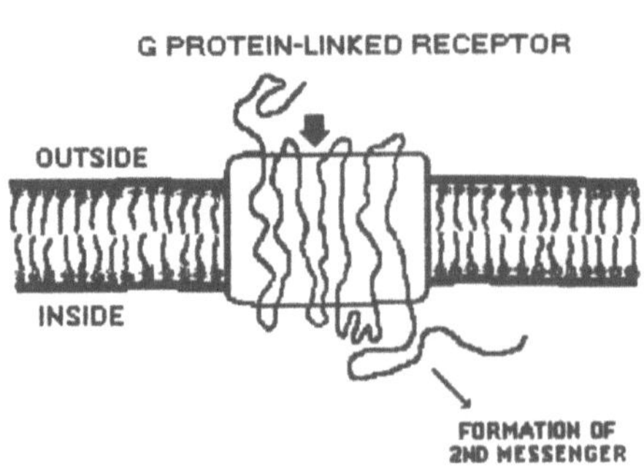

Figure 2. Model of G protein-linked membrane receptor. Arrow represents transmitter and binding site.

of gene expression. The resulting production of new protein may affect membrane receptors, the number of receptors, the anatomic form of the neuron, etc. Although such effects may be relatively slow, (not that slow, however, because they can appear within 20 to 30 minutes in some cases) they can be very long-lasting.

EXAMPLES OF RECEPTOR-TRANSDUCTION MECHANISMS THAT LEAD TO LONG-LASTING EFFECTS ON RESPIRATION

Serotonin-mediated

The first example is a long-lasting facilitation that occurs after stimulation of the carotid body or its afferents

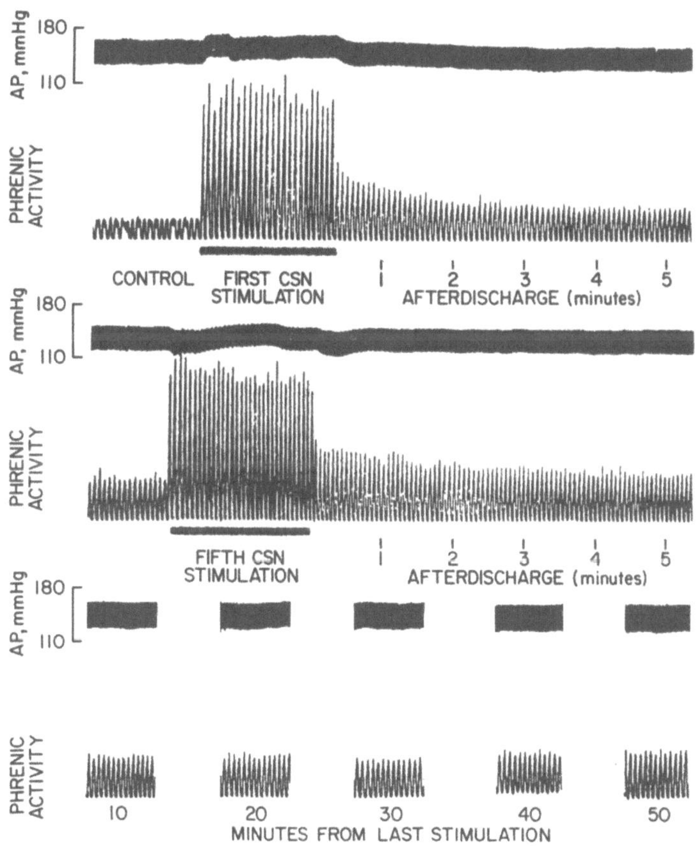

Figure 3. Long-lasting facilitatory effect of carotid sinus nerve (CSN) stimulation. Paralyzed cat with constant Pco2 of 31 mm Hg. AP, arterial pressure. Top: control, first CSN stimulation and recovery. Note increased phrenic activity even after decay of afterdischarge. 2nd, 3rd, 4th stimulations not shown. After fifth stimulation (middle panel) there is further increase of phrenic activity which remains elevated for at least 50 minutes afterwards (bottom recordings). (from Millhorn et al., 1980a)

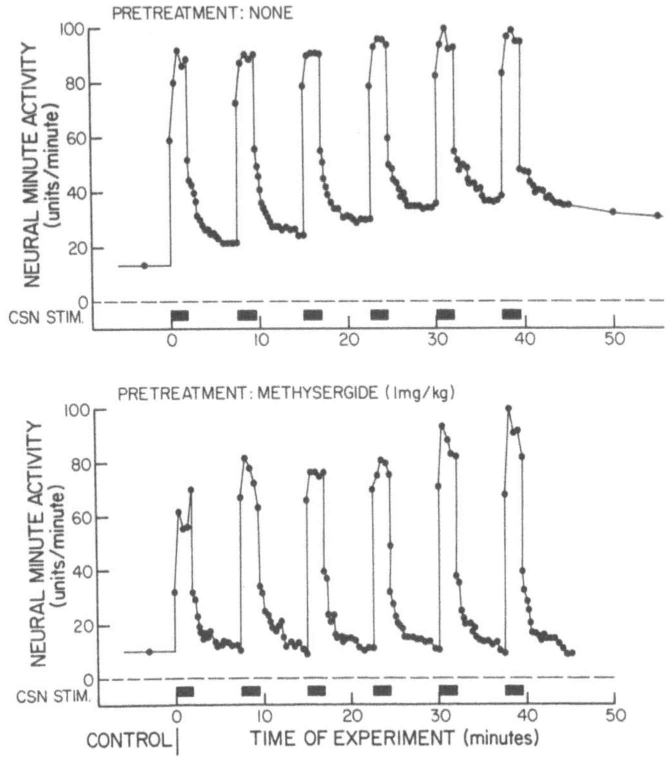

Figure 4. Effects of six successive carotid sinus n. stimulations on respiration in untreated and methysergide-pretreated cats. Long-lasting facilitation develops in untreated but is blocked in pretreated cats. (from Millhorn et al., 1980b)

(Millhorn et al., 1980a). In top recording of Fig. 3, a carotid sinus nerve stimulation (first CSN) caused respiration (phrenic activity) to increase. At cessation of stimulus an afterdischarge (AD) was apparent; however, following decay of the AD (5 min.) phrenic activity clearly remained higher than control. Subsequent stimulations (fifth is shown) had similar immediate effects but also led to further increases of respiratory output which remained stably elevated for more than 1.5 hours with no sign of return toward control. Pco2 and other traditional chemical and neural stimuli were kept constant or ablated in these paralyzed animals, so changes in them did not explain the findings. The results were similar in decerebrate animals, so the mechanism could be localized to the medulla or pons (Millhorn et al., 1980a). It was therefore concluded that the facilitation was due to a central neural mechanism that, once activated, sustained respiration for hours after cessation of the stimulus.

The mechanism involves serotinin because antagonism of serotoninergic mechanisms by a receptor antagonist (methysergide), by a synthesis blocker (parachlorophenylalanine, PCPA) and by an analog (5-7 dihydroxy-tryptamine, 5-7 DHT) that destroys serotoninergic neurons, either prevented or decreased magnitude of the long-lasting facilitatory effect (Millhorn et al., 1980b). The results with these pretreatments are shown in Figs. 4 and 5). Blockade of dopamine/norepinephrine mechanisms had no effect on the findings.

It is probable that the mechanism of long-lasting facilitation after carotid body afferent input involves release of serotonin which acts through a G protein receptor and 2nd messenger system to change membrane channels, probably by phosphorylation. A second possibility is that another transmitter, e.g., substance P, is co-released with serotonin and binds to presynaptic receptors to cause continued release of serotonin even though the original stimulus has stopped.

Progesterone-mediated

The second example is the long-lasting effect of the hormone progesterone. Progesterone has long been associated with augmented breathing in humans (Tyler, 1960). Recently, Bayliss et al., (1987) have confirmed that progesterone given intravenously (Fig. 6) or into the medulla (nucleus tractus solitarius) of anesthetized, paralyzed cats leads to prolonged stimulation of respiration. Pretreatment with RU486, a progesterone-receptor antagonist, blocked the effect, leading to the conclusion that progesterone acts centrally through a steroid receptor-mediated mechanism.

Subsequent work by Bayliss et al. (1990) leads to new insights into the mechanism. These workers studied ovariectomized cats; estrogen esposure, either natural in the estrus cycle or administered exogenously in ovariectomized animals, is a prerequisite for the respiratory response to progesterone. The response can be blocked by either the estrogen-receptor antagonist CI628 or by the progesterone-receptor antagonist RU486. Inhibitors of protein synthesis (anisomycin) and of RNA synthesis (actinomycin D) were given to ovariectomized but estrogen-treated cats (Bayliss et al., 1990). Both inhibitors markedly reduced the respiratory response to subsequently adm-

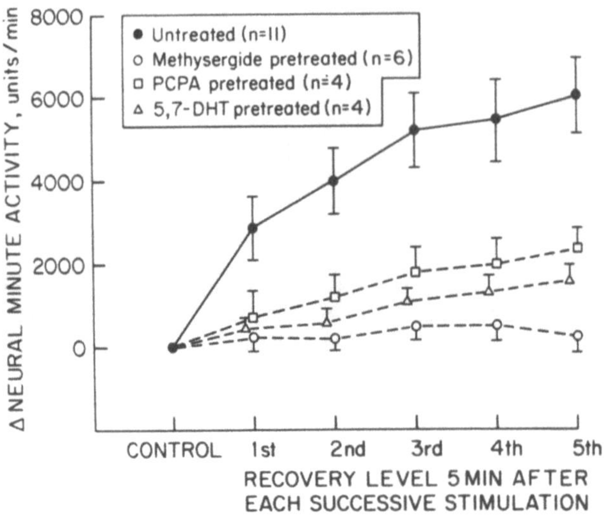

Figure 5. Comparison of long-lasting increases of respiration after carotid sinus n. stimulations in untreated cats and those pre-treated with methysergide, PCPA, and 5,7 DHT. All three serotonin antagonists block or reduce the increases. (from Moss et al., 1986)

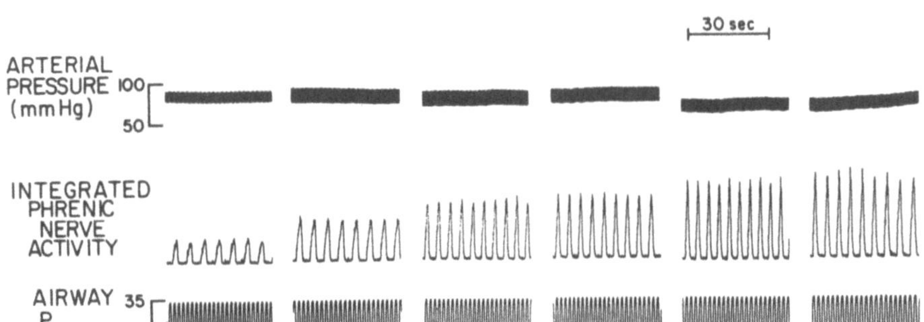

Figure 6. Effect of progressive i.v. progesterone administration on arterial pressure and integrated phrenic n. activity in paralyzed cat. Pco2 held constant at 32 mm Hg. (from Bayliss et al., 1987)

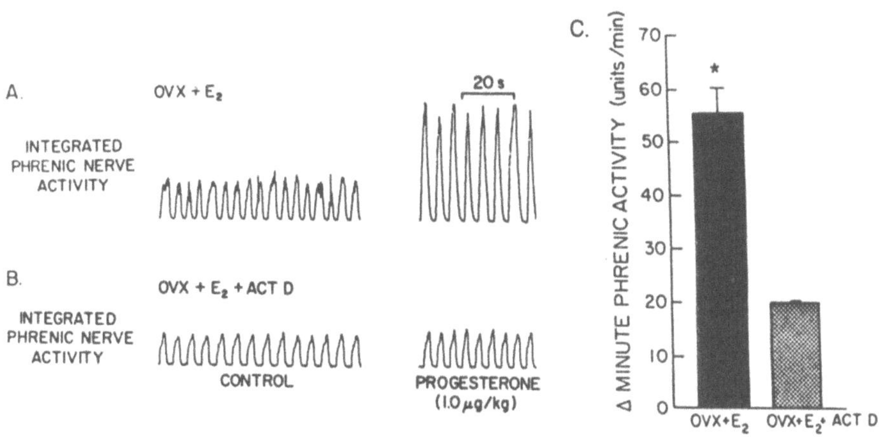

Figure 7. Effect of inhibitor of RNA synthesis, actinomycin D, on respiratory response (integrated phrenic n. act.) to progesterone in ovariectomized (OVX), estrogen-treated (E2) cats. A: control (OVX + E2). B: treated (OVX + E2 + ACT D). C: comparison of changes of respiratory response in control (n=7) and treated (n=3) groups (*p <0.05). (from Bayliss et al., 1990)

ministered progesterone (Fig. 7 shows the effect of actinomycin D), demonstrating a requirement for gene expression in the response. These workers also showed that the hypothalamus was the locus of the genomic mechanism.

ACKNOWLEDGEMENT

This work was supported by USPHS Merit Award Grant HL-17689.

REFERENCES

Bayliss, D.A., Millhorn, D.E., Gallman, E.A., and Cidlowski, J.A., 1987, Progesterone stimulates respiration through a central nervous system steroid receptor-mediated mechanism in cat, Proc. Natl. Acad. Sci. USA 84:7788.

Bayliss, D.A., Cidlowski, J.A., and Millhorn, D.E., 1990, The stimulation of respiration by progesterone in ovariectomized cat is mediated by an estrogen-dependent hypothalamic mechanism requiring gene expression, Endocrinology 126:519.

Millhorn, D.E., Eldridge, F.L., and Waldrop, T.G., 1980a, Prolonged stimulation of respiration by a new central mechanism, Respir. Physiol. 41:87.

Millhorn, D.E., Eldridge, F.L., and Waldrop, T.G., 1980b, Prolonged stimulation of respiration by endogenous central serotonin, Respir. Physiol. 42:171.

Millhorn, D.E., and Hokfelt, T., 1988, Chemical messengers and their coexistence in individual neurons, NIPS 3:1.

Moss, I.R., Denavit-Saubie, M., Eldridge, F.L., Gillis, R.A., Herkenham, M., and Lahiri, S., 1986, Neuromodulators and transmitters in respiratory control, Federation Proc. 45:2133.

Orchinik, M., Murray, T.F., and Moore, F.L., 1991, A corticosteroid receptor in neuronal membranes, Science 252:1848.

Tyler, J.M., 1960, The effect of progesterone on the respiration of patients with emphysema and hypercapnia. J. Clin. Invest. 39:34.

THE ROLE OF ENDOGENOUS OPIOIDS IN THE VENTILATORY RESPONSE TO SUSTAINED RESPIRATORY LOADS

Anthony T. Scardella, Teodoro V. Santiago, and Norman H. Edelman

Division of Pulmonary and Critical Care Medicine
Department of Medicine
University of Medicine & Dentistry of New Jersey
Robert Wood Johnson Medical School
New Brunswick, New Jersey (USA)

INTRODUCTION

The potent respiratory depressant effects of exogenous opiates are well known. Morphine and other opiates reduce ventilation, the ventilatory responses to hypercapnia and hypoxia, as well as the respiratory compensation for an acute increase in airway resistance.[1-3] The discovery of endogenous opioid peptides such as beta-endorphin and enkephalin led to investigations into the possible role of these peptides in ventilatory control in humans. In one such study Santiago and co-workers found that the opioid antagonist naloxone restored the respiratory compensation for a flow-resistive load in those patients with chronic obstructive pulmonary disease in whom it was found to be absent.[4] They postulated that in these patients endogenous opioids were elaborated in response to the stress of chronically increased airway resistance resulting in attenuation of compensation for the flow-resistive load. In a subsequent study in an unanesthetized goat model we tested the hypothesis that shorter periods of stress produced by increased airway resistance would activate the endogenous opioid system and reduce the subsequent ventilatory response.[5] We found in animals exposed to two and one-half hours of inspiratory flow-resistive loading that the reduction in tidal volume was partially reversed by naloxone given at the conclusion of the loading period and that levels of immunoreactive beta-endorphin measured in the cisternal cerebrospinal fluid were increased. Our subsequent studies have been directed at defining the effect of activation of the endogenous opioid system on central respiratory output to the respiratory muscles and the peripheral signal responsible for the activation of this system. The results of these studies are discussed in detail below.

METHODS AND RESULTS

All studies were performed in chronically instrumented male goats at least one week following surgery. General surgical procedures included implantation of an indwelling arterial catheter which was used for sampling of arterial blood and measuring of blood pressure.

Diaphragm EMG (EMGdi) was recorded using paired gold wire electrodes sutured to the abdominal surface of the costal diaphragm near the central tendon via a midline abdominal suture; external oblique EMG (EMGeo) by paired gold wire electrodes sutured directly to its surface. The EMG signals were amplified and band-pass filtered from 20 to 500 Hz, full-wave rectified, and processed by a Paynter filter with a time constant of 200 msec to generate a moving average.

The EMGdi signals were quantified as peak moving average during inspiration measured from the previous expiratory level; EMGeo signals were quantified as peak moving average during expiration. Peak moving average EMG activities were expressed as a percentage of peak amplitude at an end-tidal pCO_2 of 8%.

During each of the protocols, the goats were studied in the sternal recumbent position lightly restrained by the horns. A tight-fitting mask was placed over the snout to measure ventilation. In studies where it was measured, transdiaphragmatic pressure was measured by two thin-walled latex balloons, one positioned in the esophagus and one in the stomach and connected to separate air pressure transducers.

Only one loading protocol was performed on each day. The flow-resistive load (placed in the inspiratory line) was composed of fine wire mesh discs and was calibrated at flow rates ranging from 10-30 l/min, which corresponded to the mean inspiratory flow rates obtained in the animals studied. During all studies 25% O_2 in N_2 was given to prevent hypoxemia.

The Effect of an Inspiratory Load on the Acute Ventilatory Response and on Cerebrospinal Fluid Levels of Beta-Endorphin Immunoreactivity

In this study we tested the hypothesis that relatively short-term, intense inspiratory flow-resistive loading would be sufficient to activate the endogenous opioid system and modify the subsequent ventilatory response. Six unanesthetized goats were exposed to two and one-half hours of inspiratory flow-resistive loading at two levels, 50 (moderate load) and 80 (high load) cm H_2O/l/sec on separate days. Opioid effects were assessed by the administration of the opioid antagonist naloxone (NLX, 0.1 mg/kg) at the conclusion of the loading period. We directly tested for opioid elaboration by measuring immunoreactive beta-endorphin in the cisternal CSF. The effect of inspiratory loading (80 cm H_2O/l/sec) and NLX on tidal volume is shown in Figure 1. Tidal volume fell significantly during loading to a mean of 82.5 ± 7.5 SEM% of the baseline

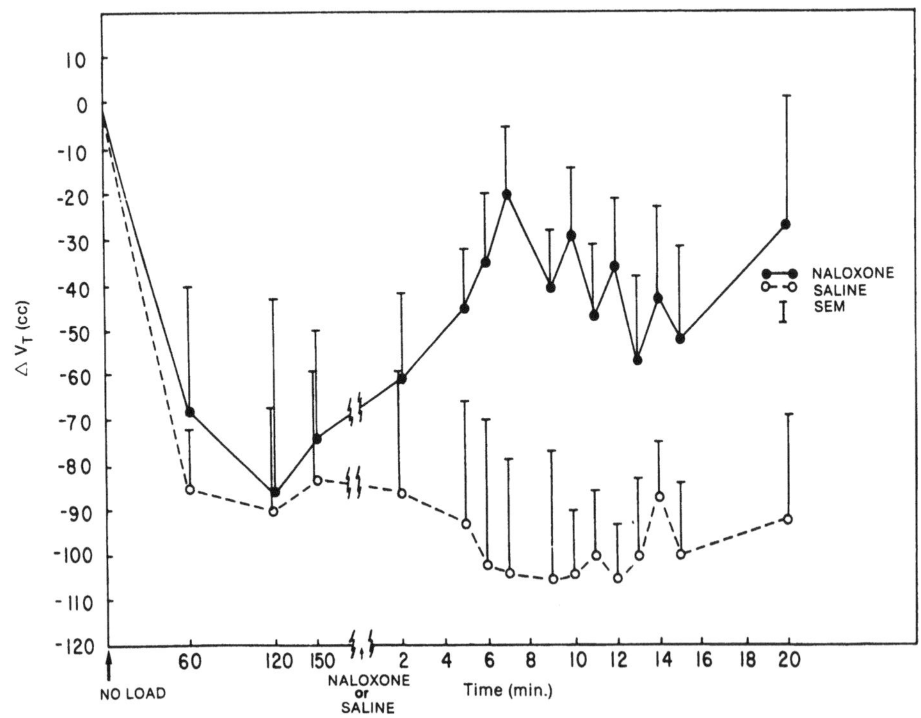

Figure 1. Tidal volume response to 2.5 hr of high inspiratory flow-resistive loading prior to and following the administration of naloxone. The tidal volume decrease associated with loading is partially reversed by naloxone. (Note the change in time scale on the x axis.) Closed circles, naloxone; open circles, saline.

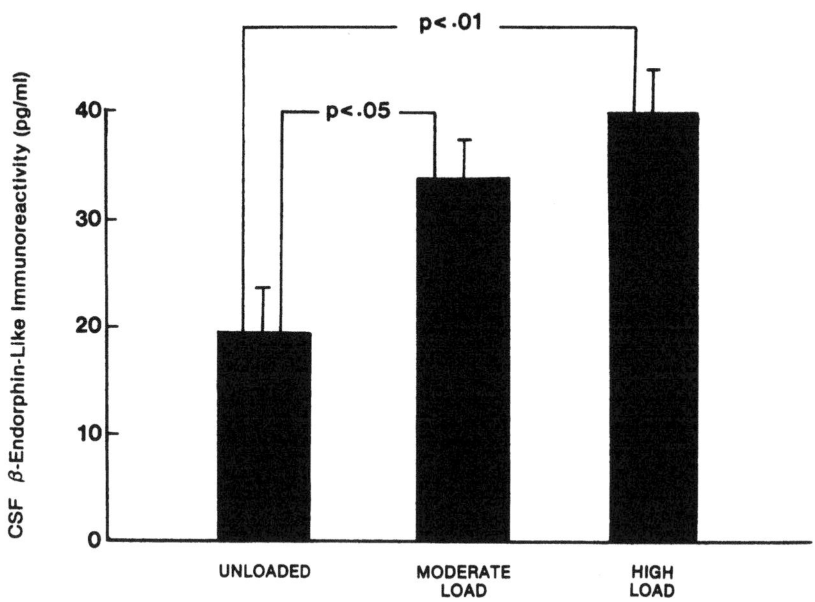

Figure 2. Beta-endorphin-like immunoreactivity in cisternal CSF in control (unloaded) and under two loading conditions. Beta-endorphin immunoreactivity was significantly increased with the moderate ($p<0.05$) and high ($p<0.01$) loads when compared to the unloaded state.

value at 2.5 hours. NLX increased tidal volume significantly but transiently after its administration, while saline administration had no effect.

Cerebrospinal fluid beta-endorphin levels in each of the unloaded, moderate, and high load experiments are shown in Figure 2. Beta-endorphin immunoreactivity was higher in the two loading conditions than in the unloaded control. Beta-endorphin levels tended to be lower with the moderate load than with the high load and were linearly related to the relative minimum tidal volume after the first thirty minutes of loading.

These data suggest that the endogenous opioid system can be activated by relatively short-term inspiratory flow-resistive loading. The increase in tidal volume immediately following NLX indicates that these potentially fatiguing flow-resistive loads reduce tidal volume prior to the onset of overt muscle fatigue by a mechanism that, in addition to the direct mechanical effect of the load, involves the endogenous opioids. These data suggest an adaptive role for activation of the endogenous opioid system under these circumstances.

The Effect of Endogenous Opioids on Abdominal Muscle Activity During Inspiratory Loading

Based on our findings above we postulated in a subsequent study that activation of the endogenous opioid system alters the ventilatory response to loading by reducing central respiratory output to the diaphragm.[6] We reasoned that, if this hypothesis were correct, a constant infusion of naloxone (NLX) during the loading period would increase ventilation relative to loading without NLX by increasing central respiratory output to the diaphragm. In addition, we hypothesized that the increase in diaphragm activity might predispose it to fatigue.

Six paired loading experiments were performed in five goats. Fifteen minutes prior to imposition of an inspiratory flow-resistive load of 120 cm H_2O/l/sec, NLX (0.1 mg/kg) was given intravenously and every 15 min thereafter. Inspiratory loading was maintained for four hours. In Figure 3 are shown the mean tidal volume and frequency responses to loading with and without (saline control) a continuous NLX infusion. In the animals given saline, the load caused a sustained reduction in tidal volume. In the presence of NLX, loading was not accompanied by a reduction in tidal volume but remained significantly above its baseline for two hours. As expected, transdiaphragmatic pressure was significantly greater than in the saline control for the same period of time. Unexpectedly, the increase in EMGdi for the saline and NLX groups was not different. However, we noted a greater end-expiratory gastric pressure during loading in the animals given NLX, implying increased activity of the abdominal muscles. An additional important finding of this study included the observation that although ventilation was greater in the NLX group for the initial two hours of loading, pCO_2 was not different in the two groups. Thus, the greater ventilation with NLX was offset by a greater CO_2 production, again suggesting an adaptive role for endogenous opioids in terms of overall energy expenditure of the respiratory muscles.

Since EMGdi did not appear to be greater during loading with a continuous NLX infusion, we investigated the possible role for endogenous opioids in selective inhibition of the abdominal muscles.[7] EMGdi and external oblique EMG (EMGeo) were monitored in seven goats during three hours of inspiratory loading (50 cm H_2O/l/sec). Mean values for EMGdi and EMGeo during loading are shown in Figure 4. EMGdi increased in response to the load, but remained constant thereafter. In contrast, after its initial increase EMGeo decreased at three hours of loading. NLX (0.1 mg/kg) at this time increased EMGdi by only 15% but EMGeo by 91%. The magnitude of change in EMGeo was greater than that for EMGdi ($p<0.05$).

These findings suggest to us that the intense activity in the respiratory muscles, especially the abdominal muscles, serves as a "noxious" stimulus resulting in activation of the endogenous opioid system. Reduction of central respiratory output appears to occur to a greater extent in the muscles receiving the greater stimulus, as the external oblique showed both greater depression during loading and a greater naloxone response. This pattern of endogenous opioid-mediated depression is similar to that which occurs in the antinociceptive pain control system.

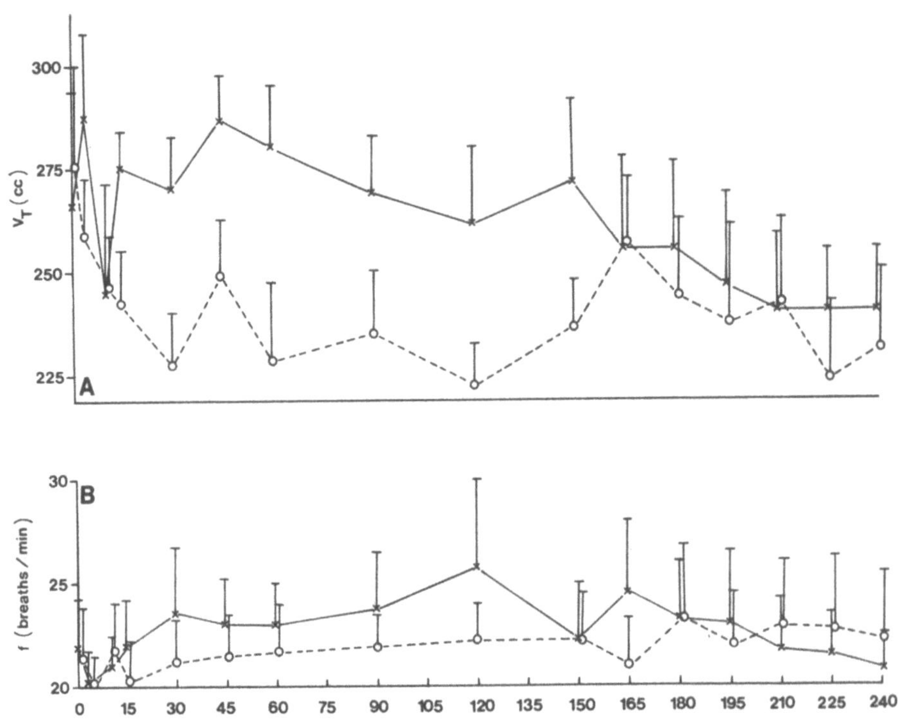

Figure 3. Tidal volume (A) and frequency (B) response during 4 hr of inspiratory loading. Loading resulted in tidal volume depression in the saline group. It did not fall during NLX and was significantly above saline up to 2.5 hr. Open circles, saline; X's, naloxone.

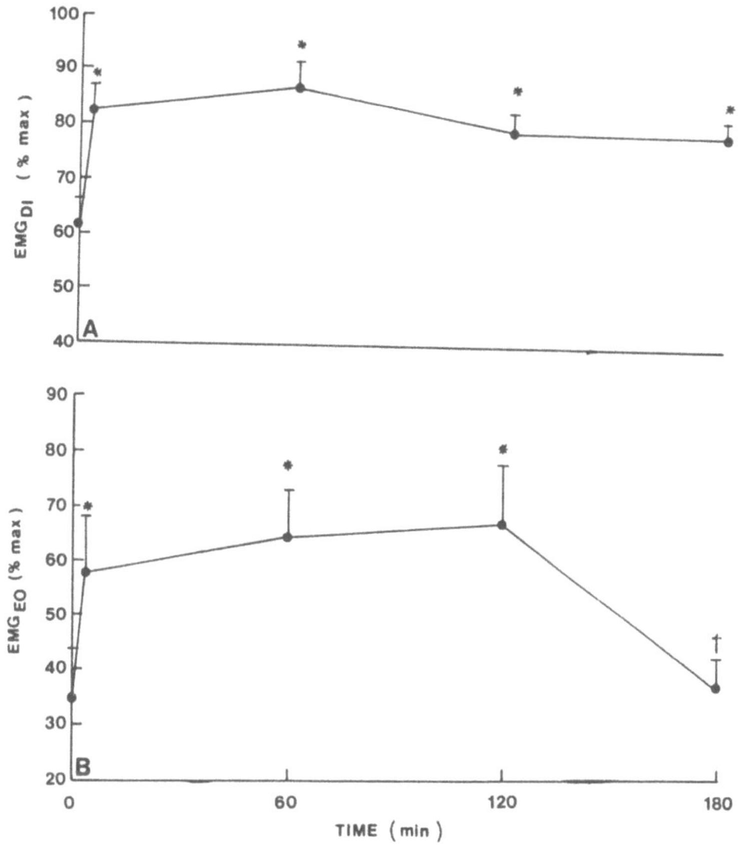

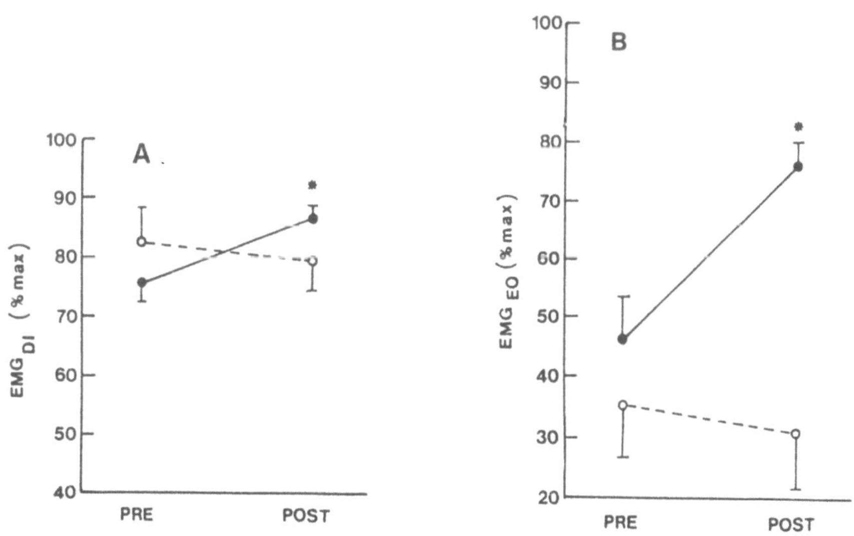

Figure 4. EMGdi (A) and EMGeo (B) during three hours of inspiratory loading. EMGdi increased in response to the load and was significantly above baseline for all three hours. EMGeo increased immediately and was significantly above baseline until 180 min when it decreased.

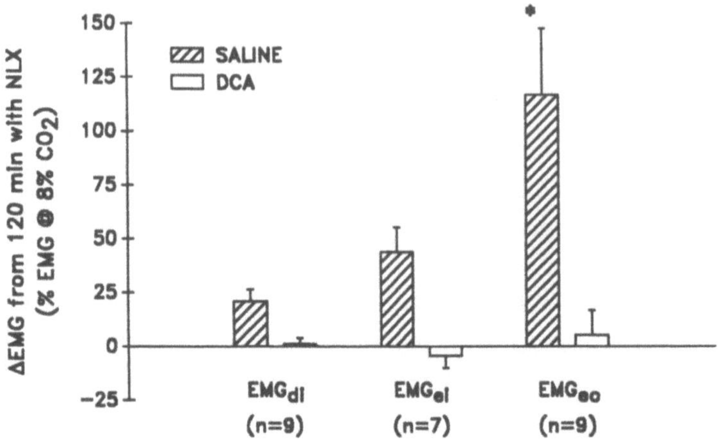

Figure 5. EMG responses to NLX after 120 min of loading with saline or DCA. With saline NLX caused an increase in all EMG's; DCA blocked the NLX response. Open bars, DCA; hatched bars, saline. * $p<0.05$ vs. EMGdi.

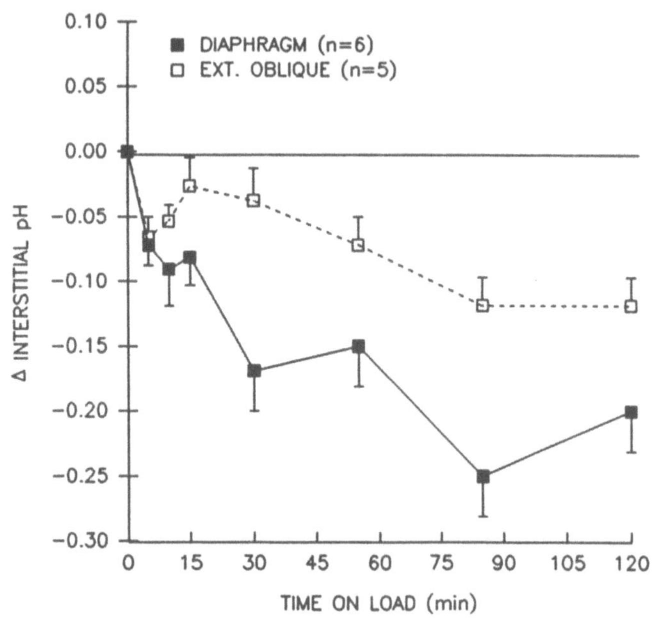

Figure 6. The change in interstitial pH of the diaphragm and external oblique during 120 min of loading. There was a greater decrease in pH_{eo} compared to pH_{di} throughout loading.

The Relationship Between Respiratory Muscle Lactic Acidosis and Endogenous Opioid Activity During Inspiratory Loading

Our previous studies had shown that the increase in motor output to respiratory muscles following NLX during inspiratory loading resulted in an increase in overall ventilation but did not reduce pCO_2 because the increased CO_2 production was associated with increased work, suggesting that the load was in the range capable of causing muscle fatigue. This suggested an adaptive role for endogenous opioids during loading, i.e., limitation of ventilation in the face of a level of airway resistance that might otherwise result in respiratory muscle fatigue. We sought to define the peripheral stimulus which would activate the endogenous opioid system but do so in a manner which would result in differential reduction in respiratory output to the diaphragm and abdominal muscles. Since lactic acid is a metabolic by-product related to intense muscle activity and is a strong stimulant of afferent (Group III and IV) fibers which can signal the central release of endogenous opioids, we hypothesized that lactic acid is the stimulus that triggers their release during loading.[8]

During two hours of inspiratory loading (50 cm H_2O/l/sec) goats were exposed to a constant infusion of either saline or dichloroacetate (DCA). DCA is a compound which enhances the activity of pyruvate dehydrogenase and thus lessens the production of lactic acid. In Figure 5 is shown the response to NLX (0.3 mg/kg) given at the conclusion of the loading period. In the goats given saline, NLX increased both EMGdi and EMGeo (along with external intercostal EMG), with the greatest response being in the EMGeo. DCA infusion competely blocked the NLX effect on respiratory activity, suggesting that lactic acid is the stimulus signaling the release of endogenous opioids.

If increased respiratory muscle lactic acid is the peripheral stimulus which activates the endogenous opioid system, then a greater decrease in pH in the external oblique compared to the diaphragm may account for the differential effect of NLX during loading. Interstitial pH was measured in the external oblique (pH_{eo}) and diaphragm (pH_{di}) using flexible glass pH electrodes. In Figure 6 is shown the change in pH_{di} and pH_{eo} from baseline during two hours of inspiratory loading. A greater decrease in pH_{eo} compared to pH_{di} was noted throughout loading. A continuous DCA infusion completely blocked the decrease in pH_{di} and significantly attenuated the decrease in pH_{eo}. From this study we conclude that the reduction in central respiratory output secondary to endogenous opioid elaboration is linked to the degree of lactic acid accumulation (pH decrease) in the respiratory muscles.

CONCLUSIONS

We have shown in this series of studies that acute, intense inspiratory flow-resistive loading activates the endogenous opioid system. Under these circumstances opioids act to reduce overall respiratory muscle activity but do so in a very specific manner, i.e., there is greater suppression of the external oblique compared to the diaphragm. The specificity of opioid-mediated suppression appears to be related to the degree of lactic acid accumulation in these muscles. Two lines of evidence support this concept. First, dichloroacetate blocks the naloxone-mediated increase in respiratory output during loading and second, the decrease in interstitial pH during loading is greater in the external oblique muscle than the diaphragm. This elaboration of endogenous opioids during inspiratory loading appears to be adaptive, since the increase in ventilation following naloxone is not accompanied by reduction in arterial pCO_2. Thus, there appears to be a pH-linked proportional control system resulting in selective reductions in central respiratory output which may serve to avoid or delay the onset of respiratory muscle fatigue.

REFERENCES

1. J.V. Weil, R.E. McCullough, J.S. Kline, and I.E. Sodal, Diminished ventilatory response to hypoxia and hypercapnia after morphine in normal man, *N. Engl. J. Med.* 292:1103 (1975).

2. T.V. Santiago, K. Goldblatt, K. Winters, A. Pugliese, and N.H. Edelman, Respiratory consequences of methadone: the response to added resistance to breathing, *Am. Rev. Respir. Dis.* 122:623 (1980).
3. M.H. Kryger, O. Yacoub, J. Dosman, P.T. Macklem, and N.R. Anthonisen, Effect of meperidine on occlusion pressure responses to hypercapnia and hypoxia with and without external inspiratory resistance, *Am. Rev. Respir. Dis.* 114:333 (1976).
4. T.V. Santiago, C. Remolina, V. Scoles, and N.H. Edelman, Endorphins and control of breathing, *N. Engl. J. Med.* 304:1190 (1981).
5. A.T. Scardella, R.A. Parisi, D.K. Phair, T.V. Santiago, and N.H. Edelman, The role of endogenous opioids in the ventilatory response to acute flow-resistive loads, *Am. Rev. Respir. Dis.* 133:26 (1986).
6. A.T. Scardella, T.V. Santiago, and N.H. Edelman, Naloxone alters the early response to an inspiratory flow-resistive load, *J. Appl. Physiol.* 67:1747 (1989).
7. A.T. Scardella, J.J. Petrozzino, M. Mandel, N.H. Edelman, and T.V. Santiago, Endogenous opioid effects on abdominal muscle activity during inspiratory loading, *J. Appl. Physiol.* 69:1104 (1990).
8. J.J. Petrozzino, A.T. Scardella, T.V. Santiago, and N.H. Edelman, Dichloroacetate blocks endogenous opioid efffects during inspiratory flow-resistive loading, *J. Appl. Physiol.* 72:590 (1992).

SIGNIFICANCE OF EXCITATORY AND INHIBITORY NEUROCHEMICALS IN HYPOXIC CHEMOTRANSMISSION OF THE CAROTID BODY

Nanduri R. Prabhakar

Department of Medicine, Physiology & Biophysics
Case Western Reserve University School of Medicine
Cleveland, Ohio 44106 U.S.A.

INTRODUCTION

Chemoreflexes arising from the carotid body are important for maintaining respiratory and cardiovascular homeostasis during hypoxic environmental stress. Conversion of hypoxic stimulus to action potential encoded signals requires transduction and transmission processes. Type I cells of the glomus tissue are considered to be the initial transducers of hypoxic stimulus. Several studies have examined the mechanisms of chemo-transduction in type I cells.[1,2,3] Some of the current ideas concerning the transduction are summarized in Figure 1. Whatever may be the transduction mechanism(s), eventually they release neurochemical(s) from the glomus cells which are necessary for transmission of the hypoxic stimulus.

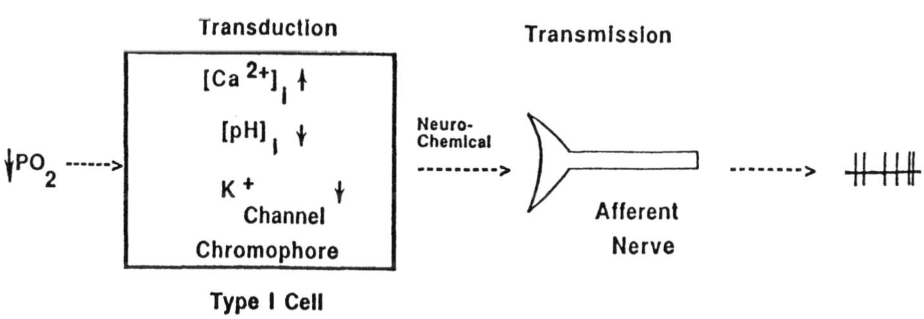

Figure 1. Schematic representation of mechanisms involved in chemotransduction and transmission in type I cells of the carotid body.

Control of Breathing and Its Modeling Perspective, Edited by
Y. Honda *et al.*, Plenum Press, New York, 1992

Till recently it was thought that transmission between type I cells and the afferent nerve ending is mediated by a single neurochemical. Recent studies, however, have shown that type I cells contain several neurochemicals including biogenic amines and neuropeptides (Table I). Often they are co-localized within the same glomus cell and perhaps co-released during hypoxia. It is believed that chemoreceptor responses to hypoxia is mediated by excitatory neurochemical(s).[4] The fact that glomus cells have several inhibitory neurochemicals suggests that the carotid body response to hypoxia is being regulated by interaction of the inhibitory with excitatory substances. The purpose of the present communication is to describe excitatory effects of Substance P (SP) and inhibitory influence of Norepinephrine (NE) on the carotid body activity and discuss their interactions with relevance to hypoxic chemo-transmission.

Table 1. Neurochemicals in Type I cells and their effects on carotid body activity.

Excitatory	Inhibitory
Substance P	Enkephalins
Acetylcholine	Dopamine
Serotonin (?)	Atrial Natriuretic Peptide
	Norepinephrine

Substance P (SP)

Localization and Metabolism. In the cat, SP-like immunoreactivity (SP-ir) is localized to many glomus cells and nerve fibres innervating the carotid body.[5,6] SP-ir seem to be confined to only those glomus cells that are in synaptic contact with the afferent nerve ending.[5] Two lines of evidence support the notion that SP is synthesized within the glomus cells. Firstly, denervation of the carotid bodies do not affect SP-ir in the glomus cells. Secondly, type I cells contain mRNA for preprotachykinin gene that encodes SP as evidenced by *in situ* hybridization histochemistry (Ringstedt and Prabhakar, unpublished observations).

In many tissues, actions of SP are terminated by the enzyme neutral endopeptidase. In the carotid body, neutral endopeptidase is present in substantial amounts and exists as membrane bound and soluble forms.[7] The latter form (i.e., soluble fraction) seems to be unique to the carotid body, because it is absent in other tissues, such as nodose and superior cervical ganglion. Phosphoramidon, an inhibitor of neutral endopeptidase, prevents degradation of SP in the carotid body.[7] These studies demonstrate that SP is synthesized in the type I cells and the enzymatic machinery responsible for its degradation is also present in the glomus tissue.

Exogenous Actions. McQueen was the first to report that exogenous administration of SP stimulates the carotid body activity in anaesthetized cats.[8] Subsequent studies confirmed the excitatory effects of SP on the carotid body activity *in vivo* and *in vitro*.[6,9,10,11] Studies on the isolated carotid body suggest that the stimulatory actions of SP are due to direct action of the peptide on the chemoreceptor tissue.

Species Variations. Before attempting to relate the effects of SP to hypoxic excitation, it is necessary to examine whether the effects of the peptide are uniform across different species. Cragg et al, have examined the chemosensory responses to SP in anaesthetized rats.[12] Intracarotid administration of SP stimulated the rat carotid bodies in a dose dependent manner. Furthermore, Capsaicin, a substance that releases endogenous SP also augmented the chemosensory discharge. SP-antagonist prevented capsaicin-induced excitation, suggesting the involvement of endogenous SP. Stimulatory effects of SP were also seen on rabbit carotid bodies.[13] The doses of the peptide required in rats and rabbits seem to be 10 to 100 fold less than that in cats. Stimulatory effects of SP seem to be uniform in all three species thus far examined.

Influence on Hypoxic Response. Maxwell et al reported that intravenous administration of SP in conscious human subjects,enhances hypoxic ventilatory drive.[14] The effects of SP were attributed to its action on the carotid body. On the other hand, peptides such as vasoactive intestinal polypeptide (VIP) were found to have no effect on hypoxic ventilatory drive.[14] We have examined the effects endogenous SP on chemoreceptor responses to hypoxia on cat carotid bodies *in vivo* and *in vitro*.[7,11] Phosphoramidon, a substance that prevents degradation of SP, significantly potentiated the chemoreceptor response to hypoxia both *in vivo* and *in vitro*. Potentiation of the low PO_2 response was blocked by SP antagonist,[11] but not by naloxone, an enkephalin blocker,[7] suggesting that the effects of phosphoramidon are due to elevated levels of endogenous SP in the glomus tissue.

Recent studies have reported a blunted hypoxic ventilatory drive in adult rats treated with capsaicin.[15,16] Cragg et al found that the carotid body responses to hypoxia were markedly attenuated in capsaicin-treated rats, compared to controls.[12] Neonatal administration of capsaicin depletes endogenous SP. Therefore, the blunted hypoxic response could conceivably be due to impaired chemo-transmission at the carotid body resulting from depletion of endogenous SP.

SP Antagonists and Hypoxic Response. Several analogues of SP have been shown to function as antagonists to SP receptors of competitive type.[17] We examined the effects of two SP antagonists (Spantide and D-Pro^2-D-Tryp^{7,9}-SP; DPDT-SP) on carotid chemoreceptor responses to hypoxia in anaesthetized cats[12,18,19] and rats. Intracarotid infusion of SP agonists at a rate of 10-15 μg/kg/min blocked the carotid body responses to SP, and attenuated or abolished the hypoxic excitation. Similar effects of the antagonist were also seen in rats. Carotid body stimulation by CO_2, however, was not affected by SP-antagonist.[19] Two factors influenced the efficacy of the antagonists. One being the dose and the other the mode of administration. For instance, in cats doses below 10 μg/kg/min were found to have no effect on chemoreceptor response to SP and hypoxia. Unlike infusion, bolus injections were found to be ineffective. Being these are antagonist peptides, it is possible that they are rapidly hydrolyzed by blood proteases, when given as bolus. These studies, thus demonstrate that SP-antagonists given in sufficient doses that block the carotid body response to SP also antagonize the chemoreceptor response to hypoxia but not that of CO_2.

SP Receptors and Hypoxic Response. Biological effects of SP are mediated by tachykinin receptors that have been classified as neurokinin-1 (NK-1), NK-2 and NK-3 subtypes (See 20 for REF). Based on the effects of various tachykinin peptides we have suggested that carotid body responses to SP are mediated by NK-1 receptors.[20] Recently Snider et al developed a non-peptide antagonist that is selective for NK-1 subtype.[21] We have examined the effects of NK-1 receptor blocker on the hypoxic response of the cat carotid bodies *in vivo* and *in vitro*[22]. Chemoreceptor responses to hypoxia but not to CO_2 were markedly attenuated by NK-1 receptor antagonist. Effective dose of the antagonist for 90% inhibition of the response was 0.3 mg/kg/min in *in vivo* studies and 0.9 μM in *in vitro* preparations. Comparable doses of an antimer had no effect on hypoxic excitation. Moreover, the antagonist effectively displaced NK-1 binding sites in the carotid body (Snider and Prabhakar, unpublished observations). From these studies it appears that NK-1 receptors exist in the glomus tissue and they are coupled to hypoxic response. Their localization in the carotid body remain to be investigated.

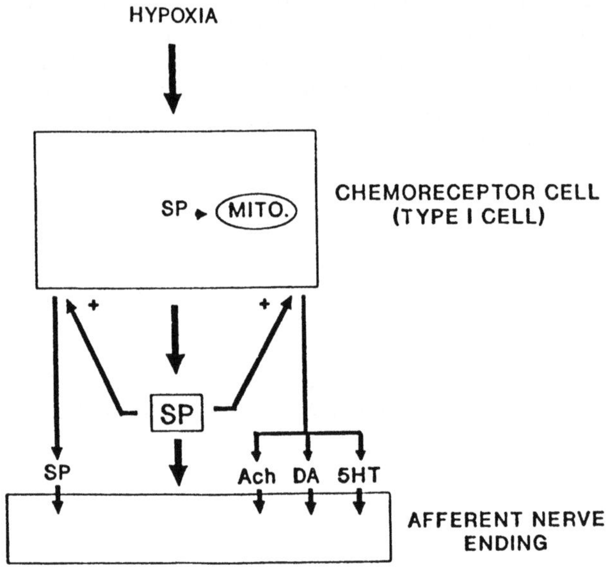

Figure 2. Schematic representation of possible actions of substance P in the carotid body.

Possible Mechanisms of SP Actions in the Carotid Body. The studies described thus far, gives an impression that actions of SP in the carotid body resemble that of a "classical" neurotransmitter or a modulator. Some of our observations are difficult to reconcile with this idea. For instance, antimycin A, a substance that inhibits mitochondrial respiration by affecting cytochrome oxidase, block not only hypoxic response but also chemoreceptor excitation caused by SP.[13] But the carotid body responses to nicotine and dopamine were not affected by antimycin. These observations suggest that the effects of SP in the carotid body may be mediated, in part, by an action on mitochondrial respiration. In fact, in a recent study it was reported that SP increases mitochondrial respiration possibly acting as a protonophore.[23] It is possible that SP participates in hypoxic chemotransmission in several ways (Figure 2). SP released by hypoxia may act on the NK-1 receptors located on the afferent nerve ending and also on the type I cells (autoreceptors). Additionally, by interfering with K^+ and/or Ca^{+2} channels in type I cells, SP enhances the release of other neuroactive agents (e.g., Ach, DA etc) as it does in other neuronal tissues. Also, acting intracellularly affects the mitochondrial metabolism, and thus exaggerates the effects of hypoxia (i.e., amplification of the hypoxic signal).

Norepinephrine (NE)

Localization. Carotid bodies resemble chromaffin tissues containing catecholamines. It is well established that norepinephrine (NE) is one of the major catecholamine present in the carotid bodies of various mammalian species.[4] In addition to type I cells, much of the NE is localized to the sympathetic nerve terminals innervating the carotid body. Being close to the carotid artery, glomus tissues are also exposed to circulating NE. In other words, two sources of endogenous NE i.e., one from the glomus tissue and the other from circulation can potentially affect the function of the carotid bodies.

Exogenous Actions and Influence on Hypoxic Response. Several studies have reported that NE causes inhibition as well as excitation of the chemosensory discharge; often inhibition preceding the excitation (See 24 for REF). It is possible that at certain doses, NE produces only inhibition, whereas at other doses, it causes excitation. We examined the effects of NE on carotid bodies of anaesthetized cats. Intracarotid administration of NE at doses between 0.1 to 5 μg/min inhibited the base line activity in a dose dependent manner. Maximum depression was seen with 5 μg/min. Increasing the dose to 10 and 20 μg/min, however, increased the sensory discharge by 25 % of the controls. The doses that inhibited the chemosensory discharge, however, had undetectable effects on arterial blood pressure. To distinguish between the NE effects on the carotid body that act directly on the chemosensing elements and indirectly via blood flow changes, parallel experiments were performed on carotid bodies *in vitro*, where the latter effect is absent. Superfusing the carotid body with NE between 10-40 μg/min depressed the base line activity, whereas higher doses (80 and 160 μg/min), augmented the sensory discharge.

The effects of NE (5 μg/min intracarotid infusion for 5 min) on hypoxic responses was examined on carotid bodies *in vivo*. Chemoreceptor responses to low PO_2 were reduced by 44% during infusion of NE. Similarly, superfusing the *in vitro* carotid bodies with 10 to 40 μg/min attenuated the hypoxic excitation by 76% of the controls.

From these studies it is evident that low doses of NE reduce the base line activity and attenuate the hypoxic excitation both in *in vivo* and *in vitro* carotid bodies. We reasoned that low doses used in our studies may represent the "physiological" concentrations of NE. Therefore, analyzed further the adrenergic receptors responsible for the inhibitory effects of NE.

Adrenergic Receptors Mediating the Inhibitory Actions of NE. It is now clear that in many tissues, inhibitory actions of NE are mediated by α_2-adrenergic receptors.[25] Kou et al[24] determined the α_2-receptor density in the glomus tissue of cats and examined the effects of α_2-receptor agonist and antagonist on chemoreceptor responses to isocapnic hypoxia. Alpha-2 receptor binding determined by [125] p-iodoclonidine averaged 10 ±2 fmol/mg of protein. Intracarotid administration of an α_2-agonist (Guanabenz; 0.5-5 μg/min; 5 min) depressed base line carotid body activity and attenuated the hypoxic excitation. Systemic administration of α_2-antagonist (SKF-86466, 0.5-2 mg/kg) prevented the effects of guanabenz on the carotid body activity, but not the responses to phenylephrine (α_1-agonist) and dopamine, suggesting its selectivity to α_2-receptors. Furthermore, α_2-antagonist alone increased baseline discharge by 68% and potentiated the hypoxic response by 46%.

In another series of experiments the effects of α_2-antagonist on carotid body responses to NE were examined *in vivo* and *in vitro*. The reduction in the baseline activity and depression of the hypoxic response by NE were prevented by α_2-antagonist.

Interaction of α_2-adrenergic Receptors with SP. Activation of α_2-receptors depress the activity of the target tissues by inhibiting the release of various neuroactive agents including SP. Two lines of evidence suggest potential interactions of α_2-receptors with SP. Firstly, α_2-antagonist potentiated the carotid body response to SP. Secondly, SP-antagonist markedly attenuated the excitatory effects of α_2-antagonist on the carotid body response to hypoxia.

Significance of Interaction of Excitatory and Inhibitory Neurochemicals in Hypoxic Chemotransmission at the Carotid Body: Hypoxia affects almost all tissues in the body. But the uniqueness of the carotid body response to hypoxia is that the sensory excitation continues during the entire period of hypoxia. So that the respiratory complex is continuously informed about the status of arterial pO_2. That is why carotid body is regarded as a sensory "receptor" for monitoring arterial O_2. Neurochemicals should not only initiate the excitation but more importantly, has to maintain the increased sensory discharge during the long periods hypoxic exposures. It is conceivable that chemicals such as SP may initiate the chemosensory response to hypoxia. The inhibitory neurochemicals on the other hand prevent the over excitation of the carotid body and thus help for the sustenance of the response. Relevant to this idea is the findings of Ponte and his co-workers,[26] who reported that hypoxic excitation of the carotid body is no longer sustained after blockade of dopaminergic receptors, which are inhibitory to chemosensory discharge.

Besides the feed back regulation, biological processes are regulated by interactions between excitatory and inhibitory (push-pull) mechanisms. The fact that glomus cells contain excitatory and inhibitory neurochemicals led to the proposal that they function as elements in a push-pull regulatory system.[27] It is conceivable that initiation and maintenance of the carotid body response to hypoxia depends on complex interplay of excitatory and inhibitory neurochemicals.

ACKNOWLEDGEMENTS

I am grateful to Professor Neil S. Cherniack for stimulating discussions and constant encouragement, and to Cheryl Diane Gilliam for secretarial assistance. The work reported here was done in collaboration with Drs. Cragg, Kumar, Kou, and Runold. The research is supported by grants from National Institutes of Health: HL-38986; HL-45780; and a Research Career Development Award, HL-02599.

REFERENCES

1. H. Acker. PO_2 chemoreception in arterial chemoreceptors, *Annu. Rev. Physiol.* 51:835-844 (1989).

2. T.J. Biscoe and M.R. Duchen. Monitoring pO_2 by the carotid chemoreceptor, *News in Physiol. Sci.* 5:229-233 (1990).

3. J. Lopez-Barneo, J.R. Lopez-Lopez, J. Urena, and C. Gonzalez. Chemotransduction in the carotid body: K^+ current modulated by pO_2 in type I chemoreceptor cells, *Science.* 241:580-582 (1988).

4. S.J. Fidone and C. Gonzalez. Initiation and control of chemoreceptor activity in the carotid body, *in*: "Handbook of Physiology - Section 3: The Respiratory System," Vol. 2, N.S. Cherniack and T.G. Widdicombe, ed., (1986).

5. I.V. Chen, R.D. Yates, and J.T. Hansen. Substance P-like immunoreactivity in rat and cat carotid bodies: light and electron microscopic studies, *Histol. Histopathol.* 1:203-212 (1986).

6. N.R. Prabhakar, S.C. Landis, G.K. Kumar, D.M. Kilpatrick, N.S. Cherniack, S.E. Leeman. Substance P and neurokinin-A in the cat carotid body: localization, exogenous effects and changes in content in response to arterial pO_2, *Brain Res.* 481:205-214 (1989).

7. G.K. Kumar, M. Runold, R.D. Ghai, N.S. Cherniack and N.R. Prabhakar. Occurrence of neutral endopeptidase activity in the cat carotid body and its significance in chemoreception, *Brain Res.* 517:341-343 (1990).

8. D. S. McQueen. Effects of substance P on carotid chemoreceptor activity in cats, *J. Physiol.* (London) 302:31-47 (1980).

9. M. Shirahata. Effects of substance P on the carotid chemoreceptor responses to natural stimuli, *in*: "Chemoreceptors and Reflexes in Breathing: Cellular and Molecular Aspects," S. Lahiri, R.E. Foster, R.O. Davies, and A.I. Pack, eds., Oxford University Press, N.Y., pp 139-145 (1989).

10. L. Monti-Bloch and C. Eyzaguirre. Effects of methionine-enkephalin and substance P on the chemosensory discharge of the cat carotid body, *Brain Res.* 338:297-307 (1985).

11. Y.R. Kou, G.K. Kumar and N.R. Prabhakar. Importance of substance P in the chemoreception of the carotid body *in vitro*, *FASEB J.* 5:A1118 (1991).

12. P.A. Cragg, Y.R. Kou and N.R. Prabhakar. Role of substance P in rat carotid body responses to hypoxia and capsaicin, *in*: "Neurobiology and Cell Physiology of Chemoreception," P. G. Data, H. Acker, and S. Lahiri, eds., Plenum Press (1992). (In press)

13. N.R. Prabhakar and N.S. Cherniack. Importance of tachykinin peptides in hypoxic ventilatory drive, *in*: "Chemoreceptors and Reflexes in Breathing: Cellular and Molecular Aspects, S. Lahiri, R.E. Foster, R.O. Davies, and A.I. Pack, eds., Oxford University Press, N.Y., pp 99-112 (1989).

14. D.L. Maxwell, R.W. Fuller, C.M.S. Dixon, F.M.C. Cuss, and P.J. Barnes. Ventilatory effects of substance P, vasoactive intestinal peptide, and nitroprusside in humans, *J. Appl. Physiol.* 68:295-301 (1990).

15. S. M. Bond, F. Cervero, and D.S. McQueen. Influence of neonatally administered capsaicin on baroreceptor and chemoreceptor reflexes in adult rat, *Br. J. Pharmacol.* 77:517-521 (1982).

16. G.T. De Sanctis, F.H.Y. Green and J.E. Remmers. Ventilatory responses to hypoxia and hypercapnia in awake rats pretreated with capsaicin. *J. Appl. Physiol.* 70:1168-1174 (1991).

17. S. Rosell and K. Folkers. Substance P antagonists: a new type of pharmacological tool, *Trends Pharmacol.* 3:211-212 (1982).

18. N.R. Prabhakar, M. Runold, Y. Yamamoto, H. Lagercrantz, and C. von Euler. Effect of substance P antagonist on the hypoxia-induced carotid chemoreceptor activity. *Acta Physiol Scand.* 121:301-303 (1984).

19. N.R. Prabhakar, J. Mitra and N.S. Cherniack. Role of substance P in hypercapnic excitation of carotid chemoreceptors, *J Appl Physiol.* 63:2418-2425 (1987).

20. N.R. Prabhakar, Y.R. Kou, and M. Runold. Effect of physalamine and eledoisin on carotid chemoreceptor activity: evidence for neurokinin-1 receptors. *Neurosci. Lett.* 120:183-186 (1990).

21. R.M. Snider, K.P. Longo, S.E. Drozda, J.A. Lowe III and S.E. Leeman. Effect of CP-96,345, a non-peptide substance P receptor-antagonist on salivation in rats, *Proc. Natl. Acad. Sci. USA.* 88:1042-1044 (1991).

22. H. Cao, R.M. Snider, N.S. Cherniack and N.R. Prabhakar. Effect of non peptide NK-1 receptor-antagonist on chemoreceptor response to hypoxia. *FASEB J.* (1992). (In press)

23. N.R. Prabhakar, M. Runold, G.K. Kumar, N.S. Cherniack, and A. Scarpa. Substance P and mitochondrial oxygen consumption: evidence for a direct intracellular role for the peptide, *Peptides.* 10:1003-1006 (1989).

24. Y-R. Kou, P. Ernsberger, P.A. Cragg, N.S. Cherniack, and N.R. Prabhakar. Role of α_2-adrenergic receptors in carotid body response to hypoxia, *Resp. Physiol.* 83:353 (1991).

25. D.B. Bylund and U.C. U'Prichard. Characterization of α_1- and α_2-adrenergic receptors, *Int. Rev. Neurobiol.* 24:343 (1983).

26. J. Ponte and C.L. Sadler. Interactions between hypoxia acetylcholine and dopamine in the carotid body of rabbit and cat, *J. Physiol (London)* 410:395-610 (1989).

27. N.S. Cherniack, N.R. Prabhakar, M.A. Haxhiu, and M. Runold. Excitatory and inhibitory influences on the ventilatory augmentation caused by hypoxia, *in*: "Response and Adaptation to Hypoxia," S. Lahiri, N.S. Cherniack, and R.S. Fitzgerald, eds., Oxford Univ. Press, pp 107-121 (1991).

ROLE OF SEROTONIN IN AIRWAY PATENCY: PHYSIOLOGICAL AND MORPHOLOGICAL EVIDENCE FOR SEROTONINERGIC INPUTS TO LARYNGEAL INSPIRATORY MOTONEURONS

Hideho Arita, and Masahiro Sakamoto

Department of Physiology, Toho University School of Medicine
Ota-ku, Tokyo 143, Japan

INTRODUCTION

Serotonin or 5-hydroxytryptamine (5-HT) is known to be widely distributed in the brain (Dahlström and Fuxe, 1964) and has been implicated in various physiological and behavioral functions, such as REM sleep (Jouvet, 1969), neuropsychiatric disorder (Peroutka and Snyder, 1980), thermoregulation (Gudelsky et al., 1986), cardiovascular control (Lovick, 1989), and so on. The present study is conducted to evaluate a role of serotonin in the control of upper airway. Using a micropressure ejection method (Kogo and Arita, 1990), we have tested responsiveness of the medullary inspiratory neurons to direct applications of serotonin, in comparison with applications of glutamate and noradrenaline. Furthermore, we have examined modes of 5-HT terminals on the medullary motoneurons projecting to the dilator muscles of the larynx, by means of combined techniques of retrograde labelling with unconjugated choleratoxin subunit-B and immunohistochemistry with an antiserum against serotonin. The present physiological and morphological results are discussed in relation to a possible role of the serotoninergic system in airway obstruction during sleep.

METHODS

Unit Recording and Application of Serotonin

Experiments were performed on decerebrate, spontaneously breathing cats. The animals were initially anesthetized with pentobarbital sodium and tracheotomized in the neck. A parietal craniotomy was performed, and the brain was transected at the precollicular level. The anesthetic was discontinued after decerebration. To expose the ventral surface of the medulla, an occipital craniotomy between the tympanic bullae was performed in the supine position, and the dura was opened along the midline.

Five-barreled micropipettes were made from borosilicate glass capillaries (1.5 mm OD). One barrel for recording was filled with sodium acetate and 2% pontamine blue dye (d.c. resistance=7-15 MΩ). Three barrels for drug ejection contained sodium L-glutamate (Glu, 10mM), serotonin-creatinine sulfate (5-HT, 5mM), and noradrenaline hydrochloride (NA, 50mM), respectively. They were dissolved in artificial cerebrospinal fluid. The remaining barrel was filled with the control artificial cerebrospinal fluid (Ctr), the pH of which was

Control of Breathing and Its Modeling Perspective, Edited by
Y. Honda *et al.*, Plenum Press, New York, 1992

adjusted to 7.35. These barrels for micropressure ejection were attached to a pneumatic micropump (WPI) by way of the tubes.

A systematic search for single units with inspiratory discharges was conducted in the region medial to the rootlets of the hypoglossal nerve and lateral to the pyramidal tract and 1.0-4.5 mm deep from the ventral surface of the medulla. The neuronal activity along with phrenic nerve activity was continuously recorded. Once activity of the inspiratory neurons was detected, Glu, 5HT, NA and Ctr were applied in the vicinity of the neuron under investigation. Effects on neuronal activity were assessed in terms of changes in firing rate.

The successful recording sites were marked by electrophoretically depositing pontamine blue dye. The site of dye deposition was determined by histological examination at the end of the physiological experiments.

Retrograde Labelling and Immunohistochemistry

In the next morphological experiments, we used the combined techniques of retrograde labelling of motoneurons with unconjugated cholera toxin subunit B (CTB), and immunohistochemistry with an antiserum against serotonin. Ten to twenty µl of 1% CTB solution was injected into two dilator muscles of the larynx, i.e., posterior cricoarytenoid (PCA) muscle and cricothyroid (CT) muscle in anesthetized cats. Three days after CTB injections, the animals were again deeply anesthetized and perfused through the left cardiac ventricle with Zamboni's fixative. The brain stem was then removed and the transverse sections of the entire medulla were made at 40 µm thickness on a freezing microtome.

For 5-HT immunohistochemistry, the free-floating sections were incubated in (1) rabbit antiserum to 5-HT in 0.01M phosphate-buffer saline (PBS), (2) biotylated goat antirabbit immunoglobulin in PBS, (3) avidin-biotin conjugated horseradish peroxidase (ABC), and (4) diaminobenzidine (DAB) with nickel ammonium sulfate in PBS. 5-HT reaction products appeared as black staining. For CTB immunohistochemistry, the sections were further incubated in goat antiserum to CTB in PBS, and in biotylated rabbit antigoat immunoglobulin in PBS. Using the ABC-DAB method without nickel, the CTB reaction products, i.e., retrogradely labelled motoneurons were visualized as light brown staining along with 5-HT positive terminals (black) in the same sections.

RESULT

Effect of Direct Application of Serotonin on Inspiratory Units

We made detailed examinations of 52 inspiratory neurons that showed clear excitation by Glu but did not react to Ctr. The excitatory response to Glu indicated that the tip of the multibarreled micropipette was situated close to the cell body under investigation. The tested sites of these inspiratory neurons were located in the caudal part of the nucleus ambiguus (0-3.5 mm rostral to the obex).

The 52 inspiratory neurons were classified into two subgroups on the basis of the firing patterns; inspiratory neurons with an augmenting firing pattern ["augmenting I units"(22/52)] and inspiratory neurons with a decrementing firing pattern ["decrementing I units"(30/52)]. The typical examples are shown in Fig. 1.

Application of NA produced predominantly inhibitory effects on both the decrementing (22/30) and augmenting I units (20/22), whereas application of 5-HT resulted in distinct or opposing effects on these two types of inspiratory neurons: the decrementing I units (25/30, 83%) were excited by 5-HT, while the augmenting I units (17/22, 77%) were inhibited by 5-HT.

Figure 2 shows a representative response of the decrementing I units to applications of 5-HT, NA, Glu and Ctr, respectively. The excitation of the decrementing I units with 5-HT was characterized by a long onset-latency of response and a prolonged recovery process. The

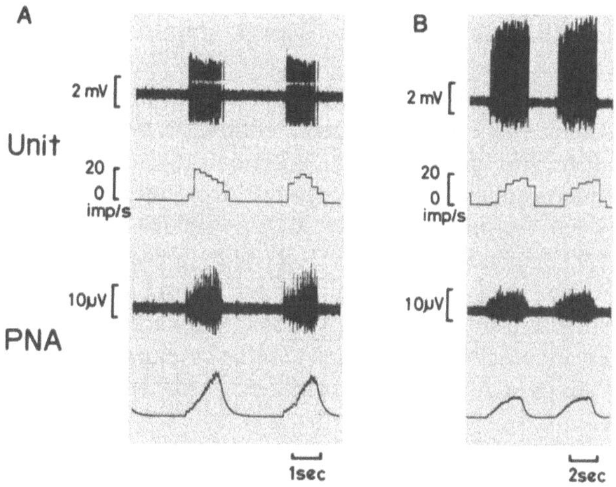

Figure 1. Examples of 2 types of inspiratory neurons with distinct firing patterns. Unit activity (Unit) is shown as a raw recording (top trace) along with its pulse-counter output (2nd trace). For comparison, phrenic nerve activity (PNA) is also shown. A: inspiratory unit with decrementing firing. B: inspiratory unit with augmenting firing. [Reprinted with permission from J. Neurophysiol.]

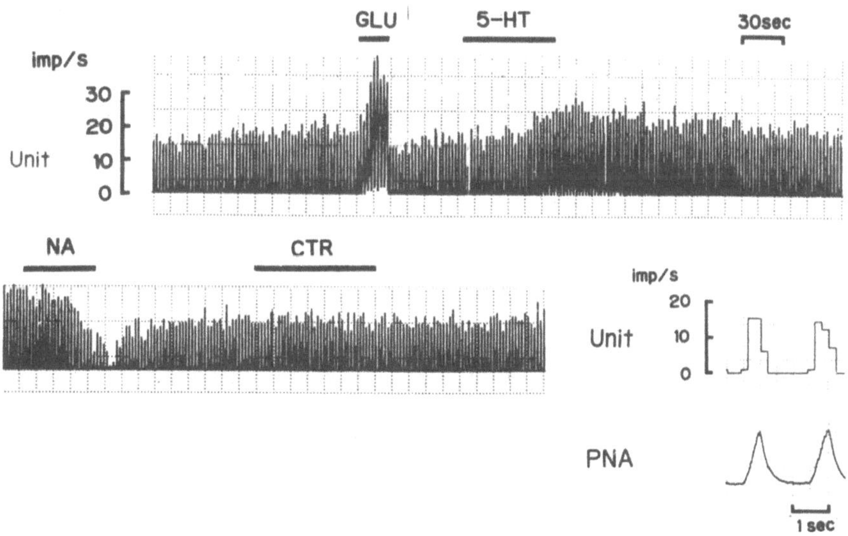

Figure 2. Typical responses of inspiratory unit with decrementing firing to micropressure applications of Glu, 5-HT, NA and Ctr. Bottom right: unit's pulse-counter output (Unit) along with integrated phrenic nerve activity (PNA) recorded at faster paper speed. Responses to applications of the various agents are shown only on the trace of unit's pulse-counter output at slower paper speed. [Reprinted with permission from J. Neurophysiol.]

increase in firing rate occurred not only during the inspiratory active phase but also during the expiratory phase (Arita and Ochiishi, 1991).

Morphological Identification of Serotoninergic Input

A typical example of light photomicrograph demonstrating retrogradely labelled motoneurons to PCA and 5-HT immunoreactive terminals is shown in Fig. 3. PCA labelled motoneurons were identified ipsilateral to the injection site within the nucleus ambiguus of the caudal part of the medulla. They were localized in a cluster of several cells. The cell bodies were usually ovoid or fusiform in shape with several neural processes.

Numerous 5-HT immunoreactive terminals were present in the area of the PCA motoneuron pool, whereas only a small number of 5-HT positive terminals were observed in the surrounding reticular formation of the ventral medulla.These 5-HT immunoreactive varicosities and intervaricose fibers were found to make intimate contacts with the cell bodies and proximal dendrites of PCA motoneurons (see arrowheads in Fig. 3).

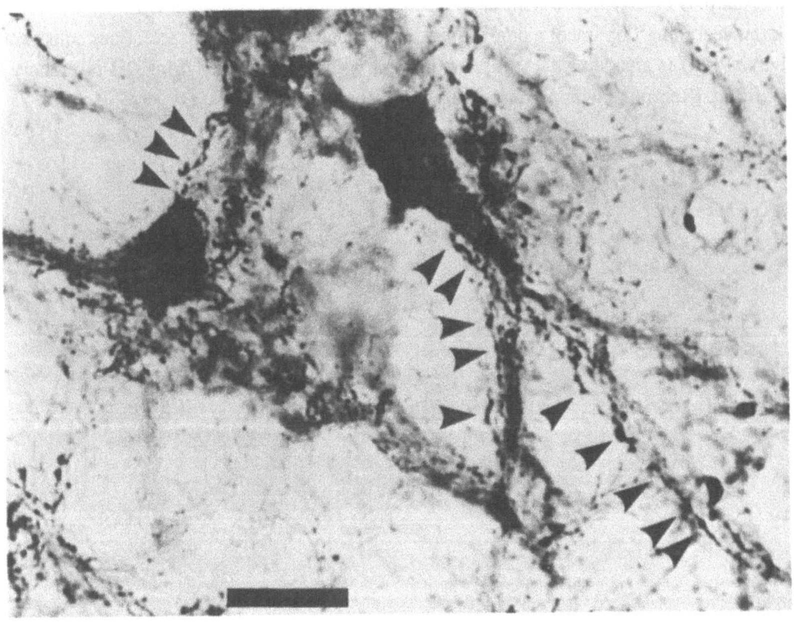

Figure 3. Photomicrographs showing retrogradely labelled motoneurons to PCA and 5-HT-immunoreactive varicosities and fibers (arrowheads) in the nucleus ambiguus at high magnification. Bar: 40µm. See text for details.

CT labelled motoneurons were also located ipsilateral to the injection site, but they appeared in two separate nuclei of the ventrolateral medulla. In the caudal part of the medulla, CT labelled cells were exclusively found within the nucleus ambiguus, whereas in the rostral part of the medulla the CT labelled cells were found not only within the nucleus ambiguus but also at the region ventrolateral to the nucleus ambiguus, i.e., the retrofacial nucleus (Fig. 4). In the nucleus ambiguus, the retrogradely labelled CT motoneurons were compactly packed and had medium-sized round cell bodies with several short dendrites (NA in Fig. 4B). In the retrofacial nucleus, the CT motoneuron pool appeared in a scattered pattern (RFN in Fig. 4B). Most of CT motoneurons were larger multipolar neurons with wide-based dendrites. These findings indicate that CT motoneurons could be classified into two distinct populations on the basis of their locations and their morphological properties.

Furthermore, the patterns of 5-HT immunoreactive terminals were also different. There was a dense accumulation of 5-HT immunoreactive terminals in the CT motoneuron area of the nucleus ambiguus. By contrast, low to medium density of 5-HT positive terminals was observed in the CT motoneuron area of the retrofacial nucleus. Although a dense network of 5-HT positive terminals was present in the nucleus ambiguus, most of 5-HT positive varicosities were found to have no contacts with the CT labelled cell bodies of this region (Fig. 4C). In the retrofacial nucleus, there existed a small number of 5-HT terminals that surrounded the cell bodies and proximal dendrites of CT motoneurons. However, the contacts were somewhat loose, as compared to the intimate contacts with PCA motoneurons described earlier.

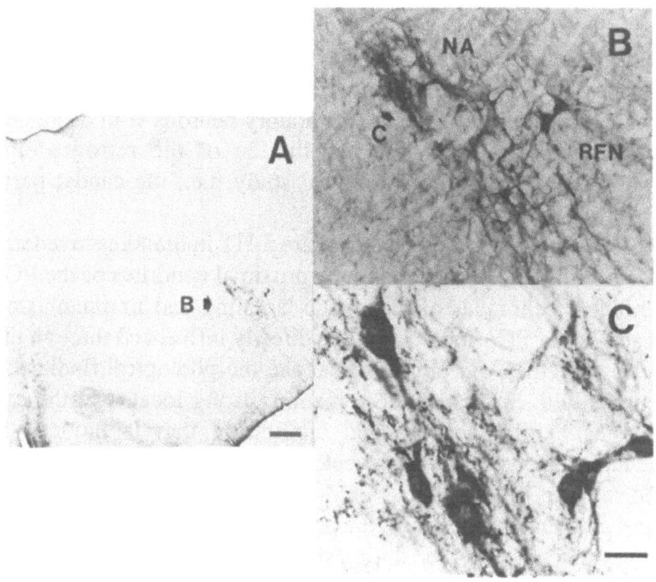

Figure 4. Photomicrographs showing retrogradely labelled CT motoneurons and 5-HT immunoreactive terminals.CT motoneurons are observed in two separate nuclei on this transverse section situated at 3.6 mm rostral to the obex, i.e., dorsally in the nucleus ambiguus (NA) and ventrally in the retrofacial nucleus (RFN). B: high magnification of A. C: higher magnification of B demonstrating 5-HT terminals in the CT motoneuron area of NA. Bars: 500µm for A, and 20µm for C.

The aforementioned data suggested that CT muscle would be innervated from two separate motoneuron pools within the ventrolateral medulla. To evaluate this aspect further, the retrograde labelling of CT motoneurons was performed in combination with surgical removal of superior laryngeal nerve (SLN) or recurrent laryngeal nerve (RLN). In the case of SLN transection, retrogradely labelled CT motoneurons were localized exclusively in the nucleus ambiguus, while in the case of RLN transection, CT labelled motoneurons were restricted to the retrofacial nucleus. These results indicated that SLN contained the axons of

CT motoneurons giving rise to the retrofacial nucleus, while RLN had the efferent fibers from the nucleus ambiguus.

DISCUSSION

The present physiological results have revealed that the direct application of 5-HT produced opposing effects on two types of medullary inspiratory neurons with distinct firing patterns. For the most part 5-HT produced excitatory effects on one group of inspiratory neurons with decrementing firing pattern and, on the contrary, inhibitory effects on the other group of inspiratory neurons with augmenting firing pattern.

Although the firing patterns themselves are not necessarily definite evidence for classification of functional subgroups of inspiratory neurons, there are accumulating data concerning the inspiratory neurons with various firing patterns in relation to synaptic inputs (Richter, 1982), axonal projections (Cohen, 1979) and so on. It is likely that the inspiratory neurons with decrementing firing are vagal inspiratory motoneurons projecting to the abductor muscles of upper airway, on the basis of the following evidence. First, it is well established that the same decrementing firing pattern can be observed in vagal inspiratory motoneurons and the dilator (PCA & CT) muscles of upper airway (Remmers and Bartlett, 1977). Second, the tested region of the nucleus ambiguus corresponds to the area where PCA motoneurons are densely packed, as demonstrated earlier by Davies and Nail (1984). We have confirmed the previous finding in the present morphological study, demonstrating that the regions where the activity of the inspiratory neurons with decrementing firing was recorded, coincided indeed with the distribution of the retrogradely labelled PCA motoneurons observed in this morphological study, i.e., the caudal part of the nucleus ambiguus.

Furthermore, we have revealed that there are 5-HT immunoreactive terminals that make intimate contacts with the cell bodies and the proximal dendrites of the PCA motoneurons, using the combined techniques of retrograde labelling and immunohistochemistry. This indicates that activity of PCA motoneurons is directly influenced through the serotoninergic terminals. Putting together, our physiological and morphological findings indicate that the PCA motoneurons with inspiratory decrementing firing located in the caudal part of the nucleus ambiguus exhibit a prolonged facilitation that is induced directly by the serotoninergic inputs. However, 5-HT receptors responsible for this facilitatory response are not determined so far.

As for the functional implications, the present results may be partly associated with the changes in airway patency observed during sleep, on the following data. First, recent studies concerning sleep apnea syndrome have demonstrated that inspiratory activity of dilator muscles of upper airway is markedly suppressed during REM sleep (Sherrey and Megirian, 1980). Second, serotonin-containing cells in the raphe nuclei show dramatic changes across the sleep-waking cycle, being active during waking, and being suppressed during REM-sleep (Trulson and Jacobs, 1979). Third, the inspiratory neurons projecting to the dilator (PCA) muscle are directly facilitated by application of serotonin, as shown in this study. Therefore, airway obstruction during REM-sleep could be caused by this serotonin-mediated disfacilitation (suppression) of the dilator (PCA) motoneurons along with other factors contributing to obstruction.

On the other hand, the distribution and the mode of 5-HT terminals of CT motoneurons are found to differ significantly from those of PCA motoneurons. Dual innervation of the CT muscle is suggested in this morphological study; one efferent route is the SLN taking its rise in the retrofacial nucleus, while the other route is the RLN originating in the rostral part of the nucleus ambiguus. This may be related to a complexity about physiological and behavioral roles of CT muscle (Woodson et al., 1989). For example, phasic respiratory contraction of CT muscle is observed in inspiration, expiration, or both phases of the cycle. The differences could be attributed to level of anesthesia, state of consciousness, respiratory drive, or routes of breathing. A phasic CT activity in inspiration and expiration might be explained by the presence of two separate populations of motor units (Rudomin, 1966). Moreover, the function of CT muscle is respiratory as well as phonatory. The notion of dual innervation of CT muscle, however, has not been established. Further studies are needed to elucidate this aspect.

The mode of 5-HT terminals in contact with CT motoneurons is found to differ from that with PCA motoneurons; the 5-HT terminal organization is characterized by loose or no contacts with the cell bodies of CT motoneurons, although there exist a dense network of 5-HT terminals in the neighboring area. There is no information concerning the direct application of serotonin in the vicinity of CT motoneuron, but such unique mode of 5-HT terminals implies a different functional role of serotonin from that undertaken by conventional neuronal system, such as "state-dependent" rather than "stimulus-response" function (Moor, 1981). In this connection, McCall and Aghajanian (1979) have demonstrated in the study of the facial motoneurons that the serotoninergic inputs set a basal level of excitability of the motoneurons that is modulated by other transmitter systems.

Finally, it has been reported that some patients with obstructive sleep apnea are successfully treated with drugs that block re-uptake of serotonin (Conway et al., 1982). This suggests that serotonin may indeed be involved in the regulation of upper airway muscle tone. On the other hand, the present study has revealed the heterogeneous modes of 5-HT terminal organization in the motoneurons projecting to the upper airway muscles. This implies diversity of serotoninergic role in the control of upper airway.

REFERENCES

Arita, H., and Ochiishi, M., 1991, Oppsing effects of 5-hydroxytryptamine on two types of medullary inspiratory neurons with distinct firing patterns, J. Neurophysiol. 66:285.

Cohen, M.I., 1979, Neurogenesis of respiratory rhythm in the mammal. Physiol. Rev. 59:1105.

Conway, W.A., Zorick, F., Piccione, P., and Roth, T., 1982, Protriptyline in the treatment of sleep apnoea. Thorax 37:49.

Dahlström, A., and Fuxe, K., 1964, Evidence for the existence of monoamine-containing neurons in the central nervous system, Acta Physiol. Scand. 62:1.

Davies, P.J., and Nail, B.S., 1984, On the location and size of laryngeal motoneurons in the cat and rabbit. J. Comp. Neurol. 230:13.

Gudelsky, G.A., Koenig, J.I., and Meltzer, H.Y., 1986, Thermoregulatory responses to serotonin (5-HT) receptor stimulation in the rat. Neuropharmacolgy 25:1307.

Jouvet, M., 1969, Biogenic amines and the states of sleep. Science 163:32.

Kogo, N., and Arita, H., 1990, In vivo study on medullary H^+-sensitive neurons. J. Appl. Physiol. 69:1408.

Lovick, T.A., 1989, Cardiovascular responses to 5-HT in the ventrolateral medulla of the rat. J. Auton. Nerv. Syst. 28:35.

Moor, R.Y., 1981, The anatomy of central serotonin neuron systems in the rat brain, in:"Serotonin Neurotransmission and Behavior," B.L. Jacobs, A. Gelperin, ed., MIT Press, Cambridge.

Peroutka, S.J., and Snyder, S.H., 1980, Long-term antidepressant treatment decreases spiroperidol-labeled serotonin receptor binding. Science 210:88.

Remmers, D.W., and Bartlett, D.Jr., 1977, Reflex control of expiratory flow and duration., J. Appl. Physiol. 42:80.

Richter, D.W., 1982, Generation and maintenance of the respiratory rhythm. J. Exp. Biol. 100:93.

Rudomin, P., 1966, The electrical activity of the cricothyroid muscles of the cat. Arch Int. Physiol. Biochem. 74:135.

Sherrey, J.H., and Megirian, D., 1980, Respiratory EMG activity of the posterior cricoarytenoid, cricothyroid and diaphragm muscles during sleep. Respir. Physiol. 39:355.

Trulson, M.E., and Jacobs, B.L., 1979, Raphe unit activity in freely moving cats: Correlation with level of behavioral arousal. Brain Res. 106:105.

Woodson, G.E., Sant'Ambrogio, F.B., Mathew, O.P., and Sant'Ambrogio, G., 1989, Effects of cricothyroid muscle contraction on laryngeal resistance and glottic area. Ann. Otol. Rhinol. Laryngol. 98:119.

EFFECTS OF TESTOSTERONE ON HYPOXIC VENTILATORY AND CAROTID BODY NEURAL RESPONSIVENESS

Koichiro Tatsumi, Bernard Hannhart, Cheryl K. Pickett, John V. Weil, and Lorna G. Moore

Cardiovascular Pulmonary Research Laboratory, University of Colorado Health Sciences Center, Denver, CO 80262

INTRODUCTION

Hypoxic ventilatory response (HVR) is known to be influenced by administration of testosterone but prior studies report conflicting results [1,2]. White et al.[1] reported an augmented HVR after testosterone replacement in hypogonadal males, whereas Matsumoto et al.[2] found HVR decreased after testosterone treatment. In addition, the site at which the hormone acts remains unclear. No studies have been undertaken to determine whether testosterone alters carotid body neural responsiveness to hypoxia and, if so, whether it acts directly on the peripheral chemoreceptors and/or on the central nervous system translation of carotid sinus nerve (CSN) output into ventilation.

Accordingly, our purpose was to determine whether testosterone altered resting ventilation and HVR and, if so, whether its effects appeared to be mediated by peripheral (carotid body) or central nervous system influences. We used neutered male cats in which measurements of resting ventilation and HVR in animals could be made before and after treatment while in the awake and anesthetized conditions. In the anesthetized animals, the ventilatory and CSN responses to hypoxia were determined simultaneously and compared with values obtained after placebo treatment to determine whether testosterone raised carotid body sensitivity to hypoxia and/or the central nervous system translation of CSN output into ventilation.

METHODS

A total of 16 neutered male cats were randomly assigned in equal numbers to either placebo or testosterone groups. Body weight was similar in the placebo and testosterone groups and did not change with treatment in either group. All animals were studied before and after one week of treatment both awake and during anesthesia. Testosterone and placebo were administered as controlled-release pellets (Innovative Research of America, Toledo, OH) implanted subcutaneously posterior to the scapula or hip region. Pellets contained either placebo or testosterone (33.3 mg/kg). Serum testosterone levels as measured by radioimmunoassay increased after hormone treatment from values <20 ng/dl to values >200 ng/dl.

The procedures used in this study were described fully in a previous publication [3] and will be outlined here briefly. Two days after a small Teflon button with an indwelling cannula for respiratory gas sampling was permanently implanted in the trachea, the awake animal was placed in a ventilated 21-liter body plethysmograph for measurements of

Control of Breathing and Its Modeling Perspective, Edited by
Y. Honda *et al.*, Plenum Press, New York, 1992

ventilation and respiratory gases. Cats were also anesthetized by the intravenous administration of a mixture of 40 mg/kg body wt chloralose and 200 mg/kg body wt urethan for the respiratory measurements. Given the invasive nature of the CSN recordings, the CSN response to hypoxia could only be measured in anesthetized animals. The CSN was carefully stripped of surrounding tissue. To minimize baroreceptor contribution to CSN activity, we stripped adventitia from the carotid sinus and also crushed the carotid sinus. Carotid body neural output was recorded by platinum bipolar electrodes from the whole desheathed nerve bundle. The amplified signal was filtered (100-3,000 Hz) and processed to produce a measure of amplitude variance of the whole nerve signals. This approach reflects a summation of action potentials proportional to both the activity of independently firing, individual fibers and the number of active fibers. The width of this distribution (amplitude variance) provides a useful index of whole nerve activity [4,5]. The amplitude variance was normalized for the value obtained at $P_{ET}O_2$=200 Torr. The CSN response to hypoxia was measured both with the nerve intact and, to remove descending central neural influences, after proximal transection of one CSN while recording from the distal (carotid body) end.

During room air breathing, minute ventilation and end-tidal O_2 and CO_2 tensions ($P_{ET}O_2$ and $P_{ET}CO_2$) were monitored until values became stable in awake and anesthetized animals. During anesthesia, expired gas was collected for determination of O_2 consumption and CO_2 production. In awake cats, HVR was determined by measuring ventilation at 12-15 $P_{ET}O_2$ values between 200 and 40 Torr. In the anesthetized animals, ventilation was recorded during progressive hypoxia. For both the awake and the anesthetized animals, $P_{ET}CO_2$ was maintained within 2 Torr of the value obtained during room-air breathing by adding CO_2 to the inspired gas.

HVR and CSN response to hypoxia were measured as the shape parameter A. The relationship between $P_{ET}O_2$ and ventilation or CSN activity is hyperbolic and can be described by the equation: V_I (or CSN activity) = $V_o + A / (P_{ET}O_2 - 26)$, where V_I is minute ventilation, V_o is the horizontal asymptote for ventilation or CSN activity, A is a measure of the curvature of the relationship, and the constant 26 is the PO_2-asymptote as determined empirically in previous studies [5]. Ventilatory and CSN responses to hypoxia are of similar shape. When these two responses are measured simultaneously and plotted in relation to each other, a linear relationship (r=0.97) results which describe the "central nervous system translation index" of peripheral chemoreceptor activity into ventilation.

Comparisons were made with and among groups using paired and two sample t-tests, and $P<0.05$ was considered as significant. Values are given as means ± 1 standard error of the mean (SEM).

RESULTS

Testosterone but not placebo treatment raised minute ventilation in the awake (1.6 ± 0.2 to 2.0 ± 0.2 l/min) and anesthetized (0.5 ± 0.03 to 0.7 ± 0.07 l/min) animals. Testosterone treatment raised O_2 consumption (22 ± 1 to 31 ± 3 ml/min) and CO_2 production (16 ± 1 to 23 ± 2 ml/min) in the anesthetized animals, but $P_{ET}CO_2$ was unchanged in either the awake (32 ± 1 to 32 ± 1 Torr) or the anesthetized (36 ± 1 to 35 ± 1 Torr) condition, implying that effective alveolar ventilation did not change. Arterial PCO_2 (Testosterone vs. placebo group: 35 ± 1 vs. 33 ± 2 Torr) and pH (7.35 ± 0.01 vs. 7.34 ± 0.02) were similar after treatment in the 2 groups.

The HVR (shape parameter A) increased 99% after testosterone treatment in the awake cats and 196% in the anesthetized animals (Fig.1). The increase was consistent, occurring in 7 of the 8 awake animals and each of the 8 anesthetized animals. The magnitude of the increase in HVR did not correlate with the change in O_2 consumption (r= - 0.18, P=NS) or serum testosterone (r= - 0.45, P=NS).

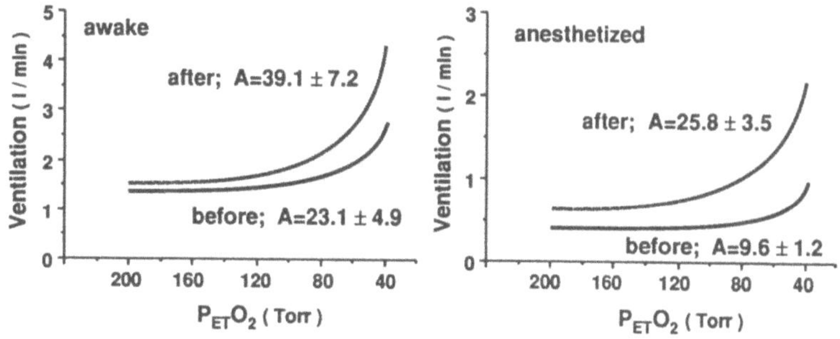

Figure 1. Testosterone treatment raised the ventilatory response to hypoxia shape parameter A (HVR_A) both in awake and anesthetized conditions. Curves represent average values before and after the hormone treatment.

The carotid sinus neural nerve response to hypoxia was greater in the testosterone-treated cats than in the placebo-treated animals. Unilateral carotid sinus nerve section reduced the A value in the testosterone-treated animals but not in the placebo group. After cutting the CSN, the CSN response to hypoxia in the testosterone group was similar to the level present in the placebo-treated animals (Fig.2).

The "central nervous system translation index" was similar in the placebo (1.02 ± 0.12) and testosterone-treated (0.82 ± 0.15) groups, suggesting that the central nervous system translation of carotid body neural output into ventilation was unaffected by testosterone treatment.

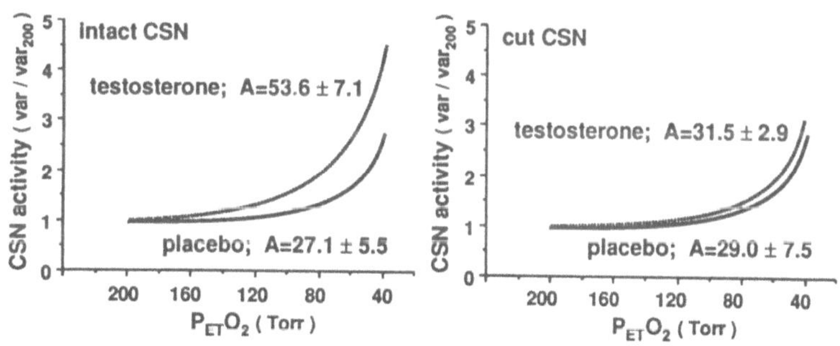

Figure 2. The carotid sinus nerve response to hypoxia shape parameter A (CSN_A) was greater in the anesthetized testosterone-treated compared with the placebo-treated cats while recording from the intact nerve ($P<0.05$). After cutting the carotid sinus nerve and measuring carotid sinus neural activity from the distal end, CSN_A decreased in the testosterone-treated animals to levels that were similar to placebo values.

DISCUSSION

The main finding of this study was that testosterone increased resting ventilation as well as ventilatory and carotid body neural responsiveness to hypoxia in cats. That the increased carotid body neural response observed in the testosterone-treated cats was eliminated by transection of the carotid sinus nerve suggested that the effect of testosterone on the carotid body was due to efferent, central nervous system influences.

Our results confirmed the previous findings of White et al.[1] which indicated that testosterone raised resting ventilation and metabolic rate. The rise in metabolic rate appeared to account for the increase in ventilation, since $P_{ET}CO_2$ did not change with treatment and arterial PCO_2 was similar in the testosterone and placebo groups after treatment.

The increase in HVR with testosterone also supported the previous observations in humans by White et al. [1] but not those of Matsumoto et al.[2]. An increase in metabolic rate has been shown to be an important determinant of HVR [6]. Therefore, the increased metabolic rate may have contributed to the elevation in HVR observed in our study, although, unlike White et al. [1], we failed to find a correlation between the increase in HVR and the change in metabolic rate among individual animals.

That testosterone increased CSN responsiveness to hypoxia suggested that the increase in HVR was the result of increased peripheral chemoreceptor sensitivity. However, since CSN responsiveness to hypoxia in the testosterone-treated animals fell to levels observed in the placebo group after cutting the CSN, it appeared that the increased peripheral chemoreceptor sensitivity was due to influence(s) descending from the central nervous system rather than to direct effects on the carotid body. The descending influence(s) was probably not due to an increased metabolic rate, given previous studies showing that an elevation in metabolic rate produced by passive hindlimb stimulation did not raise CSN responsiveness to hypoxia [7,8]. Efferent central nervous system influences on the carotid body have previously been described but, to date, all the effects identified have been inhibitory [9,10]. Because we can not determine the absolute baseline carotid body neural activity using our whole nerve preparation, we could not rule out the possibility that descending central nervous system influences exerted an inhibitory effect on baseline CSN activity which, once eliminated by CSN transection, resulted in an increase in carotid body neural responsiveness to hypoxia.

In conclusion, our study showed that in cats testosterone raised resting ventilation, metabolic rate, and the ventilatory as well as the CSN responsiveness to hypoxia. The increased HVR was not due to alterations in the central nervous system translation of chemoreceptor activity into ventilation but, rather, may have been the result of efferent central nervous system influences on carotid body. Further studies using hormone receptor blockers are required to determine whether the effects of testosterone were direct, steroid receptor-mediated and, in turn whether such hormone receptors are located in tissues involved in ventilatory control.

REFERENCES

1. D.P. White, B.K. Schneider, R.J. Santen, M. McDermott, C.K. Pickett, C.W. Zwillich, and J.V. Weil. Influence of testosterone on ventilation and chemosensitivity in male subjects. J Appl Physiol 59: 1452-1457 (1985)
2. A.M. Matsumoto, R.E. Sandblom, R.B. Schoene, K.A. Lee, E.C. Giblin, D.J. Pierson, and W.J. Bremner. Testosterone replacement in hypogonadal men: effects on obstructive sleep apnoea, respiratory drives and sleep. Clin Endocrinology 22: 713-721 (1985)
3. K. Tatsumi, C.K. Pickett, and J.V. Weil. Attenuated carotid body hypoxic sensitivity after prolonged hypoxic exposure. J Appl Physiol 70: 748-755 (1991)
4. D.E. Dick, J.R. Meyer, and J.V. Weil. A new approach to quantitation of whole nerve bundle activity. J Appl Physiol 36: 393-397 (1974)
5. M. Vizek, C.K. Pickett, and J.V. Weil. Interindividual variation in hypoxic ventilatory response; potential role of carotid body. J Appl Physiol 63: 1884-1889 (1987)
6. D.P. White, N.J. Douglas, C.K. Pickett, J.V. Weil, and C.W. Zwillich. Sexual influence on the control of breathing. J Appl Physiol 54: 874-879 (1983)
7. D. Aggarwal, H.T. Milhorn Jr., and L.Y. Lee. Role of the carotid chemoreceptors in the hyperpnea of exercise in the cat. Respir Physiol 26: 147-155 (1976)
8. R.Q. Davies, and S. Lahiri. Abscence of carotid chemoreceptor response during hypoxic exercise in the cat. Respir Physiol 18: 92-100 (1973)
9. E. Neil, R.G. O'Regan. Efferent and afferent impulse activity recorded from few fibre preparations of otherwise intact sinus and aortic nerves. J Physiol London 215: 33-47 (1971)
10. S.R. Sampson, and T.J. Biscoe. Efferent control of the carotid body chemoreceptor. Experientia 26: 261-262 (1970)

EFFECT OF THEOPHYLLINE ON BRAIN ACID-BASE STATUS DURING NORMOXIA AND HYPOXIA IN HUMANS

Masaharu Nishimura, Aya Kakinoki, Shuichi Kobayashi, Makoto Yamamoto, Yasushi Akiyama, Kenji Miyamoto, Yoshikazu Kawakami

First Department of Medicine, Hokkaido University Scool of Medicine
Sapporo, 064, Japan

INTRODUCTION

Theophylline, a methylxanthine, is known to stimulate ventilation and augment the ventilatory response to hypoxia in humans by several proposed mechanisms.[1,2] Since theophylline, an adenosine receptor antagonist, has been shown to substantially reduce cerebral blood flow (CBF) in humans,[3,4] as well as in some animal species,[5] these effects may have some relevance to the ventilation-stimulating effect of theophylline. If the acid-base status in the brain is altered as a consequence of changes in CBF, it would affect the activity of the central chemosensitivity and thus the level of ventilation. Since some animal studies have shown that effects of theophylline on CBF are apparent during hypoxia but not during normoxia,[6,7] the effect of theophylline on the PCO_2 and acid-base status in the brain may be changeable during normoxia and hypoxia. One approach by which we can approximately estimate what is occurring in the brain in humans is to measure partial gas pressures and pH in the internal jugular vein. With a technique for sampling blood from the internal jugular vein, we have previously shown that PCO_2 in the internal jugular vein significantly decreases, by 2 to 3 mmHg, under isocapnic hypoxia,[8,9] which probably reflects hypoxia-induced cerebral vasodilation. The aims of this study are first to evaluate the effect of theophylline on PCO_2 and pH in the internal jugular vein for a wide range of arterial PO_2 (PaO_2) under isocapnic conditions and second to examine the possibility that the augmented ventilatory response to hypoxia caused by theophylline is due to attenuation of hypoxia-induced cerebral vasodilation.

METHODS

Six healthy male volunteers aged 22 ± SD2 years participated in this study. All the subjects were given written information about the purpose and risks of the study and gave informed consent. The protocol of the study was approved by the Ethics Committee of the Hokkaido University School of Medicine. Before the study, catheters (Angiocath) were placed into the left bracheal artery and the right internal jugular vein under local anesthesia. Details of procedures were described in our previous publication.[10] Blood gases and pH were analyzed soon after sampling with a blood-gas analyzer (Type 1303, Instrumentation Laboratory, Lexington, MA).

The subjects were instructed to refrain from coffee, tea, or caffeinated beverages from the day before the experiment. They were placed in the supine position throughout the study and allowed to breathe spontaneously through a mouthpiece connected to a J-valve throughout each trial described below. PaO_2 and arterial PCO_2 ($PaCO_2$) were independently controlled by a servo-control system for arterial blood gases developed in our laboratory.[11] The system utilized end-tidal PO_2 ($P_{ET}O_2$) and PCO_2 ($P_{ET}CO_2$) to regulate PaO_2 and $PaCO_2$ automatically and simultaneously by changing inspiratory gas composition. Respiratory gases were monitored by a mass spectrometer (Medical Gas Analyzer MGA-1100; Perkin-Elmer Medical Instruments, Pomona, CA). Arterial oxygen saturation and heart rate were monitored with a pulse oximeter (Biox3700; Ohmeda, Boulder, CO) applied to the finger. Data were recorded on a multichannel recorder and stored at 15-sec intervals in an on-line signal processing computer (Atac-450; Nihon Kohden, Tokyo, Japan) for later analysis.

After the subject showed stable breathing in room air for at least 15 min, arterial and jugular venous blood samples were taken at the same time. $P_{ET}O_2$ controlled by the servo-control system was first gradually lowered from the baseline level to 60-65 mmHg over 5 min to get the monitored SaO_2 to around 90% (mild hypoxia) and sustained there for the next 5 min. Inspiratory oxygen concentration was then lowered to attain severe hypoxia ($P_{ET}O_2$=45 mmHg, SaO_2=80 mmHg) and sustained as well for at least 5 min. Blood samples were collected again at the end of each period of sustained hypoxia. During the hypoxic challenge, an attempt was made to maintain $P_{ET}CO_2$ at the baseline level in each subject while breathing room air. Following this, after an interval of 30 min, 6 mg/kg of aminophylline, which is a theophylline derivative, a complex of two molecules of theophylline and the simple diamine, ethylenediamine, used to increase solubility, was intravenously infused over 10 min. The blood gas measurements were made again in room air 15 min after the termination of aminophylline infusion. Then an identical protocol of hypoxic challenge was repeated for comparison. Plasma levels of theophylline were measured in all the subjects at the end of the second trial to ascertain that clinically relevant levels (10 - 20 μg/ml) were indeed achieved with the dose given in this study.

Changes of variables in three series of blood gas measurements in the hypoxia study were assessed using analysis of variance (ANOVA) for repeated measurements. Mean values in the control and aminophylline trials were compared using a paired Student's t test. P values of < 0.05 were accepted as significant. All the data are expressed as mean ± SE.

RESULTS

The data from control and aminophylline trials are shown in Table 1. While subjects were breathing room air, $PaCO_2$ was slightly, although not significantly, lower

Table. 1 Mean values of blood gas data and pH in an artery and the internal jugular vein before and after aminophylline infusion in normal volunteers (n=6)

	Room Air		Mild Hypoxia		Severe Hypoxia	
	Cont	AMN	Cont	AMN	Cont	AMN
PaO_2, mmHg	104.0	106.4	62.5	63.7	45.8	44.5
	(2.1)	(3.7)	(1.8)	(1.5)	(1.2)	(1.6)
SaO_2, %	96.6	96.5	91.2	91.1	81.3	80.0
	(0.4)	(0.3)	(0.6)	(0.4)	(1.1)	(0.9)
$PaCO_2$, mmHg	42.0	39.9	42.2	40.5 *	41.9	40.2
	(1.2)	(1.0)	(0.7)	(1.0)	(1.2)	(0.8)
$PjCO_2$, mmHg	52.3	52.3	51.5	51.2	49.5 §	50.0 §
	(1.4)	(1.0)	(1.4)	(1.5)	(1.5)	(0.6)
pHa	7.39	7.40	7.39	7.41	7.39	7.42 *
	(0.01)	(0.01)	(0.01)	(0.01)	(0.01)	(0.01)
pHj	7.34	7.34	7.34	7.34	7.36 §	7.36 §
	(0.01)	(0.01)	(0.01)	(0.02)	(0.01)	(0.01)

Numbers in parenthesis are SE values. Cont: Data from the control trial, AMN: from the aminophylline trial.

*: $p < 0.05$ compared to the value in the control trial

§: $p < 0.01$ compared to the value in room air in each study with aminophylline, which probably reflected augmented ventilation. Since $P_{ET}CO_2$ during hypoxic challenge was well controlled to the baseline value in room air in each trial, the $PaCO_2$ in aminophylline trials was again slightly lower than that of control trials, by 1.7 mmHg both in mild and severe hypoxia. The difference in mild hypoxia alone reached statistical significance. As a reflection of changes in $PaCO_2$, arterial pH (pHa) was consistently lower in control trials than that in aminophylline trials at all levels of PaO_2, although statistical significance was found only in severe hypoxia.

In contrast to $PaCO_2$ and pHa, PCO_2 and pH in the internal jugular vein ($PjCO_2$, pHj) were the same in the control and aminophylline trials (Table 1, Figure 1). In addition, $PjCO_2$ was significantly decreased from the baseline levels when subjects were exposed to severe hypoxia in both trials (by 2.8 mmHg in control trials, $p<0.01$, and by 2.3 mmHg in aminophylline trials, $p<0.01$), although $PaCO_2$ was well maintained at the baseline values in room air in both trials (Figure 1). Consequently, despite the similar levels of pHa during normoxia and hypoxia, pHj showed a significant shift from 7.34 in room air to 7.36 in severe hypoxia ($p<0.01$) in both trials (Table 1).

The mean plasma concentration of theophylline at the end of the second hypoxic trial was 14.3 ± 2.9 μg/ml.

DISCUSSIONS

This study has reconfirmed our previous findings[8,9] that $PjCO_2$ significantly decreases under isocapnic hypoxia in humans. This is probably the result of hypoxia-induced cerebral vasodilation, since the oxygen consumption is believed to remain unchanged at the level of hypoxia (PaO_2=45 mmHg)[12] used in this study, so that an increase in CBF would enhance washout of CO_2 from the tissue, thus causing a decrease in the tissue PCO_2. However, the decrease in $PjCO_2$ in response to isocapnic hypoxia was not significantly affected by theophylline (as aminophylline) infusion in this study, as is clearly shown in Figure 1. This indicates that theophylline at clinically relevant plasma concentrations does not seem to have any appreciable effect on hypoxia-induced cerebral vasodilation. However, the negative effect on hypoxia-induced cerebral vasodilation does not mean a negative effect of theophylline on CBF and the acid-base status in the brain. It

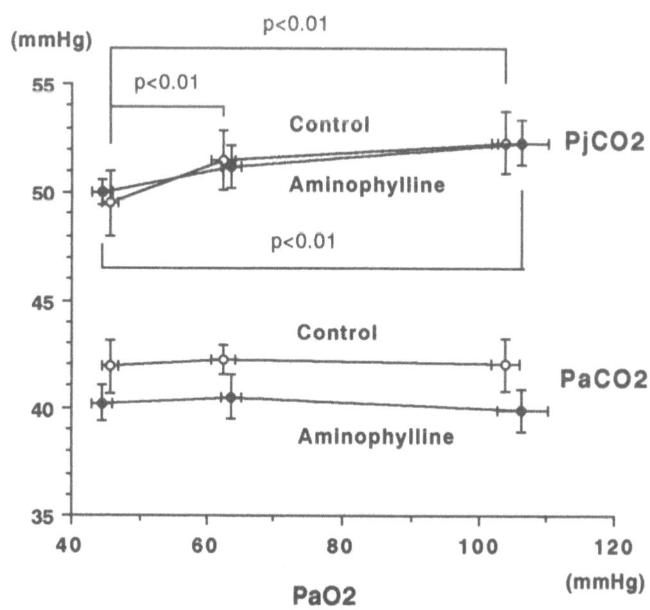

Figure 1. PCO_2 in an artery ($PaCO_2$) and in the internal jugular vein ($PjCO_2$) as a function of arterial PCO_2 (PaO_2) in control and aminophylline trials.

must be noted that the difference in PCO_2 between the artery and the internal jugular vein was apparently changed to the similar degree at all levels of PaO_2 after theophylline infusion. This probably means that theophylline in the dose given in this study decreases CBF to similar extent for a wide range of PaO_2 in humans. These results agree with the recent report in normal volunteers that aminophylline decreased CBF measured with a ^{133}Xe clearance method during both normoxia and hypoxia, but did not prevent the increase in CBF accompanying hypoxia.[13]

The pH in the artery and the internal jugular vein seemed to behave as a reflection of changes in PCO_2. Although pHa tended to more alkaline with theophylline at all levels of PaO_2 (statistical significance only in severe hypoxia), pHj was the same before and after theophylline infusion and significantly changed from the 7.34 in room air to 7.36 during severe hypoxia in both trials. Based on the assumption that $PjCO_2$ and pHj are parameters reflecting or at least influencing those of brain tissue, the significant changes in $PjCO_2$ and pHj observed during severe isocapnic hypoxia would certainly serve to modify the level of ventilation. However, as already stated, we found no significant effect of theophylline on these parameters during normoxia and hypoxia. These results are in close agreement with those of Eldridge et al[2] and Javaheli et al[14] in animal preparations. Their studies similarly showed that theophylline increased ventilation but that it was not mediated by changes in the medullary extracellular fluid PCO_2 or pH.

In summary, we reconfirm that PCO_2 and pH in the internal jugular vein significantly changed under isocapnic hypoxia, probably as a consequence of increased CBF. However, these parameters were not affected by theophylline either during normoxia or hypoxia. We therefore conclude that the ventilation-stimulating effect of theophylline reported in humans is not ikely to be mediated by changes in brain acid-base status.

REFERENCES

1. S. Lakshminarayan, S.A. Sahn, and J.V. Weil, Effect of aminophylline on ventilatory responses in normal man, Am. Rev. Respir. Dis. 117: 33 (1978).
2. F.L. Eldridge, D.E. Millhorn, T.G. Waldrop, and J.P. Kiley, Mechanism of respiratory effects of methylxanthines, Respir. Physiol. 53: 239 (1983).
3. R.L. Wechsler, L.M. Kleiss, and S.S. Kety, The effects of intravenously administered aminophylline on cerebral circulation and metabolism in man, J. Clin. Invest. 29: 28 (1950).
4. U. Gottstein and O.B. Paulson, The effect of intracarotid aminophylline infusion on the cerebral circulation, Stroke 3: 560 (1972).
5. S. Morii, A.C. Ngai, K.R. Ko, and H.R. Winn, Role of adenosine in regulation of cerebral blood flow: effects of theophylline during normoxia and hypoxia, Am. J. Physiol. 253: H165 (1987).
6. W.E. Hoffman, R.F. Albrecht, and D.J. Miletich, The role of adenosine in CBF increases during hypoxia in young and aged rats, Stroke 15: 124 (1984).
7. T.E. Emerson Jr and R.M. Raymond, Involvement of adenosine in cerebral hypoxic hyperemia in the dog, Am. J. Physiol. 241: H134 (1981).
8. M. Nishimura, A. Suzuki, Y. Nishiura, H. Yamamoto, K. Miyamoto, F. Kishi, and Y. Kawakami, Effect of brain blood flow on hypoxic ventilatory response in humans, J. Appl. Physiol. 63: 1100 (1987).
9. A. Suzuki, M. Nishimura, H. Yamamoto, K. Miyamoto, F. Kishi, and Y. Kawakami, No effect of brain blood flow on ventilatory depression during sustained hypoxia, J. Appl. Physiol. 66: 1674 (1989).
10. M. Nishimura, A. Suzuki, A. Yoshioka, M. Yamamoto, Y. Akiyama, K. Miyamoto, F. Kishi, and Y. Kawakami, Effect of aminophylline on brain tissue oxygenation in patients with chronic obstructive pulmonary disease, Thorax: in print (1992).
11. Y. Kawakami, T. Yoshikawa, Y. Asanuma, and M. Murao, A control system for arterial blood gases, J. Appl. Physiol. 50: 1362 (1981).
12. P.J. Cohen, S.C. Alexander, T.C. Smith, M. Reivich, and H. Wollman, Effects of hypoxia and normocarbia on cerebral blood flow and metabolism in conscious man, J. Appl. Physiol. 23: 183 (1967).
13. D.L. Bowton, W.S. Haddon, D.S. Prough, N. Adair, P.T. Alford, and D.A. Stump, Theophylline effect on the cerebral blood flow response to hypoxemia, Chest 94: 371 (1988).
14. S. Javaheri, J.A.M. Evers, and L.J. Teppema, Increase in ventilation caused by aminophylline in the absence of changes in ventral medullary extracellular fluid pH and carbon dioxide tension, Thorax 44: 121 (1989).

ASYMMETRICAL TRANSIENTS OF CARDIORESPIRATORY VARIABLES IN RESPONSE TO ASCENDING AND DESCENDING RAMP FORCINGS OF EXERCISE LOAD

Yoshimi Miyamoto and Kyuichi Niizeki

Department of Electrical and Information Engineering
Faculty of Engineering, Yamagata University
Yonezawa 992, Japan

INTRODUCTION

Although the response kinetics of ventilatory and gas exchange variables to incremental ramp exercise have been studied intensively in connection with the development of non-invasive techniques for detecting the AT,[1,3] analysis of responses to decremental ramp forcings is scarce except for the early work of Karlsson and Wigertz.[4] They found that with decreasing ramp slope, in both the incremental and decremental phases of the ramp exercise, the lag of the ventilation and heart rate responses behind the input stimulus became increasingly prolonged.

This paper reviews our recent work concerning the kinetics of cardiorespiratory variables during moderate ramp exercise.[5,6] Contrary to the observations of Karlsson and Wigertz,[4] we found significant asymmetry between the transient responses to the incremental and decremental phases of ramp exercise.

METHODS

Cardiac output ($\dot{Q}$) was determined by an automated measuring system based on impedance plethysmography.[7] Transthoracic impedance was measured by a constant current type impedance cardiograph (Nihon kohden, model RGA-5) and electrocardiographs (ECG) were monitored from the chest wall with bipolar electrodes to obtain heart rate (HR). The subject was allowed to breathe spontaneously during all runs. In order to eliminate the respiratory and motion artifacts from the impedance signals, an ensemble averaging technique was adopted for 10 to 20 cardiac cycles using the ECG's R wave as a trigger pulse.

Minute ventilation ($\dot{V}E$), end-tidal pressures of O_2 and CO_2 ($PETO_2$ and $PETCO_2$), O_2 uptake ($\dot{V}O_2$), CO_2 output ($\dot{V}CO_2$), and the gas exchange ratio (R) were obtained breath-by-breath.[6,8] Airflow was measured at the mouth by a hot-wire type pneumotachograph (Minato, model RF-2). The composition of expired air was continuously analyzed.

Healthy young male volunteers of average body size were studied as subjects after informed consent. No trained athlete was included in the subject group. The subject pedaled a bicycle ergometer (Lode) in the upright position at a constant rate of 60 rpm throughout all experiments. The electro-magnetic braking system of the ergometer was controlled by externally-supplied current to obtain the desired types of forcing.

The individual response variables together with timing signals for each subject were first stored in a memory disk. Group mean values for all experiments of each type were then obtained on a time-weighted basis at 5 s intervals with standard errors. Further data analyses were performed on the superimposed averaged responses of total runs.

RESULTS

The group mean responses of $\dot{V}E$ to a step forcing and three ramp forcings with different slopes are shown in Fig.1. The baseline and stimulus exercise levels were 25 and 125 W, respectively. Immediately after the step in work load, $\dot{V}E$ responded with a small abrupt rise

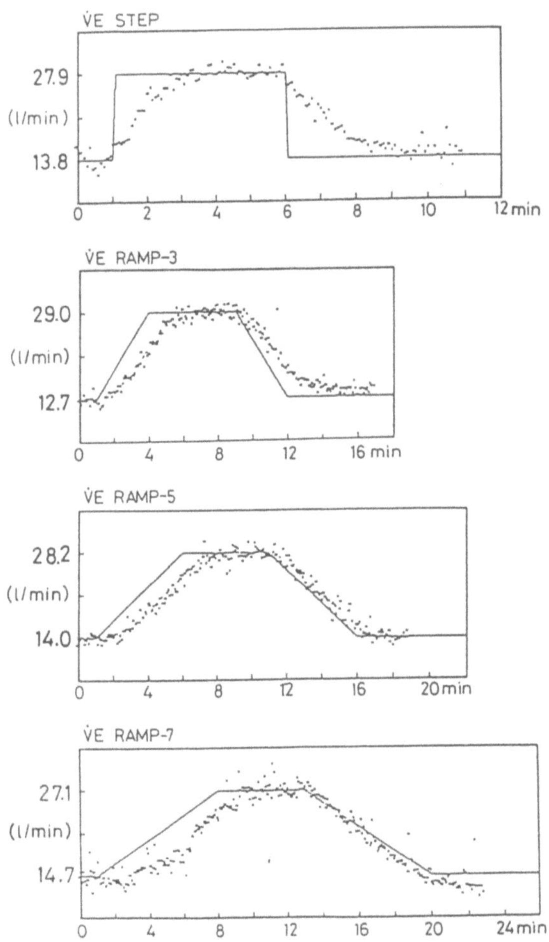

Figure 1. The responses of $\dot{V}E$ to a step forcing and ramp forcings with three different slopes (RAMP-3, 33.3 W/min; RAMP-5, 20W/min; and RAMP-7, 14.3 W/min). Points are group mean values obtained from 5 subjects. Solid lines denote the input forcing patterns (25-125 W).

(phase 1), followed by an increase (phase 2) towards the new steady-state (phase 3) that fitted well to an exponential model. Although the phase 1 response during off-transients was less prominent, the time lag between the off-response and input forcing function was similar to that of the on-response.

$\dot{V}E$ responded to incremental ramp forcings after a time delay which became progressively longer with decreasing ramp slope. On the contrary, the time delay for the decremental ramp became shortened as the slope became less steep, making the asymmetry between the incremental and decremental responses greater with decreasing ramp slope. A similar asymmetry was also observed in $\dot{V}O_2$, $\dot{V}CO_2$, HR and $\dot{Q}$.

To quantitatively describe the delayed responses of variables, the transient responses were fitted to mathematical models. The step responses fitted well to a second order exponential model consisting of two first-order components connected in parallel, each with different time constants and one with, and one without, a pure time delay. The ramp responses were fitted to a first order exponential both with or without a time delay. The mean response time (MRT), defined as the algebraic sum of the time constant and pure time delay, of variables to

Table 1. Mean response times (MRT, in sec) of variables to step and ramp forcings. Values are based on the group mean responses of 5 subjects.

	STEP		RAMP-3		RAMP-5		RAMP-7	
Variables	on	off	on	off	on	off	on	off
$\dot{V}E$	56.5	73.8	83.0	60.1	98.0	56.4	136.6	<5
$\dot{V}O_2$	34.3	39.7	51.6	23.9	45.4	<5	69.1	<5
$\dot{V}CO_2$	45.8	75.9	75.5	46.4	82.9	24.1	122.4	<5
$\dot{Q}$	26.4	39.9	13.3	<5	56.4	<5	61.1	<5
HR	40.0	40.7	79.7	26.3	97.2	30.3	115.4	25.3

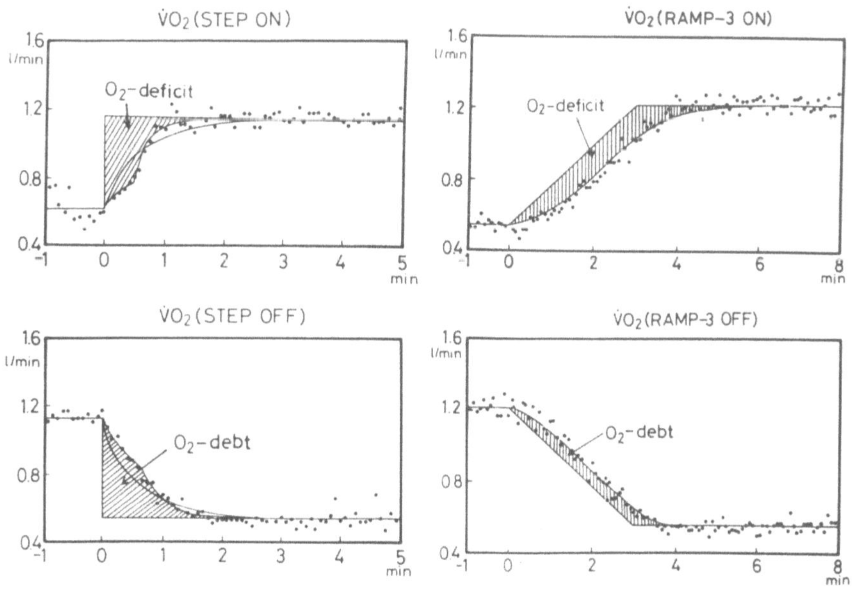

Figure 2. Curve fitting the responses of $\dot{V}O_2$ to a step and a ramp forcing pattern with a slope of 33.3 W/min by a first order (ramp) and a second order (step) model. Points are group mean values obtained from 5 subjects. Shaded areas denote the O_2 deficit and debt, respectively.

step and ramp forcings are summarized in Table 1. It is clear that the MRTs for incremental ramps are prolonged with decreasing ramp slope, while shortened for decremental ramps. The dependency of the MRTs on ramp slope is least dominant in the case of HR.

The asymmetrical ramp responses of $\dot{V}O_2$ produces a prominent discrepancy between the O_2 deficit arising during the incremental phase, and the O_2 debt during the decremental phase as shown in Fig. 2. Fig.3 shows several gas exchange parameters during the off-transitions of step and ramp exercise. At the end of step exercise, $PETO_2$ increased transiently while $PETCO_2$ decreased below the initial base level. This might be caused by hyperpnea during the period, since excessive ventilation relative to $\dot{V}O_2$ was also indicated by an overshoot of $\dot{V}E/\dot{V}O_2$. The post-exercise ventilation is matched to oxygen demand when load intensity decreases gradually. At the end of ramp exercise, these parameters return more quickly to their initial base lines as the ramp slope becomes less steep.

Asymmetrical characteristics of cardiorespiratory dynamics were also observed in response to cyclic triangular ramp exercise[5] (0-100 W) as shown in Fig. 4. In the first cycle, the time delays for the descending responses are shorter than those for the ascending responses. The delays for the ascending phase tend to prolong with decreasing ramp slope. However, the difference between the ascending and descending delays become insignificant during the second and third cycles. The same was true for $\dot{V}O_2$ and $\dot{V}CO_2$. The MRTs determined by

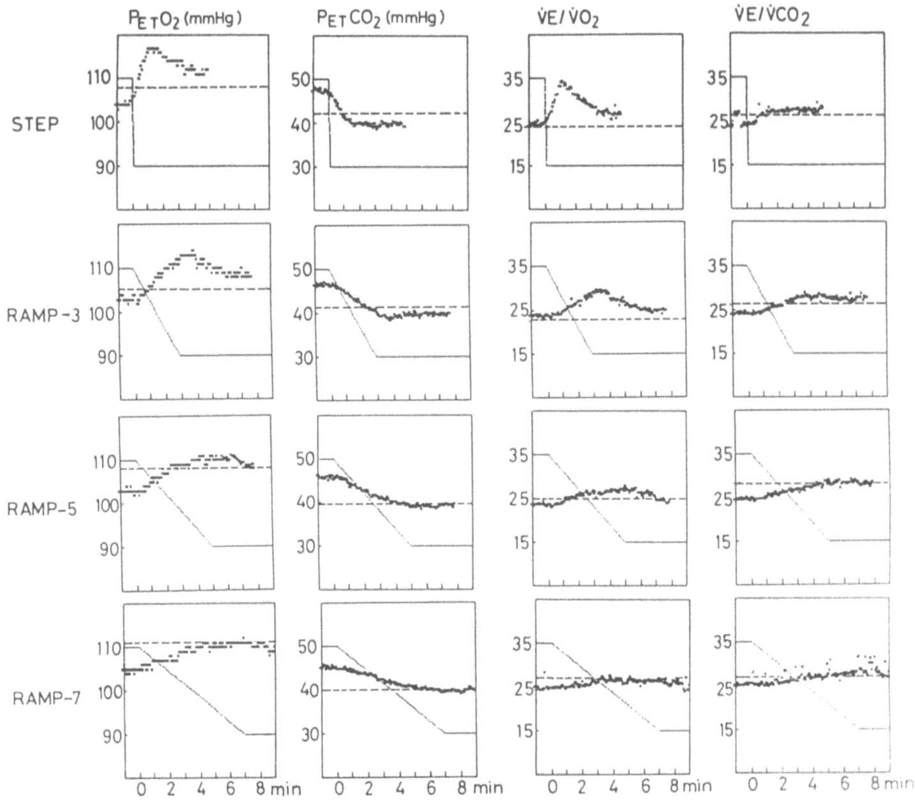

Figure 3. The $PETO_2$, $PETCO_2$, $\dot{V}E/\dot{V}O_2$ and $\dot{V}E/\dot{V}CO_2$ responses during the off-transient of a step forcing and the decremental phases of ramp forcings with three different slopes (RAMP-3, 33.3 W/min; RAMP-5, 20W/min and RAMP-7, 14.3 W/min). Points are group mean values obtained from 5 subjects. Dotted lines denote the initial control levels.

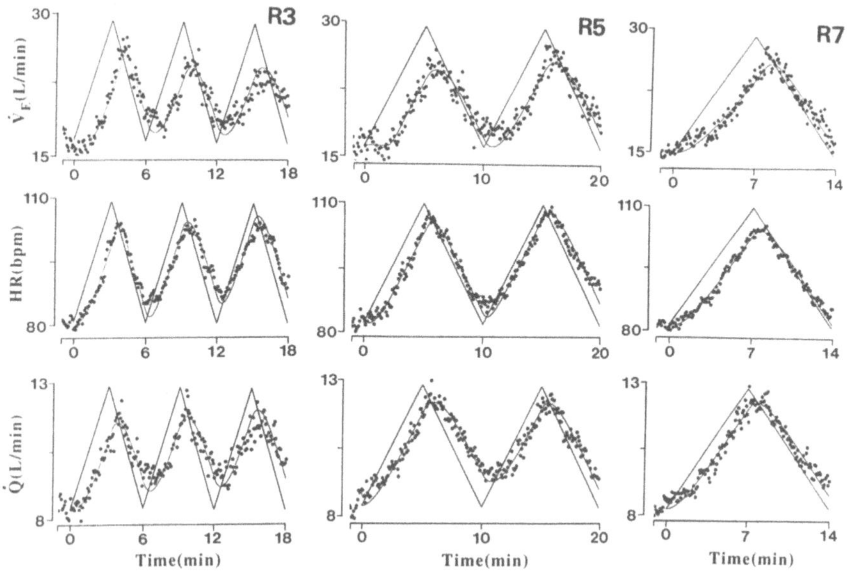

Figure 4a. The responses of $\dot{V}E$, HR and $\dot{Q}$ to cyclic ramp exercise (0-100 W). Points are averages for 6 subjects.

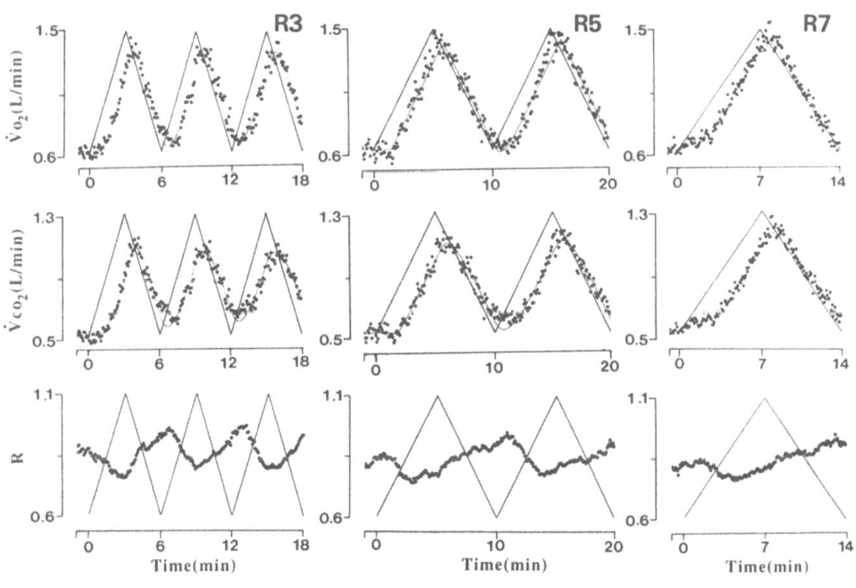

Figure 4b. The responses of $\dot{V}O_2$, $\dot{V}CO_2$ and R to cyclic ramp exercise (0-100 W). Points are averages for 6 subjects.

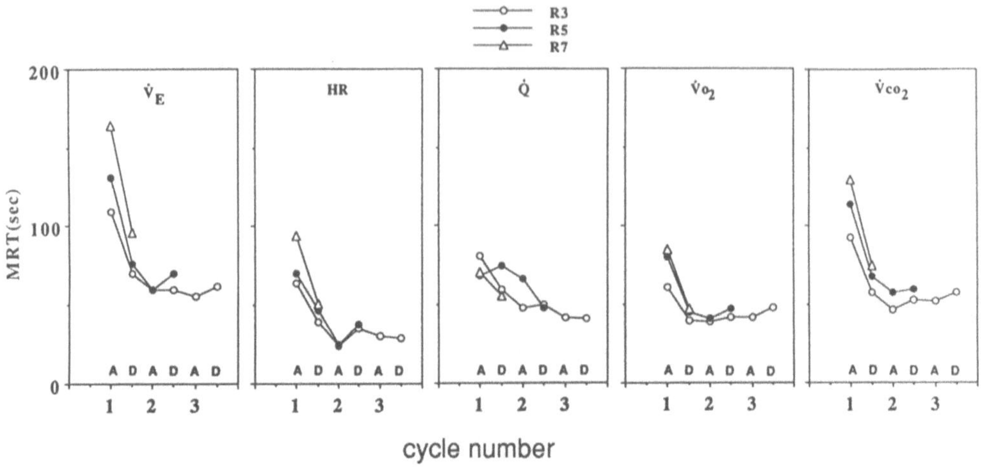

Figure 5. The relationships between the cycle number of triangular forcings and mean response time (MRT). A and D denote ascending and descending phases, respectively.

curve fitting analysis progressively decrease with subsequent cycles and attain respective asymptotical values of about 60 s for $\dot{V}E$, 30 s for HR, 40 s for $\dot{Q}$, 45 s for $\dot{V}O_2$, and 55 s for $\dot{V}CO_2$ (Fig. 5). These values are similar to the time constants determined using a sinusoidal forcing in work load.[8]

DISCUSSION

In order to examine whether anaerobic metabolism is responsible for the asymmetrical ramp responses observed in the present study, we repeated another series of similar ramp studies in a different subject group using a reduced work load range[6] of 0 W to 75 W. This provided similar results to those of the 25-125 W study. Thus, if anaerobic metabolism was present in the first study, it would appear to have little significance on the results.

The asymmetrical ramp responses of $\dot{V}O_2$ bring about a greater O_2 deficit than O_2 debt (Fig.2). We found, however, that gas exchange homeostasis had been achieved immediately after the end of decremental ramp exercise, despite the fact that most of the O_2 debt remained without repayment. The energy required for muscular contraction during the on-transient of exercise is considered to be supplied partially by anaerobic metabolism. However, the blood lactate level does not elevate significantly when the work rate remains below the AT. Thus, the accumulation of lactate can be ruled out as the origin of the O_2 debt, at least in this study. The cause of the O_2 debt is still unidentified although the release of several substances accompanying muscular contraction, such as fatty acids, norepinephrine, thyroxine, calcium and potassium ions, are thought to be required,[9] because an increase in these substances requires excess O_2 consumption by loosening the coupling between oxidation and phosphorylation (ATP production). We therefore propose that changes in the turnover rate of these substances play a role in the present finding. The rate of production of these substances would depend on the velocity or acceleration of muscular activity, which is high for a rapid contraction and low for a slow contraction. On the other hand, their rate of removal would be determined by cell metabolism, requiring more or less a constant delay. When their production rate is slow, it matches their removal rate, and thereby the net concentration remains constant and little O_2 debt is required at the off-transient of exercise, as observed in the present ramp responses.

Conversely, when their production rate is high, the removal rate is unable to keep pace, which results in more O_2 debt.

Wasserman and associates[10] have hypothesized that ventilation during exercise is regulated by CO_2 flow into the lungs, although the mechanism linking the two variables is still unclear. If this hypothesis holds true, the kinetics of ventilation should follow those of gas exchange. O_2 consumption in the muscles produces a proportional output of CO_2 but with delayed dynamics due to a high absorption rate of CO_2 into body tissues. Thus, the present ramp study may offer supplemental support to the cardiodynamic hypothesis.

However, the above explanation based on the cardiodynamic process does not explain the asymmetrical ramp response of the circulatory variables. Since the asymmetry between incremental and decremental ramp responses is observed in both ventilatory and circulatory variables, it is suspected that there may be another factor which acts in parallel with these variables. This assumes pure neural mechanisms, i.e. afferent inputs from the muscular reflex,[11] or descending inputs from the central nervous system to the respiratory and circulatory center.[12] It has been reported that medullary respiratory neurons have the property of slow potentiation.[13] It is therefore possible that the asymmetrical ramp responses of circulatory and ventilatory variables might depend on the intrinsic property of neurons in the brain stem. Observations of the responses to cyclic ramp exercise, which show considerable on-off asymmetry in only the first cycle and insignificant asymmetry in subsequent ramp cycles, might be explained by the adaptation of neural activity.

Assuming humoral factors are involved, there is an alternative explanation for the asymmetrical ramp responses. Given that changes in cardiorespiratory variables during exercise are triggered by some metabolic substances produced by muscular contraction, the time delay for the response should depend on both the transport lag of that substance traveling from active muscles to its receptor site and also the wash-in time required for its concentration to reach a threshold. In this case the transport lag and the wash-in or wash-out time would depend on the anatomical structure of blood vessels, the tissue volume surrounding the receptor and tissue blood flow. Therefore, the total time delay would be longer at the beginning of the incremental phase of ramp exercise because blood flow remains at its base level, while the total time delays would be shorter at the beginning of the decremental phase because blood flow is elevated.

Plasma catecholamine and CO_2 may be considered to be specific humoral stimuli responsible for the asymmetry of circulatory and ventilatory responses. Although arterial PCO_2 during the steady state of exercise does not increase significantly above its resting level, differences in the dynamics of $\dot{V}E$, $\dot{V}O_2$ and $\dot{V}CO_2$ would bring about a transient rise in arterial PCO_2 and a transient fall in PO_2 during the ascending phase of exercise as predicted by the model simulation of Lamarra and associates.[14] Whipp and co-workers[15] also reported that end-tidal and arterial PCO_2 increased progressively in response to incremental ramps below the AT. It is possible that these changes can be sensed by the carotid bodies at least in humans. Jacobi et al.[16] have also claimed that the CO_2 sensitivity of the respiratory center may increase during exercise near the set point. An accelerating effect of Plasma K^+ on the carotid body activities[17] seems to be important since potassium is also supposed to play a role in evoking the O_2 debt.[9]

Overall, the possible factors which might be responsible for the asymmetrical ramp responses are summarized as follows: (1) Gas exchange asymmetry as indicated by the significant discrepancy between the O_2 deficit and debt; (2) slow potentiation of the neural system to drive ventilation and circulation; and (3) non-linear behavior of the wash-in and wash-out process of some humoral stimulus substances originating in the blood flow kinetics.

The authors thank Dr. Michael Mussell for the revision of the English of the manuscript.

REFERENCES

1. K. Wasserman, B.J. Whipp, S.N. Koyal and W.L. Beaver, Anaerobic threshold and respiratory gas exchange during exercise, J.Appl.Physiol. 35:236 (1973).

2. R.L. Hughson and M.D. Inman, Oxygen uptake kinetics from ramp work tests: variability of single test values, J.Appl.Physiol. 61:373 (1986).

3. G.D. Swanson and R.L. Hughson, On the modeling and interpretation of oxygen uptake kinetics from ramp work tests, J.Appl.Physiol. 65:2453 (1988).

4. H. Karlsson and O. Wigertz, Ventilation and heart-rate responses to ramp-function changes in work load, Acta Physiol. Scand. 81:215 (1970).

5. K. Niizeki and Y. Miyamoto, Cardiorespiratory responses to cyclic triangular ramp forcings in work load, Jpn J.Physiol. 41:759 (1991).

6. Y. Miyamoto, K. Niizeki and Y. Nakazono, Responses of ventilation, circulation and gas exchange to incremental and decremental ramp forcings of exercise load, J.Appl.Physiol. (in press).

7. Y. Miyamoto, M. Takahashi, T. Tamura, T. Nakamura, T. Hiura and T. Mikami, Continuous determination of cardiac output during exercise by the use of impedance plethysmography, Med.Biol.Eng.Comput. 19:638 (1981).

8. Y. Miyamoto, Y. Nakazono, T. Hiura and Y. Abe, Cardiorespiratory dynamics during sinusoidal and impulse exercise in man. Jpn J. Physiol. 33:971 (1983).

9. G.A. Brooks and T.D. Fahey, Exercise physiology: "Human bioenergetics and its applications," John Wiley & Sons, New York (1984).

10. K. Wasserman, B.J. Whipp and R.Casaburi, Respiratory control during exercise, In:"Handbook of Physiology,". Sec.3, The respiratory system, Vol.II, Control of breathing, Part 2, N.S.Cherniack and J.G.Widdicombe, ed., Am.Physiol.Soc. Bethesda pp.595 (1986).

11. U. Tibes. Reflex inputs to the cardiovascular and respiratory centers from dynamically working canine muscles, Circ.Res. 41:332 (1977).

12. F.L. Eldridge, D.E. Millhorn, J.P. Kiley and T.G. Waldrop, Stimulation by central command of locomotion, respiration and circulation during exercise, Respir. Physiol. 59:313 (1985).

13. F.L. Eldridge and P. Gill-Kumar, Central neural respiratory drive and afterdischarge, Respir.Physiol. 40:49 (1980).

14. N. Lamarra, S.A. Ward and B.J. Whipp, Model implications of gas exchange dynamics on blood gasses in incremental exercise, J. Appl. Physiol. 66: 1539 (1989).

15. B.J. Whipp, J.A. Davis and K. Wasserman, Ventilatory control of the 'isocapnic buffering' region in rapidly-incremental exercise, Respir. Physiol. 76:357 (1989).

16. M.S. Jacobi, V.I. Iyawe, C.P. Patil, A.R.C. Cummin and K.B. Saunders, Ventilatory responses to inhaled carbon dioxide at rest and during exercise in man, Clinic. Sci. 73:177 (1987).

17. D.M. Band and R.A.F. Linton, The effect of hypoxia on the response of the carotid body chemoreceptor to potassium in the anesthetized cat, Respir. Physiol. 72:295 (1988).

DYNAMIC ASYMMETRIES OF VENTILATION AND PULMONARY GAS EXCHANGE DURING ON- AND OFF-TRANSIENTS OF HEAVY EXERCISE IN HUMANS

Brian J. Whipp[1], Susan A. Ward[2] and D.A. Paterson[3]

[1]Department of Physiology
UCLA
Los Angeles, CA 90024, U.S.A.
[2]Department of Anesthesiology
UCLA
Los Angeles, CA 90024, U.S.A.
[3]Department of Physical Education
University of Western Ontario
Canada

INTRODUCTION

Inferences for the physiological control mechanisms which couple: (a) tissue O_2 and CO_2 exchange to muscular force generation and also (b) pulmonary gas exchange to tissue gas exchange may be drawn from a precise breath-by-breath characterization of the ventilatory and pulmonary gas exchange response transients to appropriately-selected work-rate ($\dot{W}$) forcings.[1,2,3] As the components of these characterizations, in terms of delays (δ), time constants (τ) and gains (G), reflect the underlying physiological processes, this allows a physiological control model to be assembled. However, mathematical features of the model (i.e., the 'what' of parametrization) need to have as their frame of reference known, or hypothesized, physiological structures involved in the putative control scheme (i.e., the 'so what' of model formulation).

The ventilatory and pulmonary gas exchange responses to muscular exercise have been extensively parametrized, and physiological model equivalents established.[1,4] At work rates below the lactate threshold (θL), ventilatory ($\dot{V}$E) and pulmonary gas exchange ($\dot{V}CO_2$, $\dot{V}O_2$) responses exhibit mono-exponential dynamics with a delay which reflects tissue-to-lung transit.[1,2,3,4] At work rates above θL, however, a slow second component becomes evident in $\dot{V}O_2$ which has been attributed to metabolic influences of, for example, increased muscle and blood lactate and catecholamine levels.[3,5,6,7,8] This further stimulates $\dot{V}CO_2$ and $\dot{V}$E. However, the symmetry of the

Control of Breathing and Its Modeling Perspective, Edited by
Y. Honda *et al.*, Plenum Press, New York, 1992

responses at the on- and off-transient of exercise, which is an essential feature of the systems' operation, has been largely ignored especially with respect to high-intensity exercise. Consequently, as these control dynamics are frequently characterized with common parameters for both the on- and off-transient, we were interested in examining the symmetry of the on- and off-transient responses above the lactate threshold during both square-wave and ramp work-rate forcings.

METHODS

Normal males initially performed incremental ramps (15-25 $W.min^{-1}$) on a computer-controlled cycle ergometer, to determine the maximum O2 uptake ($\mu\dot{V}O_2$) and θ_L using gas-exchange criteria and also the slope ($\Delta\dot{V}O_2/\Delta\dot{W}$) of the linear component of the response. They then performed a reverse-ramp protocol; i.e., an abrupt increase in work rate to the highest value achieved on the incremental ramp, with the work rate then being decremented to "0" W at the same rate of change as that of the incremental ramp. Constant-load exercise tests were also performed at (a) 90% of θ_L and (b) ~50% of "Delta" (i.e., the difference between θ_L and $\mu\dot{V}O_2$). Tests were conducted on different days and in a randomized sequence. Respired volume was measured with a turbine transducer and gas concentrations by mass spectrometry, for on-line determination of ventilatory and pulmonary gas exchange variables breath-by-breath.

RESULTS

Below θ_L, the on- and off- transient ventilatory and gas-exchange responses to the square-wave forcing of work rate were mono-exponential, with similar time constants; i.e., the dynamics showed symmetry (Fig. 1). As shown in Figure 2, the $\dot{V}O_2$ on-transient for the supra-θ_L forcing was complicated by the addition of a slow phase of response of delayed onset. The gain of the early component of this response did not differ appreciably from that predicted from that of the sub-θ_L forcing. The delayed, slow component increased $\dot{V}O_2$ to a level above that required for wholly-aerobic energy transfer (i.e., yielding "excess" $\dot{V}O_2$), as determined from the sub-θ_L $\dot{V}O_2$-$\dot{W}$ relationship from both steady-state exercise and incremental ramps. However, the $\dot{V}O_2$ off-transient response was consistently faster than the on-, being either mono-exponential or occasionally evidencing a small ($\leq$ 10%) and slow second component (Fig. 2).

During incremental ramps, the lagged-linear behavior of $\dot{V}O_2$[9,10,11] was commonly not discernibly different above and below θ_L (Fig. 3), with a slope (10.2 $\pm$ 0.8 $L.min^{-1}.W$) which did not differ significantly from that of the steady-state relationship for the sub-θ_L square-wave forcing. However, for the decremental ramp (Fig. 3) (in which low-efficiency 'fast-twitch' fibers are likely to be recruited early in the exercise, i.e., the reverse of the incremental ramp), the descending limb of the $\dot{V}O_2$ response was ~50% steeper than that expected (14.3 $\pm$ 1.2 $L.min^{-1}.W$); i.e., either from the incremental response pattern or for system dynamics with sub-θ_L kinetics. This

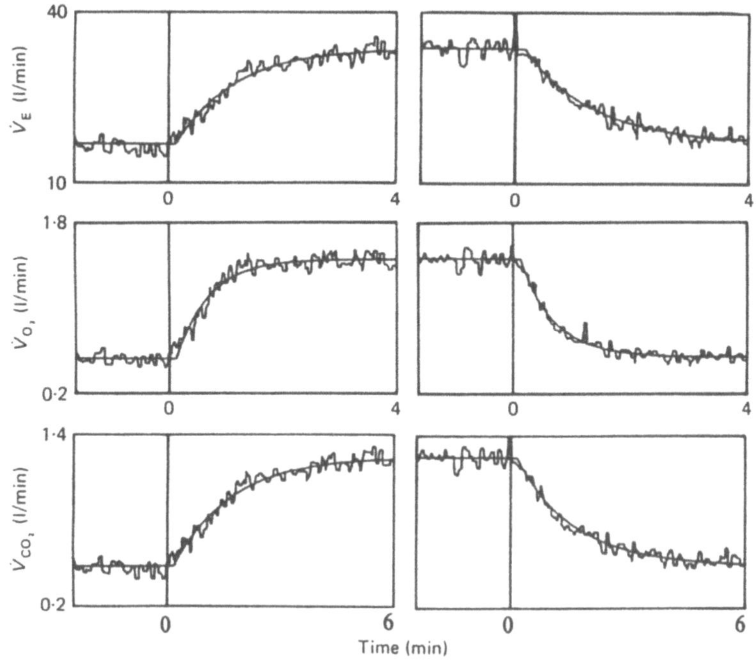

Figure 1. $\dot{V}_E$, $\dot{V}O_2$ and $\dot{V}CO_2$ responses to sub-θ_L square-wave cycling from "0" W, for a single subject. **Left**: on-transient responses, with the best-fit exponential superimposed. **Right**: off-transient responses, with the best-fit exponential superimposed. Reproduced with permission from Griffiths et al., *J. Physiol. (Lond.)* 380:387 (1986).

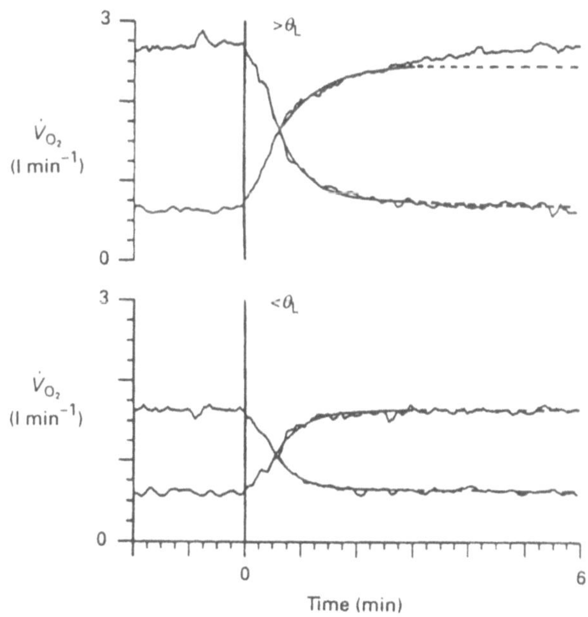

Figure 2. $\dot{V}O_2$ responses (average of 4 tests) to square-wave cycling from "0" W, for a single subject, below θ_L (110 W; **below**) and above θ_L (210 W; **above**). Superimposed on the responses are the exponential fits to 3 min (solid lines, with the dashed lines indicating the extension of the exponential to 6 min. Reproduced with permission from Paterson & Whipp, *J. Physiol. (Lond.)* 443:575 (1991).

phase was also associated with consequently-elevated $\dot{V}CO_2$ and $\dot{V}_E$ responses, consistent with the high ventilatory and metabolic costs of clearing the metabolites built up in the early phase of the decremental test. However, when the ramp was decremented from only 50% of the maximum work rate (i.e., the peak work rate was $< \theta_L$), the descending limb of the $\dot{V}O_2$ response (9.5 ± 1.2 L.min^{-1}.W) did not evidence the increased slope ; neither was there any asymmetry in the $\dot{V}CO_2$ and $\dot{V}_E$ responses.

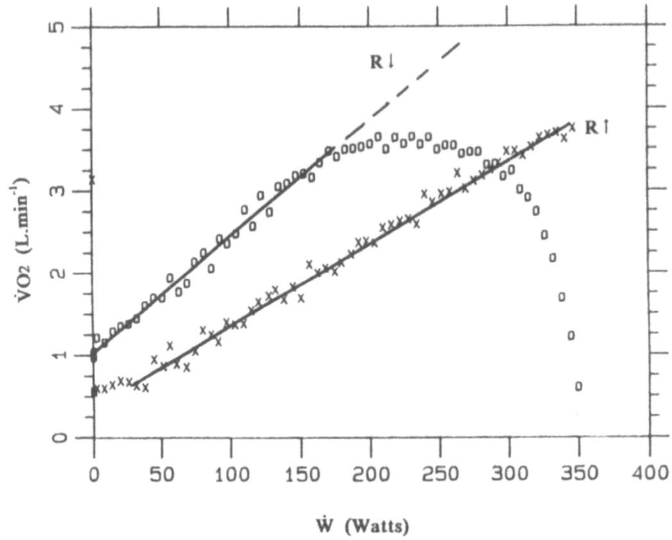

Figure 3. $\dot{V}O_2$ responses as function of work rate ($\dot{W}$) to exhausting ramp-incremental cycling from "0" W (R↑, o) and to ramp-decremental cycling to "0" W from the maximum work rate attained on the incremental ramp (R↓, x), for a single subject. Note that the slope of the $\dot{V}O_2$-$\dot{W}$ relationship is steeper for the decremental protocol than for the incremental protocol.

DISCUSSION

That the slow component of $\dot{V}O_2$ kinetics for supra-θ_L work rates in the intensity domain between the lactate threshold and approximately 50% of Delta is largely an on-transient phenomenon, as previously shown by Paterson and Whipp[12] suggests that great care should be taken in selecting work-rate forcings to investigate the dynamics of pulmonary gas exchange. Forcings which yield a single lumped parameter for both on- and off-transient responses (such as pseudo-random binary sequences and the sinusoidal function) are therefore likely to yield misleading physiological inferences for exercise in this intensity domain.

The fact that this supra-θ_L $\dot{V}O_2$ on-transient requires two components for its characterization, whereas the off-transient only requires one, has significant implications for the mechanism of the slow phase itself. For example, it suggests that the oxygen cost of respiratory and cardiac work and also the Q_{10} effect, each of which one would expect to be manifest in both the on- and off-transient responses, are likely to be of minor effect. Catecholamines may be ruled out on other grounds.[13] This observation, however, may be consistent with the hypothesis that routes of lactate metabolism may be involved. The high lactate concentrations associated with high-intensity exercise may require different mechanisms for its subsequent metabolism when the energy demands are high during the on-transient than when they are low at the off-transient, as schematized in Figure 4.

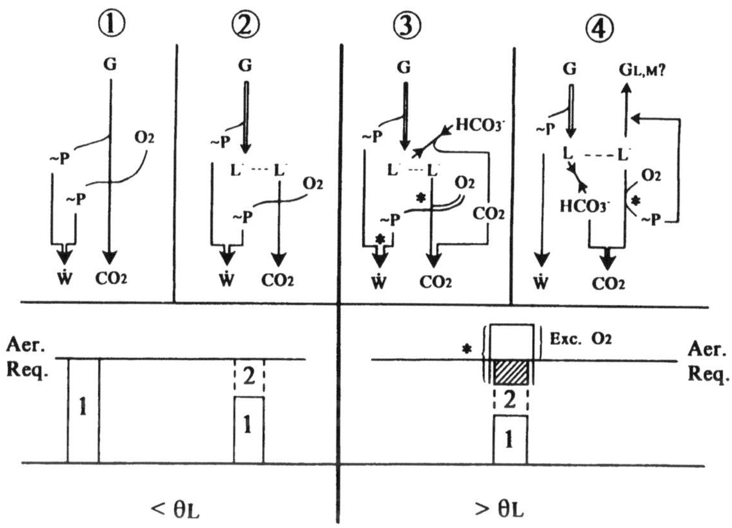

Figure 4. Schematic of sub-θL (panels 1 & 2) and supra-θL (panels 3 & 4) routes of glycogen (G) catabolism to lactate (L⁻) (**above**), and their proportional contributions to the total energy requirement for exercise (**below**). Sites of "excess" $\dot{V}O_2$ are indicated by *.

That is, with no O_2 limitation, such as during recovery, lactate can be metabolized to CO_2 and H_2O with no increased O_2 cost (Fig. 4, panel 2). The results of Roth and Brooks[14] are consistent with this hypothesis. However, with O_2 limitation in contracting units, such as during supra-threshold exercise, the lactate may now need to be metabolized in, as yet, unrecruited fibers with low ~P:O_2 ratios.[15] This would be consistent with the demonstration that the α-glycero-phosphate shuttle is prominent in 'fast-twitch' muscle fibers.[16] The hydrogen processing by this mechanism is FAD-linked rather than NAD-linked and hence, unlike the malate shuttle mechanism, bypasses an ATP-production step in the electron transport chain. The consequence is a high O_2 requirement for a given rate of ATP formation (Fig. 4, panel 3). The increased O_2 cost of glycogen resynthesis in the liver (e.g., Fig. 4, panel 4) is likely to be quantitatively small. Glycogen resynthesis from lactate in skeletal muscle could potentially lead to a large O_2 cost. However, we feel this is unlikely under the conditions of our experiments, as such glycogen resynthesis is thermodynamically improbable in fibers (or regions of fibers) which are generating lactate, and in those that are not there is likely to be only little, if any, glycogen depletion - at least in work of this duration. Were there to be significant cycling of contractile activity among motor units, however, the the O_2 cost of resynthesis during the exercise could be considerable. Further work is required, however, to establish the actual physiological basis of the slow kinetic phase of $\dot{V}O_2$ in this domain and even whether our hypothesis, based on mitochondrial shuttle mechanisms, is justifiable.

But, regardless of the mechanism of this phenomenon, the consequences of its delayed onset seems unavoidable: the conventional procedure for computing the O_2-deficit can no longer be justified in this domain, i.e., the eventual $\dot{V}O_2$ asymptote is not an appropriate frame-of-reference for the early transient events. The data of Linnarsson,[3] Whipp and Mahler[5] and Barstow and Molé[17] also support this contention.

The "excess" $\dot{V}O_2$ at high-intensity exercise is also clearly manifest in the decremental ramp (Fig. 3). The "excess" $\dot{V}O_2$ is again consistent with the increased O_2 cost of lactate metabolism under these conditions; when the peak of the decremental ramp is sub-threshold, the responses are consistent with first-order kinetics. The incremental ramp, however, commonly appears to manifest first-order behavior[4,9,10,11] both above and below the lactate threshold (very fast or very slow rates of work-rate incrementation, however, can lead to non-linear elements at high work rates[5,18]). That the "steady state" requirements for $\dot{V}O_2$ at these high work rates have a component of "excess" $\dot{V}O_2$ means that the $\dot{V}O_2$ response pattern in this supra-threshold domain of the incremental ramp manifests not actual but pseudo- first-order kinetics. That is, although the entire incremental ramp response *can* be modelled with first-order kinetics, the model is likely to be an erroneous oversimplification - i.e., the response pattern is a necessary, but not sufficient, requirement for this characterization.

In conclusion, therefore, the dynamic asymmetries of both ventilatory and pulmonary gas exchange kinetics that we have demonstrated need to be incorporated as a fundamental requirement of modelling strategies in the this intensity domain. In addition, however, they also impose stringent requirements for optimum work-rate forcings for parameter estimation and model discrimination. And, as a corollary, system characterizations which yield lumped parameters for on-and off-transient behavior are therefore likely to suggest misleading physiological correlates for high-intensity exercise.

REFERENCES

1. B.J. Whipp and S.A. Ward, Physiological determinants of pulmonary gas exchange kinetics during exercise, *Med. Sci. Sports Ex.* 22:62 (1990).
2. T.J. Barstow, N. Lamarra, and B.J. Whipp, Modulation of muscle and pulmonary O_2 uptakes by circulatory dynamics during exercise, *J. Appl. Physiol.* 68:979 (1990).

3. D. Linnarsson, Dynamics of pulmonary gas exchange and heart rate changes at start and end of exercise, *Acta Physiol. Scand.* (suppl.) 415:1 (1974).
4. R.L Hughson, Exploring cardiorespiratory control mechanisms through gas exchange dynamics, *Med. Sci. Sports Ex.* 22:72 (1990).
5. B.J. Whipp and M. Mahler, Dynamics of gas exchange during exercise, *in*: "Pulmonary Gas Exchange" Vol. II, J.B. West, ed., Academic Press, New York (1980).
6. W.L. Roston, B.J. Whipp, J.A. Davis, D.A. Cunningham, R.M. Effros, and K. Wasserman, Oxygen uptake kinetics and lactate concentration during exercise in humans, *Am. Rev. Respir. Dis.* 135:1080 (1987).
7. D.C. Poole, S. A. Ward, G.W. Gardner, and B.J. Whipp, Metabolic and respiratory profile of the upper limit for prolonged exercise in man, *Ergonomics* 31:1265 (1988).
8. R. Casaburi, T.W. Storer, I. Ben Dov, and K. Wasserman, Effect of endurance training on possible determinants of $\dot{V}O_2$ during heavy exercise, *J. Appl. Physiol.* 62: 199 (1987).
9. B.J. Whipp, J.A. Davis, F. Torres, and K. Wasserman, A test to determine the parameters of aerobic function during exercise, *J. Appl. Physiol.* 50:217 (1981).
10. G.D. Swanson and R.L. Hughson, On the modelling and interpretation of oxygen uptake kinetics from ramp work rate tests, *J. Appl. Physiol.* 65:2453 (1988).
11. T. Yoshida, Gas exchange responses to ramp exercise, *Ann. Physiol. Anthrop.* 9:167 (1990).
12. D.H. Paterson and B.J. Whipp, Asymmetries of oxygen uptake transients at the on- and off-set of heavy exercise in humans, *J. Physiol. (Lond.).* 443:575 (1991).
13. G.A. Gaesser, S.A. Ward, V.C. Baum, and B.J. Whipp, The effects of infused epinephrine on the "excess" O_2 uptake of heavy exercise in humans, *Fed. Proc.* April (1992).
14. D.A. Roth, W.C. Stanley, and G.A. Brooks, Induced lactacidemia does not affect post-exercise O_2 consumption, *J. Appl. Physiol.* 65:1045 (1988).
15. J.M. Krisanda, T.S. Moreland, and M.J. Kushmerik, ATP supply and demand during exercise, *in*: "Exercise, Nutrition, and Energy Metabolism," E.S. Horton and R.L. Terjung, MacMillan, New York (1988).
16. P.G. Schantz and J. Henriksson, Enzyme levels of the NADH shuttle systems: measurements in isolated muscle fibers from humans of differing physical activity, *Acta Physiol. Scand.* 129:505 (1987).
17. T.J. Barstow and P.J. Molé, Linear and nonlinear characteristics of oxygen uptake kinetics during heavy exercise, *J. Appl. Physiol.* 71:2099 (1991).
18. J.E. Hansen, R. Casaburi, D.M. Cooper and K. Wasserman, Oxygen uptake as related to work rate increment during cycle ergometer exercise, *Europ. J. Appl. Physiol.* 57:140 (1988).

THE EFFECT OF EXERCISE INTENSITY AND DURATION ON VENTILATION DURING RECOVERY FROM MODERATE AND HEAVY EXERCISE

Jason H. Mateika and James Duffin

Departments of Anesthesia and Physiology
University of Toronto, Ontario, Canada
M5S 1A8

INTRODUCTION

The results obtained from numerous experimental investigations suggest that the abrupt increase in ventilation ($\dot{V}_E$) that is characteristic of phase 1 of constant load treadmill exercise is elicited by neural stimuli which originate from either exercising limb afferents or higher motor centers[1,2]. Dejours and other investigators[1,2,3] have proposed that these stimuli persist throughout exercise and combine in an additive fashion with humoral stimuli from the central and peripheral chemoreceptors, which may elicit the exponential increase in $\dot{V}_E$ that is characteristic of Phase 2, to control the steady state ventilatory response observed during Phase 3.

Dejours hypothesized that these neural stimuli persist throughout exercise because the magnitude and characteristics of the ventilatory changes recorded during the on and off transition of exercise were similar[1]. However, the results obtained from other experimental investigations have demonstrated that the abrupt decrease in $\dot{V}_E$ recorded at the end of exercise ($e\dot{V}_E$) of increased duration and intensity (above the first ventilatory threshold) was smaller than the abrupt increase in $\dot{V}_E$ measured at the start of exercise ($s\dot{V}_E$)[4,5,6].

These findings suggest that the neural stimuli responsible for the control of $\dot{V}_E$ during exercise may be modified in response to alterations in exercise intensity or duration so that the magnitude of $e\dot{V}_E$ is related inversely to these two variables.

The results presented in this paper were obtained from two separate investigations that were initiated to examine the effects of varying workload and duration on the magnitude of $s\dot{V}_E$ and $e\dot{V}_E$.

METHODS

Procedure

Investigation 1. Five healthy subjects (4 males and 1 female) recruited from the

Control of Breathing and Its Modeling Perspective, Edited by
Y. Honda *et al.*, Plenum Press, New York, 1992

University of Toronto's student population completed treadmill exercise tests at each of two constant workloads that corresponded to 50% and 80% of each subject's maximum oxygen consumption ($\dot{V}O_2$ max.). The exercise tests were performed for a duration of 6 minutes (50%-6 minutes and 80%-6 minutes) and 10 minutes (50%-10 minutes and 80%-10 minutes) at each workload and the subjects completed a maximum of 2 trials at each level of exercise. A total of 38 exercise tests were completed by this group.

Investigation 2. An additional six subjects (4 males and 2 females) were recruited to perform constant load exercise tests of 6, 12 and 18 minutes duration at moderate exercise levels that corresponded to 50% and 80% of each subject's first ventilatory threshold (50%T_{vent1}- 6, 12 and 18 minutes; 80%T_{vent1}- 6, 12 and 18 minutes). In addition these subjects completed exercise tests of 2, 6 and 8 minutes duration at heavy and severe exercise workloads that corresponded to 120% and 140% of each subject's T_{vent1} (120% T_{vent1}- 2, 6 and 8 minutes; 140% T_{vent1}- 2, 6 and 8 minutes), respectively. A total of 72 exercise tests were completed by this group.

Prior to the onset of each exercise test performed during investigation 1 and 2, the subjects stood quietly on the treadmill for 3-5 minutes while ventilation was monitored breath-by-breath. The treadmill was then started abruptly without warning. Similarly, at the completion of each exercise test the treadmill was stopped abruptly and the subjects $\dot{V}_E$ was monitored for an additional 2-6 minutes.

Data Analysis

Investigation 1. The breath-by-breath ventilation data collected during the on transition of each constant load exercise test completed during investigation 1 was separated and analyzed using non-linear curve fitting techniques (Simplex algorithm) so as to obtain a measure of s$\dot{V}_E$.

A mathematical equation of the form:

$$y(t) = A^* [1-e^{-(t-D)/T}]$$

where y(t) equals the increase in V_E above the prior steady state value at time t in seconds; A equals the steady state increase in $\dot{V}_E$; t is the time after the change in work rate; T is the time constant of the response; D is the delay of the response; was fitted to the data. The magnitude of s$\dot{V}_E$ was calculated by determining the difference between y(0) and the previous steady state baseline value, which was calculated by averaging the values of the last ten breaths recorded prior to the start of exercise. However, this method did not accurately quantify the magnitude of s$\dot{V}_E$ if the abrupt change in $\dot{V}_E$ observed during Phase 1 was followed by a plateau. Therefore, if such a response was observed by visual inspection, the model was fitted to the data collected during phase 1. Using this method, the magnitude of s$\dot{V}_E$ was calculated as first described above if a negative time delay was recorded but was taken as A if the time delay was positive and the time constant was less than one second.

The breath-by-breath ventilation data collected during the off-transition of each constant load exercise test was analyzed in a similar manner so as to obtain a measure of e$\dot{V}_E$.

A general linear model, comprised of three class variables (intensity, duration and transition), with repeated measures, in conjunction with Tukey's honestly significant statistical test and least square mean values, was used to determine if statistical differences existed between the mean s$\dot{V}_E$ and eV_E values that were calculated for the separate and combined class variables (combined class variables = intensity*duration; intensity*transition; duration*transition; intensity*duration*transition).

Investigation 2. The same method described in investigation 1 was used to determine the magnitude of $s\dot{V}_E$ and $e\dot{V}_E$ from the breath-by-breath ventilation data collected during the on and off transition of each constant load exercise test, respectively.

The statistical methods described above were used to determine if significant differences existed between the mean $s\dot{V}_E$ and $e\dot{V}_E$ values that were calculated for the separate and combined class variables that characterized the exercise trials performed above T_{vent1} and below T_{vent1}. In addition, paired sample t-tests, in conjunction with Bonferroni's correction factor, were used to determine if significant differences existed between the mean $s\dot{V}_E$ values that were calculated from the exercise tests of varying intensity (T_{vent1} - 50%, 80%, 120% and 140%). The results obtained from both investigations were considered to be significant if $p < 0.05$ and weakly significant if $0.05 < p < 0.10$. The values in the text are means ± 1 SD.

RESULTS

Investigation 1

Figure 1 shows the results obtained from one exercise test completed at each level of exercise during air breathing by one of the subjects. These results were typical, and when combined with the group mean $s\dot{V}_E$ values calculated for each exercise level (figure 2) demonstrated that an abrupt increase in $\dot{V}_E$ was present at the start of each exercise test. The results obtained from the statistical analysis of the data revealed that a significant interaction existed between exercise intensity and transition ($p=0.0001$) and that the magnitude of $s\dot{V}_E$ was greater at the start of the 80% exercise tests as compared to the 50% tests ($p=0.0098$). In contrast, the mean $e\dot{V}_E$ values, which are illustrated in figure 2, showed that the magnitude of $e\dot{V}_E$ was attenuated at the higher exercise intensity ($p=0.0001$). The results also demonstrated that $e\dot{V}_E$ was less than $s\dot{V}_E$ at both exercise intensities (50%-$p=0.0064$ and 80%-$p=0.0001$).

It was also found that a weak statistically significant interaction existed between exercise transition and duration ($p=0.0547$). The least squares mean values revealed that the magnitude of $e\dot{V}_E$ calculated for the 10 minute exercise tests were significantly less than the values calculated for their 6 minute counterparts ($p=0.0008$). In contrast, the magnitude of $s\dot{V}_E$ calculated for the 10 minute exercise tests was not significantly different than the values calculated for the six minute tests ($p=0.4155$).

Investigation 2

Figures 3 and 4 show the results obtained from one exercise test completed at each level of the 50% T_{vent1} and 120% T_{vent1} workloads, respectively, by one subject. These results in conjunction with the group mean results (figure 5) demonstrated that an abrupt increase in $\dot{V}_E$ was present at the start of each exercise test. The results obtained from the statistical analysis of the data showed that the mean $s\dot{V}_E$ values which were derived from the 50% (14.44 ± 4.14) and 120% (18.18 ± 3.59) T_{vent1} exercise tests were not significantly different from the mean values calculated from the 80% (16.48 ± 4.16) and 140% (20.75 ± 5.76) T_{vent1} exercise tests, respectively. Alternatively, the mean $s\dot{V}_E$ value calculated from the 140% T_{vent1} tests was significantly greater than the mean values obtained from the 50% and 80% T_{vent1} tests ($p=0.0011$ and $p=0.0173$).

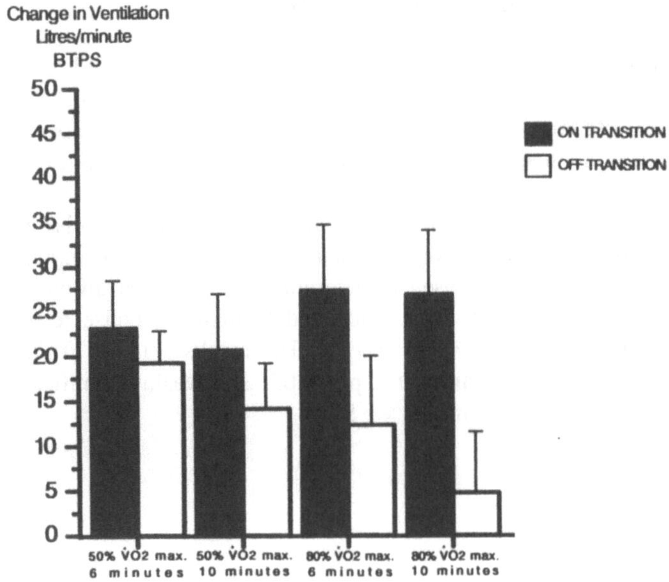

Figure 1. Plots of breath-by-breath values of ventilation versus time recorded from one subject at the four levels of exercise. The levels of exercise from top to bottom are : 80% - 10 minutes, 80% - 6 minutes, 50% - 10 minutes, 50% - 6 minutes.

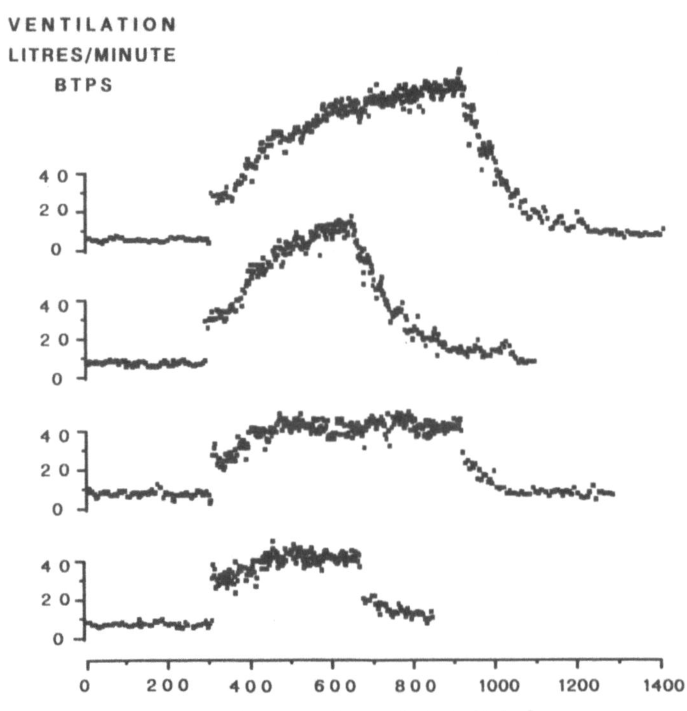

Figure 2. A histogram of the mean changes (± 1 SD) in ventilation at the on and off transition of each level of exercise completed.

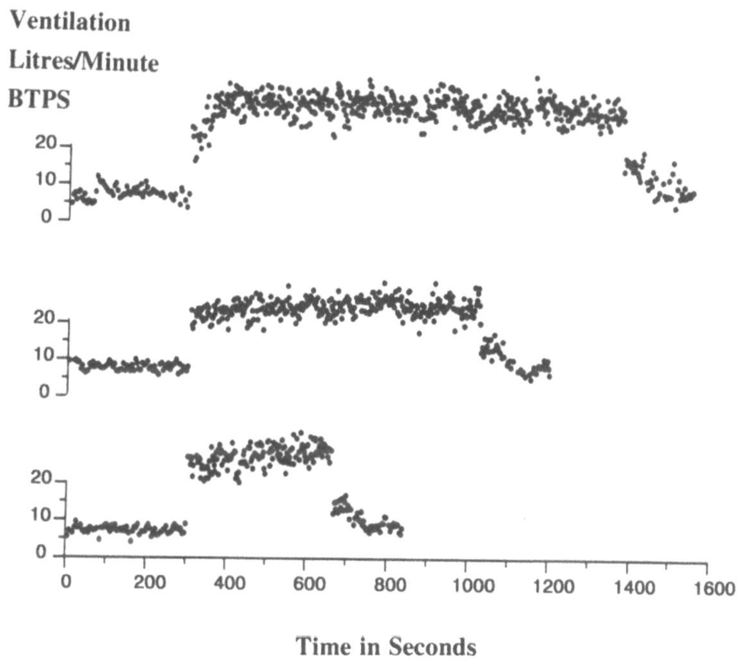

Figure 3. Plots of breath-by-breath values of ventilation versus time recorded from one subject at the three 50% T_{vent1} levels of exercise. The levels of exercise from top to bottom are: 50% T_{vent1} - 18 minutes, 50% T_{vent1} - 12 minutes, 50% T_{vent1} - 6 minutes.

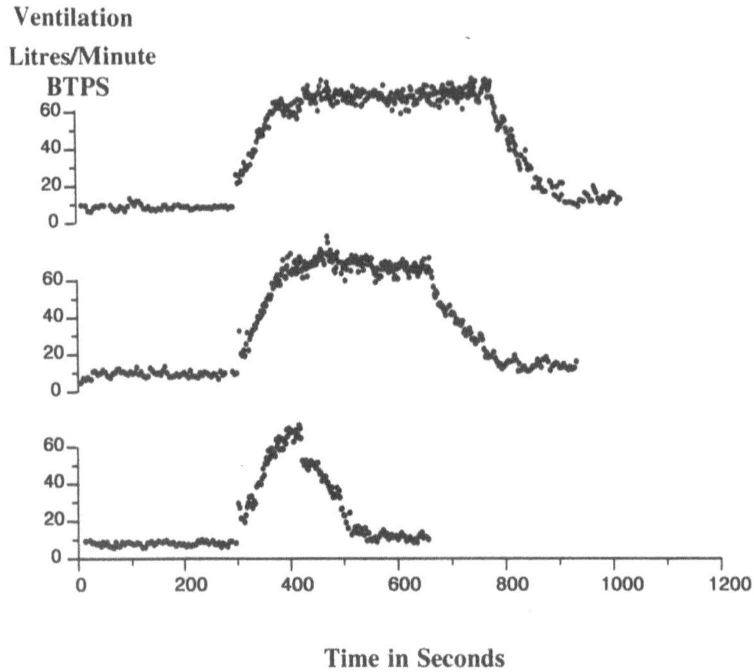

Figure 4. Plots of breath-by-breath values of ventilation versus time recorded from one subject at the three 120% T_{vent1} levels of exercise. The levels of exercise from top to bottom are: 120% T_{vent1} - 8 minutes, 120% T_{vent1} - 6 minutes, 120% T_{vent1} - 2 minutes.

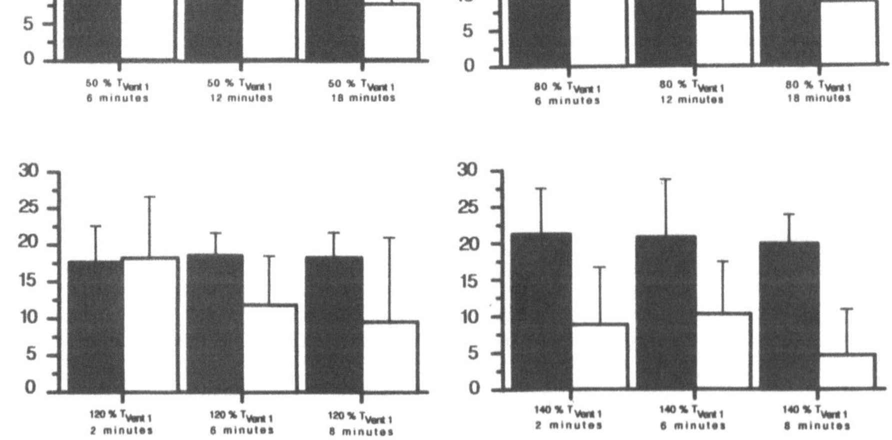

Figure 5. A histogram of the mean changes (± 1 SD) in ventilation at the on and off transition of each level of exercise completed at 50% T_{vent1}, 80% T_{vent1}, 120% T_{vent1}, 140% T_{vent1}.

It was also found that a significant interaction between exercise intensity and transition existed above T_{vent1} (p= 0.0087) because $s\dot{V}_E$ was significantly greater than $e\dot{V}_E$ at both 120% and 140% T_{vent1} (p=0.0114 and p=0.0001) (figure 5) and because the magnitude of $e\dot{V}_E$ was smaller at the end of the 140% T_{vent1} exercise tests as compared to the 120% tests (p= 0.0127) (figure 5). This significant interaction was not observed below T_{vent1} although the effect of transition existed since $s\dot{V}_E$ was significantly greater than $e\dot{V}_E$ at both 50% and 80% T_{vent1} (p=0.05) (figure 5). The effect of duration either separately or combined with other class variables did not have a significant effect on the values of $e\dot{V}_E$ calculated from the tests performed both above and below T_{vent1}.

DISCUSSION

It has been hypothesized that $s\dot{V}_E$ is elicited by neurogenic stimuli that originate from the motor cortex and/or afferents located in the exercising limbs and is proportional to limb movement frequency rather than the loading of the working muscles[1,2,7]. However, contrary to the findings of Casey et al.[7] the results obtained from investigation 1 demonstrated that the magnitude of $s\dot{V}_E$ was directly related to the intensity of exercise. This discrepancy probably exists because the findings of Casey et al.[7] were based on results that were obtained from exercise tests, of varying intensity, that were performed below the first ventilatory threshold. Consequently, the effect of higher intensity workloads (performed above the first ventilatory threshold) on $s\dot{V}_E$ was not tested.

The relationship between exercise intensity and $s\dot{V}_E$ may be non-linear, independent of variations in exercise intensity both below and above the first ventilatory threshold, but with incremented $s\dot{V}_E$ above the threshold. This hypothesis is supported by the results of the second investigation which showed that at the same walking speed, the magnitude of $s\dot{V}_E$ was not altered by the transition from light to moderate exercise (50% T_{vent} to 80% T_{vent1}), and from heavy to severe exercise (120% T_{vent1} to 140% T_{vent1}). Alternatively, a significant increase in the magnitude of $s\dot{V}_E$ accompanied the transition from light and moderate exercise to severe exercise.

It has been postulated that the stimuli responsible for $s\dot{V}_E$ persist throughout exercise and contribute, in an additive fashion with humoral stimuli, to control $\dot{V}_E$ during Phase 3 of exercise[1,2,3]. This hypothesis was formulated because Dejours reported[1] that an a abrupt change in $\dot{V}_E$ which was similar in magnitude and related directly to the intensity of exercise was observed at both the start and end of exercise. However, the results obtained from the present investigation do not support this hypothesis since the magnitude of $e\dot{V}_E$ was smaller than $s\dot{V}_E$ at each level of exercise completed in both investigation 1 and 2. This reduction may be related to the intensity of exercise since the magnitude of $e\dot{V}_E$ decreased as the exercise intensity increased during investigation 1. However, the results obtained from the second investigation show that this relationship may be alinear since the $e\dot{V}_E$ values recorded from the 50% and 80% T_{vent1} tests were not significantly different while the values recorded from the 140% T_{vent1} tests were significantly less than the 120% T_{vent1} tests. It is also possible that exercise duration may have an effect on the magnitude of $e\dot{V}_E$ since the values recorded from the 6 and 10 minute exercise tests completed during investigation 1 were significantly different. However this relationship requires further investigation since exercise duration did not have a significant effect on the magnitude of $e\dot{V}_E$ recorded from the tests performed during investigation 2.

It is possible that the reduction in $e\dot{V}_E$ that was observed occurred because the ventilatory response to the stimuli responsible for $s\dot{V}_E$ declined, in an alinear manner, with increasing exercise workload and possibly duration. This hypothesis is supported by the findings of Waldrop et al.[8] which demonstrated that the response of the respiratory controller to the simultaneous stimulation of the subthalamic locomotor region and limb afferents was smaller than would be predicted by summing the response of the respiratory controller to each individually applied stimulus. This finding suggests that the effect of two or more stimuli on the response of the respiratory controller may not be simply additive, and that the effectiveness of a stimulus to elicit a response might be reduced in the presence of other stimuli. Therefore, contrary to the traditional hypothesis, the ventilatory response to the stimuli responsible for s$\dot{V}_E$ may be reduced concomitantly with the enhancement of other stimuli which normally accompany the increase in exercise intensity and duration. If this hypothesis is correct, then the reduction of e$\dot{V}_E$ which was most prominent at the end of the heavy exercise tests in both investigations, might have occurred in conjunction with the enhancement of peripheral chemoreceptor activity. This hypothesis is supported by the findings which have demonstrated that the discharge of the peripheral chemoreceptors is enhanced by K^+ and H^+ ions which increase in concentration in response to the intensity of exercise[9,10].

It is possible that the magnitude of e$\dot{V}_E$ may be directly related to the intensity of exercise and that this relationship was not detected in the present investigations because the method used to quantify the magnitude of e$\dot{V}_E$ was employed on the assumption, which was based on the results from previous investigations, that $\dot{V}_E$ decreased abruptly at the end of exercise. However, experimental findings similar to those presented could result if short term potentiation (STP) of the ventilatory response was generated during exercise by the stimuli responsible for s$\dot{V}_E$ so that at the end of exercise the magnitude of the change in $\dot{V}_E$ had increased but the response to the removal of this stimuli was gradual and not abrupt. This hypothesis is supported by the findings of Senapati[11] and Eldridge and Millhorn[12] which showed that STP of the ventilatory response could be elicited in the dog and cat, respectively, by stimulating afferents originating from the peripheral muscles. Furthermore, Eldridge and Millhorn[12] have demonstrated that the degree of STP in the cat was related directly to the magnitude and duration of the stimulus applied to the muscle afferents.

In conclusion, the results of the present investigation have demonstrated that e$\dot{V}_E$ is reduced relative to s$\dot{V}_E$ and that the degree of reduction is inversely related to the intensity and duration of exercise. In addition, evidence has been presented to suggest that this reduction may be due either to a decline in the ventilatory response to the stimuli which produced s$\dot{V}_E$ or to a change in the dynamics of the ventilatory response such that the removal of the stimuli, which produced s$\dot{V}_E$, at the end of exercise did not result in the immediate removal of the ventilatory response.

REFERENCES

1. P. Dejours, Neurogenic factors in the control of ventilation during exercise, Circulation Res. 20 and 21 [Suppl] I:146-153 (1967).
2. T. Morikawa, Y. Ono, Y. Sasaki, Y. Sakakibara, Y. Tanaka, R. Maruyama, Y. Nishibayashi, and Y. Honda, Afferent and cardiodynamic drives in the early phase of exercise hyperpnea in humans, J. Appl. Physiol. 67(5) : 2006-2013 (1989).
3. E. D'Angelo, and G. Torelli, Neural stimuli increasing respiration during different types of exercise. J. Appl. Physiol. 30(1):116-121(1971).
4. W.L. Beaver, and K. Wasserman, Transients in ventilation at start and end of exercise, J. Appl. Physiol. 25(4) : 390-399 (1968).
5. R. Jeyaranjan, R. Goode, and J. Duffin, Changes in respiration in the transition from heavy exercise to rest, Eur. J. Appl. Physiol. 57: 606-610 (1988).
6. R. Jeyaranjan, R. Goode, and J. Duffin, Changes in ventilation at the end of heavy exercise of different durations, Eur. J. Appl. Physiol. 59: 385-389 (1989).
7. K. Casey, J. Duffin, C.J. Kelsey, and G.V. McAvoy, The effect of treadmill speed on ventilation at the start of exercise in man, J. Physiol. 391: 13-24 (1987).

8. T.G. Waldrop, D.C. Mullins, and D.E. Millhorn, Control of respiration by the hypothalamus and by feedback from contracting muscles in the cat, Respir. Physiol. 64:317-328 (1986).
9. D.J. Paterson, and P.C.G. Nye, Effect of oxygen on potassium-excited ventilation in the decerebrate cat, Respir. Physiol. 84:223-230 (1991).
10. K. Wasserman, B.J. Whipp, and R. Casaburi, Respiratory control during exercise, in: Handbook of Physiology, sect 3. The Respiratory System, vol 2., A.P. Fishman, N.S. Cherniack, and J.G. Widdicombe eds., American Physiological Society, Bethesda, MD (1986).
11. J.M. Senapati, Effect of stimulation of muscle afferents on ventilation of dogs, J. Appl. Physiol. 21(1):242-246 (1966).
12. F.L. Eldridge, and D.E. Millhorn, Oscillation, gating and memory in the respiratory control system, in: Handbook of Physiology, sect 3. The Respiratory System, vol 2., A.P. Fishman, N.S. Cherniack, and J.G. Widdicombe eds., American Physiological Society, Bethesda, MD (1986).

ON THE FRACTAL NATURE OF BREATH-BY-BREATH VARIATION IN VENTILATION DURING DYNAMIC EXERCISE

Richard L. Hughson and Yoshiharu Yamamoto

Department of Kinesiology
Faculty of Applied Health Sciences
University of Waterloo
Waterloo, Ontario N2L 3G1 Canada

INTRODUCTION

Ventilation (V_E) during rest and moderate levels of exercise is tightly coupled to the production of carbon dioxide (VCO_2) (13). This is true not only in the steady state, but also in the transient phase between work rates (13). To explain this coupling, the possible roles of carotid and central chemoreceptors, of central motor cortex irradiation, of skeletal muscle mechano- and/or metabolic receptors, and of cardiac or pulmonary mechanoreceptors have been explored (13). This stimulus-based approach to the study of respiratory control omits, as stated by Grodins and Yamashiro (4), the neural control of breathing pattern. Thus, it is important to consider the interaction of these stimuli that regulate the magnitude of the ventilatory response in conjunction with the mechanisms that produce the oscillatory pattern of respiration (3,4).

In spite of a relatively tight coupling on average, detailed investigations have shown breath-by-breath variations in V_E. Early studies characterized slow frequency oscillations occurring over approximately 25-50 breaths as well as smaller oscillations of only 2-6 breaths (5,9). Another study concluded from an examination of the run lengths and turning points in a sequence of breath-by-breath ventilation that the variations appeared to be non-random (1).

If the variations in V_E are not random, then this might prove to be valuable in the study of the interactions between the central respiratory pattern generator and the feedback and feed forward regulatory information. Recent advances in the study of chaos and fractal mathematics have been used to advantage in understanding other physiological processes such as the cardiovascular system (2).

Sammon and Bruce (11) have described the respiratory pattern of rats as chaotic. These researchers found a pattern in vagotomized rats that appeared to be a limit cycle oscillator. In contrast in rats with intact vagi, information from the pulmonary stretch receptors influenced the oscillator and induced small breath-by-breath variations characteristic of a chaotic process (11). Yamashiro (16) has questioned whether variations in the breathing pattern might represent chaos or white noise. To date, this has not been resolved for human respiration during rest or exercise.

We have developed a new method, coarse graining spectral analysis (CGSA), to

Control of Breathing and Its Modeling Perspective, Edited by
Y. Honda *et al.*, Plenum Press, New York, 1992

obtain an estimate of fractal dimension from time series data (14,15). This method is particularly suited to the current series of experiments because it extracts the fractal component independently of harmonic components. Thus, in the present study, we have been able to examine sinusoidal variations in work rate as well as steady state exercise to determine if the breath-by-breath noise in V_E and gas exchange was white, or if it was non-white noise typical of a fractal process.

METHODS

Seven healthy men (age 20-41 yr) volunteered for this study after receiving a complete description of the experimental details and signing a consent form approved by the Office of Human Research. Each subject completed an incremental cycle ergometer test to exhaustion. Work rate was at a baseline level of 25 W for 4 min before increasing as a ramp function at a rate of 15 W/min. V_E, VO_2 and VCO_2 were measured breath-by-breath (6). VO_2 and VCO_2 were processed to determine the effective lung volume factor to minimize breath-by-breath variations (6). Work rates for subsequent testing were below the ventilatory threshold.

On a separate day, all 7 subjects completed a submaximal test in which work rate varied as a sinusoid with a period of 4 min between a lower limit of 25 W and an upper limit of 125 W. 8 consecutive cycles of the sine wave were completed, with the first eliminated from analysis to provide a warm-up period.

Two constant load exercise tests at 25 W and 125 W were completed by 2 of the original subjects plus one additional subject. Subjects cycled for 5 min to attain a steady state prior to a period of 20 min data collection.

Coarse Graining Spectral Analysis

The CGSA method of Yamamoto and Hughson (14) takes the time series data of V_E, VO_2 and VCO_2 and forms a series of equally spaced samples (equivalent to an interval tachogram in heart rate variability analysis). CGSA takes advantage of the important self-similar property of fractal processes. The time scale of the original data set was changed by doubling each sample observation. Thus, when the original data was cross correlated with the rescaled version, only scale invariant components from which the spectrum of fractal (self-similar) processes was calculated were preserved (see Fig. 3 in Yamamoto and Hughson (14)). The subtraction of this component from the total spectrum yields the harmonic component. This was accomplished by dividing the total data set into three segments, doing this procedure on each segment, then taking the ensemble average to yield the final spectral data set.

The fractal dimension was obtained from the non-harmonic component of each data set by calculating the slope of the log spectral power versus log frequency. This was confined to a range of frequencies from a lower limit of the sixth data point corresponding to 0.5 breaths/min (frequency 0.08 Hz, log value -2.08) to an upper limit of 12 breaths/min (frequency 0.2 Hz, log value -0.7) to avoid encroaching on breathing frequencies during exercise. From this slope (β), fractal dimension was calculated as $D_F = 2/(\beta - 1)$ (15). Because D_F is a non-linear function of β, statistical analyses were completed only on β.

RESULTS

During the sinusoidal exercise, V_E was significantly correlated with both VCO_2

($r=0.89\pm0.03$) and $\dot{V}O_2$ ($r=0.82\pm0.05$). Inspection of the responses showed in all cases that breath-by-breath variability was observed.

CGSA analysis of the sinusoidal exercise tests showed, as expected for data where a large periodicity was intentionally introduced, that there was a high percentage of spectral power contained in the harmonic component. This was most evident for the $\dot{V}O_2$ and $\dot{V}CO_2$ where only 9-13% of the total power was in the fractal component. Perhaps somewhat surprisingly for $\dot{V}_E$, almost 24% of the spectral power was in the fractal component. The slope (ß) of the log power versus log frequency relationship for the fractal component was 1.84 ± 0.13 ($D_F=2.4$) for $\dot{V}_E$, 1.84 ± 0.07 ($D_F=2.4$) for $\dot{V}CO_2$, and 1.69 ± 0.09 ($D_F=2.9$) for $\dot{V}O_2$ (Table 1). (Fig. 1)

Steady state exercise at 25 and 125 W by three subjects showed that without the sinusoidal variation in work rate, more power (typically close to 40%) was contained in the fractal component (Table 2). The slope (ß) tended to be slightly greater during the 25 W cycling than 125 W. But, the overall pattern was similar between sinusoidal variations in work rate and constant load exercise (Fig. 1).

DISCUSSION

We have clearly demonstrated that the breath-by-breath variations in $\dot{V}_E$, $\dot{V}CO_2$ and $\dot{V}O_2$ have properties that can be described as fractal. The primary focus of this discussion will be on $\dot{V}_E$ because it is directly measured rather than calculated. We believe these are the first data to show that human ventilatory pattern during exercise could be described as fractal. The implications of this to ventilatory control will be considered below.

The term fractal refers in the current study to give an indication of the underlying complex structure that generates the time series data represented in breath-by-breath ventilation. As introduced by Mandelbrot (10), fractal was used to recognize the irregularities that can occur and the difficulties in measuring these.

We have used coarse graining spectral analysis (CGSA) (14) to obtain the non-harmonic component from which we derived the fractal dimension of the respiratory response to exercise. Kobayashi and Musha (7) and Saul et al. (12) have shown long term heart rate variability data to have a power spectrum in which log of spectral power is inversely proportional to frequency (*f*). The slope (ß) of the $1/f$ relationship is used to calculate the fractal dimension from $D_F = 2/(ß-1)$ (15).

To better appreciate the nature of the variations in breathing pattern, we have simulated different types of breath-by-breath noise and processed the data with the CGSA approach. There is reason to suspect that the ventilatory process might be described by the so-called Lorentzian spectrum. That is, if we the consider the following Markov process:

$$X(t+1) = \frac{1}{\tau_o} X(t) + \xi(t) \qquad Eq. 1$$

where $X(t)$ = ventilation at time t
τ_o = correlation time
$\xi(t)$ = Gaussian random variable.

The spectrum ($P(f)$, where f = frequency) of Eq. 1 takes the form:

$$P(f) \propto \frac{\tau_o}{1 + (2\pi\tau_o f)^2} \qquad Eq. 2$$

Table 1. CGSA analysis of fractal component during sinusoidal exercise.

	V_E		VCO_2		VO_2	
Subject	'Slope (ß)	% Fractal	Slope (ß)	% Fractal	Slope (ß)	% Fractal
1	2.33	14.2	2.13	9.1	1.98	7.7
2	1.36	34.2	1.62	7.3	1.35	6.5
3	2.05	39.6	1.87	24.2	2.00	13.6
4	2.01	23.3	1.94	10.0	1.79	7.8
5	1.35	14.8	1.52	10.9	1.35	8.6
6	1.93	19.2	1.82	16.8	1.72	10.1
7	1.88	22.4	1.99	13.9	1.63	7.8
Mean ± SEM	1.84 ±0.13	24.0 ±3.4	1.84 ±0.07	13.2 ±2.0	1.69 ±0.09	8.9 ±0.8
D_F	2.37		2.38		2.90	

Slope (ß) is taken from the log spectral power versus log frequency plot after extraction of the fractal component. The % Fractal is fractal power expressed as a percentage of total spectral power. D_F is obtained from the mean value of slope.

Table 2. CGSA analysis of fractal component during constant load exercise.

		Slope (ß)		% Fractal	
		25 W	125 W	25 W	125 W
V_E	Subj. 1	2.42	1.83	40.3	56.0
	Subj. 2	1.43	0.83	41.8	30.8
	Subj. 8	2.72	2.43	31.3	40.9
VCO_2	Subj. 1	2.27	1.50	38.7	47.0
	Subj. 2	1.27	1.05	30.5	40.2
	Subj. 8	2.60	2.02	34.8	39.2
VO_2	Subj. 1	2.26	1.42	44.1	36.3
	Subj. 2	1.13	0	34.7	33.1
	Subj. 8	2.67	1.84	39.7	34.1

Values represent individual observations in 3 subjects during exercise at 25 or 125 W. Subjects 1 and 2 are the same as in Table 2.

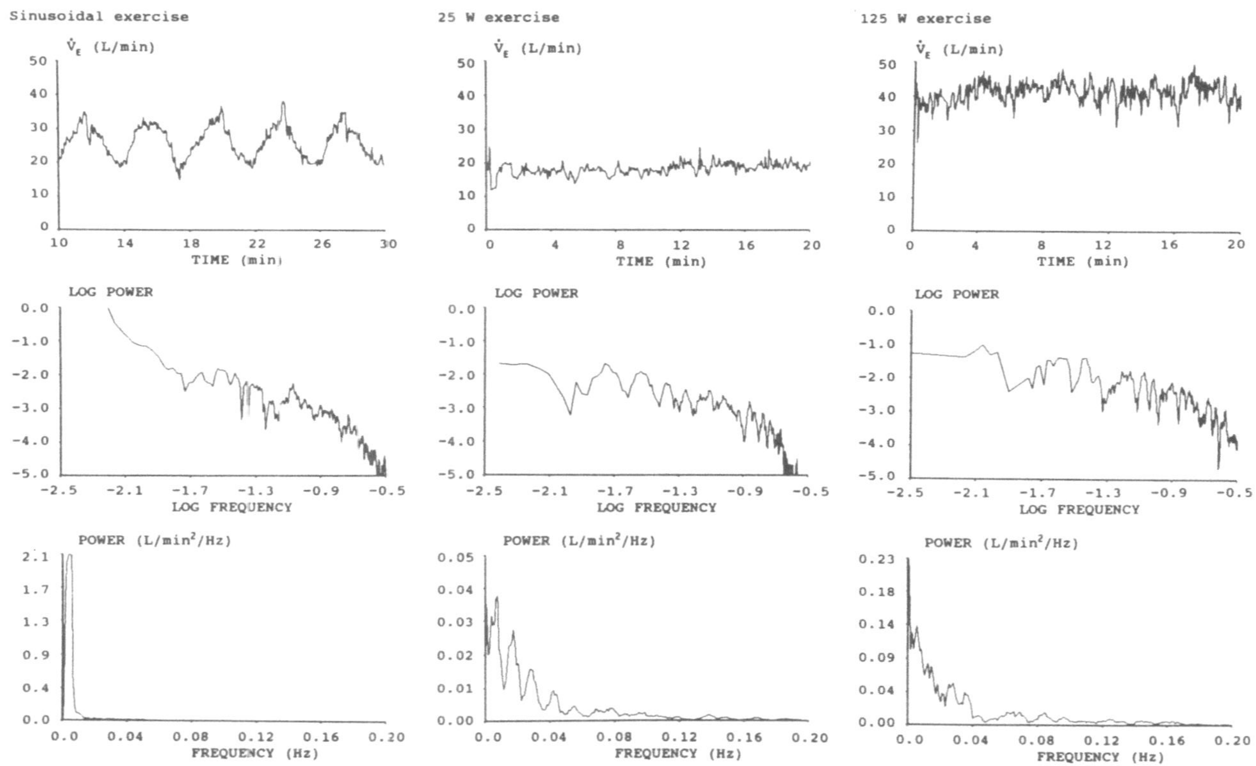

Figure 1. The original time series data of V_E for one subject (#1) are presented in the top series of figures for sinusoidal exercise (far left), 25 W constant load exercise (center), and 125 W (far right). The middle series of graphs represent the log spectral power versus log frequency relationship obtained by the CGSA method. The power spectrum of the harmonic components is presented in the bottom series.

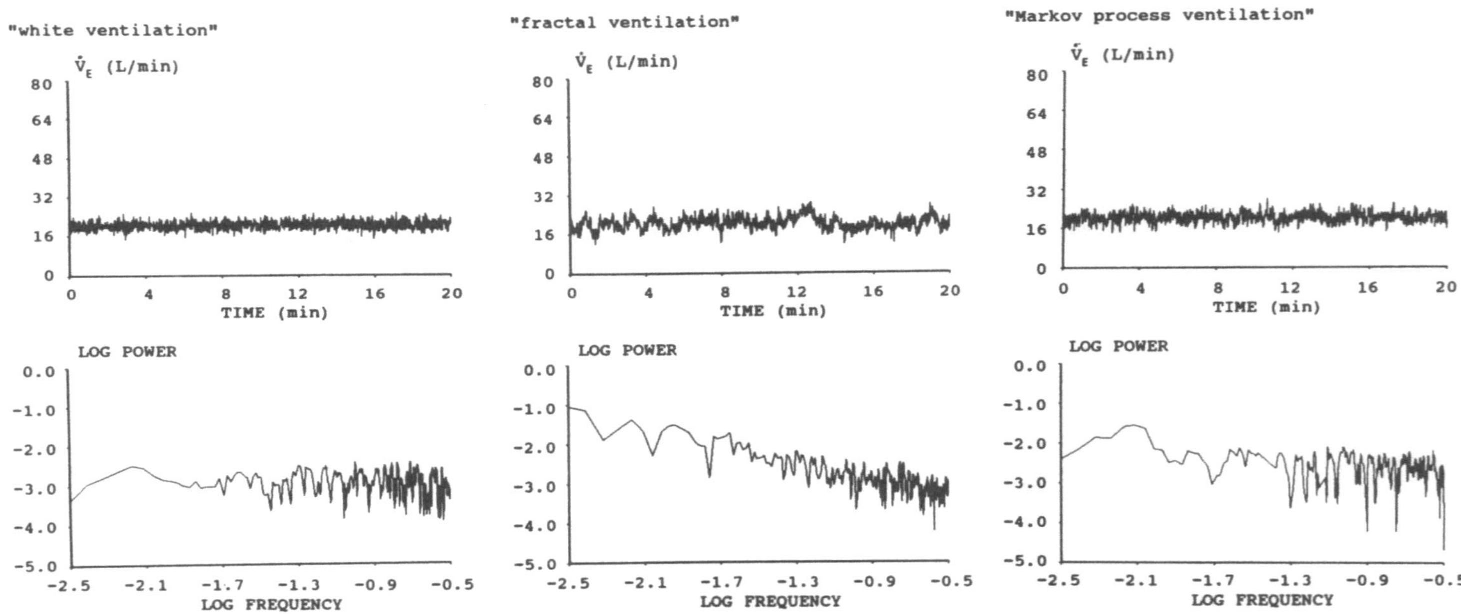

Figure 2. Simulated data for "white" (far left), "fractal" (center), and Markov process (far right) are shown as the time series (top graphs), and as the log spectral power versus log frequency relationship obtained by the CGSA method. Note the flat spectra for each of the white and Markov data sets, but the continuous decline for the fractal plot.

The log spectral power versus log frequency plot of this type of spectrum, i.e. the Lorentzian spectrum, decays with a slope of -2 for $f >> 1/2\pi\tau_o$, while it has a zero slope for $f << 1/2\pi\tau_o$.

The validity of this "linear" model (Eq. 1) was assessed by comparing this Markov process as well as "white" and "fractal" ventilation to the observed response (Fig. 2). For the Markov process, the correlation time τ_o was set to 3.0 s approximating one breath. Lamarra et al. (9) have reported that there was no substantial two breath correlation in ventilation during exercise.

Although the spectrum of the pattern simulated by Eq. 1 decayed with a slope of approximately -2 for $f > 1/2\pi\tau_o$ (approximately 0.05 Hz, data not shown), the extracted noise component from the CGSA analysis did not show such a decay. The spectrum was quite flat with properties very similar to those shown by "white" noise simulated ventilation (Fig. 2). In contrast, the noise component of the "fractal" simulated ventilation linearly decayed for the entire frequency range (Fig. 2). The difference between the Markov and the fractal simulations is largely in the manner in which the noise is added as a nonlinear process in the fractal case. The CGSA method was clearly able to differentiate this difference.

Our finding of a fractal relationship in the breath-by-breath V_E should not be too surprising in light of a recent study by Sammon and Bruce (11) that suggested the central respiratory pattern generator was modified slightly during each respiratory cycle by the feedback information from the pulmonary stretch receptors. This is consistent with the concepts of phase resetting of respiratory rhythm as shown by Eldridge et al. (3).

The meaning of the fractal dimension of V_E during exercise will be somewhat different from that observed in anesthetized rats. During exercise, there are potentially multiple inputs that could be influencing breathing pattern (13). It would be anticipated that feedback information is being provided to the respiratory pattern generator not only from pulmonary stretch receptors, but also from peripheral and central chemoreceptors, as well as skeletal muscle metabolic and mechanoreceptors, or that feed forward information is coming from the motor cortex. The fractal dimension has been suggested to represent the number of independent oscillators acting through non-linear interactions at the site of the pattern generator (15). Although we do not believe that it is possible to attribute specific physiological interpretation to the value of fractal dimension, we do suggest that multiple inputs are impinging on the central respiratory pattern generator.

During the constant load exercise at 125 W, one subject had a slope (ß) for the V_E response that was less than 1.0. Such a value is outside the range considered to be fractal (15). It is possible that in this case, the physiological control system was being driven primarily by one dominant oscillator. The work rate was selected to be below the ventilatory threshold; the respiratory exchange ratio (RER=0.94) suggested that it was. This subject had the highest absolute level of ventilation at both work rates. He also had relatively low values of ß during 25 W and during sinusoidal exercise suggesting that there are between subject differences that deserve future consideration.

In spite of the major periodicity superimposed on the ventilatory response by the sinusoidal variation in work rate, there was on average, 24% of total spectral power for V_E in the fractal component. During the constant load exercise, there was between 30 and 56% of the total spectral power in the fractal component. Thus, it appears that the complex pattern of interaction of competing information at the respiratory control centre is maintained during sinusoidal exercise.

We conclude that V_E varies on a breath-by-breath basis with a pattern that is consistent with the existence of fractal noise. The parallel of these observations with those obtained during studies of heart rate variability suggest that these two physiological processes are both regulated by complex feedback and feed forward systems. Inputs from multiple regulatory systems are integrated to yield the net driving signal. Yet, the key factor that determines whether the response is fractal is whether these input signals interact in a nonlinear manner. The breath-by-breath variability might contain information that is used in the regulatory process.

Acknowledgements

This research was supported by the Natural Sciences and Engineering Research Council of Canada.

REFERENCES

1. BOLTON, D.P.G. and J. MARSH. Analysis and interpretation of turning points and run lengths in breath-by-breath ventilatory variables. *J. Physiol. (London)* 351:451-459, 1984.

2. BUTLER, G.C., Y. YAMAMOTO, H.C. XING, D.R. NORTHEY, and R.L. HUGHSON. Heart rate variability and fractal dimension during orthostatic challenges. *Am. J. Physiol. Heart Circ. Physiol.* submitted:1991.

3. ELDRIDGE, F.L., D. PAYDARFAR, P.G. WAGNER, and R.T. DOWELL. Phase resetting of respiratory rhythm: effect of changing respiratory "drive". *Am. J. Physiol.* 257:R271-R277, 1989.

4. GRODINS, F.S. and S.M. YAMASHIRO. *Respiratory Function of the Lung and its Control.* New York: MacMillan Publishing Co., Inc., 1978.

5. HLASTALA, M.P., B. WRANNE, and C.J. LENFANT. Cyclical variations in FRC and other respiratory variables in resting man. *J. Appl. Physiol.* 34:670-676, 1973.

6. HUGHSON, R.L., D.R. NORTHEY, H.C. XING, B.H. DIETRICH, and J.E. COCHRANE. Alignment of ventilation and gas fraction for breath-by-breath respiratory gas exchange calculations in exercise. *Comput. Biomed. Res.* 24:118-128, 1991.

7. KOBAYASHI, M. and T. MUSHA. 1/f Fluctuation of heartbeat period. *IEEE Trans. Biomed. Eng.* BME-29:456-457, 1982.

8. LAMARRA, N., B.J. WHIPP, S.A. WARD, and K. WASSERMAN. Effect of interbreath fluctuations on characterizing gas exchange kinetics. *J. Appl. Physiol.* 62:2003-2012, 1987.

9. LENFANT, C. Time-dependent variations of pulmonary gas exchange in normal man at rest. *J. Appl. Physiol.* 22:675-684, 1967.

10. MANDELBROT, B.B. *The Fractal Geometry of Nature.* New York: W.H. Freeman and Co., 1982.

11. SAMMON, M.P. and E.N. BRUCE. Vagal afferent activity increases dynamical dimension of respiration in rats. *J. Appl. Physiol.* 70:1748-1762, 1991.

12. SAUL, J.P., P. ALBRECHT, R.D. BERGER, and R.J. COHEN. Analysis of long term heart rate variability: methods, 1/f scaling and implications. *Comp. Cardiol.* 14:419-422, 1988.

13. WHIPP, B.J. and S.A. WARD. Cardiopulmonary coupling during exercise. *J. Exp. Biol.* 100:175-193, 1982.

14. YAMAMOTO, Y. and R.L. HUGHSON. Coarse graining spectral analysis: a new method for studying heart rate variability. *J. Appl. Physiol.* 71:1143-1150, 1991.

15. YAMAMOTO, Y., Y. NAKAMURA, G.C. BUTLER, and R.L. HUGHSON. Fractal dimension of heart rate variability and physiological stress. *Am. J. Physiol. Heart Circ. Physiol.* submitted:1991.

16. YAMASHIRO, S.M. Distinguishing random from chaotic breathing pattern behavior. In: *Modeling and Parameter Estimation in Respiratory Control*, (ed KHOO, M.C.K. New York:Plenum Press, 1989), 137-145.

THE TIME COURSE OF MIXED VENOUS BLOOD GAS COMPOSITION FOLLOWING EXERCISE ONSET

Richard Casaburi

Division of Respiratory and Critical Care Physiology and Medicine
Harbor-UCLA Medical Center
Torrance, CA, USA 90509

INTRODUCTION

Some of our most profound insights into respiratory control mechanisms during exercise have come from the examination of dynamic responses. In particular, the events during the first few seconds after exercise begins have been scrutinized. The consideration of ventilatory and gas exchange responses early in exercise began auspiciously almost 80 years ago with the publication of "The regulation of respiration and circulation during the initial stages of muscular work"[1]. Krogh and Lindhard measured both ventilation and oxygen uptake in the first few seconds of exercise and found that both increased with no measurable latency. From this observation they drew two profound conclusions:

> *"If the excess of CO_2 produced in the muscles (at exercise onset) were responsible for the rise in ventilation there must be a latent period until the blood could reach the respiration centre... The mechanism which shall produce the abrupt changes must be a nervous mechanism."*
>
> *"During the first 6-10 seconds at least the venous blood reaching the lungs must have practically the same oxygen percentage as during rest and as the arterial blood is always practically saturated with oxygen the increase in O_2 absorption can be due only to an increase in circulation rate and must be very nearly proportional to that increase"*

The former statement is, in essence, a major foundation of "neurogenic" theories of the regulation of exercise hyperpnea. The latter statement has been used by many investigators to use the initial changes in oxygen uptake to infer changes in cardiac output.

Control of Breathing and Its Modeling Perspective, Edited by
Y. Honda *et al.*, Plenum Press, New York, 1992

Both conclusions are based on the reasonable assumption that circulatory transport from the exercising muscles to the central circulation means there is a latent period before mixed venous PCO_2 and oxygen content start to change. Though precise modelling of the circulatory transit time between the onset of exercise and the time of arrival of the products of exercise metabolism at the lung is difficult, Barstow and Mole[2] calculated that the transit time should be in the range of 12-17 seconds. However, the premise that mixed venous blood gas composition remains unchanged in the first few seconds after exercise onset has never been explicitly confirmed. To do this, a protocol was designed which allowed frequent sampling of blood from the pulmonary artery during the rest to exercise transition.

METHODS

Six healthy young male subjects participated in this study which involved abrupt transitions from rest to moderate exercise levels (averaging 170 watts) on an electromagnetically braked cycle ergometer. The reader is referred to reference 3 for full details of these procedures.

Prior to exercise testing, each subject had a catheter inserted percutaneously into an antecubital vein and advanced toward the pulmonary artery. The catheter was a thin (1.3 mm diameter) polyethylene tube whose flexibility allows it to be guided by blood flow through the venous system and cardiac chambers[4]. A pressure transducer attached to the catheter allowed confirmation of the catheter tip's progression through the right heart and into the pulmonary artery.

To obtain the desired temporal resolution in blood sampling, an anaerobic sampling manifold was designed. A roller pump withdrew blood from the catheter at a constant rate and directed it toward a series of 19 syringes. A computer sequentially activated solenoids which controlled pinch valves on each syringe and enabled sampling intervals as short as 3 seconds. Because the volume of the catheter and sampling manifold was greater than the blood sample size (averaging 2.2ml and 0.7ml, respectively), explicit correction for the time delay between blood entry into the catheter tip and arrival at the collecting syringe was made.

Each subject performed 5-7 exercise transitions. After a 3 minute period of rest, 4 minutes of exercise was performed. The manifold sampling was timed so that 2-3 samples were obtained at rest and the remainder during the first minute of exercise. Thereafter, samples were drawn manually. All blood samples were analyzed for PO_2, PCO_2 and pH; oxygen saturation was calculated by assuming a normal oxyhemoglobin dissociation curve.

RESULTS

In these six subjects we were able to record 12 studies in which the catheter was in the pulmonary artery, 5 studies in which the catheter was in the right atrium and 4 studies in which the catheter had been pulled back to the superior vena cava. In these studies the manifold sampling interval was 3 sec/sample in 6 studies, 4 sec/sample in 12 studies and 6 sec/sample in 3 studies. To obtain average time courses of response, the period between 20 seconds before and 48 seconds after exercise onset was divided into 4 second intervals and blood composition measurements from all samples whose mid-sample time fell within each 4 second interval were averaged.

Figure 1 shows the changes in mixed venous O_2 saturation and PCO_2 that occur immediately after the onset of exercise. Note that, contrary to expectations, mixed venous O_2 saturation falls with no observable delay. In fact, by 6 seconds after exercise onset, O_2 saturation has dropped from 71% to 61%, which is roughly one-third of the total steady state change. Mixed venous O_2 saturation achieves a new steady state by about 120 seconds; the half-time of decrease is approximately 32 seconds.

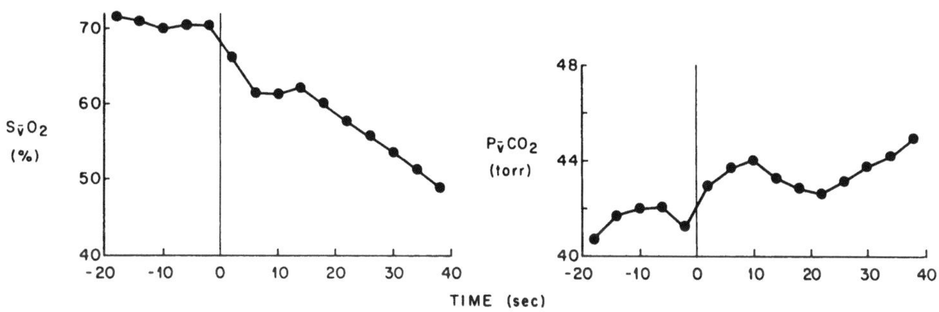

Figure 1: Averaged time course of pulmonary arterial blood composition at the transition from rest to exercise. Left: mixed venous oxygen saturation. Right: mixed venous partial pressure of carbon dioxide. (Modified from reference 3.)

Mixed venous PCO_2 also shows distinct changes in the first few seconds after exercise onset. There is an approximately 2 torr rise in the first 10 seconds, but little subsequent change through the first 30 seconds of exercise. The most rapid PCO_2 change occurred in the interval between 60 and 90 seconds after exercise onset; the half-time of response was roughly 80 seconds. The slower PCO_2 response is presumably related to the larger body stores for CO_2 than for O_2.

The time course of right atrial SO_2 and PCO_2 were virtually identical to those of mixed venous blood. Specifically, rapid changes in each variable occurred within a few seconds of exercise onset. In contrast, blood sampled from the superior vena cava showed virtually no changes until 40 seconds after exercise onset.

DISCUSSION

The finding that mixed venous blood gas composition changes with no demonstrable latency following exercise onset is startling. We have searched for possible explanations. Recently obtained data suggests that hypoxic and hypercapnic blood pooled in the legs is propelled upward with little latency by the muscle pump[5] as exercise starts[6]. Further, we have discovered that other forms of exercise transition (unloaded pedalling to exercise, supine rest to exercise and exercise to rest) do not feature an abrupt change in mixed venous O_2 saturation and PCO_2[7]. This likely explains why De Cort et al.[8], who recently inferred the time course of mixed venous SO_2 following the unloaded pedalling to exercise transition from the Fick principle, did not detect an abrupt fall in calculated mixed venous SO_2.

The findings of this study imply that the conclusions first stated by Krogh and Lindhard are in need of revision. The initial increase in oxygen uptake seen at the onset of exercise is due to a fall in mixed venous O_2 saturation as well as an increase in cardiac output. Further, a potential humorally-borne stimulus to the chemoreceptors reaches the central circulation shortly after exercise begins. However, it seems unlikely that these finding identify the CO_2-linked signal sought to explain hyperpnea early in exercise[9], since the magnitude of the early mixed venous PCO_2 change is relatively small and no convincing evidence for respiratory chemoreception in the venous circulation has been obtained[9].

ACKNOWLEDGMENT

Drs. James E. Hansen, James Daly and Richard M. Effros participated in these studies. R. Casaburi is an Established Investigator of the American Lung Association of California.

REFERENCES

1. A. Krogh and J. Lindhard, The regulation of respiration and circulation during the initial stages of muscular work, J. Physiol. 47:112-136 (1913).
2. T.J. Barstow and P.A. Mole, Simulation of pulmonary O_2 uptake during exercise transients in humans, J. Appl. Physiol. 63:2253-2261 (1987).
3. R. Casaburi, J. Daly, J.E. Hansen and R.M. Effros, Abrupt changes in mixed venous blood gas composition after the onset of exercise, J. Appl. Physiol. 67:1106-1112 (1989).
4. M.M. Scheinman, J.A. Abbott and E. Rapaport, Clinical use of a flow-directed right heart catheter, Arch. Intern. Med. 124:19-24 (1969).
5. B. Folkow, V. Haglund, M. Jodal and O. Lundgren, Blood flow in the calf muscle of man during heavy rhythmic exercise, Acta Physiol. Scand. 81:157-163 (1971).
6. R. Casaburi, W. French, W. Stringer, T. Barstow and P. Abdel-Sayed, Mechanism of the abrupt fall in mixed venous oxygen saturation at the onset of exercise, Med. Sci. Sports Exercise 23:S132 (1991).
7. R. Casaburi, C. Cooper, R.M. Effros and K. Wasserman, Time course of mixed venous oxygen saturation following various modes of exercise transition, FASEB J. 3:849 (1989).
8. S.C. De Cort, J.A. Innes, T.J. Barstow and A. Guz, Cardiac output, oxygen consumption and arteriovenous oxygen difference following a sudden rise in exercise level in humans, J. Physiol. 441:501-512 (1991).
9. K. Wasserman, B.J. Whipp and R. Casaburi, Respiratory control during exercise, in: "Handbook of Physiology. The Respiratory System. Control of Breathing.", A.P. Fishman, Ed., Am. Physiol. Soc., Bethesda, (1986).

HUMORAL ASPECTS OF THE VENTILATORY REACTIONS TO EXERCISE BY FLUX INHALATION OF CARBON DIOXIDE

Poul-Erik Paulev*, Yoshimi Miyamoto,
Michael John Mussell, and Kyuichi Niizeki

Department of Information Engineering, Faculty of Engineering
Yamagata University, Yonezawa, 992 Japan
*University of Copenhagen, Medical Physiology
Panum Institute, Copenhagen, Denmark

INTRODUCTION

The human respiratory controller (RC) accurately matches ventilation to carbon dioxide output during exercise, and often maintains the average alveolar (PACO2) and arterial (PaCO2) CO2 tension at a normocapnic level, close to that at rest (1, 7).

When a resting person inspire air with high CO2 concentration, the CO2 flux (mol/min) into the alveoli is directly dependent on ventilation, and forces PACO2 to be elevated. Instead of a rapid, initial ventilatory rise seen in dynamic exercise, there is a time delay of 15 s before ventilation starts to rise exponentially to a steady-state. The RC allows the arterial CO2 tension (PaCO2) to rise, and the steady-state hypercapnia depends on the level of the inhaled CO2 (1, 2, 3).

Metabolic CO2 diffuses continuously from the venous blood into the alveoli throughout the respiratory cycle, so during expiration PACO2 will rise, while inhaled fresh air dilutes the tension during inspiration. These alveolar oscillations are transmitted to the arterial blood as oscillations of PaCO2. Yamamoto (8) was first to consider the pulmonary oscillations in blood-gas tensions as a feed-forward factor in the control of exercise hyperpnea. Since then, several scientists have analyzed the oscillation hypothesis and found the oscillations in the alveoli phase-locked to chose of the arterial chemoreceptors (2, 5, 6). Oscillations have been detected at the carotid chemoreceptors and their neural discharge (2). The slope of the arterial oscillations is well correlated with the CO2 output. The carotid chemoreceptors respond to rapid changes in PaCO2, and exhibit CO2 oscillation rate-sensitive-properties (2).

The purpose of this analysis is to determine the ventilatory responses to constant flux CO2 inhalation and to moderate dynamic exercise in man. We avoided an early inspired CO2 bolus by the use of a large inspiratory dead space. The discussion will reveal whether the effective signal for the regulation of the steady-state ventilation is CO2 itself.

MATERIAL AND METHODS

11 healthy male subjects with no history of cardio-pulmonary disorders participated in these tests with informed consent. Their median age were 22,

Control of Breathing and Its Modeling Perspective, Edited by
Y. Honda *et al.*, Plenum Press, New York, 1992

median height was 1.71 m, and their median weight was 68 kg. Each subject participated in two CO2 inhalations tests (procedures 1 and 2), and one exercise test (procedure 3), which were performed on separate days.

Procedure 1: CO2 flux. In contrast to our first investigation (3), we used a large, inspiratory dead-space, which was filled with constant flux of CO2. We started with an exponential rise in CO2 over the first few inspirations. Thereby we avoided any artifacts due to a substantial inspiratory bolus of CO2. Each subject inhaled a constant flux of 0.3, 0.5 and 0.7 l/min STPD injected into the inspired air at rest.

Procedure 2: CO2 flux combined with 50 W exercise on a bicycle ergometer. The exercise was combined with a flux of 0.2, 0.4 and 0.7 l/min of CO2, and also here we have no early inspired pulse delivery of CO2.

Procedure 3: The work rate on the bicycle ergometer was 25, 50 and 75 W. The procedure was pure exercise, which was continued until steady-state was achieved.

RESULTS AND DISCUSSION

The correlation coefficient between ventilation and CO2 production was excellent (0.99) during the steady state of exercise. We found an additive effect of **Constant CO2 Flux** and exercise on ventilation. The human RC chooses a ventilation sufficient to eliminate metabolic CO2.

Actually, the metabolic CO2 production was found constant at around 0.3 l STPD/min for all levels of C-Flux (Fig. 1). There was no blocking effect on metabolic CO2 of the inhaled CO2. This was earlier found when inhaling a constant fraction of CO2 (3). By adding a background of 50 W exercise to the CO2 Flux, the CO2 output was constant around 0.9 l/min. However, at the high CO2 flux of 0.7 l/min a much higher metabolic CO2 production was observed, perhaps due to the increased respiratory work (Fig.1).

Ventilation increased linearly with the total CO2 delivered to the alveoli, whether we fluxed CO2 from the outside or delivered it from the working muscles during exercise (Fig. 2). Exercise and the combination of exercise and C-Flux inhalation seem to be a comparable stimulus to the RC (Fig. 2).

The classical CO2 response curves with ventilation as ordinate and alveolar CO2 tension as abscissa show a small rise in tension with increasing levels of CO2 Flux at rest (Fig. 3). The light exercise was in the normocapnic range (Fig. 3), and the combination of exercise and CO2 Flux showed

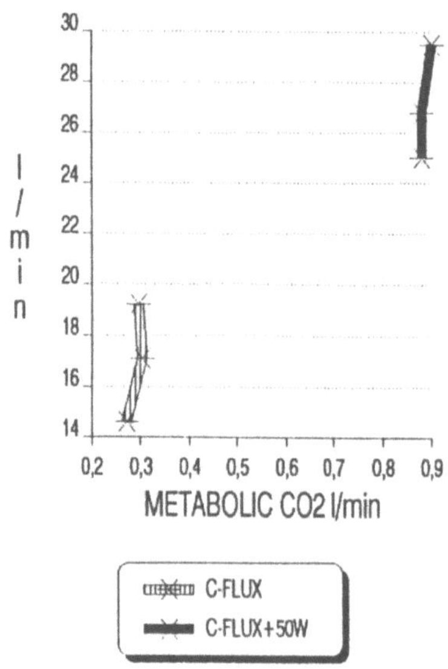

Fig. 1

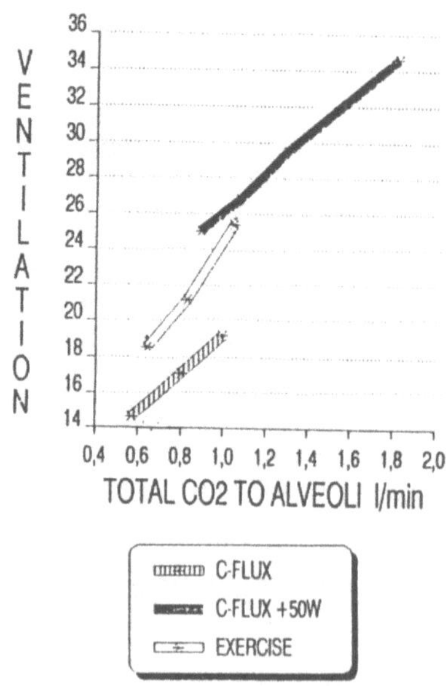

Fig.2

normocapnia and slight hypercapnia. We found no corelation between ventilation and alveolar CO2 tension during steady state exercise (Fig. 3). Thus the main control of ventilation during steady state exercise can hardly be a negative feed back system in which ventilation is regulated to give a constant PaCO2, since such a system must operate with an "error" signal. We seem to choose a ventilation that minimize any consequential rise in CO2 tension and thus the unpleasant sensation of dyspnea (Fig. 3).

Thus, the RC prefers a ventilation that eliminates both the metabolic, and the inhaled CO2. Whether we apply CO2 into the alveoli through venous blood or inhaled or as a combination of both, the RC will still try to eliminate any consequential surplus.

We confirmed previous findings that the CO2 flux to the alveolar air is an important determinant of ventilatory regulation (3, 7).

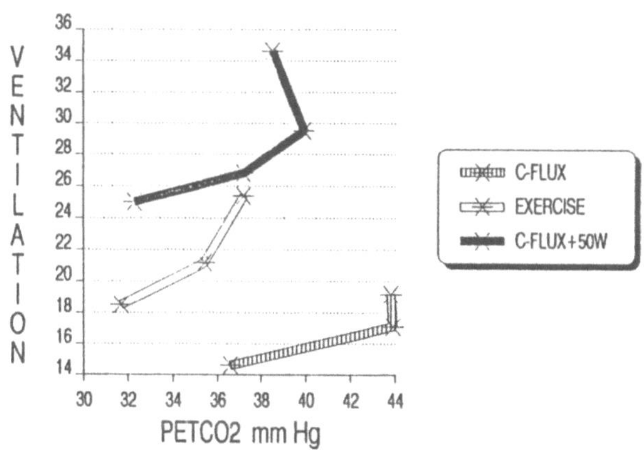

Fig. 3

Since exercise is normocapnic, RC must be considered as having "error free" feed-back regulation due to feed forward factors. Such feed forward factors may be neural factors (4) or receptors sensitive to small changes in CO2 tension.

Oscillations in PaCO2 could be the mediating pathway, which informs the RC of the amount of CO2 that is in the alveoli independent of source. But the carotid body or other CO2/dyspnea sensitive receptors may only be a potential source of information for the RC strategy.

The optimum controller hypothesis (5) implies that the RC will choose to eliminate the constant, alveolar CO2 flux by increasing ventilation, since the work of breathing is relatively small.

Another possible strategy is the concept that the subject chooses ventilation in order to minimize his conscious sensation of dyspnea by relatively small increases in ventilation.

CONCLUSIONS

1. Ventilation is matched to alveolar CO2 flux.
2. Alveolar CO2 flux seems to be a universal stimulus.
3. The mechanism linking exercise CO2 output to ventilation and inhaled, alveolar CO2 flux to ventilation is not resolved.
4. The sensing mechanism is most likely on the arterial side, and it could be related to the sensation of dyspnea.

ACKNOWLEDGEMENT

The excellent assistance of Kirsten McCord is appreciated. Dr. Paulev acknowledges the financial support from the Scandinavia-Japan-Sasakawa and the Fullbright Foundations.

REFERENCES

1. ASMUSSEN, E. 1983. Control of ventilation in exercise. Excercise and Sports Sciences Review 11: 24-54.

2. CUNNINGHAM, D.J.C.,P.A.ROBBINS, C.B. WOLF, 1986. Integration of respiratory responses to changes in alveolar partial pressures of CO2 and 02 and in arterial pH. In: Handbook of Physiology, Section 3: The Respiratory System. Vol. II. Control of Breathing, ed. by Cherniack, N.S. and Widdicombe, J.G. Am. Physiol. Soc. 475-528.

3. MUSSELL, M.J., PAULEV, P.-E., MIYAMOTO, Y., NAKAZONO, Y., AND SUGAWARA, T. 1990. A constant flux of carbon dioxide injected into the airways mimics metabolic carbon dioxide in exercise. Jpn. J. Physiol. 40: 877-891.

4. PAULEV,P.-E., 1971. Respiratory and cardiac responses to exercise in man. J. Appl. Physiol. 30: 165-172.

5. POON, CHI-SANG, 1987. Ventilatory control in hypercapnia and exercise: optimization hypothesis. J. Appl. Physiol. 62(6): 2447-2459.

6. SAUNDERS, K.B., 1980. Oscillations of arterial CO2 tension in a respiratory model: Some implications for the control of breathing in exercise. J. Theor. Biol. 84: 163-179.

7. WASSERMAN, K., B.J.WHIPP, R.CASABURI, 1986. Respiratory control during exercise. Handbook of Physiology, Section 3: The Respiratory System. Vol II: Control of Breathing, Chapter 17. Am. Physiol. Soc., Bethesda, Maryland.

8. YAMAMOTO, W.S., 1960. Mathematical analysis of the time ,course of alveolar CO2. J. Appl. Physiol. 15(2): 215-219.

NONLINEAR DYNAMICS OF CARDIOPULMONARY RESPONSES DURING EXERCISE

S.M. Yamashiro[1], P.K. Yamashiro[2], R.W. Glenny[3], and H.T. Robertson[3]

[1]Biomedical Engineering Department
University of Southern California
Los Angeles, CA 90089-1451
U.S.A.

[2]Bioengineering
[3]Dept. of Medicine
University of Washington
Seattle, WA 98195

INTRODUCTION

Interpretation of cardiopulmonary transients due to a step change in exercise in terms of fast (neural or central command) and slow (humoral or reflex) components was initiated by Krogh and Lindhard[1] and further developed by Dejours [2]. The respective roles of humoral and reflex components have been difficult to define due to the apparent lack of an error signal in the blood gases or arterial blood pressure. A recent review by Rowell and O'Leary[3] has proposed a specific hypothesis dealing with sympathetic reflexes during exercise involving muscle chemoreceptors and baroreceptors. According to their reasoning, in human exercise below a heart rate of 100 beats/min. vagal withdrawal dominates heart rate control largely by central command. Above this range, sympathetic reflexes involving muscle chemoreceptors and baroreceptors become important. Since vagal and sympathetic efferent cardiac nerve stimulation lead to different dynamics[4], a shift from vagal to sympathetic control mechanisms should lead to identifiable differences in transient response dynamics. Muscle chemoreceptor[5] and baroreceptor [6,7] stimulation can both lead to ventilatory response so ventilatory transient dynamics could be coupled to cardiovascular transients. In this way, some of the reflex or humoral component of ventilatory adjustments during exercise could be accounted for. While exercise transients have been extensively studied[8,9,10,11], the currently available experimental evidence does not provide a clear indication of whether a shift from vagal to sympathetic cardiovascular mechanisms is associated with a change in heart rate dynamics. Further, it is uncertain whether cardiovascular transients are coupled with corresponding ventilatory transients. A change in dynamics as a function of exercise level constitutes nonlinear system behavior. This study is directed at exploring whether such a nonlinearity is present.

Methods

The data were collected from 10 normal subjects (2 women and 8 men, ages 26 to 49) selected from a group of 15 subjects with a wide range of aerobic capacity. The activity levels ranged from resolutely sedentary to fit amateur athletes. Data from 5 additional

Control of Breathing and Its Modeling Perspective, Edited by
Y. Honda *et al.*, Plenum Press, New York, 1992

subjects were not analyzed because of heart rates above 100 beats/minute during loadless pedalling, or gross technical artifacts (EKG electrode or mouthpiece problems) during the transient analysis periods which did not allow at least two repetitions to be averaged. All subjects were non-smokers. Testing of the female subjects was scheduled so that all tests were performed during the follicular stage of their menstrual cycle to minimize between test variability(14). The subjects all fasted overnight and were tested during the early morning. All subjects were naive to the purpose of this study concerning comparison of transient responses.

The subjects were seated on an electromagnetically braked ergometer (MFE Ergometer 400L) connected to a Sensormedics 4400 exercise system, which recorded breath-by-breath measurements of ventilation and gas exchange. Prior to each study, the rotor volume transducer (Interface Associates) was calibrated with a syringe and the calibrations of the infrared CO_2 and oxygen fuel cell were checked. Three electrocardiographic electrodes were attached for continuous monitoring during the test. Subjects had nose clips applied and breathed through a mouthpiece to permit collection of all exhaled gas. Continuous measurements of minute ventilation, heart rate, oxygen consumption and carbon dioxide output were collected for each test.

The protocol for each subject was determined from a progressive maximal test with steady state power outputs of 25% and 50% of the maximal power output selected. At steady state, this corresponded to approximately 30% and 60% of the maximal oxygen uptake. The subsequent tests used for all data analysis consisted of a ten minute period of unloaded pedalling, ten minutes of 30% maximal power output and ten minutes of 60% maximal power output with transitions to 30% (0-1) and 60%(1-2) made in a step fashion. Ergometer load transitions to a step command input were found to occur at exactly a 11 watts/second rate due to internal controls. Each subject repeated all three exercise levels 2 to 3 times and maintained a pedalling speed of 70 cycles/minute for each exercise level.

The breath-by-breath data collected for expired minute ventilation and heart rate was analyzed in a similar way as described by Broman and Wigertz[8]. Interpolation was performed to estimate step responses with a one second sampling rate. All step response repetitions for each subject were ensemble averaged prior to analysis. One of the following equations was fitted to each measured step response over a 300 second period:

$$f_1(t)=a_1+a_2(1-e^{-t/\tau}) \qquad (1)$$

$$f_2(t)=a_1+a_2(1-e^{-t/\tau}) + a_3\, t \qquad (2)$$

Equation (1) was used for all minute ventilation responses and the 0-1 heart rate response. Equation (2) was used for the 1-2 heart rate responses which showed an increasing trend, as indicated by an estimated a_3 parameter where the null hypothesis of a zero value could be rejected ($p<0.05$). Parameters a_1,a_2,a_3 and τ were estimated using a numerical least-squares fitting algorithm of the KALEIDAGRAPH software package on a MACINTOSH II-CI computer. Population inference was based on the group averaged transient response as will now be described.

To minimize effects of inter-individual differences on the group average, individual responses were normalized by individual steady-state levels(a_1+a_2 or $a_1+a_2+300*a_3$) prior to weighting and summation to form an overall group mean response[8]. The last term in Equation (2) has no steady-state value so the value at 300 seconds was arbitrarily chosen. The weighted group mean is defined by:

$$\bar{f}=\sum_{i=1}^{N} w_i f_i \qquad (3)$$

$$\sum_{i=1}^{N} w_i = 1 \qquad (4)$$

where $w_i = K / (\text{steady state value})_i$, f_i = i th step response , and K was chosen to satisfy Equation (4). Note that if all steady-state values were the same for all subjects, Equation (3) would be equivalent to the conventional ensemble average($w_i = 1/N$). To assess variability following normalization, individually weighted terms of Equation (3) were multiplied by N prior to cumulation of sum of squares and conventional estimation of standard deviations at each point in time.

Results

A representative set of step responses of heart rate and minute ventilation and corresponding fits to equations is shown in Figures 1 and 2 for one individual. Equation 1 was used to fit all but the 1-2 heart rate response, for which Equation 2 was used instead. In this subject, both heart rate and minute ventilation time constants increased by about 10 seconds in the 1-2 step compared to the 0-1 step. The 0-1 step involved heart rates below 95 beats/min. and the 1-2 step was associated with rates which rose above 100 beats/min.. This change of response dynamics occurred with nearly a two-fold difference in heart rate and minute ventilation time constants for the 0-1 step. If Equation 1 had been used to fit all responses, the estimated 1-2 heart rate time constant increases by 5.3 seconds. Since this represents a significant fraction of the measured difference in time constants, correction for any linear trend or slow dynamics is important.

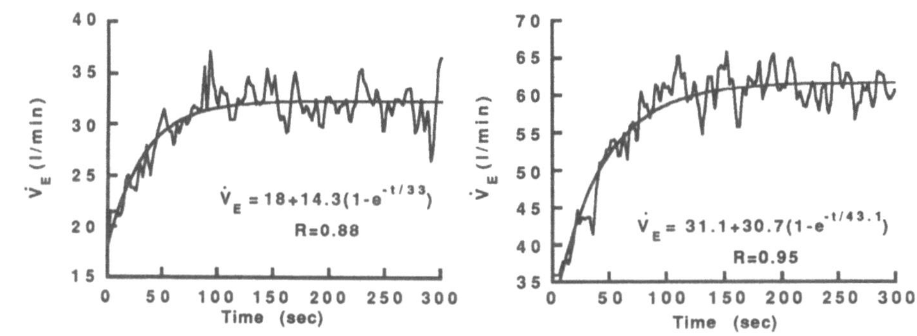

Figure 1. Individual (Subject St) heart rate response to 0-1 step (left) and 1-2 step (right). Smooth curves correspond to equation fits. Data correspond to breath-by-breath samples interpolated to form one second samples, with repetitions then ensemble averaged (n=2). R values represent the overall regression coefficients of the equation fits 87

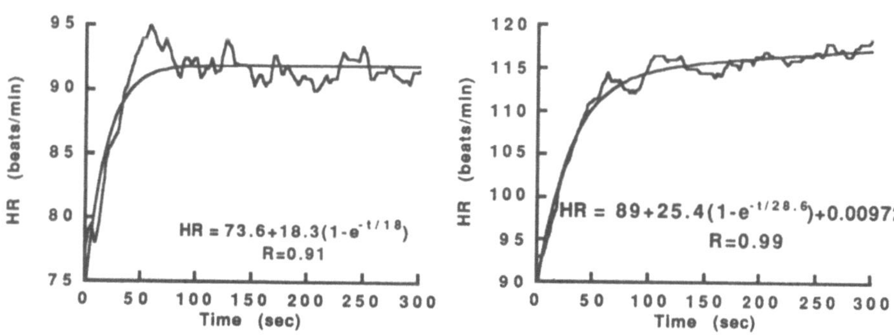

Figure 2. Individual (Subject St) minute ventilation responses to 0-1 step (left) and 1-2 step (right). (See Figure 1 caption for other details.)

Table 1. Statistical comparison of group averaged responses. Standard errors are in parentheses.

Step	Heart Rate τ (sec)	Minute ventilation τ (sec)
0-1	20.1 (0.618)	47.4 (1.51)
1-2	37.5 (0.771)	62.5 (1.75)
difference	17.4 $p<0.01$	15.1 ($p<0.01$)

Table 1 shows the results of a statistical comparison of time constants estimated from the averaged transients. The results suggest a correlated change in heart rate and minute ventilation time constants.

Discussion

Previous comparisons of heart rate time constants at different exercise levels (2,6) stressed the linearity of such responses which implies an unchanging time constant . The present results show a clear load dependency which challenges this conclusion and provides evidence for nonlinear exercise dynamics.

REFERENCES

1.A. Krogh and J. Lindhard, The regulation of respiration and circulation during the initial stages of muscular work. J. Physiol. London 47: 112-136, (1913).

2.P. Dejours, Neurogenic factors in the control of ventilation during exercise. Circ. Res. 20, Suppl. 1: 146-153, (1967).

3. L.B.Rowell and D.S. O' Leary, Reflex control of the circulation during exercise: chemoreflexes and mechanoreflexes. J.Appl. Physiol. 69: 407-418, (1990).

4. H.R.Warner and A. Cox,. A mathematical model of heart rate control by sympathetic and vagus efferent information. J. Appl. Physiol. 17: 349-355, (1962).

5.D.L.McCloskey and J.H. Mitchell, Reflex cardiovascular and respiratory responses originating in exercising muscle. J. Physiol. Lond. 224: 173-186, (1972).

6. B.Bishop, Carotid baroreceptor modulation of diaphragm and abdominal muscle activity in the cat. J. Appl. Physiol. 19: 12-19, (1974)

7. D.D. Heistad, F. Abboud, A.L. Mark, and P.G. Schmid, Effect of baroreceptor activity on ventilatory response to chemoreceptor stimulation. J. Appl. Physiol. 39: 411-416, (1975).

8. S.Broman and O. Wigertz, Transient dynamics of ventilation and heart rate with step changes in work load from different load levels. Acta Physiol. Scand. 81: 54-74, (1971).

9. Y.Fujihara, J.R. Hildebrandt, and J. Hildebrandt. Cardiorespiratory transients in exercising man. I. Tests of superposition. J. Appl. Physiol. 35: 58-67, (1973).

10 Y.Fujihara, J. Hildebrandt, and J.R. Hildebrandt. Cardiorespiratory transients in exercising man. II. Linear models. J. Appl. Physiol. 35: 68-76, (1973).

11.L.B Rowell, Human Circulation. Regulation During Physical Stress. New York: Oxford Univ. Press, (1986).

VENTILATORY RESPONSE AT THE ONSET OF PASSIVE EXERCISE DURING SLEEP IN MAN

Koji Ishida,[1] Yoshifumi Yasuda,[2] and Miharu Miyamura[1]

[1]Research Center of Health, Physical Fitness and Sports
Nagoya University, Nagoya 464-01, Japan
[2]Toyohashi University of Technology

INTRODUCTION

Although the mechanisms of exercise hyperpnea especially in phase I has been obscure, central and/or peripheral neurogenic drive is thought to be the cause for rapid increase in ventilation at the onset of voluntary exercise. Many previous investigations proved that passive exercise also produced a rapid increase in ventilation. These results may suggest that peripheral neurogenic components should be the main cause for exercise hyperpnea. However, this passive exercise was performed in an awake condition when the higher center is still active. It is considered that the higher center may influence the respiratory center in that case. So it is necessary to clarify the ventilatory response in passive exercise while the higher center is not active, for example, during sleep conditions. On the other hand, Wasserman et al.[1] proposed cardiodynamic hypotheses that ventilatory response is linked to changes in cardiac output. To confirm this theory especially in phase I, cardiac output must be measured correctly within a short time at the onset of exercise. Moreover, in a supine position, the hemodynamic response is rather different from the upright position, which affects the cardiac output. Thus, it is required to reveal the relationships between the ventilatory and cardiac response to passive exercise during sleep in a supine position. The purpose of this study is, therefore, to elucidate whether or not 1) an abrupt increase in ventilation would be observed at the onset of passive exercise during sleep in man and 2) whether these changes in ventilation are linked with concomitant changes in cardiac output in a supine position.

METHODS

Five healthy male volunteers participated in the present study. The subjects first slept in a supine position with their legs extended. Electroencephalogram (EEG), electrooculogram (EOG) and electromyogram (EMG) in the mentalis muscle were monitored to determine the sleep stage by standard criteria. Passive exercise was induced when the sleep stage was confirmed as III or IV. The subjects had both their knee joints extended and flexed

Control of Breathing and Its Modeling Perspective, Edited by
Y. Honda *et al.*, Plenum Press, New York, 1992

alternatively by pulling ropes which were connected to the subject's heels at the rate of 60 times/min for about 8s. The passive exercise began at the latter half of the expiratory period. Tidal volume (V_T), respiratory frequency (f) and minute ventilation ($\dot{V}_I$) were measured by hot-wire flowmeter. End-tidal CO_2 partial pressure (P_{ETCO_2}) and O_2 partial pressure (P_{ETO_2}) were also measured by a rapid CO_2 and O_2 analyzer, respectively. Stroke volume (SV), heart rate (HR) and cardiac output ($\dot{Q}$) were evaluated by advanced impedance cardiogram. Spot electrodes and ensemble averaging technique were adopted to effectively avoid motion artifact. We selected 4s as the average time to synchronize the time course of one breath and cardiac responses. Electrocardiogram (ECG) was used to calculate HR and R spike was the trigger signal for averaging. Phonocardiogram (PCG) was measured to compute the left ventricular ejection time correctly. All signals were converted from analog to digital at the sampling frequency of 100Hz for 80s including 30 - 40s before and after passive exercise and analyzed by computer. The passive exercise repeated 2 - 12 bouts with appropriate time intervals until the subjects became awake. After the experiments, EEG signals were analyzed by frequency analysis to exclude the data in which the δ wave of EEG changed remarkably.

RESULTS

Table 1 represents the mean and standard deviation of each parameters of all subjects. 1 response indicates one breath in terms of respiration and mean value for 4s of cardiac parameters. The pre-exercise value is the average for 5 responses preceding the exercise. The first and second responses soon after the exercise start are represented as 1st and 2nd response. Mean response represents the mean value for 1st and 2nd responses. $\dot{V}_I$ increased remarkably and significantly ($p<0.01$) at the onset of exercise. This was caused by the increase in V_T and especially in f. P_{ETCO_2} decreased and P_{ETO_2} increased slightly but significantly. HR showed significant increase ($p<0.01$) while SV decreased slightly so that $\dot{Q}$ showed no change. Figure 1 indicates the relative change of mean response value compared with pre-exercise. It is noticeable that ventilatory parameters showed remarkable increase while expired gas and cardiac parameters showed slight changes. Figure 2 represents the relationship between Δ (mean response value minus pre-exercise value) $\dot{V}_I$ and Δ $\dot{Q}$ of all data. All Δ $\dot{V}_I$ showed positive value while Δ $\dot{Q}$ sometimes showed negative value.

Table 1. Cardiorespiratory responses to the passive exercise during sleep.

	$\dot{V}_I$ (l/min)	V_T (l)	f (breaths /min)	P_{ETCO_2} (Torr)	P_{ETO_2} (Torr)	HR (beats/min)	SV (ml)	$\dot{Q}$ (l/min)
Pre-exercise	6.3 ±.65	.416 ±.045	15.4 ±2.4	40.7 ±6.2	92.5 ±4.3	55.2 ±5.9	104.2 ±13.6	5.80 ±1.3
1st response	11.2** ±2.1	.638* ±.14	18.2* ±1.7	39.7 ±5.9	94.8* ±2.7	62.8** ±7.0	98.4 ±12.5	6.15 ±.74
2nd response	11.8** ±2.7	.575 ±.13	20.8** ±2.6	38.7** ±5.5	97.5** ±3.0	65.8** ±6.5	91.0 ±20.5	5.94 ±1.1
Mean response	11.5** ±2.1	.607 ±.11	19.5** ±1.9	39.2** ±5.7	96.2** ±2.8	64.3** ±6.4	94.7 ±15.4	6.04 ±.91

Values are mean±SD. (n=5), * shows significant difference from pre-exercise values (* $p<0.05$, ** $p<0.01$)

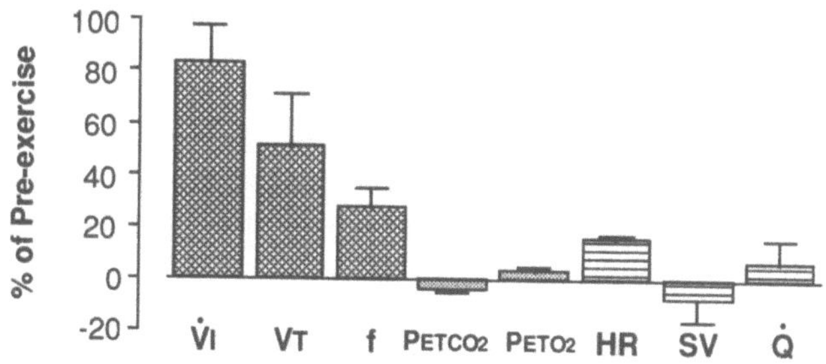

Figure 1. Relative changes of mean response value as compared with pre-exercise value.

DISCUSSION

In 1974, Wasserman et al.[1] proposed a now well-known theory that acute hyperpnea at the onset of exercise was considered to occur secondarily as the result of rapid increase in cardiac output. We attempted to apply this theory to the condition when the exercise was performed passively during sleep in a supine position. It is considered that hemodynamic response to the passive and supine exercise may be rather different from one which is voluntary and in the upright position. As shown in Figure 2, we observed that ventilation increased remarkably while cardiac output often decreased at the start of passive exercise within 8s. This was caused by a decrease in SV. Loeppky et al.[2] observed that beat-by-beat SV obtained by pulsed Doppler methods decreased at the onset of voluntary exercise within 20s in a supine position. Even though HR increase at the onset of exercise, if SV decreases because of a supine position, cardiac output doesn't always increase at the start of exercise during sleep in a supine position. Recently, Grucza et al.[3] observed no relationships between $\dot{Q}$ and $\dot{V}_E$ during the first 15s of both dynamic and rhythmic-static exercise in man. Furthermore, Morikawa et al.[4] reported that minute ventilation increased abruptly from the first breath during the voluntary and passive exercise while cardiac output increased gradually in voluntary exercise and showed no changes in passive exercise. These results, including ours, suggest that cardiodynamic theory alone cannot explain exercise hyperpnea, especially in phase I.

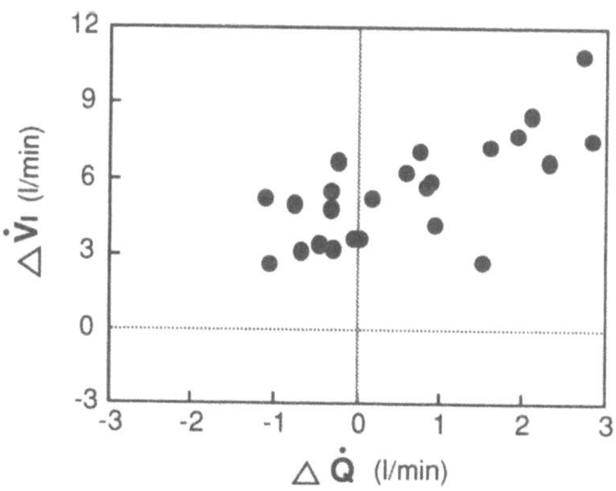

Figure 2. The relationships between $\Delta\dot{V}_I$ and $\Delta\dot{Q}$ of all data.

The neurogenic component has been thought to play an important role in exercise hyperpnea. This neurogenic component can be divided into two major factors: central and peripheral. Krogh and Lindhard[5] suggested that irradiation of the cortical motor impulse to the respiratory center might be responsible for the origin of the first component. In the present study, however, $\dot{V}_I$ actually increased at the onset of passive exercise during sleep when the motor cortex may not be active and the impulses from the motor cortex may not have reached the respiratory center. Additionally, Morikawa et al.[4] and Hida et al.[6] observed that ventilatory response to passive exercise was absent in patients having spinal cord transection or rhizotomy dogs. These results suggest that exercise hyperpnea especially in phase I was owed not to a central command from the motor cortex but to the peripheral neurogenic drive. As the passive exercise produces a change of angle of the knee joint, afferent from mechanoreceptors or other receptors in the knee joint may stimulate the respiratory center and cause the increase in ventilation. In the present study, however, the influence of the receptors and pathway of this reflex mechanism were not proven. Previous investigations postulated that group III or IV fibers play an important role in exercise hyperpnea and these thin afferent fibers may be connected to the polymodal receptors which respond to mechanical stimulations. On the contrary, in voluntary exercise in an awake condition when the higher center is active, it is possible that inhibitory and facilitatory inputs from peripheral and/or the central nervous system should complicatedly affect the respiratory center. These complex mechanisms may be integrated so that ventilation should increase abruptly at the onset of voluntary exercise. However, as the relationship between the higher center and respiratory center is still unknown, further investigation is needed.

CONCLUSION

A large increase was observed in ventilation at the onset of passive exercise during sleep in man and these changes did not match changes in cardiac output. These results suggest that exercise hyperpnea especially in phase I is mediated mainly by peripheral neurogenic drive.

REFERENCES

1. K. Wasserman, B.J. Whipp, and J. Castagna, Cardiodynamic hyperpnea: hyperpnea secondary to cardiac output increase. *J. Appl. Physiol.* 36:457-464 (1974).
2. J.A. Loeppky, E.R. Greene, D.E. Hoekenga, A. Caprihan, and U.C. Luft, Beat-by-beat stroke volume assessment by pulsed Doppler in upright and supine exercise. *J. Appl. Physiol.* 50:1173-1182 (1981).
3. R. Grucza, Y. Miyamoto, and Y. Nakazono, Kinetics of cardiorespiratory response to dynamic and rhythmic-static exercise in men. *Eur. J. Appl. Physiol.* 61:230-236 (1990).
4. T. Morikawa, Y. Ono, K. Sasaki, Y. Sakakibara, Y. Tanaka, R. Maruyama, Y. Nishibayashi, and Y. Honda, Afferent and cardiodynamic drives in the early phase of exercise hyperpnea in humans. *J. Appl. Physiol.* 67:2006-2013 (1989).
5. A. Krogh, and J. Lindhard, The regulation of respiration and circulation during the initial stages of muscular work. *J. Physiol. London* 47:112-136 (1913).
6. W. Hida, C. Shindoh, Y. Kikuchi, T. Chonan, H. Inoue, H. Sasaki, and T. Takishima, Ventilatory response to phasic contraction and passive movement in graded anesthesia. *J. Appl. Physiol.* 61:91-97 (1986).

VENTILATORY RESPONSES TO ASCENDING AND DESCENDING RAMP CHANGES IN INSPIRED CO2

Michael Mussell, Kyuichi Niizeki, Yoshimi Miyamoto

Laboratory of Biological Cybernetics
Department of Electrical and Information Engineering
Yamagata University
Yonezawa 992, Japan

INTRODUCTION

This study addresses two issues concerning ventilatory responses to ramp changes in inhaled CO2 which, unlike other forcing functions, have received relatively little attention.

(i) Miyamoto and Niizeki[1,2] observed asymmetrical cardioventilatory responses of humans to ascending and descending ramps in exercise load; ventilation closely tracked descending ramps but considerably lagged behind ascending ramps. The role of CO2 in this asymmetry was unclear, so one aim of this study was to determine if a similar asymmetry occurs in the ventilatory responses to ascending and descending ramps in inhaled CO2.

(ii) The respiratory controller closely matches ventilation to $\dot{V}$CO2 during aerobic exercise by a still-unknown mechanism.[3,4] Ventilation also closely matches a steady flow of CO2 injected into the inspired air of non-exercising subjects,[5] irrespective of the inspiratory dead space of the CO2 delivery equipment.[6] This implies ventilation is matched to alveolar CO2 flow, irrespective of whether the flow comes from venous blood or from the airways. A second aim of this study was to investigate the relationship between ventilation and dynamic changes in the flow of CO2 into inspired air.

EQUIPMENT AND PROTOCOL

Inspired CO2 was forced to follow step (control) and ramp patterns using three different inhaled CO2 delivery methods by various arrangements of equipment described elsewhere.[7] **(a)** Changes were forced in the inspired CO2 fraction - the 'FRAC' delivery - meaning the dynamics and amount of CO2 delivered to the alveoli depends on both ventilation and CO2 wash-in/wash-out. **(b)** Changes were forced in end-tidal CO2 using a closed-loop controller[7,8] - the 'FORCE' delivery - which ensures the dynamics and amount of CO2 delivered to the alveoli are independent of ventilation and CO2 wash-in/wash-out. **(c)** Changes were forced in the flow of CO2 injected into the inspiratory dead space of the equipment (estimated at 400ml) - the 'FORCE' delivery - meaning the amount of CO2 delivered to the alveoli is ventilation independent, while the dynamics of the CO2 delivery is ventilation dependent.

During all runs $\dot{V}$E, VT, f, TI, TE, FICO2 and FIO2, were recorded breath-by-breath. FECO2, FEO2, $\dot{V}$CO2 and $\dot{V}$O2 were also available for the FRAC and FLOW deliveries, but not for the FORCE delivery which entailed continuously recording FETCO2.

Control of Breathing and Its Modeling Perspective, Edited by
Y. Honda *et al.*, Plenum Press, New York, 1992

Six healthy male subjects (age range 21 to 35) participated with informed consent on 3 consecutive days: one day for the FRAC, one day for the FLOW and one day for the FORCE CO2 deliveries. Each day, subjects performed one step (control) and two ramp runs (all randomly ordered), separated by rest periods of at least 45 minutes to allow respiration to restabilize. All runs included 2 minutes before and 4 minutes after CO2 was added to the inspired air. Step runs comprised an on-step, a 5 or 7 minute long plateau of added CO2 and an off-step. Ramp runs comprised an ascending linear ramp to a 5 minutes long plateau, then a linear descending ramp back to no added CO2. Ramps took 3 and 5 minutes respectively to slew between no added CO2 and the plateau. The plateau levels of CO2 was 6.5% for the FRAC and FORCE delivery and 0.75 l/m CO2 flow for the FLOW delivery.

RESULTS AND ANALYSIS

The ensemble averaged CO2 stimulus and $\dot{V}E$ responses for the FRAC, FORCE and FLOW protocols are shown in figures 1a, 1b and 1c respectively. Figure 2 shows regression analyses of the $\dot{V}E$/CO2 flow data for the ascending and descending phases of the 3 and 5 minute ramp FLOW protocols. All correlation coefficients were excellent, being either 0.97 or 0.98. Similar analyses performed on the FRAC and FORCE ramp data (descending and ascending) yielded correlation coefficients between 0.95 and 0.98 for 5 minute ramps, and between 0.87 and 0.97 for 3 minute ramps. (Inhaled CO2 flow for the FRAC and FORCE deliveries was calculated as $\dot{V}E$ x FICO2.)

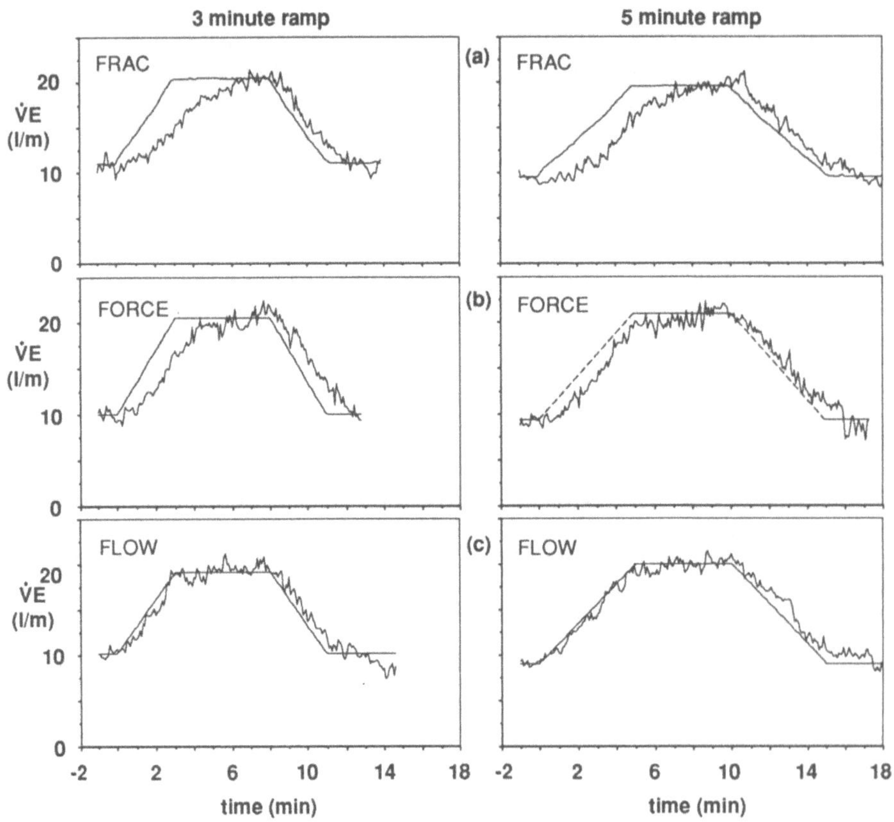

Figure 1. CO2 valve signal and ensemble averaged responses to 3 and 5 minute ramp changes in (a) the inspired fraction of CO2, (b) end-tidal CO2 and (c) the flow of CO2 into the inspired air.

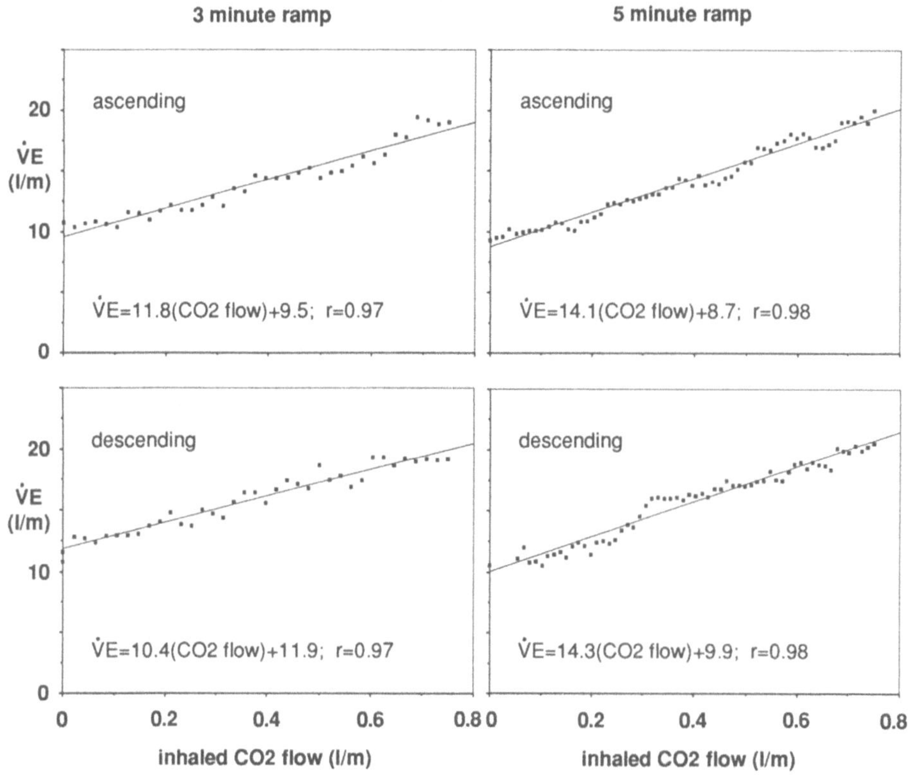

Figure 2. Linear regression analyses of the $\dot{V}E$/inhaled CO_2 flow relationship for the ascending and descending phases of the 3 and 5 minute ramp forcing functions for the FLOW CO_2 delivery. Linear regression equations are shown within each graphic frame.

DISCUSSION

Concerning Response Asymmetry

The $\dot{V}E$ responses to both 3 and 5 minute ramps in the inspired CO_2 fraction (figure 1a) are very asymmetric. The ascending responses are sluggish because resting $\dot{V}$ is low and time is needed for $PACO_2$ to wash-in. The descending responses, however, follow the forcing pattern more closely because $\dot{V}$ is elevated and CO_2 washes out faster.

The FORCE delivery overcomes the CO_2 wash-in delay and speeds up the ascending response to near that of the descending response. Consequentially, the end-tidal forcing responses (figure 1b) are more symmetric than the FRAC responses - in particular the $\dot{V}E$ response to the 5 minute ramp forcing pattern exhibits little asymmetry.

Of most interest, however, is that $\dot{V}E$ very closely tracks the flow of CO_2 into the inspired air with little deviation for both the 3 and 5 minute ramp protocols (figure 1c). Thus, since the $\dot{V}E$ responses to ramps in inhaled CO_2 flow show little or no asymmetry, we conclude that the asymmetrical exercise responses observed by Miyamoto et al. (1991) are unlikely due to the gas exchange dynamics of CO_2, though this does not discount another humoral mechanism or the influence of body CO_2 stores. Asymmetrical exercise responses are perhaps more likely, as Miyamoto suggests, to be due to the asymmetrical wash in/wash out and differing on/off concentration thresholds of some unidentified humoral substance which stimulates changes in both respiratory and cardiac control mechanisms.

Concerning the Relationship Between VE and Inhaled CO2 Flow

The ramp responses for the FLOW protocol (figure 1c) show $\dot{V}E$ accurately tracks steady and dynamic changes in inhaled CO2 flow in resting subjects. This is more formally demonstrated by the regression analyses between $\dot{V}E$ and inhaled CO2 flow (figure 2) which yield excellent correlation coefficients of between 0.97 to 0.98. This remarkable relationship, which is as profound as that seen between $\dot{V}$ and $\dot{V}CO2$ during aerobic exercise, is not only confined to the FLOW delivery. Similar regression analyses for the FRAC and FORCE deliveries also yield excellent correlation coefficients of between 0.95 and 0.98 (5 minute ramps) and between 0.87 and 0.97 (3 minute ramps). This CO2 delivery mode independence of the $\dot{V}$/inhaled CO2 flow interrelationship thus adds universality to this finding.

It is therefore apparent that the source of CO2 flow into the alveoli (be it from venous blood during exercise or from the airways at rest) is a universal stimulus to the respiratory controller (or a result of respiratory control) during both exercise and CO2 inhalation. However, the mechanism is unclear and the role of PaCO2 must still be explained.

While $\dot{V}E$ is closely tracking $\dot{V}CO2$ during exercise, PaCO2 remains near the resting level, but while $\dot{V}E$ is closely tracking an inhaled CO2 flow PACO2, and therefore PaCO2, is elevated - it is impossible for PaCO2 to remain at a resting level during any form of CO2 inhalation, but does the respiratory controller 'care' that it is elevated during CO2 inhalation.

The key issue is what the respiratory controller's priority is: is it (a) to match $\dot{V}$ to inhaled CO2 flow (which it does so well during both exercise and CO2 inhalation) regardless of the PaCO2 level, or (b) to allow PaCO2 to rise as an error signal of a poor closed-loop control system. This study supports idea (a) - that the priority of the respiratory controller is to somehow match $\dot{V}$ to CO2 flow into the alveoli, irrespective of the CO2 source and with a convenient disregard for elevated PaCO2 should it occur. By implication, our findings can be explained by the optimum controller hypothesis[9] or by the notion that some part of the ventilatory drive signal arises from a conscious sensation of dyspnea. An optimum respiratory control structure would soon discover that, unlike in the case of a FRAC or FORCE CO2 delivery, humoral imbalance can be reduced by relatively small increases in the work of breathing during the FLOW delivery. Alternatively, a subject would find that increasing ventilation can reduce their sensation of dyspnea during the FLOW delivery, but not during the FRAC or FORCE deliveries.

REFERENCES

1. Y. Miyamoto, K. Niizeki and Y. Nakazono, Asymmetrical on and off transients in gas exchange during ramp forcing in exercise load. Med. Biol. Eng. & Comp. 29 (Supp.) part 1, 457 (1991).

2. K. Niizeki, and Y. Miyamoto, Cardiorespiratory responses to cyclic triangular ramp forcings in work load. Jpn. J. Physiol. 41:759-773 (1991).

3. K. Wasserman, B.J. Whipp and J. Castagna, Cardiodynamic hyperpnea: hyperpnea secondary to cardiac output increase. J. Appl. Physiol. 36: 457-464 (1974).

4. Y. Miyamoto, T. Hiura, T. Tamura, T. Nakamura, J. Higuchi and T. Mikami, Dynamics of cardiac, respiratory and metabolic function in men in response to step work load. J. Appl. Physiol. 52(5): 1198-1208 (1982).

5. M.J. Mussell, P.E. Paulev, Y. Miyamoto, Y. Nakazono and T. Sugawara, A constant flow of CO2 injected into the airways mimics metabolic CO2 in Exercise. Jpn. J. Physiol. 40: 877-891 (1990).

6. A.R.C. Cummin, C.P. Patil, M.S. Jacobi and K.B. Saunders, Effect on ventilation of carbon dioxide delivered in early inspiration in man. Bull. Eur. Physiopathol. Respir. 23: 335-338 (1987).

7. M.J. Mussell, Y. Nakazono and Y. Miyamoto, A new CO2 inhalation system for studying regulation of breathing. Jpn. J. Physiol. 39: 635-642 (1989).

8. G.D. Swanson and J.W. Bellville, Step changes in end-tidal CO2: methods and implications. J. Appl. Physiol. 39: 377-385 (1975).

9. C.S. Poon, Ventilatory control in hypercapnea and exercise: Optimization hypothesis. J. Appl. Physiol. 62: 2447-2459 (1987).

EFFECT OF FEMORAL BLOOD FLOW OCCLUSION AND RELEASE ON VENTILATION DURING EXERCISE AND RECOVERY

Takayoshi Yoshida[1], Masahiko Ichioka[2], Mamoru Chida[2], Kouichi Makiguchi[2], Naoko Tojo[2], Ryuji Suga[2], Kouichi Tsukimoto[2] and Jun-ichi Eguchi[3]

1 Exercise Physiology Laboratory, Faculty of Health and Sport Sciences, Osaka University, Toyonaka, Osaka 560, Japan
2 Division of Respiratory Physiology and Medicine, Tokyo Medical and Dental University, Tokyo 113, Japan
3 Health and Sports Center, Komazawa University, Tokyo 154, Japan

INTRODUCTION

Since Zuntz and Geppert[11] proposed that exercise-induced hyperpnea is caused by a blood-borne substance, there has been a debate about roles of neural and humoral mechanisms in this phenomenon. Occlusion of blood flow to the leg has been used to eliminate humoral factors. However, this method has led to conflicting results regarding the mechanism of exercise-induced hyperpnea[2,4-7].

In the present study we examined the ventilatory response to exercise and its relationship to CO_2 output and to arterial K^+ ($[K^+]_a$) during and after occlusion of blood flow to the legs. By occluding blood flow, we sought to eliminate the effects of the products of metabolism on ventilation ($\dot{V}_E$) during exercise.

METHODS

Subjects: Nine healthy male subjects volunteered to participate. They were informed of the purpose and possible risks of the study, and gave their informed consent. The subjects ages were 19.6 ± 1.5 yr (mean ± SD), height 172 ± 7 cm, and weight 67 ± 7 kg.

Exercise Protocol: Each subject came into the laboratory on two different days and performed two exercise tests with circulatory occlusion during exercise. The exercise tests were performed with an electrically braked bicycle ergometer (Siemens-Elema 360), for 11 min at 80 W, which was below the lactate threshold for each subject. One thigh cuff (20 cm wide) were placed the upper part of each thigh. After 5 min of exercise, the thigh cuffs were inflated to approximately 26.7 kPa with an air compressor, and they were left inflated for 2 min. Thereafter, when the cuffs were released the subject continued to exercise for another 4 min. After the exercise (recovery period), data were collected for 4 min. During recovery, the cuffs were again iflated for 2 min. Then they were released, and data were collected for another 2 min. In the control study, there was no occlusion during recovery.

Gas Exchange Parameter Measurements: During the exercise test, $\dot{V}_E$, $\dot{V}_{O2}$ and $\dot{V}_{CO2}$ were measured with a computerized on-line breath-by-breath system (RM-300, Minato Medical

Science). Inspired and expired gas volumes were measured with a hot-wire respiratory flow system. The flow signals were electrically integrated over time to calculate $\dot{V}_E$.

Arterial Blood Sampling and Analysis: To obtain arterial blood samples, a Teflon catheter was inserted into a radial artery under local anesthesia. Arterial blood samples were taken every 30 s during occlusion and otherwise every 20 s. Blood was sampled over several breathing cycles to avoid fluctuation in blood gases due to breathing. Blood was drawn without a tourniquet or clenching of the first, to avoid raising the $[K^+]_a$, and any sample with signs of hemolysis was discarded. $[K^+]_a$ was measured in duplicate by the selective electrode method. From each 2 ml of blood sampled, a 1-ml aliquot was drawn anaerobically into a glass syringe, rinsed with heparin, for measurements of arterial partial pressure of oxygen (PaO_2) and carbon dioxide ($PaCO_2$), and pH ($[pH]_a$) (IL 813/Blood Gas Analyzer). Arterial lactate concentration ($[La]_a$) was determined enzymatically from blood supernatant (Borehinger, Mannheim, FRG).

Statistical Analysis: An analysis of variance was used to evaluate the differences among the changes associated with occlusion. Paired t-test was used to evaluate where the significance occurred. A probability level of 0.05 was accepted as significant.

RESULTS

Effect of Occlusion during Exercise: When the cuffs were inflated, during exercise, $\dot{V}_{O_2}$ decreased significantly, while $\dot{V}_E$ and $\dot{V}_{CO_2}$ increased significantly (Fig. 1). During occlusion the time courses of $\dot{V}_E$ and $\dot{V}_{CO_2}$ were similar and the relationship between $\dot{V}_E$ and $\dot{V}_{CO_2}$ remained linear (Fig. 2A). However, the slope of the curve describing this relationship during exercise with occlusion was significantly greater than that during exercise without occlusion. This suggests that during occlusion there was hyperpnea out of proportion to the increase in $\dot{V}_{CO_2}$.

During exercise with occlusion, arterial pH and PaO_2 increased, and $PaCO_2$ decreased significantly (Fig. 3). $[K^+]_a$ and $[La]_a$ also increased when the cuffs were inflated during exercise. Among the factors that might affect exercise hyperpnea, $[K^+]_a$ was strongly correlated with $\dot{V}_E$ during exercise with occlusion (for individual data; r = 0.87 - 0.99, P <

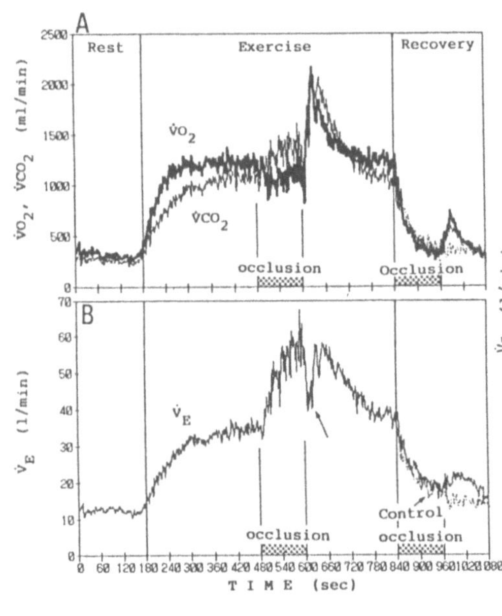

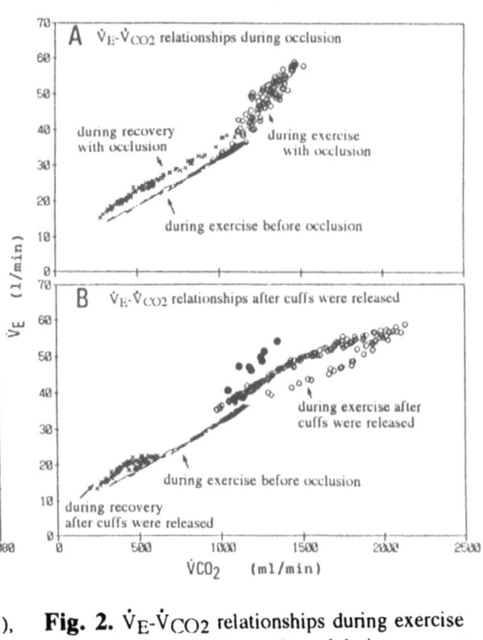

Fig. 1. Breath-by breath data for $\dot{V}_{O_2}$ and $\dot{V}_{CO_2}$ (A), and $\dot{V}_E$ (B) at rest, during exercise, and during recovery. The arrow indicates a transient reduction in $\dot{V}_E$ immediately after the cuffs were released during exercise.

Fig. 2. $\dot{V}_E$-$\dot{V}_{CO_2}$ relationships during exercise with occlusion and release (A) and during recovery with occlusion and release (B). The slope of the $\dot{V}_E$-$\dot{V}_{CO_2}$ relationship was greater during exercise with occlusion than under the others.

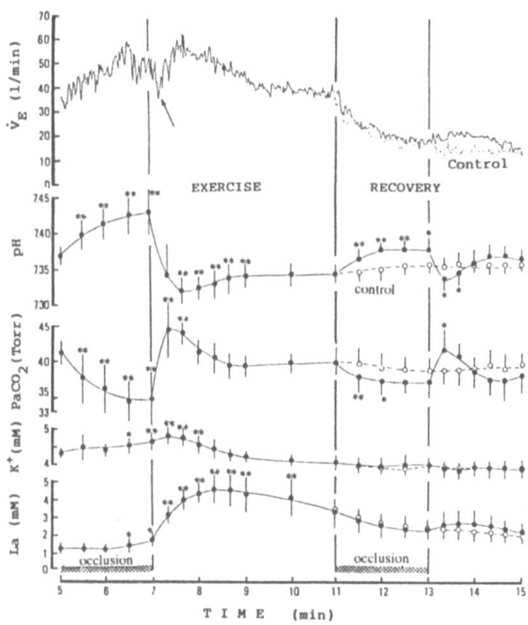

Fig. 3. The time course of $\dot{V}_E$, $[La]_a$, $[K^+]_a$, $PaCO_2$, and pH during exercise and recovery with occlusion and release. The arrow indicates a transient reduction in $\dot{V}_E$ immediately after the cuffs were released during exercise.

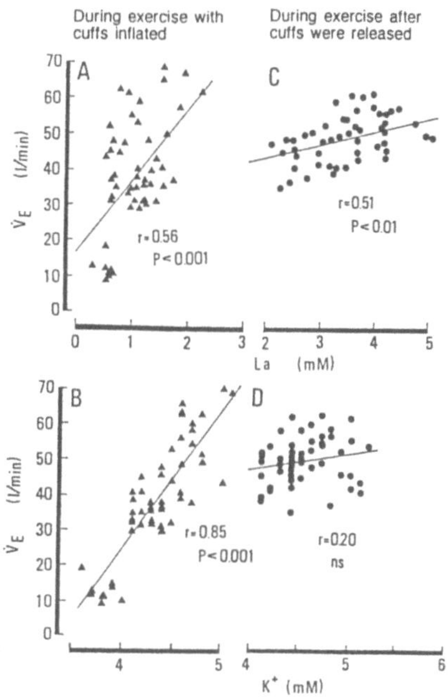

Fig. 4. The relationships between $\dot{V}_E$ and $[La]_a$ and between $\dot{V}_E$ and $(K^+]_a$ during exercise with occlusion and after the cuffs were released.

0.01) and the correlation coefficient for the pooled data was r = 0.85 (P < 0.001)(Fig. 4B). $[La]_a$ was also correlated with $\dot{V}_E$ during exercise with occlusion (r = 0.56, P < 0.01)(Fig. 4A).

Effect of Cuff Release during Exercise: After the cuffs were released, $\dot{V}_{O2}$ and $\dot{V}_{CO2}$ transiently increased and then decreased exponentially. In contrast, after the cuffs were released $\dot{V}_E$ decreased transiently (Figs. 1B & 3, see arrow), then increased and gradual decreased. For the first 10 s after the cuffs were released, the slope of the $\dot{V}_E$-$\dot{V}_{CO2}$ relationship did not change (closed circles in Fig. 2B). Thereafter it was similar to that obtained during exercise before occlusion.

After the cuffs were released during exercise, $[pH]_a$ decreased significantly, while $PaCO_2$, $[K^+]_a$ and $[La]_a$ increased significantly. PaO_2 decreased transiently, and returned to the significant elevated values within 40 sec. There was no significant correlation between $[K^+]_a$ and $\dot{V}_E$ after the cuffs were released (Fig. 4D), but there was a significant correlation between $[La]_a$ and $\dot{V}_E$ (r = 0.51, P < 0.01)(Fig. 4C).

Effect of Occlusion and Release during Recovery: Immediately after exercise, $\dot{V}_{O2}$, $\dot{V}_{CO2}$ and $\dot{V}_E$ decreased exponentially (Fig. 1, see Control). When the the cuffs were inflated, there was no significant effect on $\dot{V}_{O2}$, $\dot{V}_{CO2}$ and $\dot{V}_E$, while the changes in arterial blood gases were similar to those exercise with occlusion. The relationship between $\dot{V}_E$ and $\dot{V}_{CO2}$ during recovery was not affected bu cuff inflation (Fig. 2).

DISCUSSION

$\dot{V}_E$ increased significantly when the cuffs were inflated during exercise, and increased again after they were released. The correlation analysis indicates that hyperpnea during exercise with occlusion was largely due to changes in $[K^+]_a$ and was also moderately influenced by $[La]_a$. During exercise after the cuffs were released the $[La]_a$ also contributed to the increase in $\dot{V}_E$. There was a rapid and transient drop in $\dot{V}_E$ very soon after the cuffs were released during exercise (see arrows in Figs. 1 & 3), but this never occurred during recovery.

There were strong correlations between $\dot{V}_{CO2}$ and $\dot{V}_E$ during all phases of exercise and recovery (Fig. 2). This supports previous suggestions that CO_2 has a major role in determining $\dot{V}_E$. $\dot{V}_E$ increased immediately after an abrupt increase in the flow of CO_2 to the lungs, which was, in turn, caused by increased blood flow[8,9]. However, during exercise with the cuffs inflated, the slope of the $\dot{V}_E$-$\dot{V}_{CO2}$ relationship was significantly greater (Fig. 2A). Furthermore, this slope remained high for about 10 seconds after the cuffs were released (closed circles in Fig. 2B). Therefore, some factors other than the flow of CO_2 probably mediate the increase in $\dot{V}_E$ during exercise with the cuffs inflated.

One possible mediator is $[K^+]_a$[1,3,10]. There was a strong correlation $[K^+]_a$ and $\dot{V}_E$ during exercise with occlusion (Fig. 4B). This relationship was stronger than the one with $[La]_a$ (Fig. 4A), which would also be expected to mediate hyperpnea during exercise with occlusion. However, immediately after the cuffs were released $\dot{V}_E$ rapidly decreased, while $[K^+]_a$ remained high level. This change in $\dot{V}_E$ did not occur when the cuffs were inflated during recovery. Thus, the hyperpnea measured during exercise with the cuffs inflated may have been caused, in part, by a system that acts much more quickly than that determine $[K^+]_a$. The most likely candidate for such a system is the central nervous system (CNS). The present data indicate the CNS, via mechanoreceptos, also contributed to the hyperpnea measured during exercise with the cuffs inflated.

After the immediate post-release drop, $\dot{V}_E$ changed relatively slowly. It increased significantly and then decreased gradually. During exercise after the cuffs were released, the increased $\dot{V}_E$ was significantly correlated with $[La]_a$, but not with $[K^+]_a$. When the cuffs were released during exercise, hypercapnic blood that had pooled in the leg muscles arrived in the lungs (Fig. 3), and this increase in the flow of CO_2 flow also played a major role in the exercise-induced hyperpnea.

In conclusion, we measured ventilatory response to exercise and its relation to CO_2 output during and after occlusion of blood flow to the legs. When blood flow to the legs was obstructed during exercise, other mediators, such as $[K^+]_a$ and neurogenic reflexes, strongly influenced ventilation.

REFERENCES

1. Band, D.M.,& Linton, (1986) The effect of potassium on carotid body chemoreceptor discharges in the anaesthetized cat. J. Physiol., 381: 39-47.
2. Innes, J.A., Solarte, I., Huszczuk, A., Yeh, E., Whipp, B.J. & Wasserman, K. (1989) Respiration during recovery from exercise: effects of trapping and release of femoral blood flow. J. Appl. Physiol., 67: 2608-2613.
3. Paterson, D.J., Friedland, J.S., Bascom, D.A., Clement, I.D., Cunningham, D.A., Painter, R. & Robbins, P.A. (1990) Changes in arterial K^+ and ventilation during exercise in normal subjects and subjects with McAdle's syndrome. J. Physiol., 429: 339-348.
4. Rowell, L.B., Hermansen, L. & Blackmon, J.R. (1976) Human cardiovascular and respiratory responses to graded muscle ischemia. J. Appl. Physiol., 41: 693-701.
5. Rowell, L.B. & O'Leary, D.S. (1990) Reflex control of the circulation during exercise: chemoreflexes and mechanoreflexes. J. Appl. Physiol., 69: 407-418.
6 Sargeant, A.J., Rouleau, M.Y., Sutton, J.R. & Jones, N.L. (1981) Ventilation in exercise studied with circulatory occlusion. J. Appl. Physiol., 50: 718-723.
7. Stanley, W.C., Lee, W.R. & Brooks, G.A. (1985) Ventilation studied with circulatory occlusion during two intensities of exercise. Eur. J. Appl. Physiol., 54: 269-277.
8. Wasserman, K., Whipp, B.J. & Castagna, J. (1974) Cardiodynamic hyperpnea: hyperpnea secondary to cardiac output increase. J. Appl. Physiol., 36: 457-464.
9 Whipp, B.J. & Wasserman, K. (1980) Carotid bodies and ventilatory control dynamics in man. Fed. Proc., 39: 2628-2673.
10. Yoshida, T., Chida, M., Ichioka, M., Makiguchi, K., Eguchi, J. & Udo, M. (1990) Relationship between ventilation and arterial potassium concentration during incremental exercise and recovery. Eur. J. Appl. Physiol., 61: 193-196.
11. Zuntz N. & Geppert, J. (1986) Uber die Natur der normalen Aremreize den Ort ihrer Wirkung. Pflugers. Arch. Ges Physiol., 38: 337-338.

INTRODUCTION: DEPRESSION OF VENTILATION BY HYPOXIA - A COMPARISON OF THE PHENOMENON IN ANAESTHETIZED CATS AND CONSCIOUS HUMANS

Peter Robbins

University Laboratory of Physiology
Parks Road
Oxford OX1 3PT
U.K.

INTRODUCTION

The ventilatory response to isocapnic hypoxia is biphasic: there is an initial brisk rise in ventilation at the onset of hypoxia which is then followed by a subsequent slow decline over a period of many minutes[1]. The mechanisms underlying this slow decline in ventilation are not fully understood, but many investigators have focussed on the effects of hypoxia on the brainstem, and in particular the modulations in neurotransmitter levels that hypoxia can produce - for a review see Neubauer et al[2]. Of necessity, much of this work has been performed on anaesthetized experimental animals. The purpose of this article is to review the similarities and the differences between the hypoxic ventilatory decline observed in anaesthetized cats and that observed in conscious humans.

PHENOMENON IN ANAESTHETIZED CATS

DeGoede et al. examined the ventilatory response to the onset and relief of isocapnic end-tidal hypoxia in the anaesthetized cat[3]. They found that the degree of ventilatory decline observed during sustained hypoxia varied between cats, but that, in any single cat, this decline in ventilation during sustained hypoxia was broadly matched by an undershoot and subsequent recovery in ventilation at the relief of hypoxia. This finding has been confirmed in a further study, although there are some differences in the dynamics of the responses between the onset and relief of hypoxia[4].

In order to separate the central and peripheral effects of P_{CO2} and P_{O2}, Berkenbosch

Control of Breathing and Its Modeling Perspective, Edited by
Y. Honda *et al.*, Plenum Press, New York, 1992

et al. developed the technique of artificial brainstem perfusion[5]. Van Beek et al. used this technique to explore the effects of hypoxia when confined to the brainstem[6]. Three important observations were made in this study:

1. A reduction in the P_{O2} of the blood perfusing the brainstem caused a marked reduction in ventilation. This finding suggests that central or medullary hypoxia may play an important role in the genesis of hypoxic ventilatory decline.

2. The ventilatory sensitivity to central P_{CO2} remained unaltered.

3. The ventilatory sensitivity to peripheral P_{O2} remained unaltered.

The artificial brainstem perfusion technique also allowed deGoede et al. to examine the ventilatory response to step changes in end-tidal P_{O2} while maintaining the brainstem hyperoxic[3]. Under these conditions, they found the ventilatory response showed neither overshoot nor undershoot at the onset and relief of hypoxia.

In short, the findings above suggest that, in the anaesthetized cat, hypoxic ventilatory decline is mediated solely by the effects of hypoxia at the brainstem, and that the underlying mechanism leaves the central and peripheral chemoreflex sensitivities unaltered.

PHENOMENON IN CONSCIOUS HUMANS

As noted in the introduction, conscious humans show a biphasic ventilatory response to isocapnic hypoxia, and it might be expected that the properties observed using the artificial brainstem perfusion technique in anaesthetized cats would also hold in humans. One of the first indications that this might not be the case can be found in the data of Easton et al[7]. They observed that the magnitude of the ventilatory decline was correlated with the magnitude of the initial rise in ventilation at the onset of hypoxia. This suggests that, unlike the anaesthetized cat, the processes underlying hypoxic ventilatory decline in humans may be linked in some way to peripheral chemosensitivity.

Further evidence comes from the asymmetry of the ventilatory response to the onset and relief of sustained isocapnic hypoxia. Khamnei and Robbins observed that the magnitude of the ventilatory response at the onset of hypoxia was considerably greater than the magnitude of ventilatory response at the relief of hypoxia[8]. At the relief of hypoxia, little or no undershoot in ventilation was observed. Since the rapid ventilatory response to hypoxia is generally associated with the peripheral chemoreflex, this finding again suggests that the sensitivity of the peripheral chemoreflex has been altered.

At near normal levels of arterial P_{CO2}, one possible alternative explanation for the asymmetry of the ventilatory response at the onset and offset of hypoxia is that the "dog-leg" in the ventilation-CO_2 response curve does not allow ventilation to fall below control levels. However, if this explanation were correct, then it would be predicted that the asymmetry would disappear if the experiments were conducted at a higher P_{CO2}. This turns out not to be the case[8].

A second alternative explanation for the asymmetry could be that, whilst hypoxic ventilatory decline takes many minutes to develop, this effect of hypoxia disappears rapidly on return to euoxia[9]. If this explanation were correct, then a re-introduction of hypoxia soon after the relief of hypoxia should give a ventilatory response which matches in magnitude the ventilatory response associated with the initial introduction of hypoxia. However, the ventilatory response to the re-introduction of hypoxia is reduced in magnitude for many minutes after the relief of the initial period of hypoxia[10].

In short, both the relationship between the magnitude of the initial response to hypoxia and the magnitude of the subsequent ventilatory decline, and also the asymmetry of the ventilatory response at the onset and the relief of sustained isocapnic hypoxia, suggest that hypoxic ventilatory decline may be generated in humans via a slow reduction in peripheral chemoreflex sensitivity.

SITE OF ACTION OF HYPOXIA IN HUMANS

One possibility is that, as in the anaesthetized cat, hypoxia acts centrally to generate hypoxic ventilatory decline. However, since the fundamental mechanisms underlying hypoxic ventilatory decline appear to differ between conscious humans and anaesthetized cats, the evidence for a central site in the cat is not really applicable to humans. The author's view is that there is insufficient evidence to determine whether the site is central or peripheral in humans, but that the evidence that does exist points to an adaptation occurring within the carotid bodies.

In the anaesthetized cat, individual fibres from the carotid body respond to both hypoxia and hypercapnia[11]. If this aspect of feline physiology is also true for humans, then the simplest hypothesis is that afferents from the carotid bodies convey information concerning the overall magnitude of stimulation, but not concerning the specific type of stimulus (i.e. hypercapnia, hypoxia, etc). Now suppose sustained hypoxia acts centrally in humans to depress peripheral chemoreflex sensitivity. Then the following would be predicted:

1. The peripheral chemoreflex sensitivities to hypoxic and hypercapnic stimulation should be affected equally by sustained central hypoxia.

2. There should be a consistent relationship between peripheral chemoreflex sensitivity and hypoxic ventilatory decline when carotid body function is altered by drugs which have no central effects on respiratory chemoreflexes.

To test the first of these predictions, Bascom et al. used pulses of extra hypoxia to measure peripheral hypoxic sensitivity and pulses of extra hypercapnia to measure peripheral hypercapnic sensitivity[12]. They gave these pulses at different times during the development of hypoxic ventilatory decline in order to measure the changes occurring in these sensitivities. They found that the hypoxic sensitivity declined much more than did the peripheral hypercapnic sensitivity. Thus the first of the above predictions does not appear to hold.

To test the second of these predictions, it is necessary to compare the effects of different drugs that affect the acute ventilatory response to hypoxia on hypoxic ventilatory decline. Georgopoulos et al. reported that almitrine, a carotid body stimulant, increased the absolute magnitude of hypoxic ventilatory decline, such that the decline remained a constant fraction of the acute hypoxic response[13]. On the other hand, Bascom et al. reported that domperidone, also a carotid body stimulant, increased the magnitude of the acute hypoxic response but had no effect on the absolute magnitude of the hypoxic ventilatory decline[14]. Thus it appears that the effects of these two drugs on hypoxic ventilatory decline are not consistent, despite the fact that they both increase the carotid body's acute response to hypoxia. Consequently, the second of the predictions also does not appear to hold.

To summarize, if the medulla receives only information concerning the overall magnitude of stimulation at the carotid body, but no information concerning the nature of the stimulus, then the differential effects of sustained hypoxia on peripheral hypoxic and hypercapnic sensitivities and also the variable effects of carotid body stimulants on hypoxic ventilatory decline appear to be inconsistent with a central site of action for hypoxia in generating hypoxic ventilatory decline.

CAROTID BODY RESPONSES TO SUSTAINED HYPOXIA

If the major cause of hypoxic ventilatory decline in humans is an adaptation of the carotid body to sustained hypoxia, then it might be expected that, at least in some animals, a similar adaptation could be found in the nervous discharge in the carotid sinus nerve in response to sustained isocapnic hypoxia. In the anaesthetized cat, Barnard et al. found no effect of sustained hypoxia on carotid body discharge[15]. This finding is consistent with the other observations described here for the anaesthetized cat. In the anaesthetized rabbit, however, Kaiying et al. found that carotid chemoreceptor discharge declined over a period of 1 hr[16]. Also, Mulligan and Bhide found a decline in carotid body chemoreceptor discharge with sustained hypoxia in piglets[17]. Thus it does appear that an adaptation of the carotid body's response to sustained hypoxia can be observed in at least some mammals.

SUMMARY

In the anaesthetized cat, hypoxic ventilatory decline appears to result from a central effect of hypoxia. Peripheral and central chemoreflex sensitivities appear to be unaffected by this. In contrast, in conscious humans, the major mechanism underlying hypoxic ventilatory decline seems to involve a reduction in the peripheral chemoreflex sensitivity to hypoxia. There is some evidence to suggest that this may not be a central effect of hypoxia, but rather an adaptation at the level of the carotid bodies.

REFERENCES

1. J.V. Weil and C.W. Zwillich, Assessment of ventilatory response to hypoxia: Methods and interpretation, *Chest* 70: Supplement 124 (1976).
2. J.A. Neubauer, J.E. Melton and N.H. Edelman, Modulation of respiration during brain hypoxia, *J. Appl. Physiol.* 68:441 (1990).
3. J. deGoede, N. van der Hoeven, A. Berkenbosch, C.N. Olievier and J.H.G.M. van Beek, Ventilatory responses to sudden isocapnic changes in end-tidal O_2 in cats, *in*: "Modelling and Control of Breathing,", B.J. Whipp and D.M. Wiberg, ed., Elsevier Biomedical, New York (1983).
4. A. Berkenbosch, J. deGoede, C.N. Olievier, J.J. Schuitmaker and D.S. Ward, Dynamics of ventilation following sudden isocapnic changes in end-tidal O_2 in cats, *J.Physiol.* 394:59P (1987).
5. A. Berkenbosch, J. Heeringa, C.N. Olievier and E.W. Kruyt, Artificial perfusion of the ponto-medullary region of cats. A method for separation of central and peripheral effects of chemical stimulation of ventilation, *Respir. Physiol.* 37: 347 (1979).
6. J.H.G.M. van Beek, A. Berkenbosch, J. deGoede and C.N. Olievier, Effects of brain stem hypoxaemia on the regulation of breathing, *Respir. Physiol.* 57:171 (1984).
7. P.A. Easton, L.J. Slykerman and N.R. Anthonisen, Ventilatory response to sustained hypoxia in normal adults, *J. Appl. Physiol.* 61:906 (1986).
8. S. Khamnei and P.A. Robbins, Hypoxic depression of ventilation in humans: Alternative models for the chemoreflexes, *Respir. Physiol.* 81:117 (1990).
9. D.S. Ward, Dynamic models and parameter estimation: The hypoxic ventilatory response, *in*: "Modeling and Parameter Estimation in Respiratory Control," M.C.K. Khoo, ed., Plenum, New York (1989).
10. P.A. Easton, L.J. Slykerman and N.R. Anthonisen, Recovery of the ventilatory response to hypoxia in normal adults, *J. Appl. Physiol.* 64:521 (1988).
11. S. Lahiri and R.G. DeLaney, Stimulus interaction in the responses of carotid body chemoreceptor single afferent fibres, *Respir. Physiol.* 24:249 (1975).
12. D.A. Bascom, I.D. Clement, D.A. Cunningham, R. Painter and P.A. Robbins, Changes in peripheral chemoreflex sensitivity during sustained, isocapnic hypoxia, *Respir. Physiol.* 82:161 (1990).
13. D. Georgopoulos, S. Walker and N.R. Anthonisen, Increased chemoreceptor output and the ventilatory response to sustained hypoxia, *J. Appl. Physiol.* 67:1157 (1989).
14. D.A. Bascom, I.D. Clement, K.L. Dorrington and P.A. Robbins, Effects of dopamine and domperidone on ventilation during isocapnic hypoxia in humans, *Respir. Physiol.* 85:319 (1991).
15. P. Barnard, S. Andronikou, M. Pokorski, N. Smatresk, A. Mokashi and S. Lahiri, Time-dependent effect of hypoxia on carotid body chemosensory function, *J. Appl. Physiol.* 63:685 (1987).
16. L. Kaiying, J. Ponte and C.L. Sadler, Carotid body chemoreceptor response to prolonged hypoxia in the rabbit: Effects of domperidone and propranolol, *J. Physiol.* 430:1 (1990).
17. E. Mulligan and S. Bhide, Non-sustained responses to hypoxia of carotid body chemoreceptor afferents in the piglet. *Fed. Proc.* 3:A399 (1989).

THE EFFECTS OF DOPAMINE ON THE VENTILATORY RESPONSE TO SUSTAINED HYPOXIA IN HUMANS

Denham S. Ward and Marica Nino

Department of Anesthesiology
UCLA School of Medicine
Los Angeles, California 90024-1778

INTRODUCTION

The ventilatory response to acutely imposed sustained isocapnic hypoxia is biphasic[5]. In humans, ventilation increases immediately in the first 3-5 minutes of hypoxia (hypoxic ventilatory stimulation, HVS), then gradually declines over the next 15-20 minutes to a value intermediate between the normoxic and the peak hypoxic ventilation[5]. The initial hyperventilatory response is due to increased peripheral chemoreceptor output; however the mechanisms of the subsequent hypoxic ventilatory decline (HVD) are not well elucidated. In animals, the peripheral chemoreceptor discharge does not adapt during sustained hypoxia but the phrenic nerve efferent discharge does decrease[18] and hypoxia causes ventilatory depression when the carotid sinus nerve is cut[1]. These findings suggest a central origin for the ventilatory decline. Increases in inhibitory central neuromodulators, such as GABA[17] and adenosine[4], and the hypoxia-induced increase in cerebral blood flow washing out acid metabolites[21] have been proposed as possible mechanisms for the ventilatory decline.

Looking at alternate models of HVD to explain differences between steps into and out of hypoxia Khamnei and Robbins[11] concluded that the decline modulated the peripheral chemoreflex. This model would predict that interventions that increase or decrease HVS would also increase and decrease the magnitude of HVD. A variety of interventions have been studied but the evidence linking HVD to HVS is mixed.

Easton *et al.*[3] and Georgopoulos *et al.*[7] noted that subjects who manifest the largest increase in ventilation with early hypoxia also show the biggest decline in ventilation with sustained hypoxia. However, another human study by Suzuki *et al.*[15] found the opposite result: low responders, but not high responders, had a significant ventilatory decline during prolonged hypoxia.

Interventions that increase HVS also have not consistently caused a concurrent increase in HVD. Exercise greatly increases HVS, but both Ward and Nguyen[20] and Pandit and Robbins[13] did not find any increase in HVD. Also when HVS was increased with the peripheral dopamine antagonist, domperidone, HVD was unaffected.[2] However, increasing HVS with almitrine[8] did increase HVD.

Filuk *et al.*[6] found that somatostatin depressed both HVS and HVD in proportion, but could not determine if this was entirely a peripheral effect, since somatostatin crosses the blood-brain barrier. Intravenous administration of low-dose dopamine blunts the ventilatory increase in response to acute hypoxia[19]. Dopamine dose not cross the blood-

Control of Breathing and Its Modeling Perspective, Edited by
Y. Honda *et al.*, Plenum Press, New York, 1992

brain barrier in low doses and its effect of the hypoxic response should be confined to the carotid body. However, Bascom *et al.*[2] found that a modest reduction of HVS by dopamine did not change HVD. This study was designed to investigate the effects of dopamine on HVD. In contrast to the study by Bascom *et al.*[2] we found that dopamine did reduce both HVS and HVD.

METHODS

Ten healthy male adult volunteers age 19 to 37 years were studied with a UCLA Human Subjects Protection Committee-approved experimental protocol. Each subject had a normal physical examination, electrocardiogram, and blood glucose and hemoglobin. Before the experiment day they were familiarized with the equipment and the experimental procedure. Subjects were informed that they would experience increased breathing and possibly develop a mild headache. They were asked to refrain from stimulant and depressant substances (tea, coffee, cola, cocaine, marijuana and alcohol) for 12 hr prior to the experiment. During the experiments, subjects assumed a semi-recumbent position, listening to background music of their choice. Electrocardiogram and arterial hemoglobin saturation (Ohmeda Biox 3700 pulse oximeter) were monitored. With a noseclip in place, the subjects breathed from a mixing chamber through a mouthpiece. Inhaled and exhaled volumes were measured with a bidirectional impeller flowmeter (Sensor Medics VMM 110). Inspired and end-tidal oxygen and carbon dioxide concentrations were analyzed by mass spectrometer (Perkin-Elmer MGA 1100). Inspired concentrations of oxygen and carbon dioxide were adjusted by computer (80286-based personal computer) through evaluation of end-tidal concentrations on a breath-by-breath basis. Desired end-tidal conditions were achieved through a combination of feedback and feedforward control according to a technique described by Swanson and Bellville[16].

Heart rate, oxygen saturation, tidal volume, respiratory frequency, inspired and end-tidal oxygen and carbon dioxide concentrations were collected by computer using the TIDAL software package[9]. Minute exhaled ventilation was calculated on a breath-by-breath basis. All gas volumes were corrected to body temperature, ambient pressure, saturated (BTPS) conditions. Additionally, oxygen saturation, minute exhaled ventilation and end-tidal oxygen and carbon dioxide concentrations were monitored on a strip chart recorder.

Normal saline was infused at 50 ml/hr through a 20-gauge catheter in a large upper extremity vein throughout all experiments. Dopamine was mixed in 0.9% normal saline at 800 $mg \cdot L^{-1}$ and added to the intravenous solution via a syringe pump (Harvard Apparatus Model 22) at a rate of 3 $\mu g \cdot kg^{-1} \cdot min^{-1}$. The placement of the apparatus made it possible to turn dopamine on and off without a subject's knowledge. The subjects were unaware of the expected respiratory response to the experiment with dopamine.

Two isocapnic hypoxic tests were performed on each subject on a single day, one with and one without dopamine infusion. The subjects initially breathed room air (F_IO_2 = 0.21) for five minutes to ensure stable baseline measurements. At a constant $P_{ET}CO_2$ (approximately 40 mmHg), end-tidal oxygen tension was abruptly lowered from approximately 100 to 45 mmHg over 1 min and was followed by 20 min of constant hypoxia. This was followed by an abrupt increase of oxygen tension to hyperoxia (F_IO_2 = 0.40) for 5 min. The subject then rested while the dopamine infusion was started. The infusion ran for at least 30 minutes prior to the next experiment, to insure a constant plasma dopamine level. The sequence of breathing room air, a hypoxic mixture, then a hyperoxic mixture, was repeated. The end-tidal CO_2 was held constant by automatic adjustment of the inspired concentration throughout the experiment.

Table 1. Averages for control and dopamine experiments. $\dot{V}_E$, minute ventilation; V_T, tidal volume; f, respiratory rate. Mean $\pm$ standard deviation (N = 8). * $P < 0.05$ different from the pre-hypoxia level. # $p < 0.05$ different from the initial hypoxia level.

			Phase of Hypoxia	
		Pre-Hypoxia	Initial	Final
$\dot{V}_E$ ($L \cdot min^{-1}$)	Control	11.27 ± 2.02	$21.49^{*} \pm 5.69$	$15.58^{*\#} \pm 5.49$
	Dopamine	10.18 ± 2.36	$13.41^{*} \pm 3.02$	$11.44^{\#} \pm 2.93$
V_T (liters)	Control	0.620 ± 0.107	$0.975^{*} \pm 0.227$	$0.783^{*\#} \pm 0.223$
	Dopamine	0.578 ± 0.093	$0.708^{*} \pm 0.149$	$0.624^{\#} \pm 0.124$
f(min^{-1})	Control	18.9 ± 3.8	$22.6^{*} \pm 3.1$	$20.2^{*} \pm 3.5$
	Dopamine	17.9 ± 3.4	19.6 ± 3.2	19.0 ± 2.9
$P_{ET}CO_2$ (mmHg)	Control	40.4 ± 0.25	40.3 ± 0.07	40.4 ± 0.09
	Dopamine	40.8 ± 0.8	40.3 ± 0.25	$39.8^{*} \pm 0.86$
$P_{ET}O_2$ (mmHg)	Control	97.4 ± 0.3	$43.7^{*} \pm 0.1$	$43.6^{*} \pm 0.1$
	Dopamine	97.5 ± 0.4	$43.6^{*} \pm 0.2$	$44.3^{*\#} \pm 1.1$
T_I / T_T	Control	0.404 ± 0.028	$0.432^{*} \pm 0.022$	$0.394^{*\#} \pm 0.037$
	Dopamine	0.370 ± 0.039	0.397 ± 0.038	0.376 ± 0.034
V_I / T_I ($L \cdot sec^{-1}$)	Control	0.479 ± 0.066	$0.866^{*} \pm 0.223$	$0.663^{*\#} \pm 0.201$
	Dopamine	0.462 ± 0.081	$0.584^{*} \pm 0.113$	$0.513^{\#} \pm 0.111$

For statistical analysis the breath-by-breath data on each subject were averaged over three two-minute intervals: the final two minutes before the beginning of the hypoxic period (pre-hypoxic phase); the initial phase of the hypoxic period at three minutes into the hypoxic period; and the final phase of the hypoxic period at the end of the 20 minute hypoxic period. Since some subjects became restless at the end of the hypoxic period, the final phase was selected as a two-minute interval of regular breathing within the last five minutes of hypoxia. Statistical comparisons were performed by ANOVA and Newman-Keuls multiple comparisons test for the pre-hypoxic, initial hypoxic and final hypoxic phases. Comparisons between the control and the dopamine experiments were made with Student's paired t-test for the delta ventilations and Wilcoxon signal ranks test for the ratios of HVD to HVS and HVD to peak ventilation. Statistical analysis was performed with the SOLO computer program (BMDP Statistical Software, Inc.).

RESULTS

Because of technical problems, data from two of the ten subjects were not included in the analysis. In one subject the $P_{ET}CO_2$ could not be maintained at the pre-hypoxic value and in the other subject an incorrect dose of dopamine was used. The remaining eight subjects were used in the subsequent analysis. No subjective symptoms related to the dopamine infusion were reported by the subjects.

Table 1 gives the averages for the three time intervals in both the control and the dopamine experiments. The end-tidal oxygen and carbon dioxide levels were well controlled, although there was a small decrease in the $P_{ET}CO_2$ and increase in the $P_{ET}O_2$ levels in the dopamine group during the final phase of the hypoxic period compared to the initial phase of the hypoxic period. Although statistically significant, it amounted to less than 1 mmHg.

Table 2. Changes in variables between pre-hypoxic and initial hypoxic periods, and initial hypoxic and final hypoxic periods. Mean ± standard deviation (N = 8). * P = 0.015 different from the control change. # P < 0.01 different from the control change.

		Change Pre- to initial hypoxic phase	Change Initial to final hypoxic phase
$\dot{V}_E$ ($L \cdot min^{-1}$)	Control	10.22 ± 5.21	-5.90 ± 3.09
	Dopamine	3.24# ± 2.31	-1.97* ± 1.00
V_T (liters)	Control	0.355 ± 0.164	-0.192 ± 0.081
	Dopamine	0.131# ± 0.156	-0.084# ± 0.050
f (min^{-1})	Control	3.7 ± 2.1	-2.3 ± 1.9
	Dopamine	1.6 ± 0.9	-0.6 ± 2.4
$P_{ET}CO_2$ (mmHg)	Control	-0.09 ± .25	0.06 ± .12
	Dopamine	-0.49 ± .90	-0.53 ± .82
$P_{ET}O_2$ (mmHg)	Control	-53.8 ± 0.046	-0.08 ± 0.02
	Dopamine	-53.9 ± 0.7	0.7 ± 1.2
T_I / T_T	Control	0.028 ± 0.020	-0.038 ± 0.022
	Dopamine	0.026 ± 0.024	-0.021 ± 0.031
V_I / T_I ($L \cdot sec^{-1}$)	Control	0.387 ± 0.201	-0.202 ± 0.095
	Dopamine	0.122# ± 0.080	-0.071# ± 0.044

The ventilation in the control experiments showed the typical biphasic response with a brisk increase in the initial hypoxic phase that was not sustained; the ventilation in the final hypoxic phase was significantly less than the peak. This response was primarily in tidal volume, but there was a similar response in respiratory rate. Dividing the ventilation into the inspiratory flow (V_I / T_I) and the inspiratory time duty cycle (T_I / T_T), showed that both the increase and the decrease were primarily in the inspiratory flow, but with a similar pattern in the duty cycle.

Ventilation in the dopamine experiments showed a similar but reduced response and the ventilatory decline did not reduce the final ventilation below the pre-hypoxic value. The changes were in tidal volume and V_I / T_I, while the respiratory rate and T_I / T_T did not differ significantly in any phase during the dopamine infusion. It is important to note that throughout all experiments (both control and dopamine) inspired CO_2 was never zero. This means that had ventilation reduced still more, the servo control system would have reduced the inspired CO_2 to maintain its end-tidal level constant. Thus the ventilation could have decreased in the final hypoxic period without a counter CO_2 stimulation.

Table 2 compares the respiratory variable changes between the pre-hypoxic and initial hypoxic phases, and between the initial and final hypoxic phases, for the control and the dopamine experiments. The changes in the $P_{ET}CO_2$ and $P_{ET}O_2$ were no different between the control and the dopamine experiments. However, both the initial stimulation of ventilation with hypoxia and the subsequent decline were reduced in the dopamine experiments. The small decrease in $P_{ET}CO_2$ as well as the increase in $P_{ET}O_2$ in the final hypoxic phase of the dopamine experiments would only serve to decrease ventilation slightly and thus increase any apparent depression of ventilation.

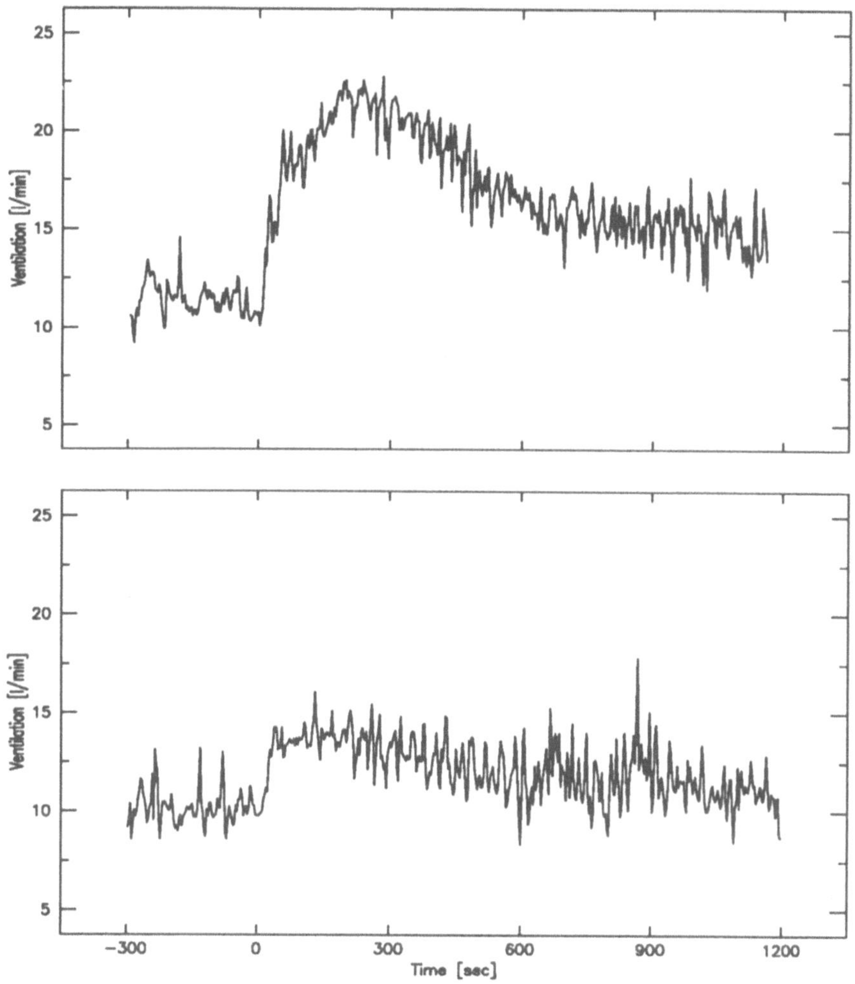

Figure 1. The average ventilatory response to the hypoxic challenge in all 8 subjects is shown. The individual breath-by-breath data were interpolated at 3 second intervals, aligned on the time of the hypoxic step, and averaged. Time zero is the start of the hypoxic period. The top panel shows the experiments without dopamine and the bottom panel with dopamine. For clarity the standard deviations are not shown, see Table 1 for the interval averages and standard deviations.

Figure 1 shows the composite ventilatory response for the control and the dopamine experiments. The breath-by-breath data sets were interpolated at 3 second intervals and ensemble averaged aligned on the time of the step transition. The reduction in both the stimulating and the depressive response to hypoxia is clearly seen.

DISCUSSION

The time course of the ventilatory response to 20 minutes of hypoxia in adult humans clearly is composed of two components (Fig. 1). The first component is a stimulation of ventilation which originates in the carotid bodies. The second component is a decline in ventilation; the origin of this response is thought to be central. Animal studies have clearly shown that there is no adaptation in the carotid body response and that the phrenic nerve firing is decreasing over this time period[1,18].

Several mechanisms have been proposed to account for this adaptation in ventilation. Since hypoxia increases cerebral blood flow there may be a wash-out of acid metabolites with sustained hypoxia[21] but Suzuki *et al.*[15] could not find any correlation between the hypoxic depression and changes in internal jugular venous PCO_2. Alternatively, there may be an increase in the central concentration of inhibitory neuro-transmitters, with GABA[17] and adenosine[4] proposed as possible candidates. Endorphins apparently do not play a role in human hypoxic ventilatory decline since naloxone does not have any effect[10]. Metabolic depression of neurons from the lack of oxygen is another possibility[12] but this is more likely at deeper levels of hypoxia.

Somatostatin has been used to investigate hypoxic ventilatory decline, since it is known to decrease the stimulating effects of hypoxia through actions on the carotid body. Our results agree with those of Filuk *et al.*,[6] who found that both the hypoxic stimulation and the decline in ventilation were decreased by somatostatin. However, somatostatin has central actions and thus the reduction in the hypoxic decline may be a result of central effects.

Dopamine does not cross the blood-brain barrier in appreciable amounts; it reduces the hypoxic stimulation by direct action on the carotid body. Our data show that dopamine reduces the peripheral drive; the stimulating effect of hypoxia was reduced by 68%, but the depression was reduced by 67%. The depression, on average, was still smaller than the stimulation and the final ventilation remained higher than the normoxic ventilation. Note that if the decline was unchanged by dopamine, the final hypoxic ventilation would have been substantially below the normoxic ventilation (Table 2).

Our results are in contradistinction to the results of Bascom *et al.*[2] who found, with a similar dose of dopamine, only a 40% reduction in HVS and no change in HVD. The protocol used by them is very similar to this protocol and the differences in the results are unexplained. They only infused dopamine for 10 minutes prior to the ventilatory tests, while we used a 30 minute equilibrium period. It is possible that, even though dopamine has a very rapid onset, 10 minutes is not a long enough period to ensure a steady-state dopamine concentration at the dopaminergic receptors in the carotid body.[14]

The role of CO_2 in the hypoxic ventilatory response must be carefully considered in interpreting our experiments. It is well known that hypoxia and hypercapnia have a potentiating effect[16]. That is, the hypoxic ventilatory response becomes much greater at higher CO_2 levels. From the study of Georgopoulos *et al.*[7], the magnitude of HVD is also increased in proportion to the increase in HVS. Thus, when comparing responses it is important that the CO_2 be held constant before and after administering the study drug. We carefully controlled the CO_2 level in our subjects with and without dopamine (Table 1). Since the hyperventilation caused by hypoxia will lower the CO_2, thus contributing to the biphasic response, it is also important that the CO_2 be held constant throughout the experiment. The change in $P_{ET}CO_2$ from the initial phase to the final phase of the hypoxic period was less than 0.5 mmHg (Table 2). Once the inspired CO_2 is reduced to zero, it is possible that a further reduction in ventilation would be stalled by a rise in CO_2. In all our experiments the inspired CO_2 never reached zero and thus the ventilation could have continued declining without a rise in end-tidal PCO_2. In a clinical situation CO_2 is, of course, not controlled. Although our data cannot be extrapolated directly to the uncontrolled CO_2 case, Easton and Anthonisen[3] found that the magnitude of hypoxic depression was similar when CO_2 was not controlled.

There are several possible explanations for our results. Dopamine may have a central effect. Dopamine may change the cerebral blood flow response to hypoxia and thus change the wash-out pattern of acid metabolites, or enough dopamine may reach the central respiratory centers to counter the hypoxic decline. However, since somatostatin gives essentially the same results, presumably through different mechanisms, it seems likely that the central depression is linked to the peripheral stimulation.

The hypoxic ventilatory decline could be dependent on peripheral input for its development, or the peripheral input could be modulated by the hypoxic depression. If the central effect of hypoxia were to modulate the peripheral input, then a reduction in peripheral input (even if the central effects of the hypoxia were the same) would result in a smaller absolute decline. Khamnei and Robbins[11] have investigated alternative models for HVD and concluded that the most likely model was a modulation of the peripheral chemoreflex. However, modulation of the total ventilatory drive (the peripheral drive plus central drive) was also a possibility. Our dopamine data can be used to address this issue. If HVD modulates the peripheral drive then the ratio of the amount of decline to the stimulation will be unchanged by dopamine. However if HVD modulates the total drive than the ratio of decline to the peak ventilation will be constant. If either of these ratios is significantly changed by dopamine, then that would be evidence against the respective model. The results of this calculation (Table 3) are inconclusive; the ratio HVD/Peak $\dot{V}_E$ did show more change but was not significant at the 0.05 level.

The evidence on the relationship between HVD and HVS remains contradictory. It will require further study with attention paid to the pattern of breathing and the experimental conditions to full elucidate the relationship between HVD and HVS.

Table 3. Comparison of the ratio of HVD to HVS and the total drive (see text). The ratios are given as the median (range) of the ratios calculated from the individual data and not from the mean data given in Tables 1 and 2. Probability of significant difference calculated by Wilcoxon Signal Rank Test.

	Control	Dopamine	
HVD/HVS	0.70 (-0.03, 1.49)	0.66 (-0.03, 8.16)	p = 0.32
HVD/Peak $\dot{V}_E$	0.30 (-0.01, 0.46)	0.14 (-0.01, 0.29)	p = 0.07

ACKNOWLEDGEMENTS

The technical help of Kamel Aqleh is greatly appreciated. Supported by a grant from the Anesthesia Education Foundation, Los Angeles, CA 90024-177820

REFERENCES

1. S. Andronikou, M. Shirahata, A. Mokashi, and S. Lahri, Carotid body chemoreceptor and ventilatory responses to sustained hypoxia and hypercapnia in the cat, Respir. Physiol. 72:361-374 (1988).
2. D.A. Bascom, I.D. Clement, K.L. Dorrington, and P.A. Robbins, Effects of dopamine and domperidone on ventilation during isocapnic hypoxia in humans, Respir. Physiol. 85:319-328 (1991).
3. P.A. Easton, and N.R. Anthonisen, Carbon dioxide effects on the ventilatory response to sustained hypoxia, J. Appl. Physiol. 64:1451-1456 (1988).
4. P.A. Easton, and N.R. Anthonisen, Ventilatory response to sustained hypoxia after pretreatment with aminophylline, J. Appl. Physiol. 64:1445-1450 (1988).
5. P.A. Easton, L.J. Slykerman, and N.R. Anthonisen, Ventilatory response to sustained hypoxia in normal adults, J. Appl. Physiol. 61:906-911 (1986).
6. R.B. Filuk, D.J. Berezanski, and N.R. Anthonisen, Depression of hypoxic ventilatory response in humans by somatostatin, J. Appl. Physiol. 65(3):1050-1054 (1988).
7. D. Georgopoulos, D. Berezanski, and N.R. Anthonisen, Effects of CO_2 breathing on ventilatory response to sustained hypoxia in normal adults, J. Appl. Physiol. 66:1071-1078 (1989).

8. D. Georgopoulos, S. Walker, N.R. Anthonisen, Increased chemoreceptor output and the ventilatory response to sustained hypoxia in normal adults, J. Appl. Physiol. 67:1157-1163 (1989).
9. J.S. Jenkins, C.P. Valcke, and D.S. Ward, A programmable system for acquisition and reduction of respiratory physiological data, Ann. Biomed. Eng. 17:93-108 (1989).
10. S. Kagawa, M.J. Stafford, T.B. Waggener, and J.W. Severinghaus, No effect of naloxone on hypoxia-induced ventilatory depression in adults, J. Appl. Physiol. 52:1031-1034 (1982).
11. S. Khamnei and P.A. Robbins, Hypoxic depression of ventilation in humans: alternative models for the chemoreflexes, Respir. Physiol. 81:117-134 (1990).
12. C.G. Morrill, J.R. Meyer, and J.V. Weil, Hypoxic ventilatory depression in dogs, J. Appl. Physiol. 38(1):143-146 (1975).
13. J.J. Pandit and P.A. Robbins, The ventilatory effects of sustained isocapnic hypoxia during exercise in humans, Respir. Physiol. 86:383-404 (1991).
14. D. Ratge, U. Steegmuller, G. Mikus, K.P. Kohse and H. Wisser, Dopamine infusion in healthy subjects and critically ill patients, Clin. Exp. Pharmacol. Physiol. 17:361-369, (1990).
15. A. Suzuki, M. Nishimura, H. Yamamoto, K. Miyamoto, and F. Kishi, No effect of brain blood flow on ventilatory depression during sustained hypoxia, J. Appl. Physiol. 66:1674-1678 (1989).
16. G.D. Swanson, and J.W. Bellville, Hypoxic-hypercapnic interaction in human respiratory control, J. Appl. Physiol. 36:480-487 (1974).
17. A.M. Taveira da Silva, B. Hartley, P. Hamosh, J.A. Quest, and R.A. Gillis, Respiratory depressant effects of GABA alpha and beta-receptor agonists in the cat, J. Appl. Physiol. 62:2264-2272 (1987).
18. M. Vizek, C.K. Pickett, and J.V. Weil, Biphasic ventilatory response of adult cats to sustained hypoxia has central origin, J. Appl. Physiol. 63:1658-1664 (1987).
19. D.S. Ward, and J.W. Bellville, Reduction of hypoxic ventilatory drive by dopamine, Anesth. Analg. 61:333-337 (1982).
20. D.S. Ward, and T.T. Nguyen, Ventilatory response to sustained hypoxia during exercise, Med. Sci. Sports. Exerc. 23(6):719-726 (1991).
21. R.B. Weiskopf, and R.A. Gabel, Depression of ventilation during hypoxia in man, J. Appl. Physiol. 39:911-915 (1975).

EFFECTS OF VARYING END-TIDAL P_{O2} ON HYPOXIC VENTILATORY DECLINE IN HUMANS

Jaideep J. Pandit, Daphne A. Bascom, Ian D. Clement and Peter A. Robbins

University Laboratory of Physiology
Parks Road, Oxford OX1 3PT, UK

It is well-established that at a given end-tidal P_{CO2}, PET_{CO2}, the magnitude of the acute ventilatory response to hypoxia increases in proportion to the degree of hypoxia, and also that if the hypoxic exposure is prolonged, ventilation falls over 20-30 min, this decline being known as hypoxic ventilatory decline or HVD[1]. Most studies have thus far concentrated on an individual's response to only one level of hypoxia. The purpose of this study was to examine the ventilatory response at five different levels of PET_{O2}.

We studied six normal young adults. Subjects were seated and breathed through a mouthpiece connected to a turbine volume measuring device and pneumotachograph. Expired gases were sampled by a mass spectrometer and analysed for P_{O2} and P_{CO2}. All experimental variables were collected by a data acquisition computer. End-tidal gas values were passed to a second computer that controlled a fast, gas-mixing system to adjust the inspired gas composition on a breath-by-breath basis and maintain end-tidal gas composition at the desired levels. Details of this gas control system have been described previously[2,3,4]. Arterial saturation was measured by continuous pulse oximetry.

Figure 1 shows the five protocols used. End-tidal P_{O2} was held at 100 Torr for 10 min, and then at one of five levels of PET_{O2} (either 45, 50, 55, 65 or 75 Torr) for 20 min, and then returned to 100 Torr for 5 min. Finally, PET_{O2} was reduced to the previous level of hypoxia for the last 5 min. End-tidal P_{CO2} was held 1-2 Torr above the subject's resting value throughout each protocol. Each protocol was attempted six times on each of the subjects, but in some subjects, we failed to reach this target and a total of 154 experimental periods was finally obtained.

Figure 2 shows the breath-by-breath results in one subject for one protocol. Three 6th degree polynomials were fitted to the breath-by-breath data for the on-response, the recovery from hypoxia and the off-response. Figure 3 shows how these polynomials were used to calculate the points V2 (peak ventilation), V4 (minimum ventilation on return to euoxia) and V6 (peak ventilation for the second hypoxic exposure). The pre-hypoxic points V1 and V5 were calculated simply as the mean ventilation over the last minute of euoxia. Similarly, point V3 was calculated as the mean ventilation over the last minute of

Control of Breathing and Its Modeling Perspective, Edited by
Y. Honda *et al.*, Plenum Press, New York, 1992

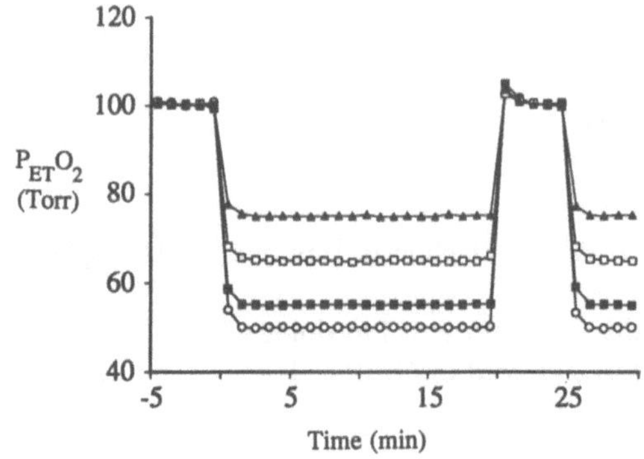

Figure 1. End-tidal values for P_{O2} for the five protocols.

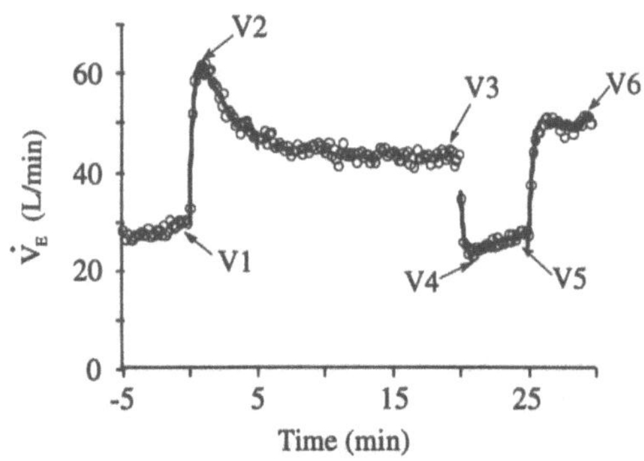

On-response-1 = V2 - V1

HVD = V2 - V3

Off-response = V3 - V4

On-response-2 = V6 - V5

Undershoot = V1 - V4

1 - (Off/On) = (V3-V4)/V2-V1)

Figure 2. Breath-by-breath values for ventilation against time for one subject. The three bold lines represent the polynomial fits for the first on-response, the recovery from hypoxia and the second on-response. See text for explanation of points V1-V6 and the calculation of parameters.

prolonged hypoxia. These six points, V1-V6, were used to calculate the five study parameters (Figure 2): On-response 1, On-response 2, HVD, Undershoot, and the ratio 1-(Off-response/On-response), which represents the change in peripheral chemoreflex sensitivity. Values for each experimental period were combined to give the mean for each subject and these subject means were combined to give the group means.

Figure 3 shows the group means for four of these study parameters plotted against arterial saturation, for each level of hypoxia. It was expected that each of the parameters would have the value zero at the euoxic value of saturation: the dotted line represents the best straight line fit of the data using least squares linear regression, that has been constrained to intercept the x-axis at the measured euoxic value of saturation of 96.4%. This is the "constrained" line. The solid line, on the other hand, represents the fit obtained when the regression is not constrained to intercept the x-axis at any particular value: this is the "unconstrained" line. The sum of squared errors from these two fits were compared using an F-ratio test to determine whether the fits were significantly different, and therefore whether the unconstrained line provided a significantly better fit than the constrained line. A value of $P < 0.05$ was considered statistically significant. For the four values in figure 4 - On-response 1, On-response 2, HVD and Undershoot - the unconstrained line does not provide a significantly better fit than the constrained line.

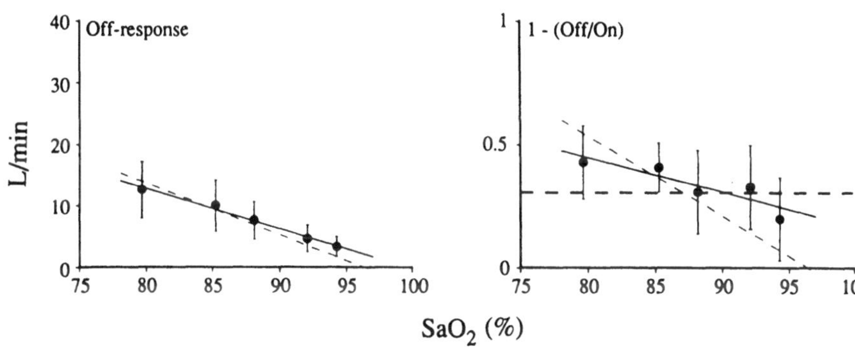

Figure 3. Mean ± SE of the on-response-1, HVD, undershoot and on-response-2 plotted against saturation for all six subjects combined. The solid line represents the results of a normal linear least squares regression. The broken line indicates a linear least squares regression with the intercept on the saturation axis held at a value of 96.4%. There is no significant difference between the fits provided by the two regressions.

However, the Off-response (Figure 4) is significantly better fit by the unconstrained line. Since the value of the Off-response should be zero in euoxia, this result suggests that in fact, the relationship between the Off-response and saturation may be a curvilinear one, intercepting the x-axis at the euoxic value of saturation.

Figure 4 also shows the relationship between the ratio 1 - (Off/On) and saturation. Two observations may be made. First, if the Off-response was always a constant fraction of the On-response, the ratio 1 - (Off/On) would be represented by the horizontal broken line, and be independent of the saturation. However, the actual, fitted, unconstrained line differs significantly from this horizontal line. Secondly, if the ratio 1 - (Off/On) was simply directly proportional to the desaturation, then the constrained, sloped, broken line would result. Again, the unconstrained line differs significantly from this.

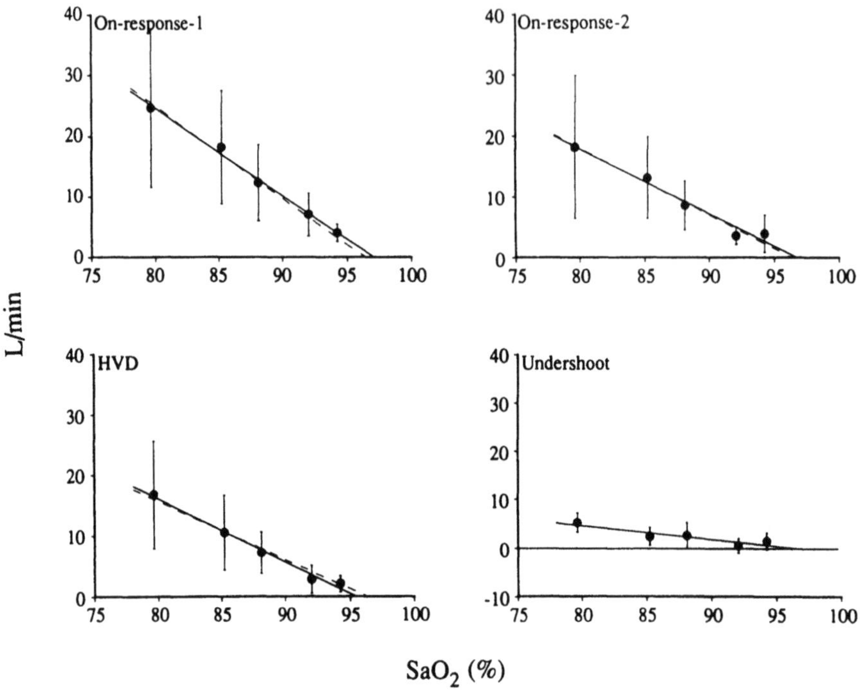

Figure 4. Mean ± SE of the off-response (left panel) and 1-(Off/On) (right panel) plotted against saturation. The solid line represents the results of a normal linear least squares regression. The broken, sloped line indicates a linear least squares regression with the intercept on the saturation axis held constant at a value of 96.4%. The unconstrained (solid) line provides significantly better fits than either of the constrained (broken lines). For further explanation, see text.

In summary:

1. The ratio 1 - (Off/On) may be thought of as a measure of the change in peripheral chemoreflex sensitivity, is not constant at different levels of hypoxia.
2. The ratio 1 - (Off/On) is not simply proportional to the degree of hypoxia.
3. There is therefore an asymmetry between the magnitudes of On-response and Off-response at all levels of hypoxia and this asymmetry increases with increasing levels of hypoxia, but not in a linear fashion.
4. This result suggests that models which assume a linear decline in chemoreflex sensitivity with sustained hypoxia may require modification.

Acknowledgments

This study was supported by the Wellcome Trust. J. J. Pandit is a Wellcome Trust Medical Graduate Fellow. D. A. Bascom is a Marshall Scholar. I. D. Clement is an MRC Student.

REFERENCES

1. J.V. Weil and C.W. Zwillich, Assessment of ventilatory response to hypoxia: methods and interpretation, *Chest.* 70: Supplement 124-128 (1976).
2. M.G. Howson, S. Khamnei, D.F. O'Connor and P.A. Robbins, The properties of a turbine device for measuring respiratory volumes in man, *J. Physiol (Lond).* 382:12P (1986).
3. M.G. Howson, S. Khamnei, M. E. McIntyre, D.F. O'Connor and P.A. Robbins, A rapid computer-controlled binary gas-mixing system for studies in respiratory control, *J. Physiol (Lond).* 394:7P (1987).
4. P.A. Robbins, G.D. Swanson and M.G. Howson, A prediction-correction scheme for forcing alveolar gases along certain time-courses, *J. Appl. Physiol.* 52:1353-1357 (1982).

DEVELOPMENTAL ASPECTS OF THE CENTRAL EFFECTS OF HYPOXIA

Teresa Trippenbach

Department of Physiology
McGill University
3655 Drummond St.
Montreal, Quebec, Canada H3G 1Y6

INTRODUCTION

Although several hypotheses have been proposed[1-3], the mechanisms of the hypoxic depression of breathing in newborns are still not well understood. A more recent study shows that the drop in ventilation coincides with a decrease in metabolic rate in newborns of several mammalian species[4]. These parallel effects imply that hypoxia does not abolish the homeostatic link between ventilation and metabolism. Nevertheless, central hypoxia could limit influences of the suprapontine structures on different control systems. These influences are independent of chemical stimuli and, via the reticular activating system (RAS), modify excitability of the respiratory neurones[5,6]. Peripheral nerve stimulation has comparable effects to the direct stimulation of the RAS, i.e. shortening of expiration and an increase in phrenic activity in newborn[7] and adult animals[6].

It is unknown how somatic stimulation affects the pattern of breathing during early development, and how hypoxia changes this interaction in newborns. Therefore, this study was undertaken to evaluate respiratory effects of saphenous nerve stimulation at different O_2 levels during the early postnatal life. The capability of the respiratory rhythm generator to respond to excitation during hypoxia-induced ventilatory depression was evaluated on the basis of respiratory effects of vagal stimulation with low frequency pulses.

METHODS

Experiments were performed on rabbits in 3 age groups: 1-3 days old (Group 1, n=13), the 2nd week of life (Group 2, n=14), and adult animals (Group 3, n=12). Rabbits were anaesthetised with urethane (0.8-2.0 g/kg), vagotomised, paralysed, and artificially ventilated with 100% 0_2 or 10% 0_2 in N_2. Blood pressure or heart rate was recorded. Integrated phrenic activity (PHR) was used as an index of the central respiratory rhythm. Stimulus parameters for both saphenous and vagus nerves were adjusted individually in each animal to produce respiratory excitation. These parameters remained unchanged until the end of the experimental run i.e.: control (C), hypoxia-induced respiratory excitation (Phase I), and respiratory depression (Phase II), and recovery (R) after hypoxia (Fig. 1). Phase I was characterised by an increase in phrenic activity and/or frequency of phrenic bursts. During Phase II the intervals between phrenic bursts increased in newborn rabbits.

Control of Breathing and Its Modeling Perspective, Edited by
Y. Honda *et al.*, Plenum Press, New York, 1992

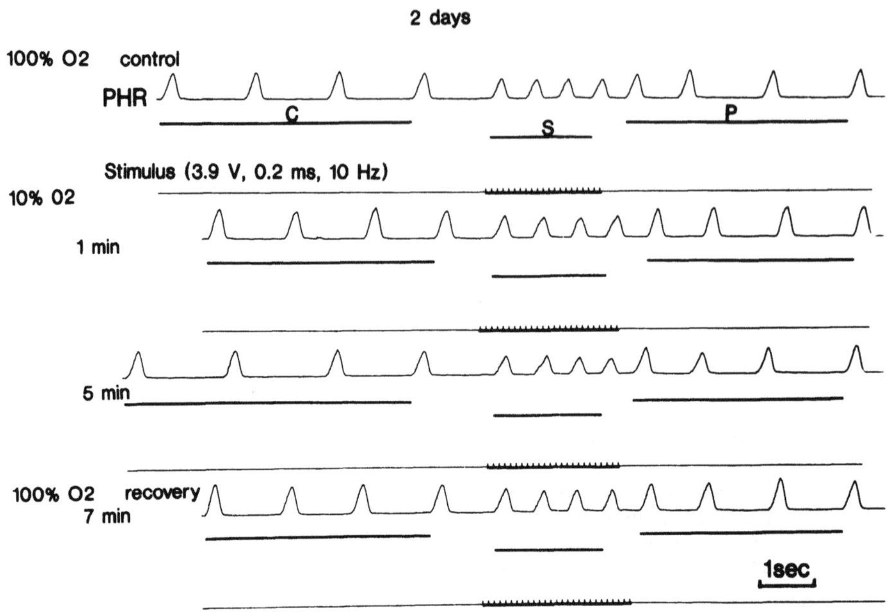

Figure 1. Effects of vagal stimulation on phrenic activity observed in a 2 day-old rabbit at control, Phase I, and Phase II of phrenic response to 10% O_2, and recovery after hypoxia. Phase I and Phase II were recorded at 2 min and 4 min of the exposure, respectively, Horizontal lines mark cycles analysed as control (C), stimulation (S), and post-stimulation (P).

RESULTS

Vagal Stimulation (VS)

At C, PHR and intervals between phrenic bursts (Te) of stimulated cycles (PHRs and Tes, respectively) always decreased from their control values. In Group 1 (Fig.1) and in Group 3 these effects were always similar. In Group 2, on the other hand, during Phase I the Te change was diminished, but remained significant. During hypoxia-evoked gasping or apnoea, VS restored phrenic activity with a frequency resembling that at C in both groups of newborns.

Saphenous Stimulation (SS)

At C, Tes was always shorter than Te. In Group 1 during Phase I, and in Group 2 during both phases of hypoxia, the Te response was not significant. In Group 3, hypoxia did not affect the Te response. Figure 2 illustrates an example of phrenic response to SS in a 12 day-old animal. Group mean increase in PHR was never significant (Fig. 3).

DISCUSSION

The respiratory effects of VS and SS observed in this study are qualitatively similar to those described in response to vagal[8] and somatic stimulation in newborn[7] and adult animals[6]. No PHR effects of SS, on one hand, and a decrease in PHR during VS, on the other, suggests different mechanisms were involved in the interaction of these two inputs with the central inspiratory neurones.

Because VS initiated effects resembling those of lung deflation, dust and histamine[9,10], most likely excitation of vagal rapidly adapting receptors and possibly C-fibres was predominant. SS probably activated only large muscle fibres. Activation of group IV of

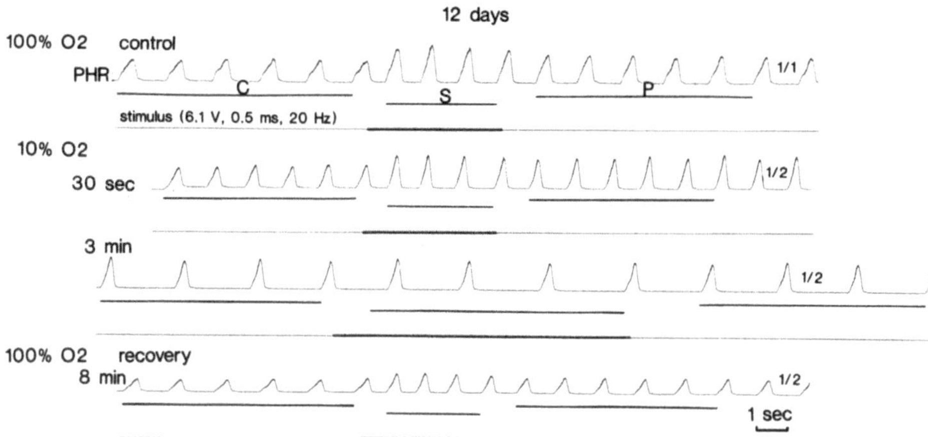

Figure 2. Effects of vagal stimulation on phrenic activity observed in a 12 day-old rabbit. For further information see legend for figure 1

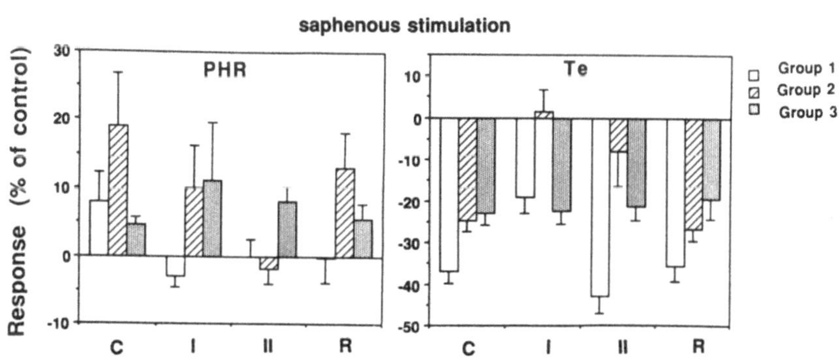

Figure 3. Effects of saphenous nerve stimulation on peak amplitude of phrenic activity (PHR) and intervals between phrenic bursts (Te) before hypoxia (C), Phase I (I), phase II (II), and recovery after hypoxia (R) in newborn (Group 1 and Group 2), and adult (Group 3) rabbits. Columns illustrate % change from control (0: no change) values recorded immediately before stimulation. Bars are ± SEM.

muscle afferents is associated with a significant increase in blood pressure in newborn kittens[11], which was not recorded in this study.

Our results show that during hypoxia-evoked respiratory depression, and even during gasping or apnoea, respiratory neurones are capable to respond with excitation to vagal excitatory input. This observation is in line with preservation of the CO_2 response during hypoxic ventilatory depression in newborn rats[12]. In contrast, the Te response to SS was abolished in Group 1 during Phase I, and in Group 2 during both hypoxic phases. Therefore it is suggested that brain hypoxia modifies the interaction between somatic inputs and the control of breathing differently immediately after birth and during the 2nd wk of life in rabbits. It is possible that at this age hypoxia depressed the RAS or blocked the inflow of this activity to the respiratory neurones. This age dependent difference may reflect the time course of maturation of the brain[13], and possibly, the predominant development of the inhibitory over excitatory neuro- transmitter systems[14]. The primary development of the inhibitory systems could explain the occurrence of SIDS only after the 1st month of life in infants.

Acknowledgments

This study was supported by the Hospital for Sick Children Foundation, Toronto, Canada. I thank Mrs. Christine Pamplin and Mr. André Duchastel de Montrouge for preparation of the manuscript.

REFERENCES

1. K. Iversen, T. Hedner and P. Lundberg. GABA concentration and turn-over in neonatal rat brain during asphyxia and recovery. Acta Physiol. Scand. 118:91 (1983).
2. H.R. Winn, R. Rubio and R.M. Berne. Brain adenosine concentration during hypoxia in rats. Am. J. Physiol. 241 (Heart Circ. Physiol. 10):H235 (1981).
3. R.L. Martin-Body and B.M. Johnston. Central origin of the hypoxic depression of breathing in the newborn Respir. Physiol. 71:25 (1988).
4. J.P. Mortola, R. Rezzonico and C. Lanthier. Ventilation and oxygen consumption during acute hypoxia in newborn mammals: a comparative study. Respir. Physiol. 78:31 (1989).
5. G.C. Salmoraghi and B.D. Burns. Notes on mechanism of rhythmic respiration. J. Neurophysiol. 23:14 (1960).
6. A. Hugelin and M.I. Cohen. The reticular activating system and respiratory regulation in the cat. Ann. N. Y. Acad. Sci. 109:586 (1963).
7. T. Trippenbach, G. Kelly and D. Marlot. Respiratory effects of stimulation of intercostal muscles and saphenous nerve in kittens. J. A. Physiol. 54:1736 (1983).
8. O.A.M. Wyss. The part played by the lungs in the reflex control of breathing. Helv. Physiol. Acta. 12 Suppl. 10:103 (1954).
9. T. Trippenbach, G. Kelly and D. Marlot. Effects of tonic vagal activity on breathing pattern in newborn rabbits. J. Appl. Physiol. 59:223 (1985).
10. T. Trippenbach and G. Kelly. Respiratory effects of cigarette smoke, dust and histamine in newborn kittens. J. Appl. Physiol. 64:837 (1988).
11. M. D. Parrish, J.M. Hill and M.P. Kaufman. Cardiovascular and respiratory responses to static exercise in the newborn kitten. Pediatr. Res. 30:95 (1991).
12. M. Seatta and J.P. Mortola. Interaction in hypoxic and hypercapnic stimuli on breathing pattern in the newborn rat. J. Appl. Physiol. 62:506 (1987).
13. J.P. Schade. Maturational aspects of EEG and of spreading depression in rabbits. J. Neurophysiol. 22:245 (1959).
14. K.L. Engish and G.D. Fischbach. The development of ACH- and GABA-activated currents in normal and target-deprived embryonic chick ciliary ganglia. Dev. Biol. 139:417 (1990).

MODULATION OF RESPIRATION BY BRAIN HYPOXIA

J.A. Neubauer, J.E. Melton, Q. Yu, L.O. Chae and N.H. Edelman

Division of Pulmonary and Critical Care Medicine
Department of Medicine
University of Medicine & Dentistry of New Jersey
Robert Wood Johnson Medical School
New Brunswick, New Jersey 08903 (USA)

The dual nature of hypoxic modulation of central respiratory activity is best appreciated in the sino-aortic deafferented, anesthetized animal during progressive reductions in the arterial oxygen content.[1] Figure 1 illustrates these two distinct central respiratory responses to brain hypoxia and their different oxygen thresholds; *respiratory depression*, which is manifested with even modest reductions in oxygenation, and *respiratory excitation (gasping)*, which occurs only with severe reductions in the arterial oxygen content (to values less than 20%).

The characteristics of hypoxic respiratory depression suggest that it is a sensitive, relatively rapid response to reduced oxygen supply[2] and mediated by mechanisms which modulate but do not impair the central respiratory circuitry from responding maximally.[3] It is tempting to explain the fall in respiratory output during hypoxia on the basis of neuronal failure due to insufficient oxygen and depletion of energy substrates. However, several observations indicate that this is not the case. First, the depression of respiratory activity even to apnea occurs without impairing the phrenic neurogram response to CO_2[3] or carotid sinus nerve stimulation.[4]

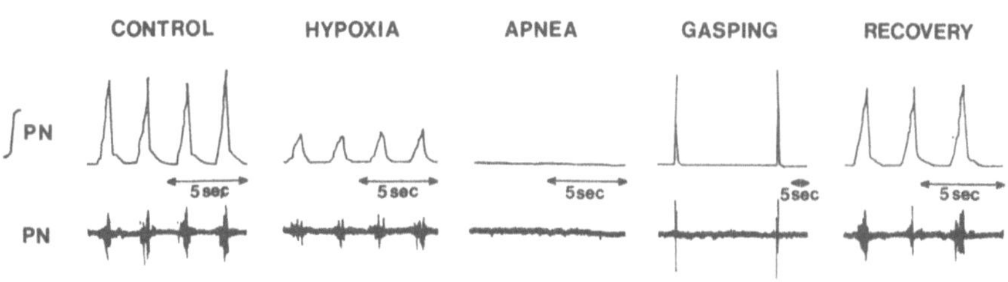

Figure 1. Phrenic neurogram (PN) response to progressive reductions in arterial oxygen content from 19.4 vol% to 4.2 vol% demonstrating initial depressant phase followed by excitatory phase (gasping) with very severe hypoxemia.

Control of Breathing and Its Modeling Perspective, Edited by
Y. Honda *et al.*, Plenum Press, New York, 1992

Figure 2A demonstrates that, even after silencing the phrenic neurogram with hypoxia (open circles), stimulation of the central chemoreceptors during CO_2 rebreathing results in the same maximal phrenic neurogram response as was obtained during baseline eupneic levels of activity (closed circles). Likewise, hypoxic depression of the phrenic neurogram does not reduce the ability of the carotid sinus nerve to augment phrenic nerve activity (Figure 2B). Secondly, neuronal failure due to energy depletion would lead to membrane instability, depolarization and ionic leakage. Using K^+-sensitive microelectrodes positioned in the medulla we have found no rise in extracellular $[K^+]$ during progressive hypoxemia until arterial oxygen contents are reduced to levels well below those associated with apnea of the phrenic neurogram.[1] These observations taken together indicate that hypoxic respiratory depression is not due to neuronal failure but is mediated by mechanisms which preserve neuronal integrity including their ability to respond appropriately to afferent input.

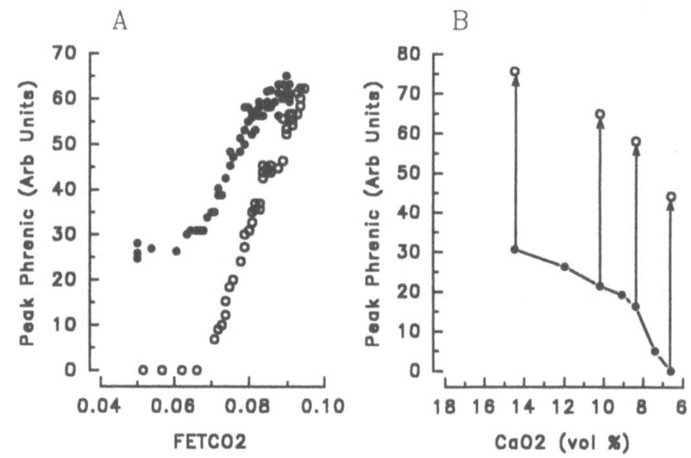

Figure 2. Phrenic neurogram responses to CO_2-rebreathing (A) and carotid sinus nerve stimulation (B). In (A), closed circles represent the control CO_2 response and open circles the CO_2 response during hypoxemia sufficient to cause apnea. In (B), closed circles represent the progressive decline in the integrated peak phrenic neurogram during hypoxia without stimulation of the carotid sinus nerve, while open circles represent the phrenic neurogram response to carotid sinus nerve stimulation at various points of phrenic depression.

The dynamic characteristics of respiratory depression have been assessed by using frequency analysis of the phrenic neurogram response to sinusoidal hypoxia (90-70% SaO_2) imposed at cycle times between 2.5-15 minutes.[2] Figure 3 illustrates that the depression of the phrenic neurogram in response to sinusoidal hypoxia (cycle time 6 minutes) has a lag of about 70 sec but is fully reversible and reproducible over successive cycles. Frequency analysis of these responses revealed that the phrenic neurogram depression has the characteristics of a first order system with a time constant of 74 seconds.[2] Potential mechanisms explaining the reduced respiratory neuronal activity which would stabilize neuronal membranes and require a time constant of several seconds are most consistent with enhanced neural inhibition due to changes in neurotransmitter mediated excitability.

A metabolically mediated shift towards greater inhibition during hypoxia could occur because of a reduced excitatory neurotransmitter influence or increased inhibitory neurotransmitter influence due to changes in release, synthesis, re-uptake or receptor kinetics of individual neurotransmitters. Several potential neuroeffectors could mediate hypoxic respiratory depression. In general hypoxia increases the extracellular concentration of most known inhibitory substances, e.g., adenosine,[5] GABA,[6,7] endogenous opioids[8] and lactic acid[9,10] and decreases in the effective extracellular concentration of most excitatory substances, e.g., acetylcholine,[11,12] aromatic monoamines,[11] and the amino acids glutamate and aspartate.[6] The importance of several inhibitory neuroeffectors in mediating the depression of phrenic nerve activity during progressive hypoxia was tested using specific antagonists (theophylline, bicuculline and naloxone) or, in the case of lactic acid, by preventing its production (sodium dichloroacetate). While adenosine and endogenous opioids contribute to hypoxic respiratory depression, their contribution appears to be minimal.[13,14] In contrast, hypoxic respiratory depression is substantially delayed by prevention of lactic acidosis[10] and reversed by GABA antagonism.[15] Both acidosis and GABA inhibition would promote neuronal membrane stabilization and hyperpolarization reducing neuronal activity without impairing the neurons from responding to a supramaximal stimulus.

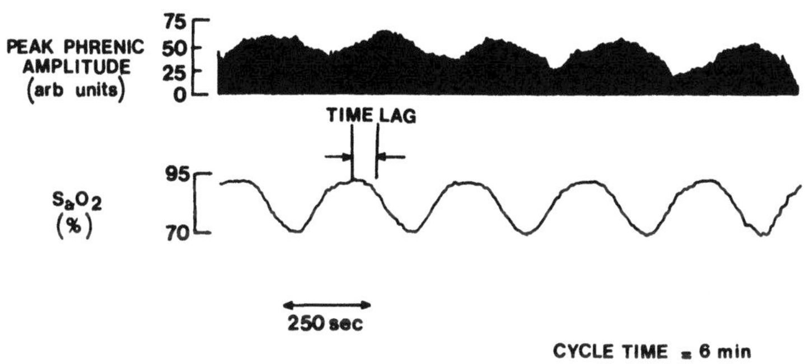

Figure 3. Phrenic neurogram response to sinusoidal hypoxia (cycle time 6 minutes). Vertical lines enclosed by arrows represent the peak-to-peak time lag between the change in SaO_2 and the change in peak phrenic amplitude.

A great deal less is known regarding the excitatory respiratory response to hypoxia (gasping). This neuronal excitation requires rather severe hypoxia and is not a unique characteristic of respiratory neurons since sympathetic neurons are also excited by severe hypoxia.[16] These excitatory events could potentially occur because of a sudden, excessive glutamergic excitation since extracellular glutamate concentrations can be significantly increased[6] when severe hypoxia causes glial reuptake mechanisms to fail. We tested this hypothesis by determining whether pretreatment with NMDA[17] and non-NMDA[18] glutamate receptor blockers could prevent respiratory and sympathetic neuronal excitation. Although antagonism of glutamate receptors altered the eupneic pattern of the phrenic neurogram and the integration of respiratory activity on the cervical sympathetic neurogram, neither receptor blocker prevented the respiratory or sympathetic excitatory responses to severe hypoxia. Blocking NMDA receptors with MK-801 and non-NMDA receptors with NBQX did not prevent the rise in tonic cervical sympathetic nerve activity or alter the frequency or amplitude of the phrenic gasp burst. These results indicate that hypoxic excitation is not dependent upon glutamergic activation and suggest that some other mechanism mediates neuronal excitation, e.g., perhaps disinhibition.

The functional values of both the depressant and excitatory responses to progressive brain hypoxia merit discussion. Although it seems counterproductive to attenuate respiratory activity during hypoxia, the depression of neuronal activity with reductions in available brain oxygen supplies may well be a generally useful and important protective mechanism functioning to stabilize the membrane and prevent neuronal failure. However, should brain oxygen levels diminish to near lethal levels excitatory responses which would promote autoresuscitation[19] and would have selective adaptive value over maintenance of respiratory and sympathetic neuronal depression.

REFERENCES

1. J.E. Melton, L. Oyer-Chae, J.A. Neubauer, and N.H. Edelman, Extracellular potassium homeostasis in the cat medulla during progressive brain hypoxia, *J. Appl. Physiol.* 70:1477 (1991).
2. Q.P. Yu, J.E. Melton, J.K-J. Li, J.A. Neubauer, and N.H. Edelman, The response of the phrenic neurogram to sinusoidal brain hypoxia, *FASEB Journal* 5:A665 (1991).
3. J.E. Melton, J.A. Neubauer, and N.H. Edelman, CO_2 sensitivity of cat phrenic neurogram during hypoxic respiratory depression, *J. Appl. Physiol.* 65:736 (1988).
4. J.E. Melton, Q.P. Yu, J.A. Neubauer, and N.H. Edelman, Effect of brain hypoxia on peripheral chemoreception and respiratory afterdischarge in the cat, *J. Appl. Physiol.* (in press).
5. H.R. Winn, R. Rubio, and R.M. Berne, Brain adenosine concentration during hypoxia in rats, *Am. J. Physiol.* 241:H235 (1981).
6. M. Erecinska, D. Nelson, D.F. Wilson, and I.A. Silver, Neurotransmitter amino acids in the CNS. I. Regional changes in amino acid levels in rat brain during ischemia and reperfusion, *Brain Res.* 304:9 (1984).
7. J.D. Wood, W.J. Watson, and A.J. Drucker, The effect of hypoxia on brain gamma-aminobutyric acid levels, *J. Neurochem.* 15:603 (1968).
8. V. Chernick, and R.J. Craig, Naloxone reverses neonatal depression caused by fetal asphyxia, *Science* Wash. DC 216:1252 (1982).
9. T.E. Duffy, S.R. Nelson, and O.H. Lowry, Cerebral carbohydrate metabolism during acute hypoxia and recovery, *J. Neurochem.* 19:959 (1972).
10. J.A. Neubauer, A. Simone, and N.H. Edelman, Role of brain lactic acidosis in hypoxic depression of respiration, *J. Appl. Physiol.* 65:1324 (1988).
11. J.N. Davis, and A. Carlsson, Effect of hypoxia on monoamine synthesis, levels and metabolism in rat brain, *J. Neurochem.* 21:783 (1973).
12. G.E. Gibson, M. Shimada, and J.P. Blass, Alterations in acetylcholine synthesis and in cyclic nucleotides in mild cerebral hypoxia, *J. Neurochem.* 31:757 (1978).
13. J.A. Neubauer, M.A. Posner, T.V. Santiago, and N.H. Edelman, Naloxone reduces ventilatory depression of brain hypoxia, *J. Appl. Physiol.* 63:699 (1987).
14. F.P. Nissley, J.E. Melton, J.A. Neubauer, and N.H. Edelman, Effect of adenosine antagonism on phrenic nerve output during brain hypoxia, *Fed. Proc.* 45:1046 (1986).
15. J.E. Melton, J.A. Neubauer, and N.H. Edelman, GABA antagonism reverses hypoxic respiratory depression in the cat, *J.Appl. Physiol.* 69:1296 (1990).
16. M.J. Wasicko, J.E. Melton, J.A. Neubauer, N. Krawciw, and N.H. Edelman, Cervical sympathetic and phrenic nerve responses to progressive brain hypoxia, *J. Appl. Physiol.* 68:53 (1990).
17. L. Oyer-Chae, J.E. Melton, J.A. Neubauer, and N.H. Edelman, Differential phrenic (PN) and sympathetic nerve responses to an NMDA blocker in progressive brain hypoxia (PBH). *FASEB Journal* 4:A405 (1990).
18. L. Oyer-Chae, J.E. Melton, J.A. Neubauer, and N.H. Edelman, Non-NMDA blockade alters eupneic but not hypoxia-induced gasping activity of the phrenic nerve, *FASEB Journal* 5:A664 (1991).
19. M.S. Jacobi, and B.T. Thach, Effect of maturation on spontaneous recovery from hypoxic apnea by gasping, *J. Appl. Physiol.* 66:2384 (1989).

RESPIRATORY DEPRESSION FOLLOWING ACUTE HYPOXIA IN VAGOTOMIZED AND CAROTID SINUS NERVE DENERVATED CATS

Teijiro Natsui and Shun-ichi Kuwana

Department of Physiology, School of Medicine, Teikyo University

Kaga 2-11-1, Itabashi-Ku, Tokyo 173, Japan

INTRODUCTION

Hypoxic depression of respiration has been observed in carotid chemodenervated animals[1,2,3]. Although there seems to be general agreement that this is a direct inhibitive effect of hypoxia on central neurones, the exact nature of this mechanism is not well understood. Recently, we observed in carotid chemodenervated cats that electrical stimulation of the carotid sinus nerve at the threshold P_{CO_2} during moderate hypoxia produced phrenic nerve activity, the peak height of which was not significantly different from that in normoxia[4]. This suggests that the neural pathway from the carotid chemoreceptors through the central respiratory neurones remains unimpaired during hypoxia. Therefore, the present study was designed to examine the possibility that respiratory depression in hypoxia might be caused by inhibition of the central CO_2 chemoreceptors.

METHODS

Adult cats weighing between 2.7 and 5.0 Kg, premedicated by inhalation of a gas mixture of 50 % N_2O in O_2, were anesthetized with chloralose (40 mg/kg) and urethane (200 mg/kg) given intraperitoneally. The head of the cat was fixed to a stereotaxic frame in the supine position. The trachea was cannulated close to the thorax for artificial ventilation. One catheter was inserted into the femoral artery for recording systemic pressure with a strain-gauge, and the other catheter into the femoral vein for drug injection. End-tidal P_{CO_2} and P_{O_2} were continuously monitored with an infrared CO_2 analyzer in series with a polaro-graphic oxygen electrode. Rectal temperature was maintained between 37 and 38 °C by placing a heating pad under the cat. Both vago-sympathetic nerves were sectioned at the mid-cervical level. The carotid sinus nerves were bilaterally cut close to the carotid body; one was used for electrical stimulation with a bipolar electrode. The phrenic nerve root (C_5) exposed on the right side was cut and the central end of the nerve was placed on a bipolar platinum recording electrode in a pool of liquid paraffin. The nerve impulses were conventionally amplified and electrically integrated. The peak height of this integrated nerve activity was used as an index of inspiratory activity. The ventral surface of the medulla was exposed, using surgical procedures previously described[5]. The dura mater was opened by a mid-line incision and its edges were fixed sideways with adhesive. The arachnoid membrane was carefully cut along the basilar artery. After the surgical preparation, the cat was initially paralyzed with pancuronium bromide, 1mg/Kg i.v., and artificially ventilated with air. Additional doses were administered as needed to prevent

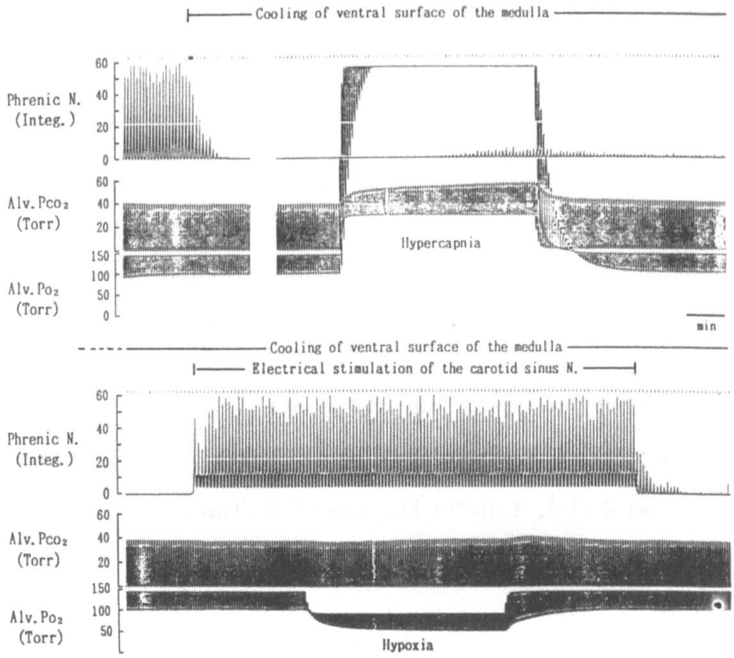

Fig. 1. Effect of hypercapnia or hypoxia on phrenic nerve activity during cooling of ventral surface of the medulla. The surface temperature was 21.7 °C during the cooling.

respiratory muscle activity. The end-tidal Pco_2 was kept at the level observed in spontaneous breathing. In order to activate the central CO_2 chemoreceptors, hypercapnic blood at a temperature of 38 °C was infused in some experments into the anterior inferior cerebellar artery (AICA). This was based on our observation[5] that the central CO_2 chemosensitive area was perfused mainly by blood from the AICA. A catheter was inserted into the basilar artery at the boundary between the pons and medulla with the tip in the caudal direction. Blood was infused with a syringe infusion pump. Heparinized arterial blood was previously sampled during inhalation of a gas mixture of 7.4 % CO_2 + 21 % O_2. The basilar artery was ligated caudal to the ramification of the AICA. Thus, any infusion flowed into both AICA. In other experiments, Ringer solution was infused to block the central CO_2 chemoreceptors because the solution contains dissolved oxygen and no carbon dioxide.

RESULTS

When the ventral surface of medulla was cooled with a cold water circulated thermode, phrenic nerve activity ceased immediately (Fig. 1). Then, when a gas mixture of 5.5 % CO_2 in O_2 was inhaled during the cooling there was almost no phrenic nerve response to hypercapnia. However, phrenic nerve activity could be induced by electrical stimulation of the carotid sinus nerve and this could not be inhibited by moderate hypoxia. These results indicate that the neural pathway from the carotid chemoreceptors is not blocked by cooling and is not suppressed by hypoxia.

Fig. 2 shows a typical example of the effect of moderate hypoxia on phrenic nerve activity before and during infusion of hypercapnic blood into AICA. Before the infusion, moderate hypoxia inhibited phrenic nerve activity (upper panel). On the other hand, the infusion produced an increase in phrenic nerve activity and this increased phrenic activity was not inhibited by moderate hypoxia (lower panel). This suggests that, as long as the central CO_2 chemosensitive area is

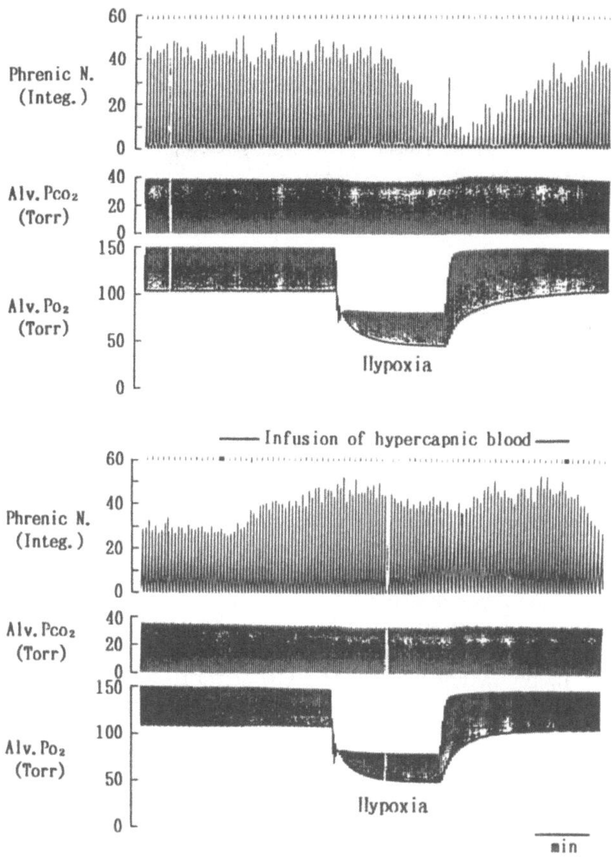

Fig. 2. Effect of hypoxia on phrenic nerve activity before and during infusion of hypercapnic blood into AICA.

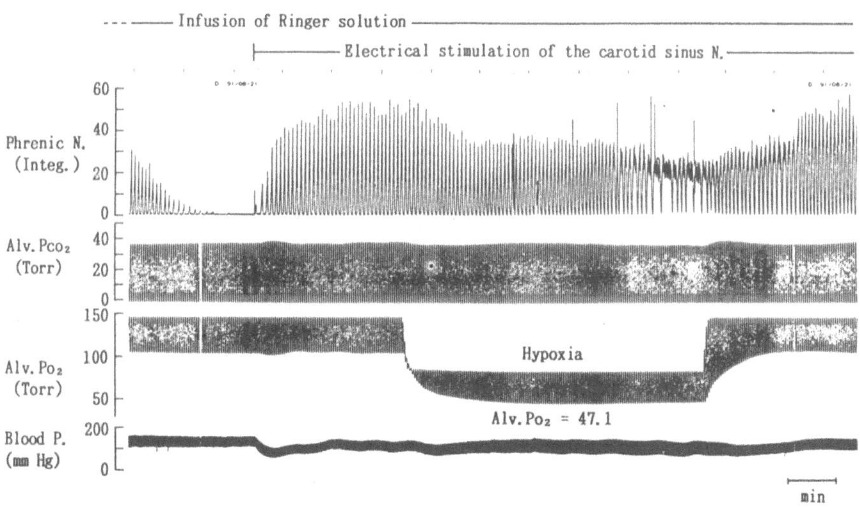

Fig. 3. Effect of hypoxia on phrenic nerve activity induced by electrical stimulation of the carotid sinus nerve during infusion of Ringer solution into AICA.

perfused with infused hypercapnic blood, the systemic hypoxia does not influence phrenic nerve activity.

Infused Ringer solution into the AICA induced phrenic nerve apnea (Fig .3), which could not be reversed by systemic hypercapnia. However, phrenic nerve activity could be induced by electrical stimulation of the carotid sinus nerve. This induced phrenic activity was not suppressed very much by moderate hypoxia during infusion. These results are similar to those obtained by cooling the medulla, thus indicating again that the neural pathway from the carotid chemoreceptors remains unimpaired during the infusion and is not suppressed by hypoxia.

DISCUSSION

The results of the present study are: 1) phrenic nerve apnea was induced by cooling the ventral surface of the medulla or by infusing Ringer solution into the AICA. Further there was no phrenic nerve response to systemic hypercapnia, but electrical stimulation of the carotid sinus nerve induced phrenic activity which could not be inhibited by moderate systemic hypoxia. 2) phrenic activity was increased by infusion of hypercapnic blood into the AICA and this was not inhibited by systemic hypoxia. These findings, suggest that respiratory depression during moderate hypoxia is mainly due to the inhibition of the central chemoreceptors and does not reflect a widespread inhibition of the respiratory neurones.

The hypoxic ventilatory depression has been demonstrated in carotid chemodenervated humans[6,7] and animals[1,2,3,4,8]; the phrenic nerve activity is inhibited by moderate systemic hypoxia, but can be restored by a sufficient increase in Pco_2, and for gradually lower values of Po_2 the CO_2-phrenic activity response curve shifts toward higher Pco_2 without changing its slope. This parallel shift of the CO_2 response curve remains true for more severe hypoxia.

This phenomenon and the results of the present study lead to the hypothesis that the central CO_2 chemoreceptors are composed of cells which have different Pco_2 threshold levels. Cells with a low Pco_2 threshold are more sensitive to hypoxia than those with a high Pco_2 threshold. At any given level of hypoxia functional arrest will occur in cells with low Pco_2 threshold and high sensitivity to hypoxia, while cells with high Pco_2 threshold and low sensitivity to hypoxia will remain intact.

In conclusion, the respiratory depression in moderate and acute systemic hypoxia can be caused by inhibition of the central CO_2 chemoreceptors.

REFERENCES

1. Katsaros, B. (1965). Die Rolle der Chemoreceptoren des Carotisgebiets der narkotisierter Katze für die Antwort der Atmung auf isolierte Änderung der Wasserstoffionen-Konsentration und des CO_2-Drucks des Blutes. Pflügers Archiv. 282: 157-178.
2. Mitchell, R. A. (1965). The regulation of respiration in metabolic acidosis and alkalosis. In: Cerebrospinal Fluid and the Regulation of Respiration, edited by C. McC. Brooks, F. F. Kao, and B. B. Lloyd. Oxford: Blackwell, 1965, p. 109-131.
3. Melton, J. E., Neubauser, J. A., and Edelman, N. H. (1988). CO_2 sensitivity of cat phrenic neurogram during hypoxic respiratory depression. J. Appl. Physiol. 65: 736-743.
4. Natsui, T. and Kuwana, S. (1989). Hypoxic respiratory depression in anesthetized, vagotomized and carotid sinus nerve denervated cats (Abstr). Proc. IUPS, 17:172-173.
5. Kuwana, S. and Natsui, T. (1990). Respiratory responses to occlusion or hypercapnic blood injection of the anterior inferior cerebellar artery in cats. Jpn. J. Physiol. 40:225-242.
6. Sørensen, S. C. and Mines, A. H. (1970). Ventilatory response to acute and chronic hypoxia in goats after sinus nerve section. J. Appl. Physiol. 28:832-835.
7. Honda, Y. and Hashizume, I. (1991). Evidence for hypoxic depression of CO_2-ventilation in carotid body-resected humans. J. Appl. Physiol. 70:590-593.
8. Natsui, T. and Kuwana, S. (1990). Central hypoxic depression in anesthetized, vagotomized and carotid sinus nerve denervated cats. Jpn. J. Physiol. 40(Suppl) S. 56.

RESPIRATORY AND POSTURAL SUPPRESSION BY STIMULATION OF THE MIDPONTINE DORSAL TEGMENTUM IN DECEREBRATE CATS

Koichi Kawahara, Kyuichi Niizeki, and Yoshiko Yamauchi

Department of Electrical and Information Engineering
Yamagata University
Yonezawa 992
Japan

INTRODUCTION

In acute precollicular-postmammillary decerebrate cats, stimulation of the dorsal part of the caudal tegmental field (DTF) in the pons results in the parallel suppression of postural tone and respiration.[1] Tonic discharges of bilateral hindlimb antigravity muscles are suppressed by DTF stimulation, and the animal becomes unable to stand upright (postural atonia). DTF stimulation also depresses diaphragmatic activity, leading to apnea for more than 20 seconds in some animals. In addition, DTF stimulation suppresses the tonic and rhythmic discharges of the external intercostal muscles[2] and of the hypoglossal nerve innervating the genioglossus muscle.[3] Previous studies[4,5] raise the possibility that DTF-elicited suppression of postural tone and respiration may result from activation of the descending axons passing through the DTF. All these findings lead us to believe that the descending inhibitory system activated by DTF stimulation may be involved in generalized motor suppression similar to the spontaneous suppression that occurs during sleep, especially during REM sleep.

In the same preparations, we have already shown that antidromic spikes are recorded in and near the nucleus reticularis pontis oralis (PoO) by stimulation of the DTF[4]. Tonic electrical stimulation of the recording sites results in the parallel suppression of postural tone and respiration. These effects are very similar to those elicited by DTF stimulation. Mitler and Dement[6] have shown that a microinjection of carbachol, a cholinergic agonist, into the rostral pontine reticular formation around or near the PoO, produces generalized motor suppression resembling that which occurs spontaneously during REM sleep. The above findings lead us to expect that the cholinoceptive or cholinergic neuronal mechanisms in the rostral pons may be involved in the parallel suppression of postural tone and respiration evoked by DTF stimulation.

In this study, we first investigated the DTF-elicited effects on cardiac parameters, such as heart rate and blood pressure. We then attempted to elucidate whether the rostral pontine neuronal structures responsible for the parallel suppression of postural tone and respiration are cholinoceptive.

Control of Breathing and Its Modeling Perspective, Edited by
Y. Honda *et al.*, Plenum Press, New York, 1992

METHODS

Experiments were done on 6 adult cats of both sexes weighing 2.8-4.3 Kg. Under halothane anesthesia, the animals were decerebrated at the precollicular-postmammillary level. The head of the animal was fixed in a stereotaxic frame and the limbs were placed on the surface of a stationary treadmill. Electromyograms (EMG) were recorded by implanting bipolar electrodes made of thin (70 μm) copper wires into the bilateral soleus muscles and the diaphragm. The CO_2 tension of expired air was monitored and recorded with an infrared gas analyzer. A catheter was inserted into the left carotid artery for monitoring arterial blood pressure. The electrocardiogram (ECG) was also monitored and recorded. Body temperature was maintained at a constant level with an infrared heat lamp.

Glass-insulated tungsten microelectrodes were used for stimulation of the DTF and for recording unit discharges. One electrode was inserted into the rostral pontine reticular formation to record antidromically-driven unit discharges extracellularly, while the DTF was stimulated with another electrode. Unit spikes with negative-positive polarity were taken to be soma action potentials. Tonic electrical stimulation of the DTF consisted of a rectangular pulse of 0.2 ms duration at 50 pulses/s with an intensity of 40-70 μA. For recording antidromically-activated spikes, a single or double shock (pulse duration, 0.2ms; intensity, 40-80 μA) was delivered to the DTF every second. A glass micropipette (tip diameter, 20-40 μm) was used for microinjection of carbachol into the sites in the rostral pontine reticular formation from which antidromically-driven spikes were recorded on stimulation of the DTF. Carbachol (4 μg/0.25 μl) was microinjected into the brain stem with the micropipette under pressure over a period of 30-40 seconds. Ringer's solution was also microinjected into the same brainstem sites by the same procedure to confirm whether the induced effects were due to chemical stimulation.

The extent of respiratory modulation of heart rate was assessed by the coherence between heart period fluctuation and respiration. ECG was amplified and then electrically differentiated. The resulting spike signal triggered a Schmitt-trigger circuit to generate a rectangular pulse of 1.0 ms duration, each representing a time position of the R wave of the ECG signal. The heart period was defined as the time interval between the onset of two consecutive pulses. The heart periods were transformed into another pulse train in which the pulses were arranged in equal intervals, with the pulse heights representing the heart period.[7,8] The coherence between the heart period fluctuation and respiration was calculated with the fast Fourier transformation (FFT) technique.

At the end of each experiment, the animals were deeply anesthetized (pentobarbital sodium, i.v.) and sacrificed. The stimulating and recording sites were marked with an electrolytic lesion (DC current of 20 μA, 20 s). The brain was fixed in 10 % formalin for later histological examination. The stimulating and recording sites were determined with reference to the stereotaxic atlas of Snider and Niemer.[9]

RESULTS

In all the animals tested, stimulation delivered to the DTF along the midline elicited simultaneous suppression of postural tone and respiration. Figure 1 shows an example of the results obtained from two different animals. DTF stimulation decreased not only the tonic discharges of the bilateral hindlimb soleus muscles but also the rhythmic discharges of the diaphragm. The tonic discharges of the bilateral soleus muscles were completely abolished during DTF stimulation, and the abolished discharges did not resume after the stimulation ended. In contrast, the once depressed rhythmic diaphragmatic activity gradually augmented despite the continuation of the stimulation. In most cases, the rebound augmentation of the diaphragmatic activity occurred immediately after the end of the stimulation causing the cats to hyperventilate.

We then analysed the DTF-elicited effects on cardiac parameters such as heart rate and blood pressure (Fig. 2). In this animal, the diaphragm showed tonic as well as rhythmic discharges before the start of DTF stimulation. The stimulation simultaneously depressed diaphragmatic and soleus muscle activities. In particular, tonic discharges of the diaphragm were completely abolished during the stimulation, as with those of hindlimb soleus muscle (Fig. 2A). DTF stimulation also produced a decrease in heart rate (- 70 beats/min from a pre-stimulus mean value of 195 beats/min), and a decrease in arterial blood pressure (-50 mmHg from a pre-stimulus value of 115 mmHg at the systolic level), as shown in Fig.2B and C. In all the animals tested, DTF stimulation not only suppressed the diaphragmatic and antigravity muscle activities but also decreased heart rate and blood pressure.

After confirmation of the effects of the DTF stimulation, another microelectrode was inserted into the rostral pons for recording. The DTF was then stimulated once per second, and antidromic spikes or field potentials were recorded from the rostral pontine reticular formation. Similar to previous findings[4,5], a systematic survey with the recording electrode showed that antidromic spikes could be recorded from the regions in and near the nucleus reticularis pontis oralis (PoO). The recording electrode was then replaced with a glass micropipette filled with carbamylcholine (carbachol). Carbachol (4 μg/0.25 μl) was injected into the site from which the antidromically driven spike was recorded over a period of about 30 s (Fig. 3). Before the injection, the hindlimb soleus muscle exhibited tonic discharges due to decerebrate rigidity (Fig. 3A). About 90 s after the start of the injection, the tonic activity of the left soleus muscle gradually decreased in parallel with a decrease in heart rate and blood pressure (Fig. 3B). The tonic activity of

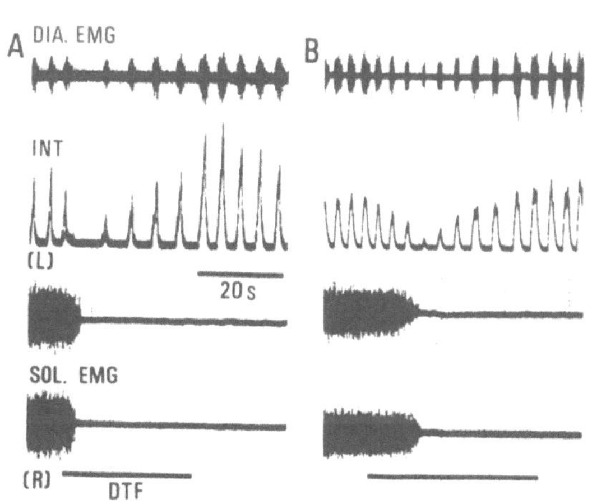

Fig. 1 Respiratory and postural depression by DTF stimulation. Tonic electrical stimulation delivered to the midline DTF depressed both the diaphragmatic and soleus muscle activities. A and B show the results obtained from two different cats. Abbreviations: DIA. EMG, diaphragmatic EMG; INT, integrated diaphragmatic EMG; SOL. EMG, soleus muscle EMG; L, left side; R, right side.

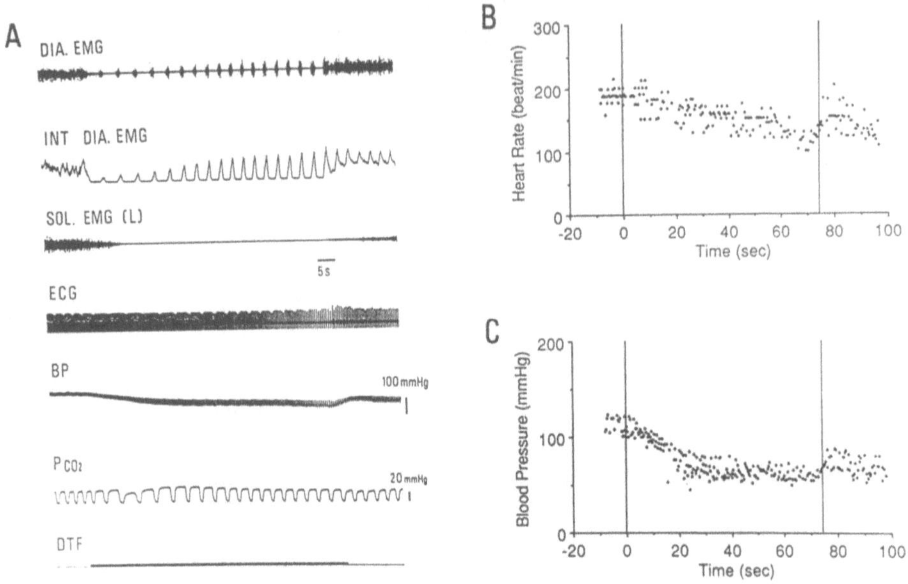

Fig. 2 Depression of heart rate and blood pressure induced by DTF stimulation. DTF stimulation induced parallel suppression of diaphragmatic and soleus muscle activities. The stimulation also produced a decrease in heart rate and in arterial blood pressure. Abbreviations: INT DIA. EMG, integrated diaphragmatic EMG; ECG, electrocadiogram; BP, arterial blood pressure; PCO2, CO2 tension of expired air; DTF, period of DTF stimulation. Other abbreviations are the same as those in Fig. 1. See text for details.

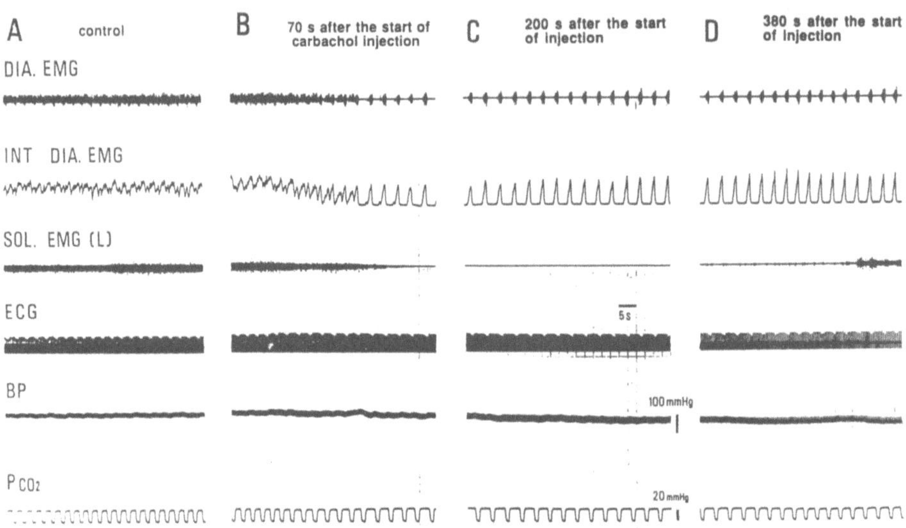

Fig. 3 Postural, respiratory and cardiac depression induced by microinjection of carbachol into the rostral pontine reticular formation. Carbachol (4μg/0.25μl) was microinjected into the pontine sites from which antidromic-driven spikes were recorded by stimulation of the DTF. Carbachol injection depressed soleus muscle activity bilaterally, although only the left soleus activity is shown in the figure. Abbreviations are the same as those in Fig. 1 and Fig. 2.

the diaphragm was gradually depressed, and the respiratory frequency also decreased. This animal became completely flaccid (postural atonia) because of the loss of decerebrate rigidity (Fig. 3C). At about 400 s after the injection, the hindlimb extensor muscles, however, started to show irregular contractions (Fig. 3D). These postural, respiratory and cardiac changes induced by microinjection of carbachol into the PoO were very similar to those elicited by electrical stimulation of the DTF and of the PoO. Microinjection of the same amount (0.25 µl) of Ringer's solution into the PoO did not induce any noticeable effects on postural, respiratory or cardiac functions.

In the other animal in Fig. 3, about 5 min after the start of the carbachol injection, rapid eye movements (REM) suddenly began to appear, and the respiratory rhythm became irregular (Fig. 4A). Three out of four animals tested with carbachol microinjection showed REM at 3 to 6 min after the end of the injection. At that time, the bilateral soleus muscles exhibited continuous irregular contractions synchronously or alternately, and the blood pressure fluctuation had a tendency to become greater. The mean arterial blood pressure level increased slightly compared with that during the carbachol induced atonic state without REM. We then analysed the extent of respiratory modulation of the heart period fluctuation (respiratory sinus arrhythmia, RSA) during the two states; one before carbachol injection, and the other during the period in which REM occurred after the injection. Before microinjection of carbachol (Fig. 4B), there was a distinct coherence peak centered at the mean respiratory frequency of about 0.45 Hz. There existed no distinct coherence peak other than that of respiratory frequency. However, the extent of RSA decreased markedly during the period in which REM occurred (Fig. 4C).

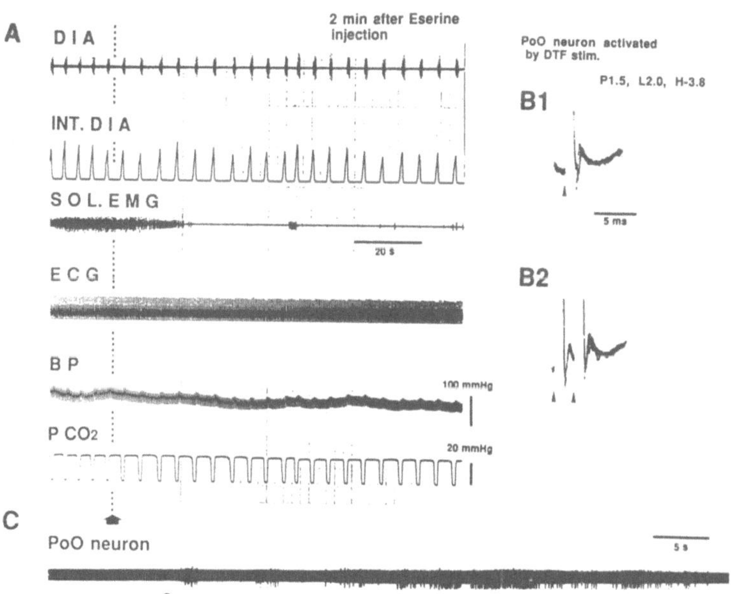

Fig. 4 Carbachol-induced rapid eye movements and irregular hindlimb muscle contractions. Upward spike artifacts observed in the top record in A (EOG) shows contaminatiof the ECG. Abbreviations: EOG, electrooculogram; RSA, respiratory sinus arrhythmia. Other abbreviations are the same as those in Fig. 1 and 2. See text for details.

We then investigated the changes in the neuronal activity of the PoO neuron antidromically driven by DTF stimulation when physostigmine (0.1 mg/Kg) was injected intravenously in the same kind of preparations. Previous studies[10,11] have shown that paradoxical sleep can be induced by intravenous administration of physostigmine (eserine) in either intact animals or those with brainstem transection. Figure 5 shows an example of the results. At about 2 min after the start of intravenous injection of eserine, the hindlimb soleus activity gradually decreased (Fig. 5A). At that time, both respiratory frequency and blood pressure also began to decrease. These generalized depressive effects seemed very similar to those elicited by electrical stimulation of the DTF or the PoO as well as those induced by carbachol microinjection into the PoO. Figure 5B shows an example of the antidromically activated PoO neurons on stimulation of the DTF. Almost all the PoO neurons identified had little or no spontaneous activity before the start of intravenous eserine injection. This neuron, however, began to fire almost coinciding with postural, respiratory and cardiac depression caused by eserine injection (Fig. 5C).

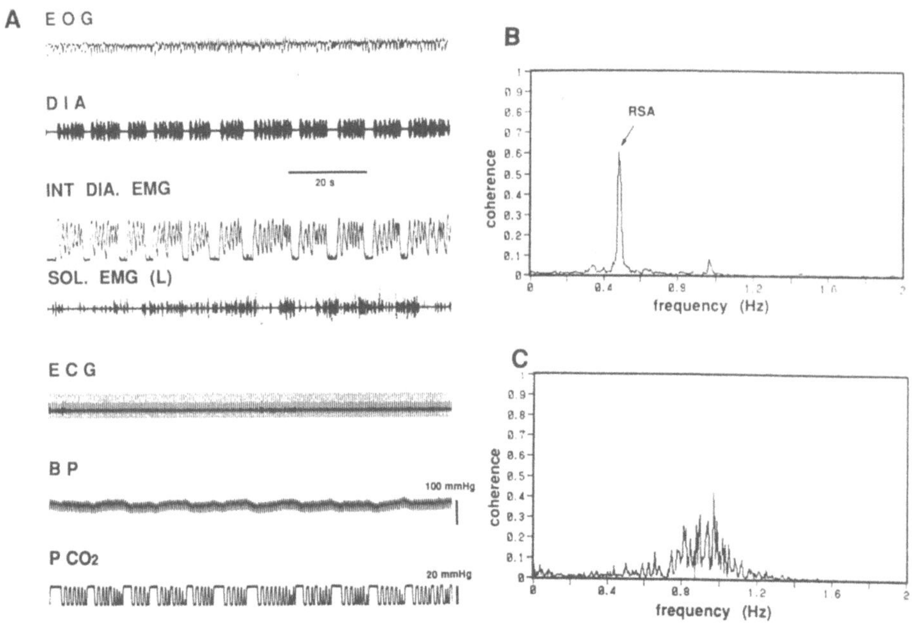

Fig. 5 Changes in the neuronal discharge of the rostral pontine reticular neuron and generalized suppressive effects induced by intravenous injection of physostigmine. The antidromically activated neuron by stimulation of the DTF (PoO neuron) was first identified (B1 and B2), and physostigmine (0.1 mg/Kg) was then administrated intravenously.

DISCUSSION

Previous findings[2,5] have led us to suggest that there are close similarities between the DTF-elicited generalized motor suppression and the motor suppression that occurs spontaneously during REM sleep. The present results seem to support this idea. The depressive effects on postural, respiratory and cardiac functions were very similar to those induced by carbachol microinjection into the PoO, from which antidromically driven units were recorded by stimulation of the DTF. Microinjection of carbachol into the reticular formation around or near the PoO produces generalized motor suppression resembling that observed during REM sleep.[6] Similar to the present results, carbachol microinjection into the rostral pontine region corresponding to the PoO produces a decrease in heart rate, as well as in arterial blood pressure.[12] Recent studies by Lydic and Baghdoyan[13], and by Gilbert and Lydic[14] have shown that cholinoceptive regions of the medial pontine reticular formation (mPRF), long known to play a role in regulating the sleep cycle, can causally mediate state-dependent changes in respiratory rate, tidal volume, minute ventilation, and respiratory cycle timing. In addition, the extent of respiratory modulation of heart rate (RSA) varies depending on the sleep-waking states, being greatest in slow wave sleep and least in REM sleep in humans.[15] The cross-spectral coherence between heart period fluctuation and respiration, reflecting the extent of RSA, greatly decreased during REM sleep. The coherence peak at the mean respiratory frequency was evident before carbachol microinjection into the PoO, but greatly decreased during the period in which REM were induced by carbachol injection (Fig. 4B and C). These results may further support the close similarities between the effects induced by carbachol microinjection into the PoO and those observed during REM sleep in intact animals or in humans.

Morales et al.[16] suggested that carbachol microinjection into the pontine reticular formation activates the same brainstem-spinal cord system responsible for the postsynaptic inhibition of motoneurons during REM sleep. Bilateral lesions of the pontine tegmentum including the PoO in cats result in REM sleep without atonia.[17] In addition, some of the identified PoO neurons without spontaneous activity, that were antidromically driven by DTF stimulation, began to fire almost coinciding with the start of postural, respiratory and cardiac depression induced by intravenous injection of physostigmine (Fig. 5C). Intravenous injection of physostigmine, an anticholinesterase, induces REM sleep, and the duration of the REM state depends on the amount of physostigmine injected.[18]

All of the above findings and our previous report on the descending inhibitory system originating from the neurons in the PoO[5] lead us to believe that the cholinoceptive neuronal structures in the rostral pons may be responsible for the generalized motor suppression. Such cholinoceptive pontine structures seem to be involved in the sleep state control. Respiratory movements are under autonomic as well as volitional control. Therefore, respiratory movements are likely to be depressed due to the generalized suppression of volitional motor systems that occurs during a REM sleep state. Respiratory movements, however, must recover under such strong inhibitory influences on respiration, because those are crucial for maintaining blood gas and acid-base homeostasis.[19] We have already demonstrated that respiratory movements recover under the strong inhibitory influences on respiration induced by DTF or PoO stimulation.[1,4] Therefore, respiratory drives must be brought about to overcome such respiratory inhibition. Chemical drives, emerging as a consequence of humoral changes due to preceding apneic states detected by central chemoreceptors, are the most probable origin of the respiratory drives.[1] If this is the case, dysfunction of central chemosensitivity would result in apnea during a REM sleep state, in which the cholinoceptive pontine neuronal structures described in this study possibly raise their activity.

Acknowledgments

We express sincere appreciations to Dr. K. Owada, Laboratory of Animal Center, Yamagata University, for preparing experimental animals. We also thank to Dr. M. Mussell for revision of the English of this manuscript.

This study was partly supported by a grant to K.K. from the Ministry of Education and Culture of Japan (03251205; 03650334).

REFERENCES

1. K. Kawahara, Y. Nakazono, S. Kumagai, Y. Yamauchi, and Y. Miyamoto, Parallel suppression of extensor muscle tone and respiration by stimulation of pontine dorsal tegmentum in decerebrate cats, *Brain Res.* 473:81(1988).
2. K. Kawahara, Y. Nakazono, and Y. Miyamoto, Depression of diaphragmatic and external intercostal muscle activities elicited by stimulation of midpontine dorsal tegmentum in decerebrate cats, *Brain Res.* 491:180(1989).
3. K. Kawahara, Y. Nakazono, S. Kumagai, Y. Yamauchi, and Y. Miyamoto, Inhibitory influences on hypoglossal neural activity by stimulation of midpontine dorsal tegmentum in decerebrate cat, *Brain Res.* 479:185(1989).
4. K. Kawahara, Y. Nakazono, S. Kumagai, Y. Yamauchi, and Y. Miyamoto, Neuronal origin of parallel suppression of postural tone and respiration elicited by stimulation of midpontine dorsal tegmentum in the decerebrate cat, *Brain Res.* 474:403(1988).
5. K. Kawahara and M. Suzuki, Descending inhibitory pathway responsible for simultaneous suppression of postural tone and respiration in decerebrate cats, *Brain Res.* 538:303(1991).
6. M.M. Mitler and W.C. Dement, Cataleptic-like behavior in cats after microinjection of carbachol in pontine reticular formation, *Brain Res.* 68:335(1974).
7. M. Kobayashi and T. Musha, 1/f fluctuation of heart period, *IEEE Trans. Biomed. Eng.* 29: 456(1982).
8. K. Kawahara and Y. Yamauchi, Spectral analysis on fluctuation of heart period in paralyzed, vagotomized, and unanesthetized decerebrate cats, *Biol. Cybern.* 63:251(1990).
9. R.S. Snider and W.T.A. Niemer, *Stereotaxic Atlas of the Cat Brain*, University of Chicago Press, Chicago(1961).
10. M. Matsuzaki, Differential effects of sodium butyrate and physostigmine upon the activities of para-sleep in acute brain stem preparations, *Brain Res.* 13:247(1969).
11. M. Matsuzaki and M. Kasahara, Induction of para-sleep by cholinesterase inhibitor in the mesencephalic cats, *Proc. Jap. Acad.* 42:989(1966).
12. Y. Katayama, D.S. DeWitt, D.P. Becker, and R.L. Hayes, Behavioral evidence for a cholinoceptive pontine inhibitory area: descending control of spinal motor output and sensory input, *Brain Res.* 296:241(1984).
13. R. Lydic and H.A. Baghdoyan, Cholinoceptive pontine reticular mechanisms cause state-dependent respiratory changes in the cat, *Neurosci. Lett.* 102:211(1989).
14. K.A. Gilbert and R. Lydic, Parabrachial neuron discharge in the cat is altered during the carbachol-induced REM sleep-like state (DCarb), *Neurosci. Lett.* 120:241(1990).
15. W.C. Bond, C. Bohs, J. Ebey, and S. Wolf, Rhythmic heart rate variability (sinus arrythmia) related to stages of sleep, *Cond. Reflex* 8:98(1973).
16. F.R. Morales, J.K. Engelhardt, P.J. Soja, A.E. Pereda, and M.H. Chase, Motoneuron properties during motor inhibition produced by microinjection of carbachol into the pontine reticular formation of the decerebrate cat, *J. Neurophysiol.* 57:1118(1987).
17. K. Henley and A. Morrison, Release of organized behavior during desynchronized sleep in cats with pontine lesions, *Psychophysiol.* 6:245(1969).
18. M. Matsuzaki, Y. Okada, and S. Shuto, Cholinergic agents related to para-sleep state in acute brain stem preparations, *Brain Res.* 9:253(1968).
19. J.L. Feldman, J.C. Smith, D.R. MaCrimmon, H.H. Ellenberger, and D.F. Speck, Generation of respiratory patterns in mammals, in: *Neural Control of Rhythmic Movements in vertebrates*, A.H. Choen et al. eds., Jhon Wiley & Sons, New York(1988).

THE ROLE OF VENTRAL GROUP NEURONS IN THE RESPIRATORY DEPRESSION DURING CARBACHOL-INDUCED REM SLEEP ATONIA

H. Kimura,[1] L. Kubin,[1] H. Tojima,[1] A.I. Pack, and R.O. Davies[1]

[1]School of Veterinary Medicine
Center for Sleep and Respiratory Neurobiology
University of Pennsylvania
Philadelphia, PA 19104, U.S.A.

INTRODUCTION

Sleep-related decreases in the tone of upper airway and respiratory pump muscles occur most frequently in association with the rapid eye movement (REM) phase of sleep and its characteristic postural muscle atonia[12]. These decreases in muscle tone may lead to upper airway obstruction and/or reduced tidal volume in patients with sleep apnea syndrome.

To study central neural mechanisms related to REM sleep per se, investigators have often used microinjections of cholinergic drugs into the dorsal pontine tegmentum of chronically instrumented cats to induce a state having many of the characteristics of natural REM sleep (see refs. 5,14 for reviews). Although the exact mechanisms whereby cholinergic stimulation within the pons leads to the REM sleep-like state are not completely understood, these studies, as well as studies using pontine lesions[4], indicate that the pons is critical for the generation of various signs of REM sleep and, in particular, the postural muscle atonia characteristic of this state. Recently, Morales et al.[10] demonstrated that microinjections of carbachol into the pons of acutely decerebrate cats produce a postural muscle atonia and inhibition of lumbar motoneurons that has the same characteristics as those observed during natural REM sleep, including state-specific inhibitory postsynaptic potentials. This finding in an acute preparation allows one to use invasive neurophysiological techniques not applicable in chronic animals, to rigidly control the experimental conditions, and to produce the postural atonia at controlled times in order to study some of the central mechanisms activated during REM sleep.

Using this acute preparation, we found that pontine microinjections of carbachol caused a powerful depression of respiratory motor output and a slowing of respiratory rate[6] (see Fig. 1A). This depression of respiratory motor output developed in parallel with the postural muscle atonia and, as in natural REM sleep, the phrenic activity was least affected, inspiratory intercostal activity intermediately depressed, and expiratory intercostal and hypoglossal activities were most depressed.

Control of Breathing and Its Modeling Perspective, Edited by
Y. Honda *et al.*, Plenum Press, New York, 1992

The mechanisms that lead to this carbachol-induced suppression of the activity of nerves innervating the respiratory pump muscles (diaphragm and intercostals) are unknown. The suppression may be due to state-specific postsynaptic inhibition at the motoneuronal level, as in lumbar motoneurons[10], or, alternatively, it may result from a disfacilitation; i.e., a state-related withdrawal of an excitatory input to respiratory motoneurons. In this study, we assessed the latter possibility. In particular, we studied the carbachol-induced changes in the activity of neurons of the ventral respiratory group (VRG) of the medulla and their primary targets, phrenic and intercostal motoneurons[3,8], in order to determine whether a suppression of the respiratory-modulated excitatory drive at this premotoneuronal level may underlie reductions in ventilation during the atonia of REM sleep.

METHODS

Cats (n=22) were preanesthetized with ketamine (80 mg, i.m.) and diazepam (2 mg, i.m.), anesthetized with halothane and then decerebrated at a pre-collicular level. Anesthesia was then discontinued. The surgical methods and techniques of recording from peripheral nerves and pontine microinjections were previously described[6]. In all experiments, phrenic activity and, to monitor postural muscle tone, a motor branch of the fourth cervical nerve (C_4) were recorded. In some experiments, the internal and external branches of intercostal nerves (T_{8-10}) were recorded. Medullary VRG cell activity was recorded extracellularly with tungsten microelectrodes. Penetrations were made from 1.0 mm rostral to 4.0 mm caudal to the obex, from 2.4-4.0 mm lateral to the midline, and at a depth of 2.0-5.0 mm from the dorsal surface of the medulla. Only cells showing a incrementing, ramp-like activity phase-locked to either inspiration or expiration and not antidromically activated by electrical stimulation of the vagus nerve were selected for study. VRG cell action potentials were amplified and processed using standard techniques, and counted with a digital moving average counter (window width of 200 ms). Peripheral nerve activities were amplified and fed to analog moving averagers having time constants of 200 ms.

The drugs were injected from glass pipettes (tip diameter 20-40 μm) using pressure pulses. The pipettes were placed in the dorsal pontine tegmentum, 2-4 mm posterior to stereotaxic zero, about 1.5-3.0 mm lateral to the midline and at a depth of -3 to -5 according to Berman's atlas[1]. The volume injected was determined by measuring the movement of the meniscus in the pipette. First, carbachol was injected (mean: 230 nl) and the steady-state response determined. The effects of this injection were subsequently reversed by an atropine injection (mean: 292 nl) into the same pontine site. Then, a second carbachol injection was made in the equivalent site in the contralateral pons; its effects were then reversed by a second atropine injection. Each microinjection was made while recording the activity of a single VRG cell; thus, four neurons could be studied in each experiment.

Carbachol produced a parallel suppression of the moving averages of both C_4 and respiratory activities and this was reversed by atropine. The accompanying changes in the activity of each VRG cell studied were characterized by the change in the peak firing rate that occurred on the transition from the preinjection state (before carbachol or atropine) to the new steady conditions following the microinjection. The changes in peak cell firing were related to the changes in the peaks of the moving averages of the phrenic and intercostal activities. The changes in the activities of the nerves and VRG neurons were expressed relative to either the precarbachol or post-atropine states, referred to as control conditions (cf. ref. 6). A two-tailed Student's t-test was used for statistical comparisons; differences were considered significant when $p<0.05$.

RESULTS

Data are reported for 57 VRG neurons; 34 were inspiratory (I) and 23 expiratory (E). Since the changes in cell and nerve activities resulting from the carbachol and atropine microinjections were not different when referred to the control conditions (defined as either the initial precarbachol state, or the state following the reversal of the response by atropine), the data from the two procedures were pooled.

In 42 of the 57 neurons studied, the cell peak firing rate was reduced during the carbachol-induced postural and respiratory depression as compared to control conditions. As shown for an E neuron in Fig. 1A, this reduction displayed a similar time course to that of the depression of the peripheral nerve activities. The remaining 15 cells, however, showed either no change (5) or an increase (10) in firing in response to carbachol, while the simultaneously recorded phrenic and intercostal nerve activities never increased. An example of an increase in the peak firing of an I neuron following pontine carbachol is shown in Fig. 1B. The peak firing of the cell during the inspiratory burst increased to 117% of control, whereas peak phrenic activity was reduced to 87% and peak inspiratory intercostal to 60%.

For the total population of cells studied, both I and E neurons, carbachol induced a small depression in peak firing rate that was quantitatively similar to the average depression of the phrenic nerve activity, but considerably smaller than the reduction in either inspiratory or expiratory intercostal nerve activity. This was particularly apparent for E neurons, since expiratory intercostal activity was depressed drastically (often abolished), whereas the neuron always maintained a strong rhythmic discharge, albeit at a reduced peak frequency (cf. Fig. 1A). For all 57 neurons, I and E, the peak firing following carbachol was reduced to 88.5% ± 16.3 (SD) of control (range: 48-127%) and the simultaneously recorded peak phrenic activity was reduced to 77.9% ± 11.5 (SD) (range: 50-98%) (cf. Fig. 3).

For I neurons, cell activity was reduced by 7% ± 14 (SD) and for E neurons it was reduced by 17.9% ± 17.7 (SD) ($p<0.02$, t-test). The corresponding reductions in phrenic nerve activity were 20% ± 10.4 (SD) for I cell experiments and 25.9% ± 13.5 (SD) for E cell experiments. This difference in phrenic response may have been due to the more extensive dissections of the intercostal nerves during E cell experiments. Consequently, at least a part of the difference between I and E cell depressions could be related to different magnitudes of general depression, as characterized by the phrenic nerve depression in those respective experiments. These data for I and E cells, and the corresponding phrenic nerve depressions, are shown in Fig. 2A.

A proportional relationship between the phrenic nerve and VRG cell depressions, regardless of cell type, is also suggested by the scatter plot shown in Fig. 2B. In this plot, each data point shows the change in peak firing rate of individual I (●) and E (○) neurons with respect to the corresponding change in peak phrenic nerve activity. We found a highly significant ($r=0.492$, $p<0.001$) linear correlation for these data. This and the intermingling of data points for I and E cells suggests that a common and quantitatively similar mechanism affects both I and E cell activity during the carbachol-induced atonia. Moreover, the distribution of the majority of points below the identity line in Fig. 2B suggests that, in addition to disfacilitation from VRG cells, there is another mechanism that also contributes to the depression of phrenic nerve activity during the carbachol-induced atonia.

Figure 3 shows the average effects of carbachol on inspiratory and expiratory intercostal nerve activity in relation to the average changes seen in VRG cell and phrenic nerve activities. While cell activity was reduced to 88.5% ± 16.3 (SD) of control and phrenic activity to 77.9% ± 11.5(SD), the reduction in inspiratory intercostal activity was to 63.4% ± 21.6(SD) and in expiratory intercostal activity to

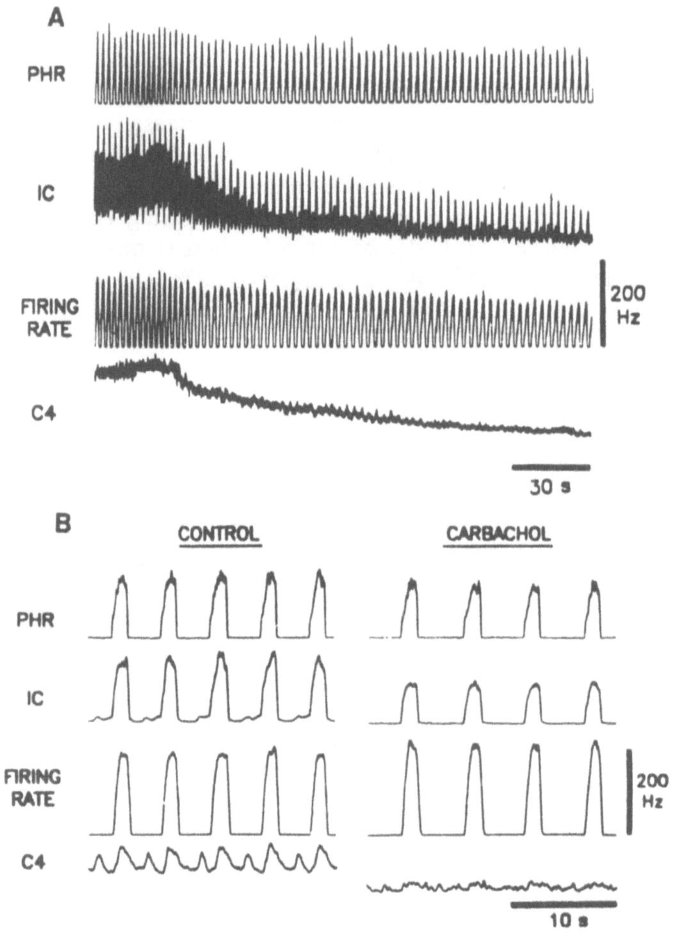

Figure 1. A. Polygraph record showing the transition from control state to carbachol-induced atonia and the effect of carbachol on the firing of an E neuron. Carbachol (1.0 μg) was injected 3.5 min before the beginning of the record. Traces show, from top to bottom, the moving averages of the neural activities of the phrenic (PHR) and intercostal (IC, both inspiratory and expiratory activities present) nerves, the cell firing rate and the moving average of the neural activity of a cervical motor nerve branch (C4). Note the similar time courses of changes in all the activities and the maintenance of E neuron firing in the presence of almost total suppression of expiratory activity (lower peaks) in the IC signal. **B.** Effect of carbachol on the firing of an I neuron. Traces labeled as in A. Note the increase in peak firing of the cell while PHR, IC and C4 were depressed.

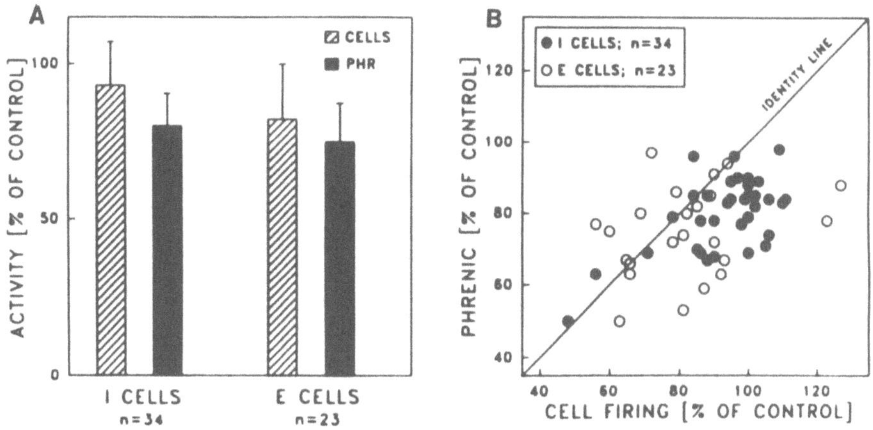

Figure 2. A. Mean I and E neuron peak firing rates and mean activity of the simultaneously recorded phrenic nerve during carbachol-induced atonia. All activity levels are normalized to the control, precarbachol, level (100%). Hatched bars, peak cell firing rates: filled bars, amplitude of the moving average of the phrenic activity. Data are means ± SD. **B.** Relationships of changes in peak cell firing rates to changes in the simultaneously recorded magnitudes of the moving averages of phrenic nerve activity during carbachol-induced atonia. Responses of the phrenic nerve are plotted against the simultaneously recorded individual VRG cell activity changes for both inspiratory (I) and expiratory (E) cells. Note the roughly proportional relationship between the cell and phrenic activity changes regardless of the cell type and that most of the responses fall below the line of identity.

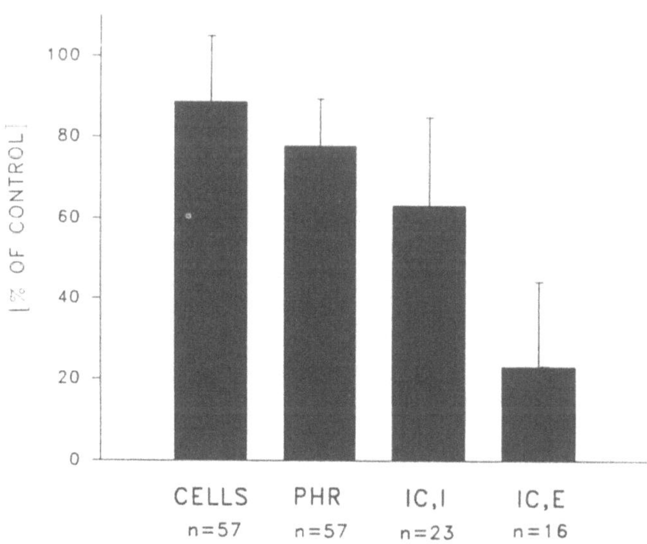

Figure 3. Relationship of the mean change in peak cell firing rate (I and E cells combined) to the mean changes in the moving averages of the phrenic (PHR), inspiratory intercostal (IC, I) and expiratory intercostal (IC, E) activities during carbachol-induced atonia. All activity levels are normalized to the control, precarbachol, level (100%). Data are means + SD.

23.2% ± 21.2(SD). The difference between the two intercostal nerve activities was significant ($p<0.05$). In addition, inspiratory intercostal nerve activity was significantly ($p<0.001$) more depressed than phrenic nerve activity. The differences between the depression of I cell and inspiratory intercostal nerve activity, and between the depression of E cell and expiratory intercostal nerve activity, were both statistically significant ($p<0.005$). As reported previously[6], in individual responses to carbachol, phrenic and intercostal activities consistently showed the same relative pattern of depression as that seen in the average data presented in Fig. 3. Thus, there seems to be a mismatch between the relatively weak reduction in VRG cell peak firing and the depression of intercostal nerve activities, especially expiratory.

DISCUSSION

We have previously shown that microinjections of carbachol into the dorsal pontine tegmentum of decerebrate cats suppress the activity of motoneurons to various respiratory muscles, both pump and upper airway[6]. This depression develops in parallel to a postural muscle atonia and has a pattern that is similar to that observed during natural REM sleep in cats. The magnitude of this depression is larger than that seen during carbochol-induced REM sleep in chronically instrumented cats[9] and highly exaggerated when compared to natural REM sleep. Both the postural muscle atonia and suppression of respiratory motor output are evoked in decerebrate cats by cholinergic stimulation of the same small region of the pontine tegmentum (Kim, Kubin, Kimura and Davies, unpublished observations) as that producing REM sleep in chronic cats[15]. In addition, in this same region, the concentration of acetylcholine rises during natural REM sleep in cats[7]. Thus, local pontine microinjections of carbachol are likely to activate some of the cholinergic mechanisms that also are activated during REM sleep.

Our study shows that cholinergic stimulation of this pontine site leads, on average, to a depression of the activity of VRG premotor respiratory neurons, together with the depression of their target motoneurons. However, the depression of VRG cells was relatively weak compared to the depression of respiratory motor output. In addition, despite the fact that the dominant effect at the motoneuronal level was depression, an excitatory input, presumably also activated by pontine cholinergic stimulation, was evident in some VRG cells during the atonia.

Although the bulbospinal VRG neurons have been shown to provide direct mono- and polysynaptic drive to both phrenic and intercostal motoneurons[3,8], the input-output characteristics of the synaptic connections between VRG cells and these respiratory motoneurons have not been determined. Thus, it is not clear whether a relatively small average depression of VRG cell activity can significantly contribute to the suppression seen at the motoneuronal level by the withdrawal of a portion of the excitatory input (disfacilitation). Nevertheless, as explained below, our data lead us to doubt that disfacilitation from VRG cells provides a sufficient explanation for the depression of either phrenic or intercostal nerve activity during the carbachol induced atonia.

As for the phrenic nerve activity, we found that, on average, I cell activity was reduced by 7% whereas phrenic activity was reduced by 20%. This difference could reflect the input-output characteristics of the synaptic connectivity between VRG cell population and phrenic motoneurons. However, the fact that the majority of data points in Fig. 2B are below the identity line indicates that phrenic activity will be depressed even when the average VRG activity remains unchanged following pontine carbachol. (We estimate that, when VRG firing is unchanged, phrenic activity will be depressed by 6-18%.) The source of this additional phrenic depression (either inhibition or disfacilitation) is not known.

Some I and E cells increased their activity at the same time that phrenic activity was substantially depressed. This suggests that the response of an individual VRG neuron depends on the relative balance of excitatory and inhibitory inputs that may be simultaneously activated by carbachol (or activated during REM sleep). Important in this regard is the fact that in our experiments the rise in CO_2 in response to a decreased ventilation (decreased phrenic activity) was eliminated by artificial ventilation. If chemical feedbacks were left intact, VRG cells would probably show less depression than that observed in this study, with a net facilitation possibly being the dominant response pattern.

The carbachol-induced depression of intercostal activity was always greater than that of phrenic (cf. Figs. 1 and 3). Again, this difference may be due, in part, to differences in the input-output characteristics of the synaptic connections between VRG cells and the different motoneuronal pools. However, based on the data for VRG cells and phrenic nerve shown in Fig. 2B, it is clear that a depression of intercostal nerve activity, even more prominent than that for phrenic, may occur even in the absence of carbachol-induced changes in VRG cell activity. Thus, we again conclude that additional inhibitory and/or disfacilitatory inputs must play a role in the carbachol-induced depression of the activity of the intercostal motoneuronal pool (cf. refs. 2,12,13).

Our findings are consistent with those of Orem[11,12] who concluded, based on studies of VRG neurons during REM sleep in cats, that both excitatory and inhibitory inputs impinge on VRG neurons during REM sleep, with the inhibitory effects being dominant during "tonic" REM sleep (no pontogeniculooccipital waves). He also postulated that REM sleep-dependent effects on phrenic and intercostal motoneuronal activity are mediated by both the medullary respiratory premotor neurons and non-respiratory, REM sleep-related neurons of the brain. Thus, our data (from carbachol-induced atonia in decerebrate cats) and Orem's (from natural REM sleep in behaving cats) lead to similar qualitative predictions as to the character of state-specific inputs to VRG cells and motoneurons during REM sleep.

Our finding that cholinergic stimulation within the pontine region that has been implicated in the generation of various signs of REM sleep[14] gives origin to a depression of both VRG premotor cell and peripheral motoneuronal activities suggests that pathways and mechanisms similar to those activated in our experimental conditions may underlie the pathological depression of breathing seen in central sleep apnea syndrome.

ACKNOWLEDGMENTS

This study was supported by a National Heart, Lung and Blood Institute Specialized Center for Research Grant, HL-42236. H. Kimura was on leave from the Department of Chest Medicine, Chiba University, Japan.

REFERENCES

1. A. L. Berman. "The Brainstem of the Cat: A Cytoarchitectonic Atlas with Stereotaxic Coordinates," University of Wisconsin Press, Madison (1968).
2. B. Duron and D. Marlot. Intercostal and diaphragmatic electrical activity during wakefulness and sleep in normal unrestrained adult cats. *Sleep*, 3:269-280 (1980).
3. K. Ezure. Synaptic connections between medullary respiratory neurons and considerations on the genesis of respiratory rhythm. *Progr. Neurobiol.*, 35:429-450 (1990).

4. K. Henley and A.R. Morrison. A re-evaluation of the effects of lesions of the pontine tegmentum and locus coeruleus on phenomena of paradoxical sleep in the cat. *Acta Neurobiol. Exp.*, 34:215-232 (1974).
5. J.A. Hobson, R. Lydic, and H.A. Baghdoyan. Evolving concepts of sleep cycle generation: from brain centers to neuronal populations. *Behav. Brain Sci.*, 9:371-448 (1986).
6. H. Kimura, L. Kubin, R.O. Davies, and A.I. Pack. Cholinergic stimulation of the pons depresses respiration in decerebrate cats. *J. Appl. Physiol.*, 69:2280-2289 (1990).
7. T. Kodama, Y. Takahashi, and Y. Honda. Enhancement of acetylcholine release during paradoxical sleep in the dorsal tegmental field of the cat brainstem, *Neurosci. Lett.*, 114:277-282 (1990).
8. S. Long and J. Duffin. The neuronal determinants of respiratory rhythm. *Prog. Neurobiol.*, 27:101-182 (1986).
9. R. Lydic and H.A. Baghdoyan. Cholinoceptive pontine reticular mechanisms cause state-dependent respiratory changes in the cat. *Neurosci. Lett.*, 102:211-216 (1989).
10. F.R. Morales, J.K. Engelhardt, P.J. Soja, A.E. Pereda, and M.H. Chase. Motoneuron properties during motor inhibition produced by microinjection of carbachol into the pontine reticular formation of the decerebrate cat. *J. Neurophysiol.*, 57:1118-1129 (1987).
11. J. Orem. Medullary respiratory neuron activity: relationship to tonic and phasic REM sleep, *J. Appl. Physiol.*, 48:54-65 (1980).
12. J. Orem. Neuronal mechanisms of respiration in REM sleep. *Sleep*, 3:251-267 (1980).
13. J.E. Remmers. Control of breathing during sleep, *in*. "Regulation of Breathing," T.F. Hornbein, ed., Dekker, New York (1981).
14. J.M. Siegel. Brainstem mechanisms generating REM sleep, *in*. "Principles and Practice of Sleep Medicine," M.K. Kryger, T. Roth and W.C., Dement, eds., Saunders, New York (1989).
15. K. Yamamoto, A.N. Mamelak, J.J. Quattrochi, and J.A. Hobson. A cholinoceptive desynchronized sleep induction zone in the anterodorsal pontine tegmentum: locus of the sensitive region. *Neurosci.*, 39:279-293 (1990).

CONTROL OF BREATHING IN COPD-PATIENTS WITH SLEEP HYPOPNEA; EFFECTS OF ACETAZOLAMIDE AND CHLORMADINONE ACETATE

H. Folgering, P. Vos, C. van Herwaarden

Dept. Pulmonology Dekkerswald, Univ. of Nijmegen, The Netherlands

INTRODUCTION

Hypoxemic patients with chronic obstructive pulmonary disease (COPD) may develop more profound oxygen desaturations during sleep due to alveolar hypoventilation(1).

Drugs that stimulate ventilation may ameliorate hypoxemia and carbon dioxide retention in certain patients with COPD and may thereby provide a possible alternative for treatment with oxygen suppletion.

Chlormadinone acetate (CMA) raises arterial oxygen tension and reduces carbon dioxide tension during wakefullnes and sleep (2).

Another respiratory stimulant is acetazolamide (ACET). It is a strong carbonic anhydrase inhibitor which increases ventilation by inducing metabolic acidosis, possibly acting at both the peripheral and the central chemoreceptors (3).

The aim of the current study was to determine if CMA and ACET are clinically relevant in the treatment of hypoxia and hypercapnia in addition to, or instead of oxygen therapy, in COPD-patients.

METHODS

Patients

Fifty-three out-patients with chronic obstructive pulmonary disease, as defined by the American Thoracic Society, participated in this study. Admission criteria were: FEV1 value less than 65% of the reference value, and daytime PaO_2 values of 8.5 kPa or less. At the time of the study all patients were in a stable clinical condition for at least four weeks and received optimal bronchodilatory treatment. Exclusion criteria were: thrombo-embolic events in the past, and indices for possible obstructive sleep apnea. The characteristics of the patients are shown in table I.

Control of Breathing and Its Modeling Perspective, Edited by
Y. Honda *et al.*, Plenum Press, New York, 1992

Protocol

All patients were studied during three nights; two consecutive nights during which they received either room air or oxygen 1 l/min via a nasal cannula in a random order, and single blind. The second part of the study was designed as a randomised, double blind and placebo controlled trial with pharmacological respiratory stimulants. After the two first nights the patients took during one week twice a day orally, either CMA (25 mg), or ACET (250 mg) or placebo, all in identical capsules. The patients were randomised for FEV1 value (above or below 35% ref. value) and daytime PaO_2 (above or below 7.5 kPa). During the third night at the end of the one week treatment, room air was administered via a nasal cannula in all patients, in a single blind way.
After each night the patients answered the St. Mary's Hospital sleep questionnaire concerning subjective sleep quality (4). After the first night and after one week "drug treatment", the ventilatory responses to hypercapnia and hypoxia were measured.

Table 1

PATIENTS

	CMA	ACET	Plac
sex	14M 4F	11M 6F	14M 4F
age	64 (9)	64 (6)	65 (9)
FEV_1 %pred	31 (11)	31 (9)	29 (9)
IVC %pred	78 (14)	75 (14)	70 (10)
TLC %pred	99 (26)	100 (18)	103(21)
BMI W/L^2	23 (4)	23 (4)	25 (5)

Techniques

The nocturnal oxygen saturation was measured by a pulse-oximeter (Oxyshuttle, SensorMedics). Desaturation was defined as a decrease in oxygen saturation from the asleep baseline SaO_2 of more than 4%. The total desaturation time was calculated.

The chest wall movement was assessed by respiratory inductance plethysmography (Vitalog or Respitrace).

The $P_{ET}CO_2$ was determined by introducing a cannula through the nose in the oral pharyngeal cavity. An air sample was withdrawn (150 ml/min), via a perma-pure drying tube to a sampling capnograph (Mijnhart capnolyser).

An electrooculogram was performed to distuinguish sleep in the REM and the NONREM sleep stage.

Statistical analysis

Statistical analysis for paired comparisons was performed using the signed rank test. The Kruskal-Wallis test was used to determine statistical significant differences between the three groups. Correlations were established with the Spearman rank analysis. p-values of 0.05 or less were considered to be significant.

RESULTS

Daytime laboratory measurements

The effects of CMA, ACET, placebo and oxygen therapy on daytime bloodgas values are shown in table 2. Arterial PCO_2 was significantly lowered during CMA and ACET therapy. During supplemental oxygen therapy, $PaCO_2$ was elevated by 0.6 kPa. Daytime hypoxemia improved during all treatments ($p < 0.05$). During ACET and oxygen treatment, the improvement was significantly higher than during placebo and CMA.

ACET therapy induced a metabolic acidosis as indicated by a mean decrease in base excess of 6.7 ± 3.3 mmol/l ($p < 0.05$).

Ventilation

The hypercapnic ventilatory response and minute ventilation were significantly increased during CMA therapy. After ACET administration both the hypercapnic and the hypoxic ventilatory responses were augmented.

Subjective parameters

In the CMA group 7/18 patients (39%) said to benefit from this therapy. Seven of the 16 patients (41%) who had received ACET reported a favourable reaction. In the placebogroup, 8/18 patients (39%) claimed to feel better. In all three treatment regimens, the same amount of side effects were presented.

Table 2

	ACET before	ACET during	CMA before	CMA during	Plac before	Plac during
DAYTIME						
PaO2 kPa	7.5	9.4 °'	7.2	7.8 °	7.6	8.0 °
PaCO2 kPa	6.4	5.9 °	6.5	5.8 °	6.3	6.6 °
pHa	7.42	7.35°'	7.42	7.43	7.41	7.40
BE	+6.6	-0.1°'	+6.2	+4.2	+4.8	+5.1
(A-a)DO2	5.4	4.1 °'	5.6	5.9	5.5	4.6 °
HCVR	2.7	3.7 °	2.3	3.3 °	2.2	2.0
HVR	.22	.34 °	.20	.20	.19	.24 °
Ve l/min	10.5	10.5	9.6	11.8°	9.4	9.5
NOCTURNAL						
SaO2 %	85	89°'	82	84°	86	87
PetCO2 kPa	5.3	5.0°	5.2	4.7°	5.7	5.5
TST hrs	4.6	6.7	5.4	4.7	5.3	4.4

°: $p<0.05$ Wilcoxon; 'during' vs 'before'. ': $p<0.05$ ACET vs CMA

Subjective sleep quality

The results of the St. Mary's Hospital sleep questionnaire showed that CMA had no effect, whereas ACET therapy significantly improved the subjective quality of sleep.

Nocturnal parameters

Mean nocturnal oxygen saturation improved significantly during CMA, ACET and oxygen therapy, as compared to room air and to placebo (table 2). The fall in SaO_2 during the course of the night was not significantly changed by CMA and ACET, but O2 suppletion significantly reduced the fall in SaO_2. Both CMA and ACET lowered the baseline $P_{ET}CO_2$ values during sleep, whereas oxygen increased the $P_{ET}CO_2$.

The mean desaturation time in the whole group of patients in the control situation was 47 minutes. Oxygen suppletion decreased the time which was spent desaturated to 10 minutes on average.

The results of the total sleep time showed an increase of two hours total sleep time in the ACET treated patients (table 2).

DISCUSSION

The results of this study suggest that progesterone stimulates ventilation via central chemoreceptors, and that ACET apparently had its effect on both the peripheral and on the central receptors. The alveolar-arterial oxygen difference in the ACET group was substantially reduced. Therefore the correction of bloodgas values during ACET administration could also partly be explained by improved ventilation-perfusion ratio.

The patients who received ACET reported a significantly improved subjective sleep quality that was comparable with their oxygen night. Both oxygen and ACET relieved nocturnal desaturations and less 4% dips occurred as compared with the room air night. Less 4% dips might cause less arousals and therefore better objective sleep quality resulting in improved subjective sleep quality. This confirms the results of Kearly et al who assessed an increased sleep time in COPD during oxygen suppletion (5). Moreover, Fleetham and colleagues have shown that 40% of the arousals were associated with oxygen desaturations (6). Nevertheless, they did not find a decrease in arousal frequency during O2 therapy.

Skatrud et al, studied the effects of medroxyprogesterone and ACET. They divided their patients in correctors and non correctors by using the criterion of a decrease of more or less than 0.67 kPa in daytime $PaCO_2$ during treatment (7). They suggested that patients who increased tidal volume rather than respiratory rate would react more favourably on the respiratory stimulants. Neither in our CMA group nor in the ACET group could a correlation be established between the reponse of $PaCO_2$ and the severity of the obstruction or the hyperinflation. Apparently the response in our patients to the used respiratory stimulating drugs was not dependent on mechanical limitations. Indeed, all the CMA patients and 89% of ACET patients were able to decrease their $PaCO_2$ by at least 0.7 kPa with voluntary ventilation. Furthermore, no significant correlation was found between the initial PaO_2, $PaCO_2$ or any of the other patient characteristics and the response of the $PaCO_2$ to either drug.

It is concluded that CMA as well as ACET decrease awake and asleep $PaCO_2$ whereas additional inspiratory O_2 increases the $PaCO_2$. The decrease in $PaCO_2$ induced by CMA and ACET has been shown to be of the same magnitude as the increase when breathing oxygen (8). Therefore, CMA and ACET may be added to oxygen therapy in order to correct the CO_2 retention. ACET improved also day and nighttime oxygenation and total sleep time. Hence, ACET may have beneficial value in COPD patients with mild hypoxemia who do not fulfill -yet- the criteria for home oxygen therapy (9).

References

1. Douglas NJ, Flenley DC. Breathing during sleep in patients with chronic obstructive lung disease. State of the art. Am Rev Respir Dis 1990; 141:1055-1070
2. Tatsumi K, Kimura H, Kunitomo F, Kuriyama T, Watanabe S, Honda Y. Effect of chlormadinone acetate on sleep arterial oxygen desaturation in patients with chronic obstructive pulmonary disease. Chest 1987; 91:688-692
3. Sutton JR,Houston CS, Marsell AL et al. Effect of acetazolamide on hypoxemia during sleep at high altitude. N Engl J Med 1979; 301:1329-1331
4. Ellis BW, Murray WJ, Lancaster R, Raptopoulos P, Angelopoulos N, Priest RG. The St. Mary's Hospital sleep questionnaire: a study of reliability. Sleep 1981; 4(1):93-97
5. Kearly R, Wynne JW, Block AJ, Boysen PG, Linsey S, Martin C. Effects of low flow oxygen on sleep disordered breathing and oxygen desaturation. Chest 1980; 78:682-685.
6. Fleetham J, West P, Mezon B, Conway W, Roth T, Kryger M. Sleep, arousals, and oxygen desaturation in chronic obstructive pulmonary disease. The effect of oxygen therapy. Am Rev Respir Dis 1982; 126:429-433.
7. Skatrud JB, Dempsey JA, Bhansali P, Irvin C. Determinants of chronic carbon dioxide retention and its correction in humans. J Clin Invest 1980; 65:813-821
8. Medical Research Council Working Party Report. Long term oxygen therapy in chronic hypoxic cor pulmonale complicating chronic bronchitis and emphysema. Lancet 1981; 1:681-686
9. Stradling JR, Controversies in sleep related breathing disorders. Lung 1986; 164:17-31

SERIAL SUBMENTAL STIMULATION DURING SLEEP IN PATIENTS WITH OBSTRUCTIVE SLEEP APNEA

Wataru Hida, Shinichi Okabe, Satoru Ebihara, Tatsuya Chonan, Yoshihiro Kikuchi, and Tamotsu Takishima

First Department of Internal Medicine
Tohoku University School of Medicine
Sendai, 980, Japan

INTRODUCTION

We previously reported that submental electrical stimulation may be an effective treatment for obstructive sleep apnea (Miki et al., 1989a, Hida et al., 1991). The mechanism of relief of apneic episodes during submental stimulation may be the contraction of genioglossus and/or geniohyoid muscles, which would open the upper airway by pulling the tongue forward (Miki et al., 1989b, Gottfried et al., 1983). However, the effects of submental stimulation during consecutive nights on apneic episodes in patients with obstructive sleep apnea (OSA) have not been studied. In the present study in order to assess whether submental stimulation is necessary for treatment every night, we examined the effects of submental stimulation for a serial four or five sleep nights on apneic episodes and changes in apneic episodes on stimulation-off nights after the stimulation nights using a portable submental stimulator which we have developed.

METHODS

Eight male patients with OSA previously diagnosed by polysomnography according to the definitions proposed by Guilleminault ct.al., (1980) were studied. Average age, height and body weight were 51.4 ± 3.2 yr, 165.1 ± 1.2 cm and 76.8 ± 4.1 kg (means ± SE), respectively. They were all snorers and had mild excessive daytime sleepiness and slight morning headache. All were in a clinically stable condition and free of such complication as lung disease or heart failure. Each patient gave informed consent to the protocol, which had received prior approval from the Human Research Committee at our institution.

Overnight sleep studies were performed in a quiet darkened room using standard polysomnographic techniques, including an electroencephalogram, electrooculograms, and submental electromyogram with surface electrodes to determine the sleep stages. Respiratory movements of the rib cage and abdomen with tidal volume summation were measured using induced plethysmography (Respitrace; Ambulatory Monitoring, Ardsley, NY) placed at the level of the midchest and umbilicus. Arterial oxygen saturation was measured continuously using a pulse oximeter (Biox 3700; Ohmeda, Boulder, CO). Airflow at the nose and mouth was recorded with specially designed thermisters. All variables were

Control of Breathing and Its Modeling Perspective, Edited by
Y. Honda *et al.*, Plenum Press, New York, 1992

recorded continuously on a multi-channel thermal polygraph (Model 360; NEC San-ei, Tokyo, Japan) and data recorder (A-109; SONY, Tokyo, Japan). Apnea defined as the cessation of airflow at the nose and mouth lasting longer than 10 s was classified as obstructive, central, or mixed type using the Respitrace and airflow signals in the manner previously described (Miki et al., 1989a, Guilleminault et al., 1980).

The airflow demand-type portable submental stimulator which we have developed consists of two parts : apnea detection and a stimulating system. This apparatus weighs approximately 300 g, and is operated for two night by a rechargeable 9 DC volt battery. Airflow signals were recorded by thermisters placed near the nose and mouth. The presence of apnea was defined as any point where the thermisters fell to less than 10% of their respective intensities during tidal breathing. Two silver chloride ECG electrodes (10 mm in diameter) were used as stimulation electrodes and were attached 10 mm apart on the skin in the submental region. Stimulation was performed with 0.2 msec pulses, repeated at 100 Hz, and at an intensity of approximately 5-30 volts. Electrical stimulation was started when apnea lasted for five seconds, and was stopped immediately after breathing was detected by oronasal airflow or after twenty seconds at the longest. The preset stimulation intensity was determined for each patient while awake, by choosing an intensity just below the threshold of mild pain. If the oronasal airflow did not appear within 5 sec of stimulation, the intensity automatically increased stepwise; at a 50% increase from initial intensity for the following 5 sec, and a 75% increase from the initial intensity for the next following 10 sec. The stepwise increased stimulation intensity always returned to the initial level.

Patients with OSA underwent a polysomnographic study in an all-night session on two nights without stimulation, a serial stimulation-on four or five nights and a following stimulation-off three nights. "Control night (C)" was defined as the second night without submental stimulation, "Stimulation night (S)" was defined as a night studied using the portable stimulator and "Stimulation-off night (S-off)" was defined as the following night without submental stimulation. In five patients, upper airway resistance (Ruaw) in the supine position and during wakefulness was obtained before and after the serial four or five stimulation-on nights. Inspiratory and expiratory Ruaw was determined by simultaneously recording mouth flow and pressure difference between mouth and lower pharynx. Pharyngeal pressure was measured at the level of the epiglottis using a low bias flow catheter (1.4 mm ID) and a Validyne differential pressure transducer (MP-45, ±50 cmH_2O). Mouth flow was obtained with a pneumotachometer and a pressure transducer (MP-45, ±2 cmH_2O).

RESULTS

Table 1 shows mean values of the apnea index expressed as the frequency of apneic episodes expressed per hour (AI), desaturation index expressed as the apneic episodes in times per hour that SaO_2 dropped below 85% (SaO_2<85%/hr) and the apnea time expressed as percent of sleep time (AT) on C, the first S, the second S, the last S, the first S-off, the second S-off and the third S-off. The values of these three parameters on stimulation nights decreased significantly compared with the values of C. Furthermore, the values of these three parameters still decreased significantly on the first S-off, and SaO_2<85%/hr and the apnea time decreased significantly on the second S-off. On the third S-off, the values of these three parameters returned to the level of C. The quality of the sleep on S and S-off improved in comparison with that of C. That is, stage 1 and 2 decreased significantly, whereas stage 3 and 4 increased significantly on S and the following S-off. Inspiratory and expiratory Ruaw before submental stimulation night were 0.75 ± 0.12 and 0.60 ± 0.11 cmH_2O/L/sec, respectively. Inspiratory and expiratory Ruaw after serial submental stimulation nights were 0.71 ± 0.12 and 0.62 ±0.10 cmH_2O/L/sec, respectively. Both inspiratory and expiratory Ruaw did not differ significantly before and after serial submental stimulation nights.

Table 1, Apnea index, SaO2<85%/hr and apnea time on the control night, the stimulation nights and the following stimulation-off nights.

	C	S (1)	S (2)	S (Last)	S-off (1)	S-off (2)	S-off (3)
AI	53.8±7.0	31.3±5.5**	33.2±7.0*	27.3±5.7**	36.8±7.5*	38.8±7.5	47.0±10.2
SaO2<85%/hr	32.5±7.0	18.6±6.1**	19.0±6.8*	11.3±3.3**	17.1±6.4*	17.4±6.1*	26.1±8.5
AT (%)	43.2±7.1	25.2±5.5**	24.1±6.3**	23.7±5.6**	25.4±6.3**	26.2±7.7*	35.2±10.5

Definition of abbreviations: S(1), S(2) and S(Last), the first, second and last stimulation night, respectively; S-off (1), S-off (2) and S-off (3), the first, second and third stimulation-off night, respectively. AI, apnea index; SaO2<85%/hr, desaturation index expressed as the number of times per hour that oxygen saturation dropped below 85%; AT(%), apnea time expressed as percent of sleep time. Significantly differnet from the control night: ** P<0.01; * P<0.05.

DISCUSSION

We studied the effects of serial submental stimulation on apneic episodes during and following stimulation nights in patients with OSA using a portable submental stimulator which we have developed. We found an improvement of apneic episodes during submental stimulation nights with good reproducibility and on at least two stimulation-off nights after four or five submental stimulations. The effectiveness of submental stimulation on apneic episodes is determined by several factors. The first is the stimulation frequency of the pulse repetition rate. In the present study, we used 100 Hz frequency, because in our animal experiments, stimulation over 50 Hz decreased and plateaued upper airway resistance (Miki et al. 1989b). The second is the intensity of stimulation. Submental stimulation at this intensity did not open the upper airway during some apneic episodes. We first set the intensity of stimulation for each patient below the pain threshold while awake. If we used a high intensity to stimulate, this would induce arousal, and this intensity would not be adequate. The third is the sleep stages. In fact, the stimulation threshold used in this study was only effective the apneic episodes during non-REM sleep, and not on apneic episodes during REM sleep. The fourth is the apnea type. Submental stimulation is effective in the obstructive or mixed type of apnea but not on the central type (Miki et al. 1989a) The fifth is the skin thickness in the submental area. Submental skin thickness in obese patients tended to be thicker than in non-obese patients. The intensity of stimulation set while awake tended to be greater in the former than the latter. A final factor is adhesive force in the pharynx during upper airway obstruction. If the adhesive force is strong, it becomes hard to open the upper airway and a higher intensity of stimulation would be required to open the obstructed upper airway (Miki et al., 1992).

We found an improvement in apneic episodes during the stimulation-off nights. This improvement continued for at least two nights. Several mechanisms for this can be explained. The first is the possibility that the upper airway submucosal edema before treatment may disappear during stimulation-on and the following stimulation-off nights. We did not find any significant differences in the upper airway resistance before and after the serial submental stimulation nights. Therefore, it is unlikely that a local change, such as a decrease in subcutaneous edema of the upper airway wall during submental stimulation, explained the improvement of the apneic episodes during stimulation-off nights. The second is the possibility that improvement in sleep-related respiration, oxygenation, sleep fragmentation or alertness may produce an improvement of the apneic episodes on the following stimulation-off nights. As we did not find an improvement of the apneic episodes after the two serial stimulation-on nights in additional studies, this possibility is unlikely. An alternative explanation is that a conditioning effect formed in the brain during sleep on the serial stimulation nights may improve the apneic episodes of the following stimulation-off nights. The formation of new neural connections, that is, a form of associated plasticity may be induced during the submental stimulation nights. Several investigators have argued that the conditioning hypothesis is not only relevant to normal respiratory functions but also various respiratory dysfunction such as the anticipatory change in breathing pattern before muscular exercise (Tobin et al., 1986) asthmatic attack (Dekker et al., 1957) dyspnea (Dudley et al., 1980) and ventilatory response (Gallego et al., 1991). Therefore, we cannot discard the possibility of such conditioning.

REFERENCES

Dekker, E., Pelser, H.E., and Groen, J., 1957 Conditioning as a cause of asthmatic attack, J Psychosom Res. 2:97.

Dudley, D.L., and Pitts-Porch A.R., 1980, Psychophysiologic aspects of respiratory control, Clin Chest Med. 1:131.

Gallego, J., and Perruchet, P., 1991 Classical conditioning of ventilatory responses in humans, J Appl Physiol. 70:676.

Gottfried, S.B., Strohl, K.P., Van de Graaff, W., Fouke, J.M., DiMarco, A.F., 1983, Effects of phrenic stimulation on upper airway resistance in anesthetized dogs, J Appl Physiol. 55:419.

Guilleminault, C., Cummiskey, J., and Dement, W.C., 1980, Sleep apnea syndrome; recent advances, in "Advances in Internal Medicine, Vol. 26," Stollerman, G.H., ed., Year Book Medical Publishers, Chicago.

Hida, H., Miki, H., Kikuchi, Y., Chonan, T., and Takishima, T., 1991, Treatment of obstructive sleep apnea with airflow demand-type submental stimulator, in "Control of Breathing and Dyspnea," T.Takishima and N.S. Cherniack, ed., Pergamon Press, Oxford.

Miki, H., Hida, W., Chonan, T., Kikuchi, Y., and Takishima, T.,1989a, Effects of submental electrical stimulation during sleep on upper airway patency in patients with obstructive sleep apnea, Am. Rev. Respir. Dis. 140:1285.

Miki, H., Hida, W., Kikuchi, Y., Chonan, T., Satoh, M., Iwase, N., and Takishima, T., 1992, Effects of pharyngeal lubrication on the opening of obstructive upper airway, J Appl Physiol. In press.

Miki, H., Hida, W., Shindoh, C., Kikuchi, Y., Chonan, T., Taguchi, O., Inoue, H., and Takishima, T., 1989b, Effects of electrical stimulation of the genioglossus on upper airway resistance in anesthetized dogs, Am. Rev. Respir. Dis. 140:1279.

Tobin, M.J., Perez, W., Guenther, S.M., D'Alonzo, G., and Dantzker, D.R., 1986 Breathing pattern and metabolic behavior during anticipation of exercise, J Appl Physiol. 60:1306.

EFFECTS OF VISCOELASTIC PROPERTIES OF RESPIRATORY SYSTEM ON RESPIRATORY DYNAMICS

J. Milic-Emili and E. D'Angelo

Meakins-Christie Laboratories, McGill University
Montreal, Quebec, H2X 2P2, Canada
Istituto di Fisiologia Umana, Università di Milano, 20133 Milan, Italy

INTRODUCTION

In 1955 Mount assessed the dynamic work per breath on the lung (Wdyn,L), as given by dynamic volume-pressure loops, in mechanically ventilated open-chest rats. To explain the relatively high values of Wdyn,L found at low respiratory frequencies and the progressive decrease in dynamic pulmonary compliance with increasing respiratory frequency, he proposed a two-compartment viscoelastic model of the lung which "confers time-dependency of the elastic properties". In 1967, Sharp et al. proposed a similar viscoelastic model for both lung and chest wall. For a long time these models were ignored. Recently, however, it has been recognized that the viscoelastic properties of the respiratory system play an important role in respiratory dynamics. In this review we will discuss the implications of viscoelastic behaviour in terms of (a) time-dependence of pulmonary and chest wall elastance and resistance; and (b) work of breathing. Understanding respiratory mechanics is important for understanding the control of breathing.

VISCOELASTIC MODEL

The necessity to study respiratory mechanics in physiology and clinics has generated the need for relatively simple models that can mimic to some degree the mechanical behaviour of the respiratory system. Based on the pioneering studies by Mount (1955), Bates et al. (1989) have recently proposed an 8-parameter spring-and-dashpot model of the respiratory system which applies to experimental animals (Similowski et al., 1989; Shardonofsky et al., 1991). In humans, the model simplifies to seven parameters (Fig. 1) because the human chest wall does not exhibit any appreciable standard (newtonian) resistance (D'Angelo et al., 1991).

The model in Fig. 1 consists of two submodels, the lung and chest wall, which are arranged mechanically in parallel because the lung and chest wall undergo the same volume changes. The lung submodel consists of a dashpot representing airway resistance (Raw) arranged in parallel with a Kelvin body. The latter consists of a spring representing standard static elastance (Est,L) in parallel with a Maxwell body, i.e. a spring (E_2,L) and a dashpot (R_2,L) arranged serially. E_2,L and R_2,L, and the corresponding time constant (τ_2,L=R_2,L/E_2,L) account for the viscoelastic properties of the lung. In contrast, the chest wall submodel is represented solely by a Kelvin body consisting of standard static chest wall elastance (Est,w) and corresponding viscoelastic parameters (E_2,w, R_2,w and τ_2w).

Control of Breathing and Its Modeling Perspective, Edited by
Y. Honda *et al.*, Plenum Press, New York, 1992

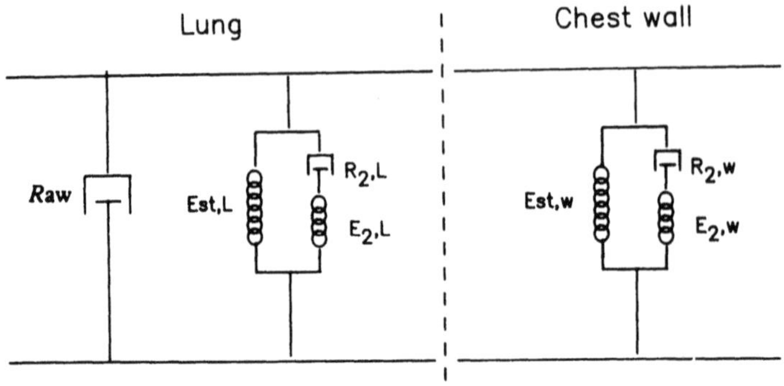

Figure 1. Scheme of spring-and-dashpot model of human respiratory mechanics. Respiratory system consists of airway resistance (Raw) and static elastance of lung (Est,L) and chest wall (Est,w) in parallel with in series spring-and-dashpot bodies (E_2 and R_2, respectively) of the lung and chest wall that represent viscoelastic (stress adaptation) units. Distance between two horizontal bars is analogue of lung volume (V), and tension between these bars is analogue of pressure at airway opening (Pao). Modified from D'Angelo et al. (1991).

The viscoelastic model in Fig. 1 has been validated in experiments on anesthetized paralyzed normal humans during constant-flow inflation by D'Angelo et al. (1991). The values of the viscoelastic parameters obtained by these authors are provided in Table 1. Over the volume range studied ($\Delta V=1$ L), Est,L and Est,w amounted to 8.2±1.7 and 6.3±1.1 cmH_2O/L, respectively. Thus, on average, E_2,L and E_2,w amounted to 39 and 26% of the corresponding static elastances. The viscoelastic time constants of the lung and chest wall were considerably longer than the standard mechanical time constant of the respiratory system [Raw/(Est,L + Est,w)] which amounted to about 0.15 s.

TABLE 1. Average values (±SD) of viscoelastic parameters of Fig. 1 obtained on 18 anesthetized paralyzed humans by D'Angelo et al. (1991).

	R_2 ($cmH_2O/L/s$)	E_2 (cmH_2O/L)	τ_2 (s)
Lung	3.44±0.97	3.21±1.14	1.13±0.36
Chest wall	2.12±0.58	1.66±0.37	1.30±0.34

At the moment, the structural basis of the viscoelastic elements in Fig. 1 remains to be elucidated. In spite of its simplicity, the model is useful for understanding the dynamic behaviour of the respiratory system (Fung, 1981). Indeed, results obtained in anesthetized dogs (Bates et al., 1989; Similowski et al., 1989), cats (Shardonofsky et al., 1991), and humans (D'Angelo et al., 1989; D'Angelo et al., 1991) could be adequately explained in terms of simple viscoelastic models.

FREQUENCY DEPENDENCE OF ELASTANCE AND RESISTANCE

The viscoelastic elements within the lung and chest wall confer time-dependency of elastance and resistance. Indeed, at high respiratory frequencies (f), the springs E_2 in Fig. 1 will oscillate so fast that there will be insufficient time for their tension to be dissipated through the dashpots R_2. Under these conditions, dynamic elastance (Edyn) will approximate Est + E_2 while Raw will approximate total respiratory resistance (Rrs). By contrast, at low frequencies the dashpots R_2 are given time to move and dissipate the elastic energy stored in E_2. In the limit, as frequency tends to zero the springs E_2 should remain at fixed length (i.e. at their resting length at which tension is zero). In this case, Edyn will be close to Est while Rrs will approach Raw + R_2,L + R_2,w. This implies that Rrs decreases with increasing f while the opposite is true for Edyn,rs. Frequency dependence of resistance and elastance occurs in both components of the respiratory system, namely the lung and chest wall. An increase in pulmonary elastance with increasing f will concomitant decrease in pulmonary resistance can also be caused by time constant inequalities within the lungs (Otis et al., 1956). In normal subjects, however, such contributions appear to be negligible. In contrast, in normal humans the viscoelastic properties have been found to confer marked time dependence of elastance and resistance both to lungs and chest wall (D'Angelo et al., 1991).

WORK OF BREATHING

In open-chest rats, Mount (1955) measured the relationship between Wdyn,L and f during sinusoidal cycling at fixed tidal volume. His results suggested that in rats at low frequencies the viscoelastic component of Wdyn,L was relatively large. The same is true in normal humans for both lung and chest wall (D'Angelo et al., 1991). This is shown in Fig. 2 (*top*) which depicts the relationships between the viscoelastic work per breath (Wvisc) of the lung and chest wall. Viscoelastic work per breath was computed at fixed tidal volume (V_T) of 0.47 L, according to the following equation (Milic-Emili, 1991):

$$\text{Wvisc} = 0.5\pi^2 R_2 V_T^2 f/(1 + 4\pi^2 f^2 \tau_2^2) \qquad (1)$$

Fig. 2 (*top*) shows that, at f >0.15 Hz, the viscoelastic work per breath for both lung and chest wall decreases with increasing f. This is predictable in view of the progressive decrease of pulmonary and chest wall resistance with increasing f (D'Angelo et al., 1991). Fig. 2 (*bottom*) shows the frequency dependence of Wdyn,L which is the sum of Wvisc and the work due to airway resistance (Wres,L). The latter was computed according to the classic equation of Otis et al. (1950):

$$\text{Wres,L} = 0.5\pi^2 K_1 V_T^2 f + 1.33\pi^2 K_2 V_T^3 f^2 \qquad (2)$$

where K_1 and K_2 are Rohrer's constants (Raw = $K_1 + K_2\dot{V}$). In computing Wres,L we used the average values of K_1 and K_2 obtained by D'Angelo et al. (1991) on 18 anesthetized paralyzed normal subjects, namely 1.85 $cmH_2O/L/s$ and 0.43 $cmH_2O/L^2/s^2$.

In line with the observations on rats by Mount (1955) and on awake humans by Cavagna et al. (1962), at low frequencies Wvisc,L contributes substantially to Wdyn,L but its relative contribution becomes negligible at f>0.6 Hz. Fig. 2 (*bottom*) also shows that at low frequencies the relationship between Wdyn,L and f cannot be described simply in terms of Eq.2. According to Fig. 1, the viscoelastic work of the chest wall in Fig.2 (*top*) should represent the total dynamic work of the chest wall. It should be noted, however, that the dynamic volume-pressure loops may include a component due to the so-called "quasi-static hysteresis". However, this work component becomes important only with large tidal volumes (Shardonofsky et al., 1990).

While the viscoelastic properties of the lung and chest wall appear to contribute appreciably to Wdyn only at relatively low frequencies, their contribution to inspiratory elastic work increases progressively with increasing f, reflecting the increase of Edyn,L and Edyn,w with increasing frequency. This implies that with increasing f (e.g.

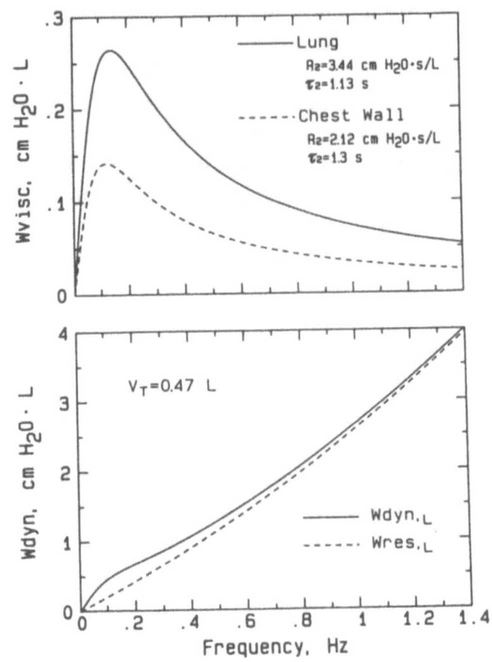

Figure 2. *Top*: Relationship of viscoelastic work per breathing cycle (Wvisc) of lung and chest wall with respiratory frequency computed according Eq. 1 for tidal volume of 0.47 L using average values of constants R_2 and τ_2 in Table 1. *Bottom*: Relationship of total dynamic pulmonary work per breath (Wdyn,L) and airway resistive work (Wres,L) with respiratory frequency. Wres,L was computed according Eq. 2 using the average values of K_1 and K_2 of D'Angelo et al. (1991). Wdyn,L was obtained by adding Wvisc,L to Wres,L. From D'Angelo et al. (1991).

muscular exercise) the inspiratory elastic work per breath is not given by 0.5 Est,rs V_T^2 as commonly described (Otis et al., 1950) but it is greater because, as a result of viscoelastic behaviour, Edyn,rs is higher than Est,rs. At f>0.6 Hz, Edyn,rs is about 30% higher than Est,rs. Although the dynamic work due to the viscoelastic properties of the lung and chest wall represents energy that is dissipated as heat, part of the elastic energy stored during inspiration in springs E_2 should be available as expiratory driving pressure. Thus, during increased ventilation (e.g., muscular exercise), some of the requirements for increased expiratory flow rates are intrinsically met by the increase in Edyn,L and Edyn,w due to increased f. The augmentation of the expiratory driving pressure through a viscoelastic mechanism has been previously proposed by Mortola et al. (1985) in studies of newborn kittens.

A more detailed account of the role of viscoelastic properties on respiratory dynamics can be found elsewhere (D'Angelo et al., 1991; Milic-Emili et al., 1990; Milic-Emili, 1991).

REFERENCES

Bates, J.H.T., Brown, K.A., and Kochi, T., 1989, Respiratory mechanics in the normal dog determined by expiratory flow interruption, J. Appl. Physiol. 67:2276-2285.

Campbell, E.J.M., 1958, The respiratory muscles and the mechanics of breathing. London:Lloyd-Luke.

Cavagna, G., Brandi, G., Saibene, F., and Torelli, G., 1962, Pulmonary hysteresis, J. Appl. Physiol. 17:51-53.

D'Angelo, E., Calderini, E., Torri, G., et al., 1989, Respiratory mechanics in anesthetized paralyzed humans: effects of flow, volume, and time, J. Appl. Physiol. 67:2556-2564.

D'Angelo, E., Robatto, F.M., Calderini, E., et al., 1991, Pulmonary and chest wall mechanics in anesthetized paralyzed humans, J. Appl. Physiol. 70:2602-2610.

Fung, Y.C., 1981. Biomechanics, New York: Springer-Verlag, chapt. 2, p. 41-56.

Milic-Emili, J., Work of breathing, in: The Lung: Scientific Foundations, R.G. Chrystal, J.B. West, P.J. Barnes, N.S. Cherniack, E.R. Weibel, eds., Raven Press, New York, chapt. 5.1.2.8, p. 1065-1075.

Milic-Emili, J., Robatto, F.M., and Bates, J.H.T., 1990, Respiratory mechanics in anaesthesia, Br. J. Anaesth. 65:4-12.

Mortola, J.P., Magnante, D., and Saetta, M., 1985, Expiratory pattern of newborn mammals, J. Appl. Physiol. 58:528-533.

Mount, L.E., 1955, The ventilation flow-resistance and compliance of rat lungs, J. Physiol. Lond., 127:157-167.

Otis, A.B., Fenn, W.O., and Rahn, H., 1950, The mechanics of breathing in man, J. Appl. Physiol. 2:592-607.

Otis, A.B., McKerrow, C.B., Bartlett, A., et al., 1956, Mechanical factors in distribution of pulmonary ventilation, J. Appl. Physiol., 8:427-443.

Shardonofsky, F.R., Sato, J., and Bates, J.H.T., 1990, Quasi-static pressure-volume hysteresis in the canine respiratory system in vivo, J. Appl. Physiol., 68: 2230-2236.

Shardonofsky, F.R., Skaburskis, M., Sato, J., et al., 1991, Effects of volume history and vagotomy on pulmonary and chest wall mechanics in cats, J. Appl. Physiol., 71: 498-508.

Sharp, J.T., Johnson, F.N., Goldberg, N.B., and Van Lith, P., 1967, Hysteresis and stress adaptation in the human respiratory system, J. Appl. Physiol. 23:487-497.

Similowski, T., Levy, P., Corbeil, C., et al., 1989, Viscoelastic behaviour of lung and chest wall in dogs determined by flow interruption, J. Appl. Physiol., 67:2219-2229.

INFLUENCE OF AGE ON VENTILATORY PATTERN DURING EXERCISE

Yoshinosuke Fukuchi, Shinji Teramoto, Takahide Nagase
Takeshi Matsuse, Kiyoshi Ishida and Hajime Orimo

Department of Geriatrics, Faculty of Medicine
University of Tokyo
Hongo, Tokyo, Japan

INTRODUCTION

Previous investigators have reported that exercise capacity is decreased in the elderly with the reduction in maximal oxygen uptake[1]. We have previously demonstrated that healthy elderly achieve greater $\dot{V}E/\dot{V}O_2$ for a given exercise load than the young during exercise[2]. However, the pattern of ventilation on exercise has not been extensively evaluated and little is known as to the changes in the fractional contribution of abdomen vs thorax in relation to total ventilation during exercise in old age.

The most notable changes of mechanical properties with aging of the respiratory system are found in the loss of elastic recoil of the lung and the decrease of the chest wall compliance. We then hypothesized that thoracic contribution on exercise to total ventilation in the elderly may become less than the young due to greater mechanical impedance of the chest wall.

We conducted this study in order to test this hypothesis and found significant decrease of the thoracic contribution to total ventilation during exercise in healthy old subjects.

SUBJECTS AND METHOD

We conducted this study in two parts, Study 1 and Study 2.

Study 1

The subject of the study 1 were 12 young volunteers (Mean age 25±4.9, %VC 100.7±8.8%, FEV1.0% 93.2±4.2%) and 12 elderly (Mean age 71.9±5.3, %VC 98.3±6.5%, FEV1.0% 91.5±3.2%). All the subjects underwent a steady state exercise tests on a cycle ergometer (Collins 220 watt model) for three minutes at 20 and 40 watts. A face mask with an oneway valve was tightly fitted to each subjects and expirates were monitored through a pneumotachymeter breath by breath.

Control of Breathing and Its Modeling Perspective, Edited by
Y. Honda *et al.*, Plenum Press, New York, 1992

Total ventilation was measured by integrating the expiratory flow (VPn). Measurement were made of minute ventilation ($\dot{V}E$), oxygen consumption ($\dot{V}O_2$) and respiratory frequency (f) by mass spectrometry (WLCS–1400, Westron Corp, Japan)–pneumotacyhmeter –computer (PC–9801, NEC Corp, Japan) system. An inductance plethysmograph (Respitrace, RIP) was applied to the thorax at mid-sternal level (rib VT) and over the abdomen (abd VT). The RIP (VRIP) was calibrated with a spirometer by least square method to fit a best regression line within 10 % difference of both measurements. The data were obtained in one minutes of each exercise and they were accepted for further analysis only when VPn and VRIP agreed within 10% of difference in estimating total ventilation. The relative contribution of abdomen vs rib cage in total tidal ventilation was evaluated as the changes in the ratio of the two measurements (abd VT/rib VT).

Study 2

In order to simulate stiffer chest wall in the elderly, chest straps were applied to the young subjects. Ten out of twelve young subjects in study 1 were enrolled into study 2 (Mean age; 26.4±4.2, %VC 99.6±4.4%, FEV1% 93.6±4.1%). The chest strap was adjusted by an elastic band to reduce the peak flows on three flow–volume curves by 20 %. The calibration of the equipments and the protocol of the exercise tests were exactly same as in study 1.

The data were expressed as Mean±Standard Deviation and statistical significance was tested by student t test.

RESULTS

Study 1

The average difference between VPn and VRIP was 4.7±0.6% to 8.3±1.4% from resting to 40 watts exercise in the young and 5.3±1.0% to 8.8±1.5% in the elderly.

The relative abdominal contribution to total tidal volume were 52.4, 50.0, and 48.5 % at rest, 20 watts and 40 watts of exercise respectively in the young subjects. In the elderly, the abdominal fraction contributed 53.5±4.6% at rest, 57.2±5.6% and 59.4±5.7% at 20 and 40 watts of load respectively. The elderly showed progressive increase in abdominal contribution on exercise in this range of load. On the contrary, young subjects had decreased contribution by the abdominal fraction to total tidal volume for the same exercise. In Figure 1, the changes in abd VT/rib VT is plotted at rest and during exercise. Both young (1.17±0.16) and elderly (1.18±0.06) subjects ventilated similarly at rest with greater abdominal fraction. This was followed in the old by the marked increase in the ratio of 1.38±0.07 at 20 watts and 1.52±0.08 at 40 watts of exercise, but it was decreased significantly in the young at 20 watts (1.02±0.09) and 40 watts (0.97±0.08).

Study 2

The changes in abd VT/ rib VT after the chest strap is shown in Figure 2 as compared to the data for non–strap tests. The ratio was increased to 1.35±0.07 at rest, 1.52±0.06 at 20 watts and 1.55±0.06 at 40 watts respectively. This pattern of tidal ventilation after chest strapping among young subjects follows that of the elderly rather than the young without strapping.

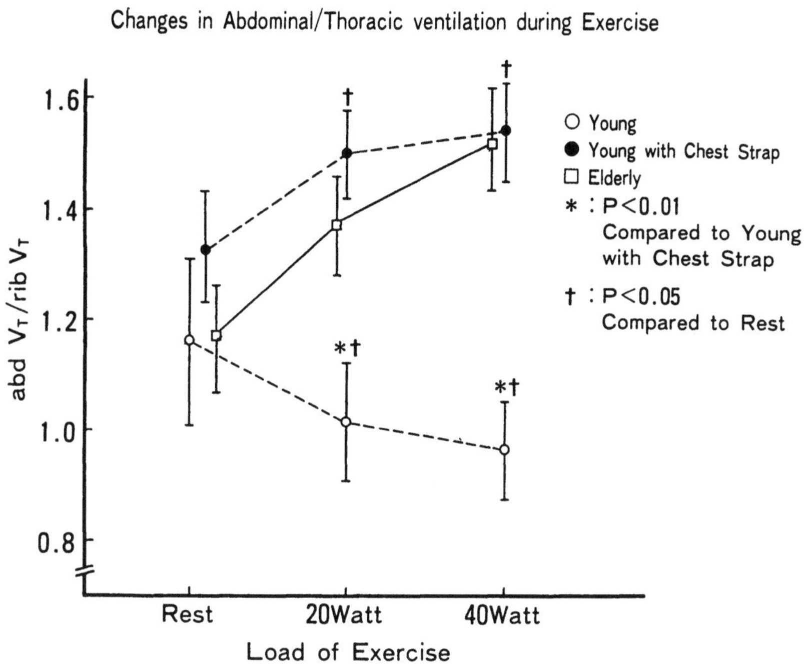

Figure 1

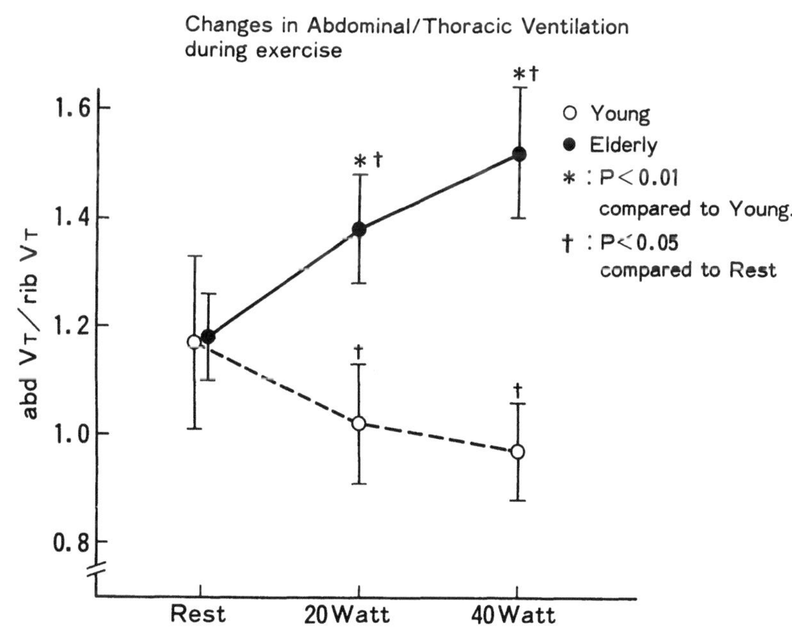

Figure 2

DISCUSSION

The resting tidal breaths of the old and young subjects in this study are characterised by greater contribution of the abdominal component than that of the rib cage. The calibration of the VRIP validated each of abd VT and rib VT to be truly reflecting the fractional contribution of corresponding total VT. The abdominal fraction represents the diaphragmatic participation and the rib cage fraction is proportional to the expansion of the chest wall other than the diaphragm.

The ventilatory equivalents for oxygen were greater in the elderly than in the young at 20 and 40 watts of the exercise load. The old subjects achieved this greater increase in minutes ventilation by recruiting greater participation of the diaphragmatic motion. The young behaved in different fashion and they rather increased the rib cage contribution for a given exercise load. But this response is not fixed in that they increased the abdominal fraction with significant chest strap applied during the exercise. This results suggest that there is an adaptive mechanism in man by which the greater minute ventilation can be reached with selective increase of the less impeded fraction of the tidal breath. Subsequently, the hypothesis of stiffer rib cage in the elderly resulting in greater contribution of diaphragmatic motion seems to be acceptable within the range of the magnitude of exercise as described.

REFERENCES

1. N. L. Jones and E. J. M. Cambell, "Clinical Exercise Testing" W. B. Saunders, Philadelphia (1982).
2. K. Sou, Y. Fukuchi, I. Nishi and M. Harasawa, Influence of age on the function of the respiratory center, Jpn. J. Geriat. 22: 399 (1985).

ABDOMINAL MUSCLE FUNCTION WITH CHANGE IN POSTURE OR CO_2 REBREATHING IN HUMANS.

T. Abe[1], K. Kobayashi[1], N. Kusuhara[1], T. Tomita[1] and P.A. Easton[2]

[1]Kitasato University, Kanagawa, Japan and
[2]University of Calgary, Alberta, Canada

INTRODUCTION

The relative respiratory function of the individual abdominal muscles in humans is not well defined. Abdominal muscles have been reported to act as a single unit during breathing[1], yet others have reported differential activation of specific abdominal muscles during respiration[2].

Previous investigators have employed surface[3] or needle[1,2] electrodes to record electromyogram(EMG). However, with surface electrodes, interference from external oblique (EXTERN) disturbs the recording from internal oblique (INTERN) and transversus abdominis (TRANSV). Needle electrodes have been used to record from TRANSV, EXTERN, and rectus abdominis (RECTUS)[2], but needles can be unstable during stimulated breathing or posture change.

In awake humans, the differential activity with change in posture or chemical stimulation, of the abdominal muscles, including EXTERN, TRANSV, RECTUS and INTERN, can be determined by EMG recordings from fine wire electrodes.

METHODS

A pair of fine wire electrodes, approximately 10 mm apart, were inserted into RECTUS, EXTERN, INTERN, and TRANSV muscles in six awake, normal young male volunteers under direct vision provided by high resolution ultrasound echograph. Site of insertion of the guide needle for RECTUS, EXTERN, INTERN, and TRANSV was 2 cm to the right of the umbilicus, right midclavicular line at the level of the umbilicus, 2 cm below the level of umbilicus on the midclavicular line, and 1 cm below the right costal margin on the anterior axillary line, respectively. Ultrasound provided clear visualization of the individual muscles layers and the guide needle used for electrode insertion (Figure 1).

With subjects breathing on a mouth piece, airflow, end tidal CO_2 ($ETCO_2$), and moving average EMG signals were sampled directly to computer during: 1) resting ventilation in supine and standing postures, and 2) CO_2 stimulated ventilation through a

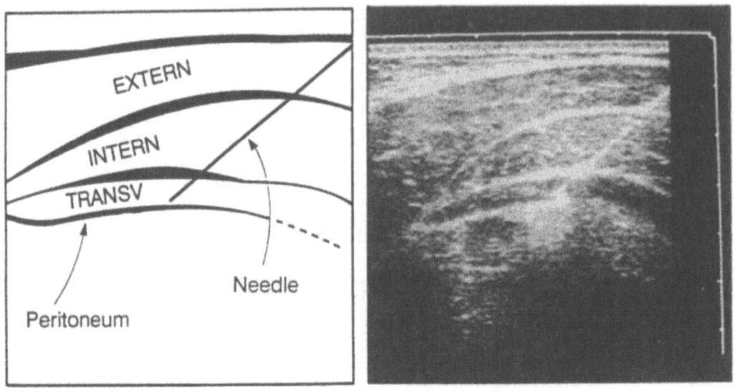

Figure 1. Ultrasound echogram of the abdominal muscles and guide needle.

rebreathing circuit, in the supine position. Whole breath tidal values and normalized intrabreath profiles of flow and EMG were calculated.

RESULTS

Electrical activity of TRANSV, INTERN, and EXTERN increased from 0.10 to 1.06, 0.14 to 0.75, and 0.15 to 0.66 Volts with change in posture from supine to standing, without change in RECTUS. During resting supine ventilation, EMG activity during expiration was slightly greater than during inspiration (Figure 2).

During resting ventilation, while standing, phasic activity was observed in TRANSV, INTERN, and EXTERN (Figure 3). The TRANSV activity was greater than INTERN or EXTERN, while RECTUS was unchanged from supine breathing. Phasic activity increased through end expiration and persisted into the following inspiration as post expiratory expiratory activity (PEEA).

During supine, CO_2 stimulated ventilation, all muscles including RECTUS showed phasic activity (Figure 4). Phasic activity of individual muscles was not identical to the

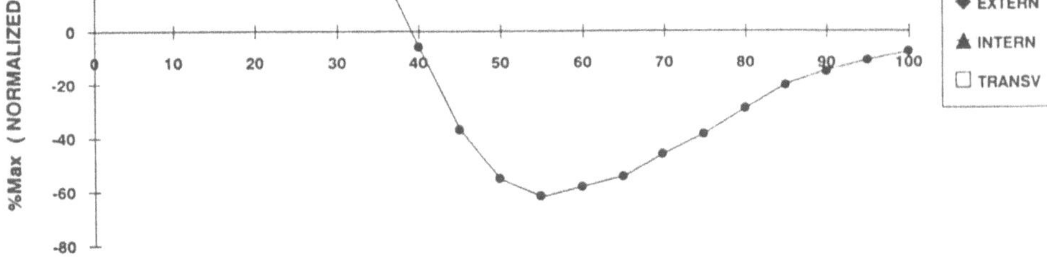

Figure 2. Intrabreath EMG activity of the abdominal muscles at rest in supine posture. Intrabreath flow profile is superimposed.

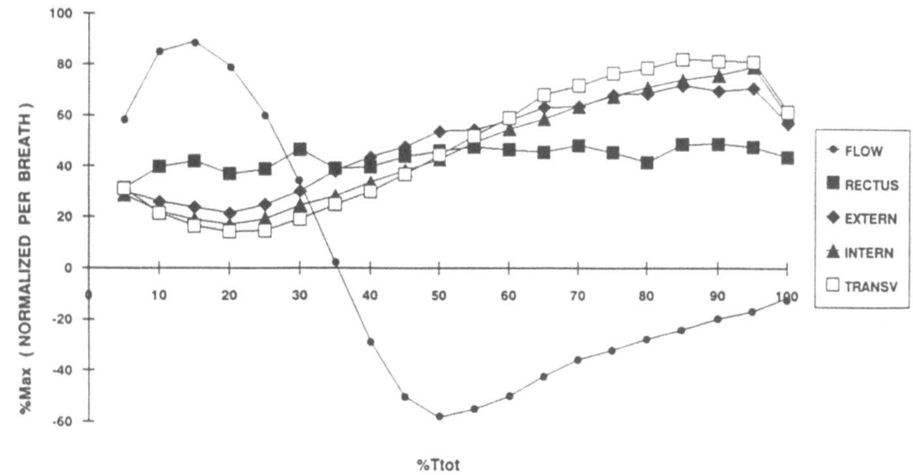

Figure 3. Intrabreath EMG activity of the abdominal muscles at rest in standing posture.

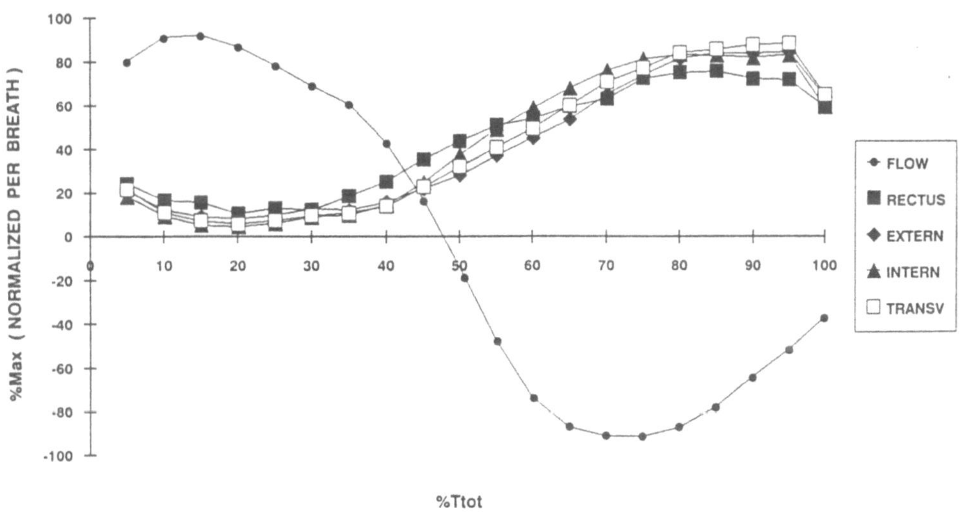

Figure 4. Intrabreath EMG activity of the abdominal muscles during CO_2 stimulated breathing in supine posture. Mean $ETCO_2$ = 69 mmHg.

activity during standing. PEEA was present in each muscle. CO_2 response curves for individual abdominal muscles are shown for a typical subject (Figure 5). During rebreathing, TRANSV was recruited most prominently, followed by INTERN, then EXTERN, and finally RECTUS.

For the group, the mean increase in electrical activation per change in CO_2 was unique per muscle. The relative CO_2 response for TRANSV, INTERN, EXTERN, and RECTUS was 0.172, 0.156, 0.145, and 0.07 Volts/mmHg, respectively. During stimulated ventilation, initial activation of each muscle occurred at different $ETCO_2$. TRANSV EMG doubled its baseline value earliest, at $ETCO_2$ of 53.8 mmHg. INTERN, EXTERN, and RECTUS followed at 54.5, 56.2, and 60.8 mmHg, respectively.

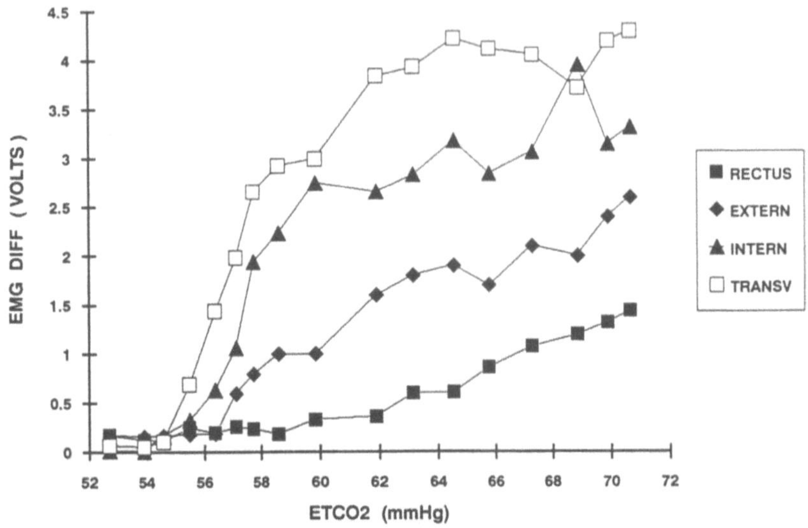

Figure 5. CO_2 response curves for abdominal muscles of a typical subject.

DISCUSSION

In awake humans, EMG recorded from fine wire electrodes demonstrated that abdominal muscles, TRANSV, EXTERN, INTERN, and RECTUS, are activated differentially, in a consistent pattern with changes in posture or CO_2 stimulated ventilation. With postural change from supine to standing, EMG activity increased most in TRANSV, then INTERN and EXTERN, without change in RECTUS. Post expiratory expiratory activity (PEEA) was noted with both CO_2 stimulated ventilation and postural change. With increasing CO_2, there was a consistent order of activation; TRANSV first, followed by INTERN, then EXTERN, and finally RECTUS. The CO_2 response, and threshold of initial activation, was specific per muscle.

The accuracy of electrode placement is crucial for investigation of function of individual abdominal muscles. Some investigators have determined electrode placement by using various maneuvers[4], while others have inserted electrodes to premeasured length according to computed tomographic or ultrasonographic measurements of the muscle layers[1,3]. Here, we used real time display to confirm the site of the electrodes during insertion.

Needle electrodes have been used by some authors[1,2]. However, they create more pain, which may inhibit recruitment of the muscle of interest, compared to fine wire electrodes. Moreover, it is difficult to avoid movement or dislocation of the needle electrodes, especially during stimulated breathing or posture change. Fine wire electrodes maintain stable placement even under strenuous effort, with minimal pain.

REFERENCES

1. J.M. Goldman, R.P. Lehr, A.B. Millar, and J.R. Silver, An electromyographic study of the abdominal muscles during postural and respiratory manoeuvres, J. Neurol. Neurosurg. Psychiatry, 50:866 (1987).
2. A. De Troyer, M. Estenne, V. Ninane, D. Van Gansbeke, and M. Gorini, Transversus abdominis muscle function in humans, J. Appl. Physiol. 68:1010 (1990).
3. K.P. Strohl, J. Mead, R.B. Banzett, S.H. Loring, and P.C. Kosch, Regional differences in abdominal muscle activity during various maneuvers in humans, J. Appl. Physiol. 51:1471 (1981).
4. D.J. Carman, P.L. Blanton, and N.L. Biggs, Electromyographic study of the anterolateral abdominal musculature utilizing indwelling electrodes, Amer. J. Phys. Med. 51:113 (1972).

SUBACUTE EFFECTS OF DENERVATION ON RAT HEMIDIAPHRAGM

C. Shindoh, W. Hida, Y. Kikuchi, S. Ebihara, and T. Takishima

First Department of Internal Medicine
Tohoku University School of Medicine
Sendai 980, Japan

INTRODUCTION

Phrenic nerve damage, which is frequently the result of cervical trauma or spinal nerve diseases, may cause respiratory muscle failure. Previous studies have shown that the acetylcholine receptors on the surface of skeletal muscles increased after denervation[1], and the resting membrane potential changed to depolarization within two days after denervation[2]. Another study showed that transformation of muscle fibers occurs in experimental hyperthyroidism, i.e., soleus (slow-twitch fiber) changes to fast and plantaris muscles (fast-twitch fiber) changes to slow within 40 days[3]. However, it is still unclear whether diaphragm muscle, which is a combination of fast and slow twitch muscle fiber, changes to slower or faster after denervation compared to intact diaphragm. Therefore, we examined the subacute effect of denervation on the physiological changes in muscle contractility and fatigability.

METHODS

Experiments were performed on 40 Wister rats (250 - 300 g). The denervation was carried out by cutting an unilateral cervical phrenic nerve about 10 mm long from the phrenic nerve root of the lower cervical nerve plexus to the insertion at the thoracic cavity under anesthesia of Ketamine (50 mg/kg). After surgery, animals recovered in about 30 minutes, and given food and water ad libitum. The animals were divided into five groups (n = 8 each) consisting of 4 denervated groups (1 day, 3 days, 7 days and 14 days after denervation) and a sham operated group as control.

The animals were killed by decapitation on each experimental day, and the entire diaphragm muscle was resected and immersed in Krebs - Henseleit solution oxygenated with 5% CO_2 and 95% O_2. Muscle strips (3 - 4 mm wide) were dissected from the costal region of both sides of the diaphragm (i.e. intact and denervated sides). The origin of the muscle with rib bone was mounted in an organ bath with pins. The central tendinous insertion was tied with 4 - 0 surgical thread and connected to a tension transducer (UL - 100GR, Mineba Co.). The Krebs - Henseleit solution in the organ bath was oxygenated with a 5% CO_2 and 95% O_2 gas mixture and maintained at a temperature of 23.5 ± 0.5 °C and a pH of 7.40 ± 0.05. Strips were stimulated electrically (SEN - 320, Nihon Kohden Co.) with 0.2 ms impulses delivered via two platinum electrodes placed on both sides of the muscle. Supramaximal stimulation (i.e. 1.2 to 1.5 times of the current required to elicit maximal twitch tension, 200 - 250 mA) was used throughout the experiment. The muscle organ bath was mounted on a fixed chassis. Muscle strip length was altered by moving the position of the tension transducer with a micrometer - controlled rack and pinion gear. The optimal length of the muscle (Lo) was defined as the fiber length at which active twitch tension (0.2 ms pulse delivered at 1 Hz) was maximally generated. The response of the muscle to increasing stimulus frequency was assessed at Lo by applying 1, 10, 20, 30, 50, 70 and 100 Hz pulses applied in 1 - s trains using a functional generator. Muscle tension was calculated as isometric force per unit of cross sectional area.

To assess muscle fatigability, rhythmic isometric contractions were performed for a 5 - min period, and the rate of fall in peak isometric tension was measured. Stimulation frequency was 20 Hz, contraction rate was 90/min, and duty cycle (i.e. the ratio of contraction times [0.22 s] over total cycle time [0.66 s]) was 0.3. Muscle fatigability was calculated by the following formula; Fatigability (%) = (Ti - Tf) / Ti x 100, in Ti = initial tension and Tf = final tension at 5 min during a fatigue trial. Statistical differences in group mean data for sham and four denervated groups were examined by one - way analysis of variance followed by Student's *t* test. Significance was $p < 0.05$.

RESULTS

Table 1 summarizes the mean changes of twitch tension, contraction time, half relaxation time in twitch kinetics of sham and denervated diaphragm.

Table 1 . Denervation effect on twitch contraction.

	T.T (kg/cm^2)	C.T (msec)	1/2RT (msec)
Control	0.49 ± 0.08	46.6 ± 4.0	55.6 ± 9.2
1 day	0.50 ± 0.09	54.1 ± 8.5	64.4 ± 10.9
3 days	0.51 ± 0.09	82.1 ± 10.7**	96.5 ± 11.6**
7 days	0.54 ± 0.05	64.8 ± 4.0**	87.6 ± 10.6**
14 days	0.67 ± 0.05**	88.0 ± 9.5**	135.9 ± 26.6**

Mean ± SD, T.T; twitch tension, C.T; contraction time, 1/2RT; half relaxation time, ** $p < 0.001$ compared to control.

The twitch tension of 14 days after denervation (0.67 ± 0.05 kg/cm^2) was significantly increased compared to sham (0.49 ± 0.08 kg/cm^2, $p < 0.001$). Both contraction time and half relaxation time from 3 days to 14 days after denervation were significantly elongated compared to control diaphragm muscle ($p < 0.001$). These results indicated that the contraction curve evoked by single pulse stimulation became slower than that of control.

Table 2 summarizes the force - frequency relationships in control and denervated diaphragm muscle. The maximal tensions were significantly decreased from 1.93 ± 0.14 kg/cm^2 (control) to 1.62 ± 0.24 (1 day, $p < 0.05$), 0.95 ± 0.15 (3 days, $p < 0.001$), 0.97 ± 0.06 (7 days, $p < 0.001$), and 1.19 ± 0.11 (14 days, $p < 0.001$). The frequency of maximal tension at 14 days after denervation was decreased to 30 Hz from 70 Hz of control. The leftward shift of force - frequency curve indicates that the characteristics of the diaphragm muscle changed to those of slow fiber muscle.

The fatigability value obtained in our stimulation mode was 85.9 ± 0.5 % of control, and that was significantly decreased at 7 days (82.8 ± 0.8 %, $p < 0.01$) and 14 days (74.5 ± 1.2 %, $p < 0.001$). This decrease in fatigability indicates that the muscle fibers changed to fatigue resistant, which characteristic is found in slow muscle fiber.

Table 2 . Force - frequency relationships after denervation.

	10 Hz	20 Hz	30 Hz	50 Hz	70 Hz	100 Hz
Control (kg/cm^2)	0.66	1.15	1.54	1.85	1.93	1.80
1 day (kg/cm^2)	0.69	1.10	1.37	1.59	1.62*	1.35
3 days (kg/cm^2)	0.74	0.90	0.95	0.95**	0.86	0.67
7 days (kg/cm^2)	0.71	0.90	0.95	0.97**	0.90	0.68
14 days (kg/cm^2)	1.07	1.19	1.19**	1.10	0.89	0.75

Values are means. * $p < 0.05$, ** $p < 0.001$ compared to maximal values of control.

DISCUSSION

In the present study, the twitch contraction of denervated diaphragm muscle became slower than that of control diaphragm, and its maximal tension of the force - frequency relationship was reduced from 3 to 14 days, and the frequency of maximal tension shifted to the lower frequency. The fatigability of denervated muscle changed to fatigue resistant compared to control. These physiological characteristics seem to be compatible with slow muscle which suggests that the denervated diaphragm change to slow - twitch fiber.

It is a well known phenomenon of muscle plasticity that the fiber composition of skeletal muscle changes as a result of various kinds of stimulation with and without neural activity. The effect of denervation on diaphragm muscle has been assumed to be simply a slow muscle to fast muscle transformation[4], because the fiber composition in diaphragm changed to type II in long term denervation. Furthermore, the myosin heavy chain (MHC) isoform composition in soleus (slow - twitch) and extensor digitorum longus (fast - twitch) muscles changed after denervation in reverse way, and recovered to their original fiber compositions after reinnervation[5]. Generally speaking, in the early stage of denervation fast - twitch fiber atrophy is found as a result of neurogenic atrophy in the skeletal muscle, therefore, it seems possible that fast - twitch fiber atrophy and slow fiber relative increase were induced in denervated diaphragm, and their muscle contractility changed to slower.

Our results seem to be at variance with the long term denervation model. It is obvious that we could not compare the early stage and later stage (4 or 6 months) of degenerative process in skeletal muscle without additional experiments. Probably, there are two possibilities of type I fiber atrophy, or type I to type II fiber transformation to explain the previous studies. Regardless of unknown degenerative mechanisms, we might conclude at least that 14 days after denervation diaphragm muscle contractility becomes slower than intact diaphragm. Furthermore, the previous studies in the chronic cerebellar[6] and spinal neuronal degeneration[7] showed that muscle fiber reduction and fatty degeneration occur in the diaphragm tissue. Our results also suggest that these degenerative changes in the diaphragm muscle start relatively earlier than 14 days after denervation.

REFERENCES

1. D. Elmqvist and S. Thesleff, A study of acetylcholine induced contractures in denervated mammalian muscle, Acta. pharmacol. et toxicol. 17:84 (1960).
2. T. Lomo and R.H. Westgaard, Further studies on the control of ACh sensitivity by muscle activity in the rat, J. Physiol. 252:603 (1975).
3. C.D. Ianuzzo, N. Hamilton and B. Li, Competitive control of myosin expression: hypertrophy vs.. hyperthyroidism, J. Appl. Physiol. 70:2328 (1991).

4. J.F.Y. Hoh and C.J. Chow, The effect of the loss of weight-bearing function on the isomyosin profile and contractile properties of rat skeletal muscles, in :"Molecular Pathology of Nerve and Muscle", A.D. Kidman, ed, Clifton, NJ:Hamana (1983).
5. A. Jakubiec-Puka, J. Kordowska, C. Catani and U. Carraro, Myosin heavy chain isoform composition in striated muscle after denervation and self-reinnervation, Eur. J. Biochem. 193:623 (1990)
6. A. Mier-Jedrzejowicz and M. Green, Respiratory muscle weakness associated with cellellar atrophy, Am. Rev. Respir. Dis. 137:673 (1988).
7. H. Haas, L.R. Johnson, T.H. Gill and T.S. Armentrout, Diaphragm paralysis and ventilatory failure in chronic proximal spinal muscular atrophy. Am. Rev. Respir. Dis. 123:465 (1981).

ß-ENDORPHINS-BASED INHIBITORY COMPONENT OF THE MECHANISM OF VENTILATORY RESPONSE TO HYPOXIA

Shu-Tsu Hu, Mei-Chun Gong, Dong-Hai Huangfu,
Tie Xu, Fa-Di Xu and Ren-Qi Huang

Physiological Laboratory of Hypoxia Study
Shanghai Institute of Physiology
Chinese Academy of Sciences
Shanghai 200031, China

INTRODUCTION

The hyperventilatory response to continuous hypoxic exposure in adult animals typically comprises an initial sharp rise of ventilation, a following precipitous fall to a lower level of hyperventilation (so-called "ventilatory depression"), and finally a gradual re-rising of ventilatory acclimatization. In contrast, the hypoxic ventilatory response in newborn animals and infants[1,2,3] is predominantly hypoventilation instead of hyperventilation although an initial brief sharp rise of ventilation does appear as a rule.

The complexity of these phasic variations seems to imply that the mechanisms underlying the hypoxic ventilatory response are likely to be involved rather than simple.

That the overall mechanism of the hyperventilatory response to hypoxia includes an inhibitory component aside from the generally known chemoreflexive ventilatory excitation as another component was clearly shown in bilaterally chemodenerveted adult rabbits and cats by Gong et al[1]. With the chemoreflex removed, the ventilation was seen to be "depressed" as long as the hypoxic inhalation lasted. Since the effect may be obtained in both cats and rabbits, anesthetized and awake as well, it may be inferred that in intact animals hypoxia elicits two parallel activities: a dominating ventilatory drive due to increased chemoreflexive activity, and a subordinate inhibitory event, which normally is masked and unnoticed.

ENDORPHIN AS POSSIBLE MECHANISM OF THE INHIBITORY COMPONENT OF THE VENTILATORY RESPONSE TO HYPOXIA

Enlightened by Grunstein's view that endorphins play a role causing the hypoxic hypoventilation in newborns, Gong et al[1] hypothesized that the hypoxic ventilatory inhibition detectable in the chemodenervated model of adult animals could also be due to effect of endorphins which is likely to be mobilized during the hypoxia stress. Their experiments using the specific opioid antagonist naloxone as a pharmacological tool provided

Control of Breathing and Its Modeling Perspective, Edited by
Y. Honda *et al.*, Plenum Press, New York, 1992

support to this idea. Naloxone (0.4 mg/kg b.w. in 1 ml physiological saline, intracarotid infusion within 3 min) reversed the inhibition in unanesthetized, chronically chemodenervated rabbits (n=10) which were hypoventilating under hypoxia. The above observation was confirmed in lightly anesthetized cats.

A series of studies has been carried out by Hu and his group at Institute of Physiology in Shanghai with the view of gathering evidences of involvement of endorphins as the basis of the mechanism of hypoxic ventilatory inhibition[4-12].

INVOLVEMENT OF CENTRAL ENDORPHIN SYSTEM

ß-Endorphins in CSF

To detect central endorphin activity under the given condition of hypoxic exposure as used in the experiment, ß-endorphins in CSF was measured by radioimmunoassay. There was a tremendous increase of ß-endorphins content in the CSF of the chemodenervated rabbits during hypoxia, detectable 10-20 min following the advent of hypoxemia and hypoxic hypoventilation (Fig.1). There exists an inverse curvilinear relationship between the CSF level of ß-endorphins and the minute ventilation during hypoxic inhalation in chemodenervated rabbits (Fig.2).

Hypothalamic Level—Role of Nucleus Arcuatus (ARC)

The authors suggested that hypoxia, being a stressor, would stimulate some central neuronal structures to liberate ß-endorphins, thereby effecting an inhibition of ventilation. The hypothalamic nucleus arcuatus(ARC) which is known to be the main pool of endorphinergic neurons in the central nervous system[13], was considered the first candidate.

In an electrophysiological study[4] in which phrenic activity of 7 chemodenervated cats was observed to represent ventilation, the depres-

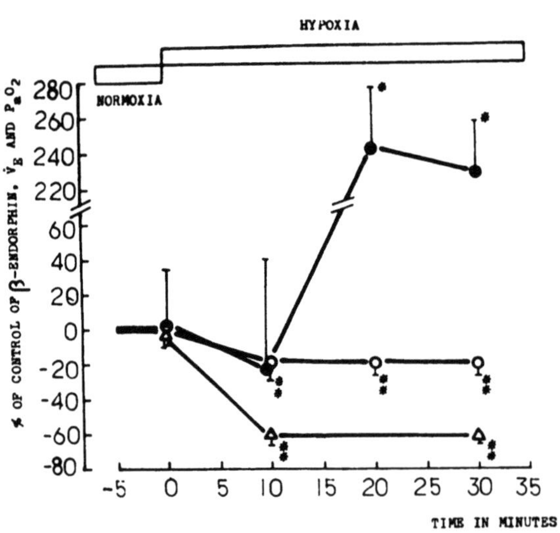

Figure 1. Increased level of ß-endorphins in CSF of chemodenervated rabbits during hypoxia with parallel changes of $\dot{V}_E$ and Pao_2 $\dot{V}_E$=open circles. ß-endorphins in CSF=solid circles. Pao_2=solid trangles. *: $p<0.05$ **: $p<0.01$

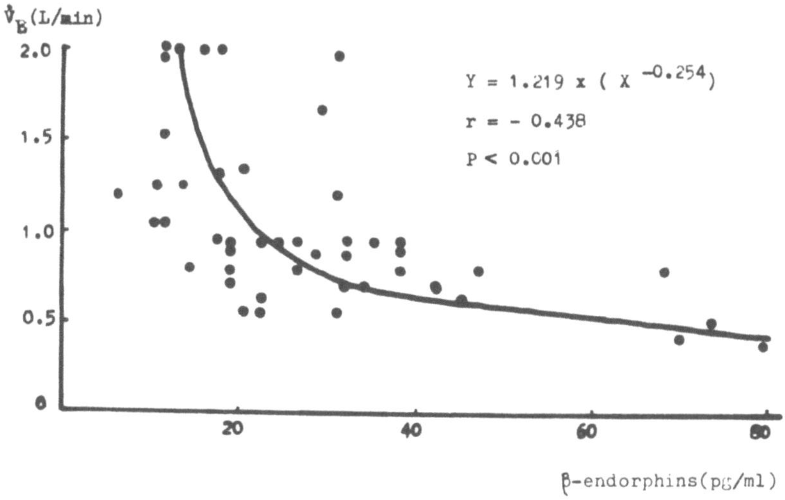

Figure 2. Inverse relationship between CSF level of β-endorphins and minute ventilation during hypoxic inhalation of gas mixtures of various concentrations of oxygen in chemodenervated rabbits.

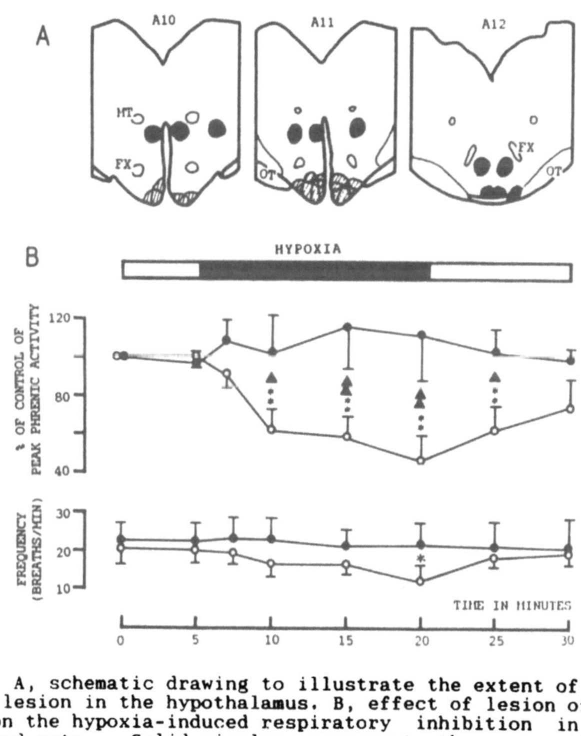

Figure 3. A, schematic drawing to illustrate the extent of electrolytic lesion in the hypothalamus. B, effect of lesion of the ARC area on the hypoxia-induced respiratory inhibition in chemodenervated cats. Solid circles represent the group with lesion of the ARC area (striped area in A); open circles represent the group with lesion of control area (blackened area in A). Asterisks show statistically significant difference from its own control; triangles denoted difference between groups (*, ▲: p<0.05; **, ▲▲: p<0.01).

sion of the peak phrenic activity under hyoxia became much less marked (-53%, p<0.01) or even reversed to excitation in 4 of 7 cats (Fig.3). Microinjection of 0.04 μmol DL-homocysteic acid, a nonspecific excitant amino acid into ARC area produced depression of phrenic activity which can be blocked by simultaneous administration of naloxone injected intravenously (0.4 mg/kg) or intracisternally (100 μg in 20 μl) (Fig.4).

In another study[11] on 16 adult male bilaterally chemodenervated rabbits under urethane anesthesia, bilateral electrocoagulative lesions placed at ARC when ventilatory depression became manifest under hypoxic exposure promptly terminated the hypoventilation 1.52 ± 0.15 L/min, resulting in a re-rising of the ventilation level to 1.82 ± 0.17 L/min (P<0.01).

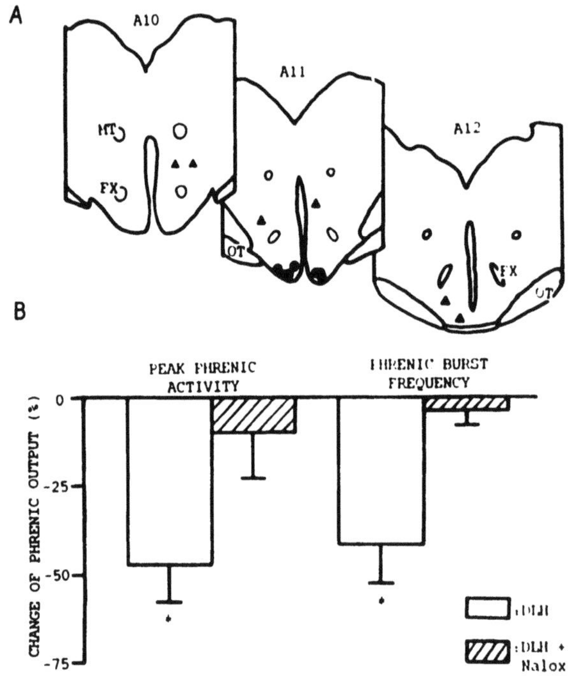

Figure 4. A: Schematic drawing to illustrate the microinjection sites in the ARC area (circles) and control area (triangles). B: Effect of microinjection of DLH or DLH+naloxone into the ARC area on respiratory output in chemodenervated cats. Note the blocking effect by naloxone.

Medullary Level—Role of Ventro-lateral Neucleus Tractus Solitaris

The authors further studied[11] the medullary ventro-lateral nucleus tractus solitaris (vlNTS), a site where respiration-related neurons are most densely concentrated. In 11 bilaterally chemodenervated rabbits naloxone (5 μg) microinjected into vlNTS abolished the hypoxia-induced hypoventilation (Fig.5). On the other hand, microinjected ohmefentanyl, a potent, specific opioid μ-receptor agonist (0.002 μg) into vlNTS aggravated the existing hypoxic hypoventilation (Fig.6).

Thus it seems likely that hypoxia inhibits ventilation by stimulating hypothalamic ARC to liberate ß-endorphins which bind μ receptors in the respiration-related neurons of vlNTS, eventullay resulting in ventilatory depression.

The ventro-lateral surface of the medulla (vlM) seems to be a potentially important area with regard to central endorphin system. Topical

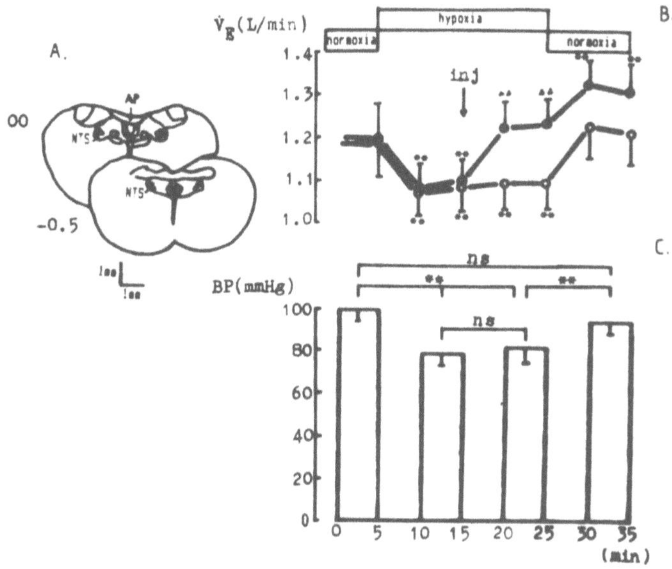

Figure 5. Effect of microinjected naloxone into vlNTS on hypoxic ventilatory inhibition chemodenervated rabbits.

A. Sites of microinjection. B. Change of minute ventilation. Solid circles show hypoxic ventilatory inhibition. **=p<0.01 (compared with prehypoxia level). Open circles show the antagonistic effect of naloxone (injected at ↓) on the inhibition. ΔΔ=p<0.01 (compared with preinjection value). ** as for solid circles. C. Columns showing arterial hypotension during hypoxia (**=p<0.01), but the hypotension is not affected by naloxone and thus irrelevant to endorphins action.

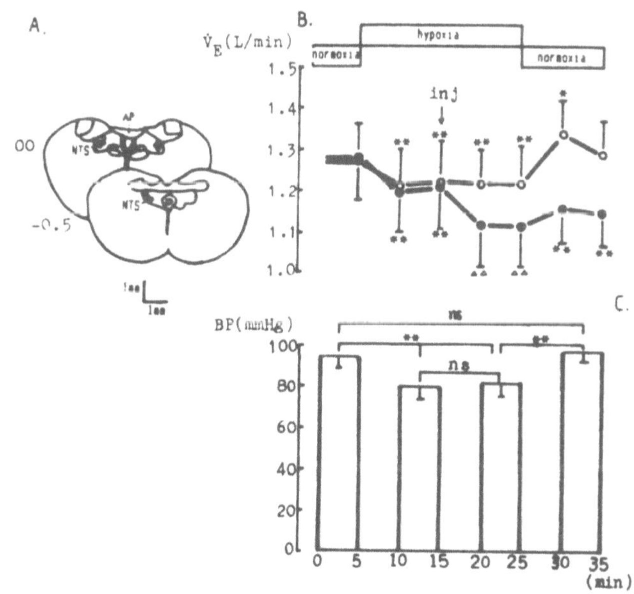

Figure 6. Effect of microinjected ohmefentanyl into vlNTS on hypoxic ventilatory inhibition

Explanatory caption as for figure 1 except in B: Open circles show hypoxic ventilatory inhibition. Solid circles show the aggravating effect of ohmefentanyl (injected at ↓) on the inhibition.

infusion of naloxone solution (27 mM in 10 µl infused at ↓) markedly attenuated the hypoxia-induced respiratory inhibition[4] (Fig.7).

Pontine Level—Role of Nucleus Parabrachialis (NPB)

In a recent study[12], NPB was examined for its relevance in the central endorphin system involved in the hypoxic respiratory depression. Unilateral microinjection of naloxone into NPB area blocked the respiratory depression caused by hypoxic inhalation, and attenuated the respiratory depression caused by electric stimulation of nucleus arcuatus.

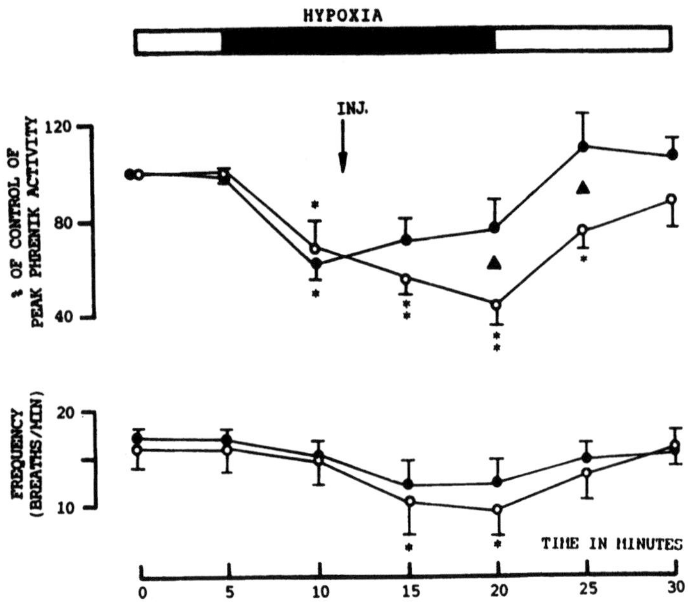

Figure 7. Influence of application of naloxone onto the ventral surface medulla upon hypoxia-induced respiratory in chemodenervated cats. Solid circles: Naloxone was applied; open circles: mock CSF was applied. (*,▲: p<0.05; **,▲▲:P<0.01)

CONCLUSION

With all the evidences so far obtained, the authors are convinced that the ventilatory response to acute hypoxia is operated by a bicomponent mechanism in which inhibitory activity as one component is set in motion from the very beginning of the hypoxic exposure and runs its course parallel to the excitatory driven by chemoreflex. The inhibitory mechanism is based on a central endorphin system in which neural structures at various levels are involved (Fig.8). Among them n. arcuatus and n. tractus solitaris play important role. Functionally the inhibitory endorphin system seems to act in ventilatory control during hypoxia as a brake to prevent the hyperventilation from excessive overshooting. But its bearing in ventilatory control under pathological conditions should not be neglected.

Moreover, a recent study from this laboratory found that hypoxia, beside producing respiratory depression, has an analgesic effect also mediated by endorphins. Thus the endorphin system when activated by hypoxia seems to serve as a common basis for a number of functional inhibitory activities among which respiratory depression is but one.

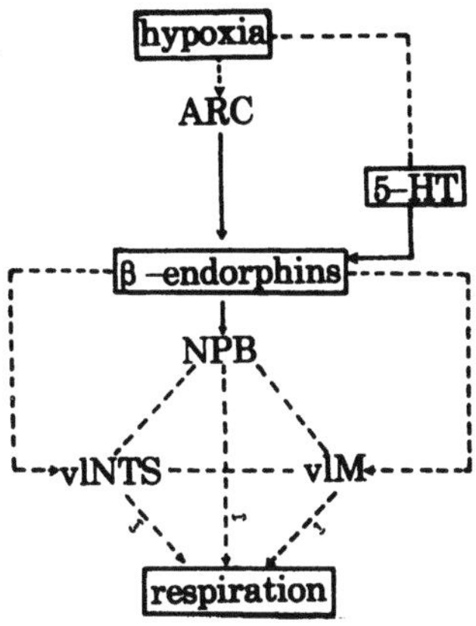

Figure 8. An outline of the pathways of β-endorphin responsible for hypoxic ventilatory inhibition based on our studies

β-endorphin released from ARC during hypoxia activates opiod receptors in NBP and in turn causes ventilatory inhibition via unknown pathways. It may also bind with μ-receptor in vlNTS to inhibit ventilation. Central 5-HT system acts as a neuromodulator on β-endorphin system to affect hypoxic ventilatory inhibition. Solid line represents the confirmed pathways; dashed line stands for possible pathways. "(-)" means inhibition.

Acknowledgements

The authors would like to thank Mr. L.Q. He, Mr. C.M. Li, Ms. X.F. Wu and Ms. X.Y. Zhou for their skillful technical asssistance. The authors greatly appreciate the permission of *Chinese Journal of Physiological Sciences* and *Acta Physiologica Sinica* to use materials from their publications as appear in this article.

REFERENCES

1. M.-C. Gong, S.-T. Hu, S.-Y. Huang and F.-D. Xu, Inhibitory component in the hyperventilatory response to acute hypoxia in adult rabbits and cats and evidence of participation of endorphins in the underlying mechanism, *Acta Physiologica Sinica* 37:107(1985)
2. M.M. Grunstein, T.A. Harinski, and M.A. Schueter, Respiratory control during hypoxia in newborn rabbits: Implied action of endorphins, *J. Appl. Physiol., Respirat., Environ., Exercise Physiol.* 51:122(1981)
3. J.P. Brady and E. Ceruti, Chemoreceptor reflexes in the newborn infant: Effect of varying degrees of hypoxia in a warm environment, *J. Physiol. (Lond.)* 184:631(1966)
4. D.-H. Huangfu, R.-Q. Huang, M.-C. Gong and S.-T. Hu, Central mechanism of hypoxia-induced respiratory depression, *Chinese Journal of Physiological Sciences* 6:166(1990)
5. D.-H. Huangfu and S.-T. Hu, Changes of β-endorphins and monoamine transmitter metabolites in cat CSF during hypoxia, *Chinese Journal of Physiological Sciences* 6:201(1990)
6. D.-H. Huangfu, T. Xu, H.-J. Du and S.-T. Hu, Response of neurons in the nucleus of solitary tract to transient deep hypoxia in rat medullary slices, *Ibid* 7:34(1991)
7. T. Xu, and S.-T. Hu, Relationship between β-endorphins and serotonin system in mechanism of hypoxic ventilatory regulation, *Acta Physiologica Sinica* 43:589(1991)

8. R.-Q. Huang, T. Xu and S.-T. Hu, Evidence for endorphin mediated analgesia induced by hypoxia in rabbits, *Chinese Journal of Physiological Sciences* 8:68(1992)
9. T. Xu and S.-T. Hu, ß-endorphins as basis of mechanism of hypoxic ventilatory inhibition, *Ibid* 8:95(1992)
10. T. Xu and S.-T. Hu, Role of central monoamine in hypoxic ventilatory inhibition, *Ibid* 8:103(1992)
11. T. Xu and S.-T. Hu, Central localization of hypoxia-induced inhibitory effect of ß-endorphin on ventilation in rabbits. *Ibid* (submitted)
12. R.-Q. Huang and S.-T. Hu, The role of nucleus parabrachialis in hypoxic respiratory inhibition in rabbits. *Ibid* (submitted)
13. F.E. Bloom, J. Rossier and E.L. Battenberg, ß-endorphins: Cellular localization of electro-physiological and behavioral effects, *in:*"Advances in Biochemical Psychopharmacology," E. Costa and M. Trabuchi ed., Raven Press, New York

PHOSPHOINOSITIDES AND SIGNAL TRANSDUCTION IN THE CAT CAROTID BODY

Mieczyslaw Pokorski and Robert Strosznajder

Department of Neurophysiology, Polish Academy of Sciences Medical Research Center, Warsaw, Poland

INTRODUCTION

The phosphoinositides, namely phosphatidylinositol-4,5-bisphosphate (PIP_2) and phosphatidylinositol (PI) play an influential role in intracellular signaling. Phospholipase C (PLC) hydrolyzes PIP_2 and PI to produce the signaling molecules inositol 1,4,5-trisphosphate (IP_3) and diacylglycerol (DAG) which act to mobilize intracellular calcium and to stimulate protein phosphorylation, respectively.[1] Quantifying PLC changes is thus one way to assess the phosphoinositide-dependent signal transduction process. We have previously reported that there are measurable amounts of PLC activity in the cat carotid body in the normoxic condition and that PLC activity is increased by hypoxia.[2] Since activation of PLC may be part of the general mechanism by which the carotid body stimuli induce physiologic effects, we now extended that study by characterization of the effects on phosphoinositide metabolism of two further natural carotid body stimuli, respiratory and metabolic acidosis.

METHODS

[2-^{3}H]PIP_2 (1.0 and 7.1 Ci/mmol) and [2-^{3}H]PI (9.2 and 18.9 Ci/mmol) were purchased from Amersham, UK.

Cats were anesthetized with α-chloralose and urethane (35 and 800 mg/kg, i.p., respectively), paralyzed, ventilated, and the carotid bodies were prepared for excision on both sides of the neck. The cats were then exposed for 20 min to normoxia ($PaO_2 \approx 90$ mmHg; $PaCO_2 \approx 36$; pH≈7.42), hypoxia ($PaO_2 \approx$ 20 mmHg; $PaCO_2 \approx 34$ mmHg; pH≈7.39), respiratory acidosis ($PaCO_2 \approx 69$ mmHg; pH≈7.22; $PaO_2 > 400$ mmHg), or metabolic acidosis (1 M lactic acid in a dose of 7 mmol/kg, iv; pH≈7.13; $PaCO_2 \approx 37$ mmHg; $PaO_2 \approx 86$ mmHg). The cats were sacrificed by perfusing them through the aorta with an icy cold Krebs solution and the carotid bodies were removed and frozen to -20°C.

Control of Breathing and Its Modeling Perspective, Edited by
Y. Honda *et al.*, Plenum Press, New York, 1992

Each pair of carotid bodies was homogenized in 400 µl of 10 mM Tris-HCl (pH 7.4) in a Dounce homogenizer. PLC activity in the carotid body tissue was assayed by measuring the formation of [^{3}H]inositol phosphate from the exogenous labeled substrate. 13 nmol PIP_2 (30 000 cpm) or 20 nmol PI (20 000 cpm) was added to each incubation vial. The organic phase was dried under N_2 and the content dissolved in 10 mM Tris-HCl buffer (pH 6.6 for PIP_2-PLC and pH 7.8 for PI-PLC) with the addition of 0.1% sodium deoxycholate and 10 mM LiCl in a final volume of 200 µl. The mixture, containing 20-30 µg protein, was vortexed and incubated for 15 min (PIP_2-PLC) and 30 min (PI-PLC) at 37°C. The reaction was terminated

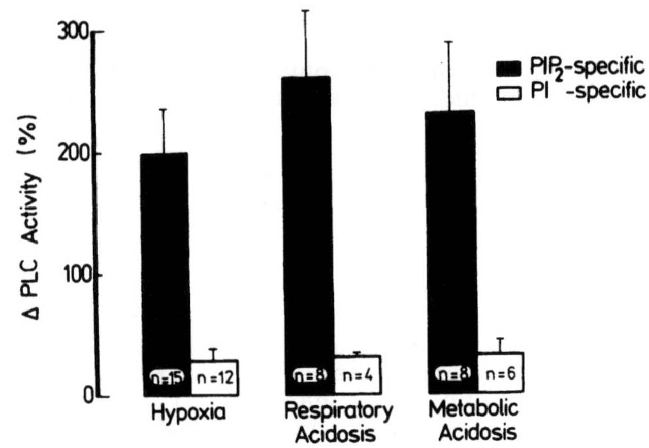

Figure 1. Percentage changes of PLC activity in the stimulated carotid body over the basal, normoxic level. The proportions of PLC activities acting on PIP_2 and PI are shown by closed and open bars, respectively. Data are means ±SE. The n refers to the number of vials.

with 1 ml chloroform/methanol/concentrated HCl (100:100:0.6 by vol) and 0.3 ml of water were added. The assay was performed on endogenous Ca^{2+} for PIP_2 breakdown and on 0.5 mM $CaCl_2$ for PI breakdown. After centrifugation, 0.4 ml of the upper aqueous phase were added to 8 ml of Bray scintillation fluid and counted for radioactivity.

All experiments were performed in duplicate. Changes of PLC activity are expressed as the percentage of basal activity, taken as that in normoxia. These changes are expressed in nmol/mg protein/min for the comparison between the effects of respiratory and metabolic acidosis. Data are means ±SE. Statistical analysis was done with a t-test or one-way analysis of variance. $P<0.05$ was considered significant.

RESULTS

Figure 1 compares the average changes in PIP_2-PLC and PI-PLC activities under the stimulatory conditions of hypoxia, respiratory acidosis, and metabolic acidosis. In each condition, the PLC acting against either phospholipid increased significantly ($P<0.05$; t-test) over the basal level found in the unstimulated carotid body. The PIP_2-PLC was severalfold greater than that acting against the PI. Analysis of variance revealed no interstimulus differences between the increments of either PLC .

Some studies have suggested that CO_2 affects, at least in part, the carotid chemoreceptor discharge in an H^+-independent manner.[3] We therefore further analyzed the results to see if this physiological phenomenon might be reflected in phosphoinositide metabolism. Figure 2 shows the average PIP_2-PLC changes for comparable increases in arterial $[H^+]$

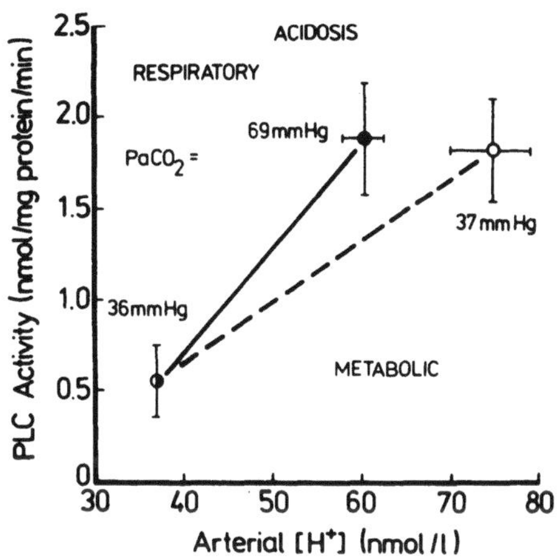

Figure 2. Incerases in PIP_2-PLC activity as a function of arterial $[H^+]$ induced by respiratory acidosis (solid circles) and by metabolic acidosis (open circles). Lines are corresponding mean regression slopes.

induced by increases in $PaCO_2$ alone and by injection of lactic acid at a constant $PaCO_2$. Respiratory acidosis caused greater increases in PLC than metabolic acidosis for the same increases in arterial $[H^+]$. The mean slopes of the response curves were 0.06 and 0.03 mg protein$\cdot$min$^{-1}\cdot$l, respectively.

DISCUSSION

This study shows that the natural carotid body stimuli low O_2, low pH, and high CO_2 increased the activity of phosphoinositide specific PLCs in the carotid body tissue. The stimulus-induced phosphoinositide hydrolysis by a PLC is bound to increase the formation of IP_3 and DAG, the key messengers for eliciting cellular responses. PLC may also be a mediator of neural effects of dopamine,[4] a transmitter ultimately involved in the generation of carotid body responses.[5] The description of PLC changes may thus have relevance in understanding the mechanisms of signal transduction in the carotid body.

The pronounced increase in PIP_2-PLC we found conforms to other studies which show that PIP_2 is the main target for PLC hydrolysis in the stimulated tissue and that changes in PIP_2 precede those in PI.[6] The small but evident increase in PI-PLC implicates that this phospholipid not only plays a precursory role due to its rapid phosphorylation to PIP_2 but also may be a target for PLC hydrolysis.

The results showed that the phosphoinositide breakdown was enhanced more by respiratory acidosis than by metabolic acidosis. This finding gives biochemical support for those physiological studies which point to a greater CO_2 excitatory effect on the neural chemosensory responses than that ascribable solely to the pH change.[3] The determinants of the CO_2 effect are unclear. CO_2 may increase access of PLC to its substrate in the plasma membrane by modifying the enzyme. It has been found in another tissue that the intracellular HCO_3^- derived from CO_2 is capable of activating the adenylylcyclase molecule in the plasma membrane.[7] This leads to increased cyclic AMP formation amd activation of protein kinases, which in turn may activate PLC by its phosphorylation.[1] Although hypoxia increases cyclic AMP in the carotid body,[8] such a chain of events has yet to be established.

In conclusion, the results of this study warrant an ascription of an intrinsic role of phosphoinositides, via the action of phospholipase C, in the transduction of the signal in the carotid body.

REFERENCES

1. E. Meldrum, P.J. Parker, and A. Carozzi, The PtdIns-PLC superfamily and signal transduction, Biochim. Biophys. Acta 1092:49 (1991).
2. M. Pokorski and R. Strosznajder, PO_2-dependence of phospholipase C in the cat carotid body, in: "Neurobiology and Cell Physiology," P.G. Data, H. Acker, and S. Lahiri, ed., Plenum Press, London, (1992).
3. M. Pokorski and S. Lahiri, Aortic and carotid chemoreceptor responses to metabolic acidosis in the cat, Am. J. Physiol. 244:R652 (1983).
4. P.S. Rodrigues and J.E. Dowling, Dopamine induces neurite retraction in retinal horizontal cells via diacylglycerol and protein kinase C, Proc. Natl. Acad. Sci. 87:9693 (1990).
5. A. Gomez-Nino, B. Dinger, C. Gonzalez, and S.J. Fidone, Differential stimulus coupling to dopamine and norepinephrine stores in rabbit carotid body type I cells, Brain Res. 525:160 (1990).
6. R.H. Michell and C.J. Kirk, Studies of receptor-stimulated inositol lipid metabolism should focus upon measurements of inositol lipid breakdown, Biochem. J. 198:247 (1981).
7. N. Okamura, Y. Tajima, S. Onoe, and Y. Sugita, Purification of bicarbonate-sensitive sperm adenylylcyclase by 4-acetamido-4'-isothiocyanostilbene-2,2'-disulfonic acid-affinity chromatography, J. Biol. Chem. 266:17754 (1991).
8. W.J. Wang, G.F. Cheng, B.G. Dinger, and S.J. Fidone, Effects of hypoxia on cyclic nucleotide formation in rabbit carotid body in vitro, Neurosci. Lett. 105:164 (1989).

INTRODUCTION: OPTIMIZATION HYPOTHESIS IN THE CONTROL OF BREATHING

Chi-Sang Poon

Harvard-MIT Division of Health Sciences and Technology
Massachusetts Institute of Technology
Cambridge, MA 02139

INTRODUCTION

The idea that living organisms may be optimally adapted to their environments owes its roots to Darwinian theory. It is clear that, inasmuch as all life processes are sustained by some form of energy, those organisms that can efficiently procure and exploit the needed energy are likely to enjoy a better chance of survival in a competitive environment. This notion has recently received considerable attention in the study of evolutionary adaptation.[1] Although not a universal criterion, the optimality condition may be manifest at many different levels of organization in a variety of ways. For example, optimality of biological functions has been variously implicated in the design of biochemical pathways,[2] regulation of plant metabolism,[3] control of animal gaits,[4] and control of ventricular ejection.[5]

In the study of the respiratory system, the possibility of an optimal pattern in breathing was suggested as early as 1925 by Rohrer.[6] For several decades, this proposition was mostly ignored despite repeated advances by noted physiologists including (cited in ref. 7) Otis, Fenn and Rahn; Agostoni et al.; Milic-Emili and Petit; Mead; Widdecombe and Nadel; Purves;[8] and Grodins and Yamashiro.[9] Weibel, Taylor and Hoppeler[10] recently demonstrated that the morphometric and physiologic designs of the respiratory system are optimized for the aerobic demand. Specifically, comparative studies of respiration in different mammalian species revealed a close matching between peak aerobic loads and capacities at all steps of the respiratory pathway - from mitochondrial oxygen uptake to pulmonary gas exchange - over a wide range of aerobic capacity.

Our investigation of respiratory optimization[11-14] was part of a general quest for a coherent framework for understanding respiratory control under diverse physiological conditions. If optimality may be considered as a motif in respiratory design, then it could provide a conceptual glue for the varied aspects of respiratory control. What evidence can be drawn for (or against) such an optimization hypothesis? How does it relate to classical concepts of respiratory control? Why is it a relevant (albeit controversial) issue? What are the implications for future research? It is the purpose of this overview to address these questions.

Control of Breathing and Its Modeling Perspective, Edited by
Y. Honda *et al.*, Plenum Press, New York, 1992

MODELING METHODOLOGY

Despite the long legacy of the optimization hypothesis, until recently it has received relatively little attention in the literature. Why was there such a general reluctance toward the optimization hypothesis? For one thing, there is a long-held belief that respiratory control is but a simple reflex phenomenon that is always characterized by certain stimulus-response relationships. This "reflex model" appears to offer a simple and adequate explanation for a range of chemo- and mechanoreflexes. Implicit in this model is the assumption that the respiratory controller represents a simple relay unit in the afferent-to-efferent path. This is the simplest form of neural information processing - so speciously simple that it seems almost irrefutable.

By contrast, the optimization model postulates a complex transformation from the afferent inputs to efferent outputs. What is the basis of this postulation? In the reductionist's view such a complex model would hardly be tenable, for without any compelling evidence in the microstructure of the respiratory controller the rule of parsimony (Ockham's razor) favors the reflex model.

The reductionist's approach holds that all models start with microscopic data and the validation of the models lies in their ability to explain macroscopic data (Fig. 1). In this "bottom-up" approach the model does not by itself constitute a hypothesis; any discrepancy in the macroscopic prediction of the model is corrected by postulating alternative micro-mechanisms in addition to, but not in lieu of, the original model. As a consequence, the preoccupation of the chemoreflex model of ventilatory control has led to the general postulation of a putative "exercise stimulus" as an explanation for exercise hyperpnea based on a similar reflex scheme.[15,16]

The optimization model, on the other hand, represents a "top-down" approach based on *inductive* reasoning. The model integrates macroscopic phenomena and predicts unknown microscopic behavior. It is, therefore, both an *empirical* model of macroscopic data and a *hypothetical* model of microscopic mechanisms. The inductive approach, of course, is unorthodox in physiology research and this difference in methodology contributed to the lack of general appreciation of the optimization hypothesis over the past several decades. In the physical science, inductive methods have led to grand theories like thermodynamics, quantum mechanics, and relativity which are powerful models for studying molecular and subatomic structures. In the biological science, macroscopic observations inspired Mendel's heredity theory which paved the way for modern molecular genetics. Could the optimization model render such usefulness in respiratory physiology?

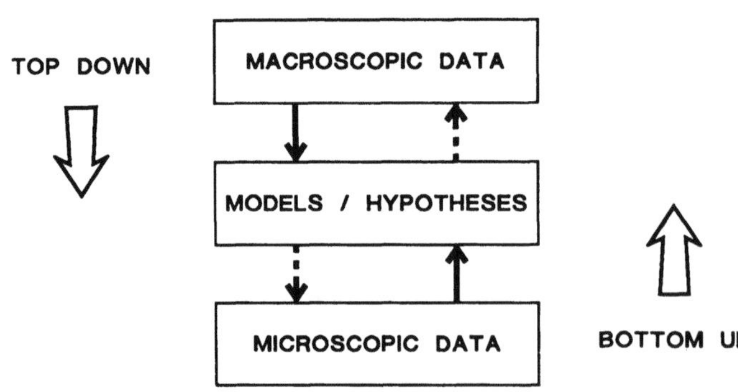

FIG. 1. Reductionist's (*bottom-up*) and inductionist's (*top-down*) approaches to modeling and hypothesis testing. Model is fully validated if microscopic data predict (*broken arrow*) macroscopic data, and *vice versa*.

To answer this question, we first note that a hypothetical model is worthwhile if it meets the following criteria: 1) It integrates and generalizes a wide range of macroscopic data; 2) It makes specific predictions of microscopic behavior as meaningful and testable hypotheses; 3) It constitutes a coherent and parsimonious framework for relating microscopic and macroscopic events. These are the *generality*, *testability* and *coherency* properties. Although Rohrer's original idea was brilliant, it was never clear that it satisfied any of these criteria. Granted, all models are not perfect and the validity of any model must be evaluated relative to other competing models. However, even in comparison with the reflex model Rohrer's proposition seemed to fall short: it pertained to only a single macroscopic event, namely, respiratory frequency at rest; it gave no clues as to the possible underlying microscopic (neural) event; it did not afford a complete and consistent explanation of respiratory control, either on its own or in conjunction with other models.

Interests in the optimization hypothesis were recently rekindled by the growing realization that the optimality principle could in fact be much more general and coherent than previously thought. The following section presents some recent advances in the optimization model. In addition, I propose that the optimization model could shed important light on the underlying neuronal mechanisms by serving as a general guiding principle for developing novel and testable hypotheses.

OPTIMIZATION OF NEURAL PATTERN

A *modus operandi* of the reductionist's approach is to divide and conquer. For a long time, this fateful tenet spawned the general assumptions that the mechanisms of ventilatory and breathing pattern control were distinct and that by attacking each problem separately and then combining the outcomes, a complete account of the respiratory controller would emerge. Such a dichotomy led to the widespread postulations of a host of purportedly reflexogenic mechanisms for separate control of ventilatory drive[16,17] and pattern.[18] However, although it is convenient to attribute different respiratory components to different control mechanisms, it remains unclear as to how these discrete mechanisms may be integrated at the controller to produce the overall respiratory response under differing physiological conditions (Fig. 2).

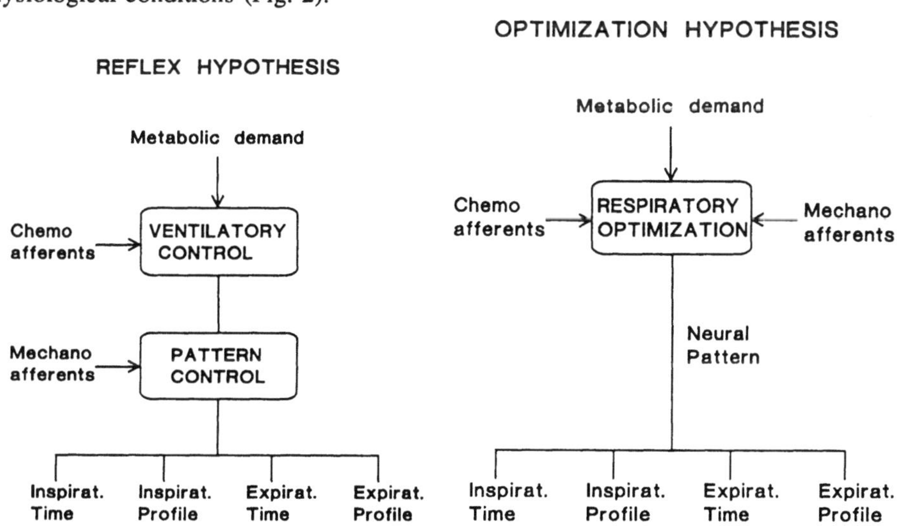

FIG. 2. Reductionist's *vs.* inductionist's view of respiratory control. The reflex (reductionist) hypothesis assumes a hierarchy of discrete commands for different components of ventilatory pattern. The optimization (inductionist) hypothesis assumes a single optimality principle in the neurogenesis of rhythmic respiratory movement which then translates into a hierarchy of ventilatory patterns.

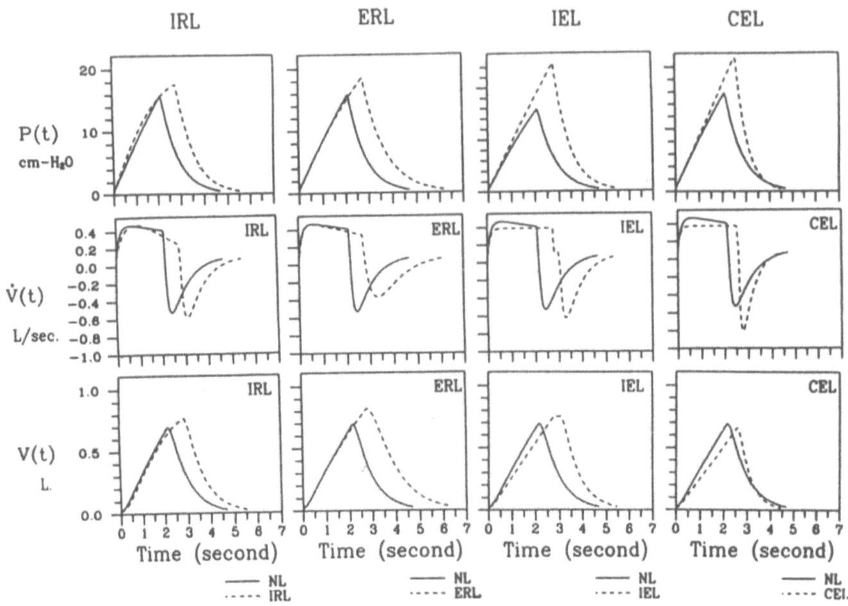

FIG. 3. Optimal waveshapes of inspiratory neural output (P), respiratory flow ($\dot{V}$) and volume (V) in resting state (*solid lines*) and under various types of ventilatory loading (*dotted lines*) with expiratory drive being suppressed. Results are strikingly similar to experimental observations. IRL/ERL, inspiratory/expiratory resistive load; IEL/CEL, inspiratory/continuous elastic load. (From ref. 14).

Early versions of the optimization model suffered from a similar drawback in that they generally assumed that ventilatory drive was always preset by chemical and metabolic demands while certain aspects of breathing pattern were being optimized. Thus, the controller was presumed to operate under separate reflex and optimization principles in a hierarchical manner although it was not certain how these disparate principles might work in harmony to generate an integrated response.

The ventilatory optimization model I previously proposed[11,13] overcame this difficulty by postulating that ventilatory drive *per se* might also be subject to optimization by the controller. By combining the optimization principles for ventilatory and breathing pattern control, it is possible to obtain a composite optimization criterion that applies to both ventilatory drive and timing components.[12]

In a recent study[14] we extended the combined optimization criterion to include an explicit description of the rhythmic respiratory neural drive. For any given chemical and exercise inputs and respiratory-mechanical load there is an optimal respiratory neural pattern that minimizes the work rate of breathing without causing excessive disturbances in arterial chemical homeostasis (i.e., it minimizes the sum of the mechanical and chemical costs of breathing). This pattern consists of three distinct phases:[19] a rising (inspiratory) phase, a declining (postinspiratory) phase, and a reversal (expiratory) phase. This basic neural pattern represents the integrated, instantaneous output of the controller within a complete respiratory cycle. It also determines all other variables that are customarily used to characterize tidal respiratory pattern (e.g., tidal volume, frequency, inspiratory duty cycle, etc.) and total ventilation. Thus, unlike previous models this general optimization model does not assume a hierarchy of control *mechanisms*; rather, it produces a hierarchy of control system *outputs* that stem from the same optimization principle (Fig. 2).

During the inspiratory phase the inspiratory muscles perform mechanical work against the viscous and elastic forces of the respiratory system. During the postinspiratory phase the inspiratory muscles perform negative work against the elastic recoil of the lung and chest wall; it is likely that a portion of the returned energy may be recovered as elastic energy in stretched tendons and cross-bridges of the muscles.[20,21] Thus, as in other skeletal muscles[20] the pliometric contraction during the postinspiratory phase may serve to enhance the mechanical efficiency of the inspiratory muscles during the inspiratory phase. As shown in Fig. 3 the optimal neural waveshapes are consistent with experimental findings under resting conditions and with various types of ventilatory loads. The distinctive respiratory neural patterns predicted by the model under different ventilatory loads suggest that respiratory load compensation is strongly influenced by mechanical work optimization.

Furthermore, with increased stimulation by exercise and airway CO_2 inputs the model predicts an augmentation of peak inspiratory activity, decreases in inspiratory and postinspiratory durations, and a more rapid decay of postinspiratory inspiratory activity. The predicted optimal responses in ventilatory pattern and ventilatory output derived from the corresponding neural patterns are similar to those resulting from previous models[11-13] and are generally compatible with experimental findings under similar conditions.[14]

These results suggest that the optimization model is a much stronger hypothesis than the conventional reflex hypothesis (Table 1). In terms of generality, both models are applicable to a wide range of inputs, but the optimization model is more comprehensive in that it predicts the complete neural pattern rather than selected ventilatory components. The optimization model is also more coherent in that the same optimization principle is consistently applied to all aspects of respiratory control whereas in the reflex model a myriad of reflex mechanisms are invoked. Finally, as expounded in the following section, the optimization model points to the possible existence of neuroplasticity as a means of optimization and adaptation in the respiratory neural network, in sharp contrast to the hard-wired network structure generally assumed in the reflex model. This unique prediction presents a critical test of the optimization model. In contrast, whereas the reflex model has received wide acceptance (by default), its validity remains questionable as many of its critical implications such as the origins of the exercise stimulus and of the load compensation reflex have so far defied experimental verifications.

TABLE 1. Comparison of reflex and optimization hypotheses.

	Reflex Hypothesis	Optimization Hypothesis
Generality	Applicable inputs: Chemical, mechanical, exercise	Applicable inputs: Chemical, mechanical, exercise
	Applicable outputs: $\dot{V}_E$, T_I, T_E	Applicable outputs: $\dot{V}_E$, T_I, T_E, V_T, P(t)
Coherency	Multiple principles: Chemoreflex Mechanoreflex Exercise reflex	One unifying principle: Chemical-mechanical optimization
Testability	Hard-wired neural network ? Exercise stimulus ? Load compensation reflex ?	Neural plasticity/adaptation ?

$\dot{V}_E$, ventilation; T_I/T_E, inspiratory/expiratory time; V_T, tidal volume; P(t), neural output.

NEURAL LEARNING AND OPTIMIZATION

Respiratory optimization signifies a complex mode of neural information processing - the significance of which is only beginning to be appreciated. It is well recognized that mammalian neurons are capable of multifarious computations.[22] These computational abilities are most evident in higher brain structures such as the neocortex, hippocampus, and cerebellum. Recently, there is increasing evidence that brainstem neurons might also be responsible for certain computations involved in motor learning and control.[23] The pervasiveness of sophisticated neural computations in different regions of the mammalian CNS (including the brain stem) should not be surprising, since the same have been demonstrated even in invertebrates.[24]

Neurons in the brainstem respiratory regions are known to be endowed with certain computational abilities such as long-term and short-term memory and temporal gating.[25] There is increasing evidence that such memory effects may result from some forms of synaptic long-term (LTP)[26] and short-term potentiation (STP)[27] similar to those found in other brain structures.[28] In vitro studies of neurons in the solitary complex have demonstrated prolonged alterations of postsynaptic excitability following afferent stimulations.[29]

These observations suggest that certain respiratory neurons may exhibit *plasticity*, a phenomenon characterized by alterations of synaptic transmissibility following neuronal excitation.[24,28,30] If so, then the classical model of the respiratory controller as a hard-wired neural network that effects reflexogenic control of breathing and respiratory rhythmogenesis would be an oversimplification, since the connectivity and connection strengths of the network might vary under different patterns of afferent stimulation.

Although neuroplasticity has been well documented in many other neural systems, its nature and possible role in the neural control of respiration in the brain stem have not been well studied. A primary mechanism of learning and adaptation in nervous systems, neuroplasticity is an important means of neural computation for many complex tasks. This type of neural learning has been suggested to underlie the adaptive control of movement and certain physiological functions.[31]

The plausibility of neural learning as a means of optimization is suggested by a recent study[32] in which the neural control of breathing in the brain stem was emulated in the cerebral cortex as an equivalent task of visually guided motor tracking. In that study subjects were asked to manually control a marker so as to dynamically minimize a displayed "cost" signal, much as the respiratory controller would minimize the chemical and mechanical costs of breathing. Not surprisingly, optimization was achieved in most cases through some trial and error. This mode of optimization is reminiscent of a cognitive phenomenon called operant conditioning, a form of associative learning in which the control action is modified by positive/negative reinforcement until an optimal outcome is attained.[33] The importance of associative learning in modulating exercise hyperpnea is also indicated in a recent experiment in goats in which exercise ventilation was found to be potentiated after pairing with prolonged chemoreceptor stimulation.[34] The pairing-specific enhancement (PSE) effect is an example of classical (Pavlovian) conditioning, another form of associative learning in which the response to a conditioned stimulus is sensitized after pairing with a powerful unconditioned stimulus.[33]

One possible neuronal mechanism of associative learning is associative long-term potentiation of Hebbian synapse.[35] In general, a modifiable synapse is called Hebbian (after D.O. Hebb[36]) if it is strengthened by simultaneous occurrence of presynaptic and postsynaptic activity. In this event, an associative learning response to presynaptic (conditioned) input is obtained when postsynaptic activity is modulated by a separate synaptic input from an unconditioned stimulus. There are two possible ways in which the synapse may be associatively strengthened. In a *conjunctional* Hebbian synapse (Fig. 4a) associative LTP in the conditioned pathway is induced by the simultaneous appearance of

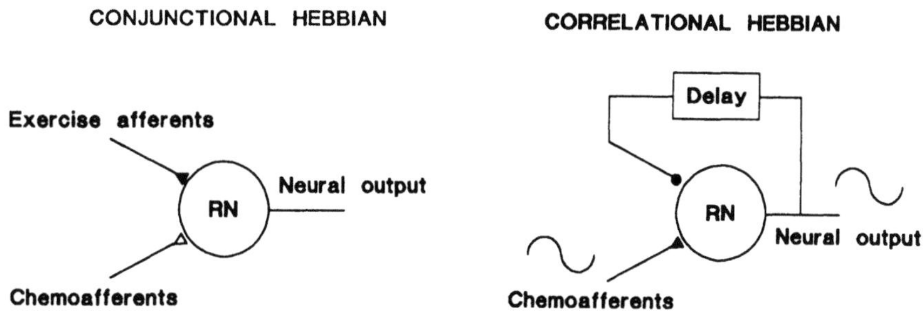

FIG. 4. Associative learning by conjunctional (*left*) and correlational (*right*) Hebbian synapses (*filled triangles*). A conjuctional synapse cooperatively enhances the exercise (conditioned) signal and chemoreceptor (unconditioned) signal by associative LTP. A correlational synapse produces similar effects by virtue of the temporal coupling between the chemical input and neural output, without the need for an explicit exercise signal. Associative conditioning is achieved by recurrent inhibition of the respiratory neuron (RN) and temporal alignment of the inputs is adjusted by variable memory delay in the recurrent path (See ref. 38).

the conditioned and unconditioned stimuli. Alternatively, in a *correlational* Hebbian synapse (Fig. 4b) temporal fluctuations of the conditioned input may result in synaptic potentiation or depression depending on whether the input is positively or negatively correlated to the unconditioned stimulus.

Conjunctional Hebbian learning has been suggested to be a possible neuronal mechanism for classical conditioning[37] and as such may be responsible for the reported PSE response. On the other hand, a correlational Hebbian synapse could be an effective means of operant conditioning in that fluctuations in respiratory neural output (trial signal) and chemosensory feedback (error signal) may reinforce/weaken the mean neural output by increasing/decreasing neuronal transmissibility in the controller. Thus, it is possible for such a synapse to optimally regulate respiratory neural output in direct proportion to the metabolic CO_2 load even in the absence of an explicit exercise induced afferent input.[38]

Conjunctional and correlational Hebbian learning are two alternative hypotheses regarding the mechanism of ventilatory control during exercise and chemoreceptor stimulation. Both models predict the exercise and chemoreflex responses as well as the multiplicative ventilatory CO_2-exercise interaction reported in goats[39] and healthy subjects.[40] Furthermore, both models are compatible with respiratory optimization since a multiplicative controller is optimal for CO_2 and exercise inputs.[11,13,41,42] The correlational synapse may also be compatible with PSE if the correlational event induces LTP as in the hippocampus.[43] However, in a correlational synapse the "exercise signal" is encoded in the synapse as a memory element whereas in a conjunctional synapse an explicit exercise input is needed. Correlational enhancement of synaptic transmissibility may offer a possible neuronal explanation for the putative "PCO_2 oscillation" signal which has been suggested as the cause of exercise hyperpnea.[44] Although the latter remains controversial, the significance of such a signal is evidenced by the recent findings that both ventilatory exercise-CO_2 interaction[45] and the degree of load compensation[46] in humans are significantly augmented when the PCO_2 oscillation is altered by increased respiratory dead space.

A multiplicative controller may also be engendered by nonadaptive neuronal mechanisms such as shunting inhibition and excitation.[47] However, such a controller again requires an explicit exercise input as an excitatory shunt to the chemosensory input (or *vice versa*). Furthermore, it does not explain the PSE response and the reported modulation of the multiplicative effect by changes in PCO_2 oscillation. It is important for future research to elucidate the nature of neuronal learning in the brain stem and its possible role in respiratory optimization, the induction of exercise hyperpnea, and the adaptive ventilatory adjustments under varied conditions.

BREATHING AS AN ADAPTIVE BEHAVIOR

If neuronal learning is indeed a mechanism of ventilatory optimization, then the respiratory controller resembles a self-tuning adaptive regulator.[48] This is consonant with Priban and Fincham's conjecture[49] that the respiratory controller may adaptively "*keep the operating point of the blood at the minimum while using a minimum of energy.*" The optimization model provides a quantitative account of this conjecture. Furthermore, as discussed below, it offers a general and unifying conceptual framework for integrating many other aspects of respiratory control that are seemingly disconnected otherwise.

Adaptive Compensation and Structural Redundancy

Like most physiological systems, the respiratory system exhibits considerable redundancy in its afferent[50] and neuronal[51] architecture. Although stimulation of each chemoreceptor (or mechanoreceptor) pathway may elicit a reflex response, the combined effect of all redundant pathways is often hypoadditive.[52,53] The competition in afferent traffic is unlikely a simple saturation effect since hypoadditivity could occur even at relatively low stimulation levels. Another possibility is adaptive competition of convergent pathways, perhaps by associative LTD/LTP in a Hebbian/anti-Hebbian synapse.[35,38] If so, then interruption of any afferent or neuronal projections would likely lead to adaptive modulations of the remaining redundant pathways. There is some evidence that failure of certain mechanosensory pathways may be adaptively compensated by experience-dependent sensitization of other redundant pathways so that homeostasis is restored over time.[34] Also, successive lesions of the dorsal and ventral respiratory groups in the medulla of decerebrate cats resulted in only slight progressive decreases in phrenic motor outflow until both neuronal groups had been extensively destroyed.[54] Thus, structural redundancy could be an important fail-safe mechanism provided the controller is self-adaptive.

Furthermore, an adaptive controller may also recruit surrogate sensory pathways in the event that all normally functional redundant pathways have failed. Such functional substitution can be seen in certain perceptive dysfunctions such as blindness (which may be partially compensated by sensitization of other non-visual inputs like hearing, touch, etc.). In a similar fashion, deprivation of chemosensitivity in congenital central hypoventilation syndrome[55] could be partially compensated by recruitment or sensitization of corticosensory pathways that may otherwise be torpid in healthy individuals or in sleep. In this event, the reductionist technique of functional identification by structural elimination would be misleading if the system is self-adaptive.

Rhythm and Pattern Generation

The central pattern generator (CPG) of respiratory rhythm has been suggested to be composed of either an oscillating network[19] or, more recently, pacemaker cells.[56] The lingering issue as to which modality plays a more important role in rhythmogenesis may be oversimplified, since the same could also result from a combination of both, as in many invertebrate CPG's.[57] An often overlooked question is how the basic rhythm and the associated pattern produced by either or both modalities are modulated by different patterns of phasic chemo- and mechanoafferent inputs. By demonstrating the prevalence of optimality, the optimization hypothesis suggests a *sine qua non* that any proposed mechanism must satisfy in explaining such a modulatory process.

The reductionist's approach assumes that the basic mechanisms of respiratory rhythmicity and pattern formation are similar in reduced preparations and intact animals, and that any discrepancies may be explained in terms of systematic transformations from simple to more complex organizations.[58] Although extensive data is presently available in various reduced states, the transformations leading to more complex behaviors *in situ* remain unclear. It is important to note that neural networks may possess "emergent" properties such that the collective behavior of the network could be much more intricate than the totality of individual neurons or neuronal groups. Two important emergent properties are the adaptive modification of the network by associative learning[33,35,37] and combinatorial optimization of neural states in a Hopfield network,[59] both of which have been implicated in the optimization of ventilatory drive (see above).[38]

The optimization character of the respiratory neural pattern, which is evidenced under a wide range of modulatory inputs[38] and in all vertebrate classes,[60] suggests that respiratory rhythm and pattern (or drive) may be an integral and collective outcome of the respiratory neural network in the intact animal. This is in contrast to the suggestions that rhythm and pattern generation are separate functions in the reduced state[58] and represent different components of the respiratory controller in the intact state.[61] Furthermore, it is possible that such a collective behavior may arise from similar emergent properties of the network as proposed above for ventilatory control. There is evidence that the CPG may undergo phasic adaptive changes (with memory-like effects) in response to vagal[62,63] and pontine stimuli.[64] It is therefore important for any model of the respiratory CPG to include the phasic modulations by direct chemo- and mechanoafferent inputs as well as the adaptive modification of the CPG consequent to these inputs (Fig. 5).

Behavioral and Defensive Interactions

The optimization models so far have only considered the interaction of chemo- and mechanoafferent inputs. However, the respiratory pump is employed in many other behavioral and defensive tasks (vocalization, posture, coughing and sneezing, panting, etc.) which compete with the chemical drive for the control of the respiratory efferent apparatus. Defensive responses evoked by various sensory receptors may assume different forms ranging from apneusis to explosive expulsion; in many instances these patterns are ineffective for the sake of pulmonary ventilation but may be optimal for the alleviation of their respective provocations. In respiratory fatigue, the adoption of a shallow and rapid breathing pattern may spare the respiratory muscles from excessive ventilatory work thereby delaying the onset of respiratory failure.[65] Similarly, the central ventilatory depression induced by sustained hypoxia - which is clearly counterproductive in restoring PaO_2 - may be an important protective mechanism to enhance survival of brainstem respiratory neurons in anoxic environments.[66]

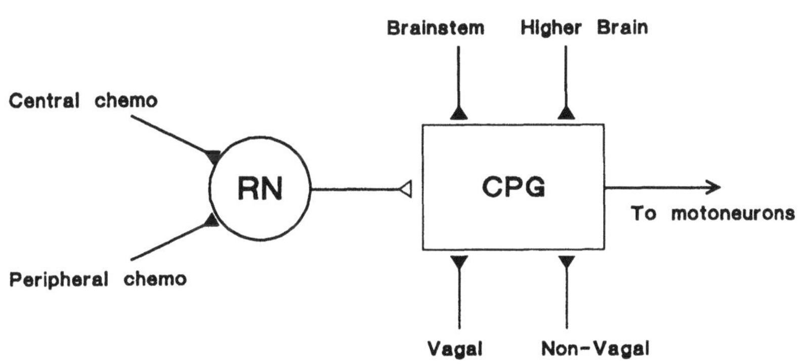

FIG. 5. Simplified model of a self-adaptive respiratory controller. Neuroplasticity might exist in the CPG, the integrative neuron(s), and the synaptic connections (*filled triangles*) from various chemoafferent, mechanoafferent, brainstem, and suprapontine inputs.

Furthermore, cortical events may also directly modulate[67] or entrain[68] the respiratory drive, and certain cortical inputs may be associatively conditioned by the automatic ventilatory drive[69] (or *vice versa*[70]) to exert independent influence on ventilatory output. The associative conditioning indicates that neuroplasticity may exist in the corticomedullary pathway. It has been pointed out that the condition required for minimizing obnoxious respiratory sensations such as dyspnea is different from the optimization criterion for the physiological control of breathing.[42] It is possible that the respiratory controller might switch from a mechanical optimization mode at low ventilatory levels to a sensation minimization mode at dyspneic levels. Thus, a truly "optimal" respiratory response must be considered in the broader context of multiple physiological, behavioral, and defensive functions of the respiratory apparatus, as well as the reflex and adaptive interactions of the respective sensory pathways (Fig. 5).

Oscillation, Randomness, and Chaos

A prerequisite for any adaptive system is that the transmitted signals must be "persistently exciting[48]," i.e., they must constantly be innovative. In the respiratory system, phasic oscillations in chemical and vagal inputs associated with tidal breathing provide some naturally self-innovative signals for neural adaptation. In addition, ventilatory patterns exhibit both cyclic and random fluctuations over several time scales ranging from seconds and minutes to hours and days.[71] In vagal intact animals some of these fluctuations may be due to deterministic chaos.[72] It has been suggested that chaotic neural patterns may be crucial for the brain's ability to adapt to changes[73] and to perceive and recognize distinct objects.[74] Thus, an array of persistently exciting signals is (theoretically) available to the respiratory controller for associative adaptation. Further studies are needed to establish the possible significance of these fluctuating signals in respiratory optimization and adaptation.

CONCLUDING REMARKS

The assumption that respiratory control is a simple reflex phenomenon has permeated the physiology literature for over a century. In the past, the optimization hypothesis has not received equal attention owing to methodological and conceptual difficulties. In this report I propose that the optimization hypothesis, in a broad sense, could be a general and cohesive model for describing a wide range of seemingly unrelated respiratory phenomena. This is not to suggest, however, that the hypothesis would prevail in all circumstances; indeed it is conceivable that many respiratory behaviors may depart from optimality under specific physiological conditions. While this certainly indicates that the optimization model has intrinsic limitations (as do all models), one should not discount on this basis its potential usefulness - as an alternative to the reflex model - in designing and interpreting new experiments. As J.M. Diamond nicely puts it:[75] "*Many biologists react instinctively to optimality models as do bulls to red flags.... Optimality criteria are not proposed in the belief that animals' bodies will prove economically designed throughout. Instead, they are proposed as a null hypothesis of biological design in order to detect when these simple economy-based expectations break down, and thus to be able to analyze the reasons for the breakdown. Without such a null hypothesis, one would not even know that there was a miseconomy of design to be explained.*"

One useful application of such a null hypothesis is in providing insights into the nature of various biological adaptations that underlie many physiological and evolutionary processes. The respiratory optimization model should be a valuable guide for future experimental studies of the neuronal adaptation mechanisms. As emphasized by Parker and Maynard Smith,[1] "*Optimization models... serve to improve our understanding about adaptations, rather than to demonstrate that natural selection produces optimal solutions.*"

It is perhaps only with this broad perspective that the physiological mechanisms underlying the integrative control of breathing may be better understood someday.

ACKNOWLEDGEMENT

This work was supported in part by National Institutes of Health grant HL30794.

REFERENCES

1. G.A. Parker and J. Maynard Smith, Optimality theory in evolutionary biology, *Nature* 348:27 (1990).
2. M.A. Savageau, "Biochemical Systems Analysis," Addison-Wesley, Reading, MA (1976).
3. D. Cohen and H. Parnas, An optimal policy for the metabolism of storage materials in unicellular algae, *J. Theoret. Biol.* 56:1 (1977).
4. J.R. Merkt, J.E. Peters, and C.R. Taylor, Running cheaply: an adaptation for desert life? *Physiologist* 33:A61 (1990).
5. S.M. Yamashiro, J.A. Daubenspeck, and F.M. Bennett, Optimal regulation of left ventricular ejection pattern, *Appl. Math. Comput.* 5:41 (1979).
6. F. Rohrer, Physiologie der Atembewegung, *in*: "Handbuch der normalen und path. Physiologie," Vol. 2, A.T.J. Bethe et al., eds., Springer, Berlin (1925).
7. E.J.M. Campbell, E. Agostoni, and J. Newsom Davis, "The Respiratory Muscles: Mechanics and Neural Control," Saunders, Philadelphia (1970).
8. M.J. Purves, What do we breathe for? *in*: "Central Nervous Control Mechanisms in Breathing," C. von Euler and H. Langercrantz, eds., Pergermon, NY (1979).
9. F.S. Grodins and S.M. Yamashiro, What is the pattern of breathing regulated for? *ibid.*
10. E.R. Weibel, C.R. Taylor, and H. Hoppeler, The concept of symmorphosis: a testable hypothesis of structure-function relationship, *Proc. Natl. Acad. Sci. USA* 88:10357 (1991).
11. C.S. Poon, Optimal control of ventilation in hypoxia, hypercapnia and exercise, *in*: "Modelling and Control of Breathing," B.J. Whipp and D.M. Wiberg, eds., Elsevier, NY (1983).
12. C.S. Poon, Optimality principle in respiratory control, *Proc. 2nd Am. Control Conf.* 2:36 (1983).
13. C.S. Poon, Ventilatory control in hypercapnia and exercise: optimization hypothesis, *J. Appl. Physiol.* 62:2447 (1987).
14. C.S. Poon, S.L. Lin, and O.B. Knudson, Optimization character of inspiratory neural drive, *J. Appl. Physiol.* 72(5): (1992). (in press)
15. F.S. Grodins, Analysis of factors concerned in regulation of breathing in exercise, *Physiol. Rev.* 30:220 (1950).
16. K. Wasserman, B.J. Whipp, and R. Casaburi, Respiratory control during exercise, *in*: "Handbook of Physiology: The Respiratory System," Sect. 3, Vol. II, N.S. Cherniack and J.G. Widdicombe, eds., Am. Physiol. Soc., Bethesda, MD (1986).
17. D.J.C. Cunningham, P.A. Robbins, and C.B. Wolff, Integration of respiratory responses to changes in alveolar partial pressures of CO_2, O_2 and in arterial Ph, *ibid.*
18. M.K. Younes and J.E. Remmers, Control of tidal volume and respiratory frequency, *in*: "Regulation of Breathing," Part I, T.F. Hornbein, ed., Marcel Dekker, NY (1981).
19. D.W. Richter, D. Ballantyne, and J.E. Remmers, How is the respiratory rhythm generated? *News Physiol. Sci.* 1:109 (1986).
20. N.C. Heglund and G.A. Cavagna, Mechanical work, oxygen consumption, and efficiency in isolated frog and rat muscle, *Am. J. Physiol.* 253:C22 (1987).

21. D.A. Syme, Passive viscoelastic work of isolated rat, Rattus Norvegicus, diaphragm muscle, *J. Physiol.* 424:301 (1990).
22. T.J. Sejnowski, C. Koch, and P.S. Churchland, Computational neuroscience, *Science* 241:1299 (1988).
23. S.G. Lisberger and T.A. Pavelko, Brain stem neurons in modified pathways for motor learning in the primate vestibulo-ocular reflex, *Science* 242:771 (1988).
24. E.R. Kandel, "A Cell-Biological Approach to Learning," Grass Lecture Monograph 1, Soc. Neuroscience, Bethesda, MD (1978)
25. F.L. Eldridge and D.E. Millhorn, Oscillation, gating, and memory in the respiratory control system, *in*: "Handbook of Physiology," Sect. 3, Vol. II, N.S. Cherniack and J.G. Widdicombe, eds., Am. Physiol. Soc., Bethesda, MD (1986).
26. D.E. Millhorn, Stimulation of *raphe* (*obscurus*) nucleus causes long-term potentiation of phrenic nerve activity in cat, *J. Physiol.* 381:169 (1986).
27. P.G. Wagner and F.L. Eldridge, Development of short-term potentiation of respiration, *Respir. Physiol.* 83:129, 1991.
28. K.L. Magleby, Synaptic transmission, facilitation, augmentation, potentiation, depression, *in*: "Encyclopedia of Neuroscience," G. Edelman, ed., Vol.2, Biekhauser, Boston (1987).
29. J. Champagnat, M. Denavit-Saubié, K. Grant, and K.F. Shen, Organization of synaptic transmission in the mammalian solitary complex, studied *in vitro*, *J. Physiol.* 381:551 (1986).
30. R.G.M. Morris, Synaptic plasticity, neural architecture and forms of memory, *in*: "Brain Organisation and Memory," J.L. McGaugh, N.M. Weinberger, and G. Lynch, eds., Oxford Univ. Press (1990).
31. J.C. Houk, Control strategies in physiological systems, *FASEB J.* 2:97 (1988).
32. C.S. Poon, Optimization behavior of brainstem respiratory neurons: a cerebral neural network model, *Biol. Cybern.* 66:9 (1991).
33. J.H. Byrne, Cellular analysis of associative learning, *Physiol. Rev.* 67:329 (1987).
34. G.S. Mitchell, M.A. Douse, and K.T. Foley, Receptor interactions in modulating ventilatory activity, *Am. J. Physiol.* 259:R911 (1990).
35. T.H. Brown, E.W. Kairiss, C.L. Keenan, Hebbian synapses: biophysical mechanisms and algorithms, *Annu. Rev. Neurosci.* 13:475 (1990).
36. D.O. Hebb, "The Organization of Behavior," Wiley, NY (1949).
37. M.A. Gluck and R.F. Thompson, Modeling the neural substrates of associative learning and memory: a computational approach, *Psych. Rev.* 94:176 (1987).
38. C.S. Poon, Adaptive neural network that subserves optimal homeostatic control of breathing, *Ann. Biomed. Engr.*, Special issue in honor of Dr. F.S. Grodins (in press).
39. G.S. Mitchell, C.A. Smith, and J.A. Dempsey, Changes in the $\dot{V}_I$-$\dot{V}CO_2$ relationship during exercise in goats: possible role of carotid bodies, *J. Appl. Physiol.* 57:1894 (1984).
40. C.S. Poon and J.G. Greene, Control of exercise hyperpnea during hypercapnia in humans, *J. Appl. Physiol.* 59:792 (1985).
41. G.D. Swanson and P.A. Robbins, Optimal respiratory controller structures, *IEEE Trans. Biomed. Engr.* BME33:677 (1986).
42. Y. Oku, G.M. Saidel, T. Chonan, M.D. Altose, and N.S. Cherniack, Sensation and control of breathing: a dynamic model. *Ann. Biomed. Engr.* 19:251 (1991).
43. P.K. Stanton and T.J. Sejnowski, Associative long-term depression in the hippocampus induced by hebbian covariance, *Nature* 339:215 (1989).
44. W.S. Yamamoto, Mathematical analysis of the time course of alveolar CO_2, *J. Appl. Physiol.* 15:215 (1960).
45. C.S. Poon, Potentiation of exercise ventilatory response by airway CO_2 and dead space loading, *J. Appl. Physiol.* (in press).
46. D.A. Sidney and C.S. Poon, Adaptive behavior in the respiratory chemoreflex response, *Proc. 17th Northeast Bioeng. Conf.* 17:66 (1991).

47. T. Poggio, Biophysics of computation, *in*: "Neuroscience in the Twenty-first Century: New Perspectives and Horizons," Georgetown Univ. Bicentennial Symp., Washington, DC (1989).
48. K.J. Åström and B. Wittenmark, "Adaptive Control," Addison-Wesley, Reading, MA (1989).
49. I.P. Priban and W.F. Fincham, Self-adaptive control and the respiratory system, *Nature* 208:339 (1965).
50. G.D. Swanson, Redundancy structures in respiratory control, *in*: "Control of Breathing and its Modeling Perspective," Y. Honda, Y. Miyamoto, K. Konno, and J. Widdicombe, eds. (1990). (this volume)
51. J. von Neumann, Probabilistic logics and the synthesis of reliable organisms from unreliable components, *in*: "Automata Studies," C.E. Shannon and J. McCarthy, eds., Princeton Univ. Press, Princeton, NJ (1956).
52. F.L. Eldridge, P. Gill-Kumar, and D.E. Millhorn, Input-output relationships of central and peripheral respiratory drives involved in respiration in cats, *J. Physiol. London* 311:81 (1981).
53. J.M. Adams and M.L. Severns, Interaction of chemoreceptor effects and its dependence on the intensity of stimuli, *J. Appl. Physiol.* 52:602 (1982).
54. D.F. Speck and E.R. Beck, Respiratory rhythmicity after extensive lesions of the dorsal and ventral respiratory groups in the decerebrate cat, *Brain Research* 482:387 (1989).
55. J. Oren, C.J.L. Newth, C.E. Hunt, R.T. Brouillette, R.T. Bachand, and D.C. Shannon, Ventilatory effects of almitrine bismesylate in congenital central hypoventilation syndrome, *Am. Rev. Respir. Dis.* 134:917 (1986).
56. J.C. Smith, H.H. Ellenberger, K. Ballanyi, D.W. Richter, and J.L. Feldman, Pre-Bötzinger complex: a brainstem region that may generate respiratory rhythm in mammals, *Science* 254:726 (1991).
57. A. Selverston and P. Mazzoni, Flexibility of computational units in invertebrate CPGs, *in*: "The Computing Neuron," R. Durbin, C. Miall, and G. Mitchison, eds., Addison-Wesley, Reading, MA (1989).
58. J.L. Feldman, J.C. Smith, H.H. Ellenberger, C.A. Connelly, G. Liu, J.J. Greer, A.D. Lindsay, and M.R. Otto, Neurogenesis of respiratory rhythm and pattern: emerging concepts, *Am. J. Physiol.* 259:R879 (1990).
59. J.J. Hopfield, Neural networks and physical systems with emergent collective computational abilities, *Proc. Natl. Acad. Sci. USA* 79:2554 (1982).
60. W.K. Milsom, Mechanoreceptor modulation of endogenous respiratory rhythms in vertebrates, *Am J. Physiol.* 259:R898 (1990).
61. J. Milic-Emili and M.M. Grunstein, Drive and timing components of ventilation, *Chest*, 70 Suppl.:131 (1976).
62. W.A. Karczewski and J.R. Romaniuk, Neural control of breathing and central nervous system plasticity, *Acta Physiol. Pol. Supl.* 20:1 (1980).
63. M.A. Grippi, A.I. Pack, R.O. Davies, and A.P. Fishman, Adaptation to reflex effects of prolonged lung inflation, *J. Appl. Physiol.* 58:1360 (1985).
64. M. Younes, J. Baker, J.E. Remmers, Temporal changes in effectiveness of an inspiratory inhibitory electrical pontine stimulus, *J. Appl. Physiol.* 62:1502 (1987).
65. C. Roussos, Ventilatory muscle fatigue governs breathing frequency, *Bull. Eur. Physiol. Respir.* 20:445 (1984).
66. J.A. Neubauer, J.E. Melton, and N.H. Edelman, Modulation of respiration during brain hypoxia, *J. Appl. Physiol.* 68:441 (1990).
67. K. Murphy, A. Mier, L. Adams, and A. Guz, Putative cerebral control involvement in the ventilatory response to inhaled CO_2 in conscious man, *J. Physiol. (London)* 420:1 (1990).

68. F. Haas, S. Distenfeld, and K. Axen, Effects of perceived musical rhythm on respiratory pattern, *J. Appl. Physiol.* 61:1185 (1986).
69. J. Gallego and P. Perruchet, Classical conditioning of ventilatory responses in humans, *J. Appl. Physiol.* 70:676 (1991).
70. J. Gallego, J. Ankaoua, M. Lethielleux, B. Chambille, G. Vardon, and C. Jacquemin, Retention of ventilatory pattern learning in normal subjects, *J. Appl. Physiol.* 61:1 (1986).
71. M.J. Tobin, M.J. Mador, S.M. Guenther, R.F. Lodato, and M.A. Sackner, Variability of resting drive and timing in healthy subjects, *J. Appl. Physiol.* 65:309 (1988).
72. M.P. Sammon and E.N. Bruce, Vagal afferent activity increases dynamical dimension of respiration in rats, *J. Appl. Physiol.* 70:1748 (1991).
73. A.L. Goldberger, D.R. Rigney, J. Mietus, E.M. Antman, and S. Greenwald, Nonlinear dynamics in sudden cardiac death syndrome: heart rate oscillations and bifurcations, *Experientia* 44:983 (1988).
74. C.A. Skarda and W.J. Freeman, How brains make chaos in order to make sense of the world, *Behav. and Brain Sci.* 10:161 (1987).
75. J.A. Diamond, The red flag of optimality, *Nature* 355:204 (1992).

EFFECTS OF CHANGES IN VENTILATORY PATTERN DURING ALTITUDE ACCLIMATIZATION

Michael C.K. Khoo

Biomedical Engineering Dept., University of Southern California, Los Angeles, CA 90089

INTRODUCTION

The frequent occurrence of periodic breathing (PB) during sleep in sojourners to high altitude has been well documented[1-6]. PB is thought to be most pronounced during the first few nights at altitude and to diminish over time as the process of acclimatization takes its course[4]. However, in recent studies conducted at extreme altitudes[3,5,6], it was found that PB and recurrent apneas persisted even after the subjects tested had been at the altitudes in question for more than 3 weeks. Weil et al.[2] reported the common occurrence of "undulating respirations of varying amplitude but without true apnea" in long-term high-altitude residents. Thus, it is probably safe to conclude that there is a tendency for the regularization of breathing pattern with altitude acclimatization, although in some individuals PB is never completely eliminated.

Although it is generally believed that the alternating hyperpneas and apneas of PB are detrimental to gas exchange, this notion is based on anecdotal comparisons with uniform tidal breathing without careful matching of the average ventilation levels. Thus, while there is no doubt that large fluctuations in blood gases accompany PB, it is less clear whether on average gas exchange is impaired. Furthermore, during sleep, the work of breathing probably accounts for a substantial portion of total oxygen (O_2) consumption. Therefore, it is also pertinent to ask whether the occurrence of PB is associated with increased mechanical costs that place greater demands on O_2 consumption. Finally, it would be useful to determine how these changes in chemical and mechanical costs are influenced by the process of altitude acclimatization.

To address the above questions, we employed a computational model to compare the gas exchange and mechanical consequences of uniform tidal breathing with those of a variety of PB patterns. The latter group ranged from patterns with hyperpneic and hypopneic phases (without apnea) to those with very short hyperpneas and long apneas.

METHODS

The model employed in a previous study[7] was extended to include O_2 effects. Gas exchange was assumed to occur in a homogeneous alveolar chamber connected in series to a dead space that consisted of five equal-sized well-mixed compartments. The volume of the alveolar chamber increased and decreased in synchrony with the breathing cycle. The number of dead-space compartments was chosen so that a simulated Fowler washout maneuver

Control of Breathing and Its Modeling Perspective, Edited by
Y. Honda *et al.*, Plenum Press, New York, 1992

produced a volume estimate that approximated the value of 150 ml employed in the model. Instantaneous and complete equilibration of pulmonary end-capillary and alveolar gas partial pressures were assumed. The blood CO_2 and O_2 dissociation relations, which included Bohr and Haldane effects, were the same as those described by Khoo[8]. Arterial P_{CO_2} (P_{aCO_2}) and arterial O_2 saturation (S_{aO_2}) at the level of the peripheral chemoreceptors were computed from pulmonary end-capillary values after accounting for the effects of convective transport and mixing in the heart and vasculature. The differential equations of the model were solved numerically on a 80386-based microcomputer. Parameter values were chosen to represent a sleeping healthy adult in the first week of acclimatization at 4300 m. The effective lung volumes for CO_2 and O_2 at end-expiration were assumed to be 3 and 2.5 liters, respectively. Body stores for CO_2 and O_2 were taken to be 15 and 6 liters, respectively. Cardiac output was assumed to have been restored to the sea-level value of 6 l/min after 1 day of acclimatization, while bicarbonate level was decreased by 12%.

Simulations with the above model were performed in the following way. In each case, a the model was "driven" by a uniform tidal breathing pattern or a PB pattern until the predicted blood gases attained the same values at the same phase in each cycle. The intra-breath flow pattern assumed a sinusoidal time-course; this is a reasonable assumption for breathing under resting conditions. The tidal volumes (V_T) of consecutive breaths in the ventilatory phase of each PB cycle were modulated in the form of a half-sinusoid. Respiratory cycle duration (T) was assumed constant from breath to breath. The PB cycle duration (T_c) was also assumed to be constant. PB patterns were classified according to:

(a) the number of breaths, N_B, contained in the ventilatory phase of each PB cycle;

(b) the cycle-averaged ventilation, $\dot{V}_E$, defined in the following manner:

$$\dot{V}_E = (60/T_c)V_T(i) \tag{1}$$

In the case of regular breathing, $\dot{V}_E$ was equal to the minute ventilation.

A previous study[9] has shown that the oxygen consumption rate per liter of air ventilated is more strongly correlated with mean pressure (P_I) of the inspiratory muscles than with work-rate of breathing. Thus, as an index of the mechanical cost of PB breathing, we employed the "cycle-averaged pressure cost", defined as:

$$Jp = \{P_I(i) + \cdots + P_I(N_B)\}T/T_c \tag{2}$$

where T_c represents the duration of one PB cycle. P_I for each breath was computed from the following expression:

$$P_I = 0.5KV_T + 2K_1V_T/T + \pi^2K_2V_T^2/2T^2 + 2RV_T/T(1 + 4\pi^2\tau^2/T^2) \tag{3}$$

The first term in the right-hand side of the above expression represents the elastic contribution, with K being the respiratory system elastance (= 14.5 cm H_2O/liter). The second and third terms reflect the laminar and turbulent flow-resistive contributions. The respective Rohrer constants, K_1 and K_2, were 1.9 cm H_2O.s/liter and 0.52 cm H_2O.s/liter^2 at sea level. However, at 4300 m, K_1 and K_2 became 1.83 and 0.24, respectively, the reduced values reflecting the decrease in air viscosity and density that accompany the decrease in barometric pressure[10]. Viscoelastic pressure losses in the lung tissues and chest wall were accounted for by the last term[11]. Here, R represents the resistance associated with thoracic viscoelasticity, while τ is the viscoelastic time constant for the lungs and chest wall combined. R and τ were assigned values of 5.86 cm H_2O.s/liter and 1 s, respectively[11].

The model was constrained to follow the same time-course of acclimatization as that described previously by White et al.[4]. During the first night at altitude (N1), mean ventilation was changed from the normoxic level of 7.2 l/min to 8.2 l/min. On the fourth night (N4),

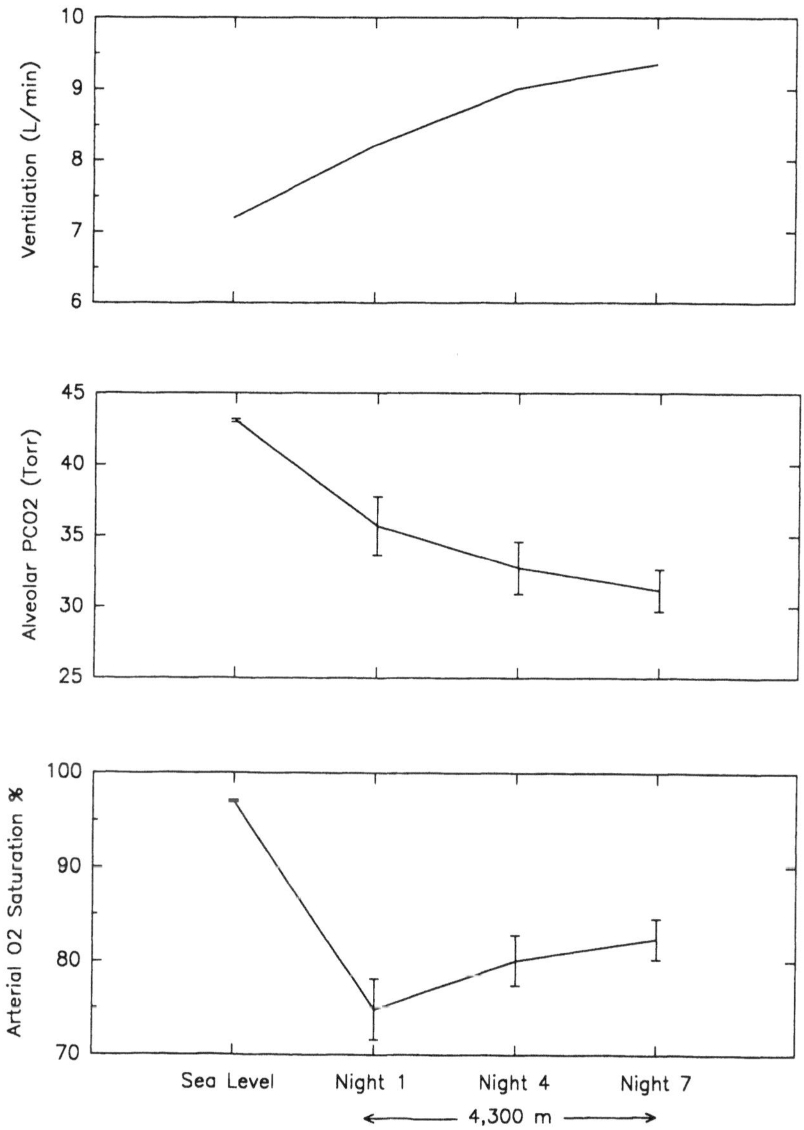

Fig.1. Changes in ventilation, P_{aCO_2} and S_{aO_2} accompanying acclimatization to 4300 m, as predicted by the model.

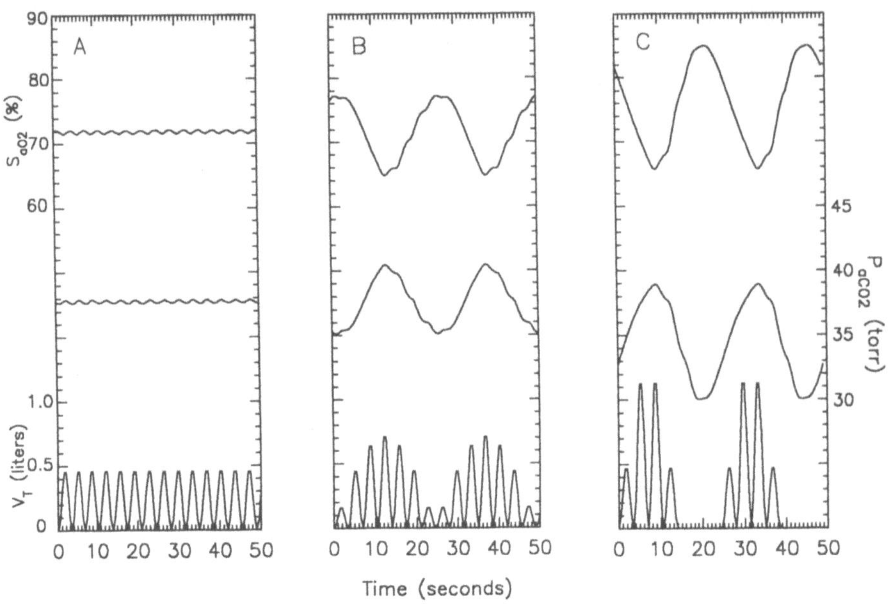

Fig.2. Effects on gas exchange of different ventilatory patterns: (A) tidal breathing; (B) non-apneic periodic breathing (N_B=7); and (C) periodic breathing with apnea (N_B=4).

$\dot{V}_E$ was assumed to have increased to 9 l/min. $\dot{V}_E$ during the seventh night (N7) was further increased to 9.35 l/min. At each of these nights, comparisons were made among the various patterns of PB and tidal breathing.

RESULTS

As Fig.1 clearly indicates, the increase in ventilatory drive accompanying altitude acclimatization led to a concommitant decrease in P_{aCO_2}. This ventilatory adjustment also led to a continual improvement of mean S_{aO_2} following the abrupt drop that accompanied the initial ascent to 4300 m. These changes in blood gas levels are consistent with those reported by White et al.[4]. The error bars in the figure reflect the different values obtained from simulations with various types of breathing patterns.

A comparison of the gas exchange consequences of ventilating with three types of breathing patterns at 4300 m is shown in Fig.2. With tidal breathing, the fluctuations in blood gas levels produced by the breathing cycle were greatly attenuated as a result of cardiovascular mixing and smearing (Fig.2A). Mean S_{aO_2} and P_{aCO_2} were approximately 72% and 38 torr, respectively. For the same cycle-averaged ventilation, the PB pattern with no apnea (ie. N_B=7) produced swings in S_{aO_2} that exceeded 12 points (Fig.2B). Mean S_{aO_2} was slightly lower and mean P_{aCO_2} slightly higher than the corresponding values in regular breathing. The swings in S_{aO_2} and P_{aCO_2} associated with the PB pattern with apnea (N_B=4) were even larger (Fig.2C). However, mean S_{aO_2} was about 4% higher than the case with regular breathing, even though mean ventilation was maintained at the same level.

Figure 3A shows the average S_{aO_2} obtained with each of the breathing patterns at each of the 3 nights of study (N1, N4 and N7). Starting with regular tidal breathing (labelled 'U'), progressing along the horizontal axis to the right implies increasing temporal nonuniformity of ventilation (ie. decreasing N_B). Although the non-apneic PB pattern (N_B=7) led to a decrease in mean S_{aO_2}, the apneic patterns actually produced higher mean saturation levels. At each of the 3 nights studied, the highest mean S_{aO_2} was achieved with a PB pattern with N_B=3 or 4.

A comparison of the values attained at the nadir of the S_{aO_2} time-course is shown in Fig.3B. As one would have expected, the highest minimum S_{aO_2} levels were achieved with the regular breathing pattern. The minimum S_{aO_2} levels associated with the PB patterns were significantly lower. However, it was somewhat surprising to find that the patterns with N_B=4 and 5 produced the highest minimum S_{aO_2} values. Figure 4 shows the pressure costs resulting from the various breathing patterns and nights of acclimatization. The numerical value of the pressure cost for each case has been normalized with respect to the value of Jp during regular breathing at sea level. Although $\dot{V}_E$ at N1 (8.2 l/min) was significantly higher than $\dot{V}_E$ at sea level (7.2 l/min), the increase in Jp was quite small because of compensatory effects due to the reductions in air density and viscosity. In each of the 3 nights of acclimatization, increasing temporal nonuniformity of ventilation led to increasing pressure cost. However, Jp was by far more sensitive to changes in $\dot{V}_E$ that accompanied the progression of altitude acclimatization.

DISCUSSION

It was initially intriguing to find that mean S_{aO_2} actually improved with the apneic type of PB patterns. However, we realized later that this result was not difficult to explain. With uniform tidal breathing, a portion of the ventilation is 'wasted' on dead space. However, with PB, dead space ventilation is minimized by increasing the tidal volumes of the breaths in the hyperpneic phase and not ventilating at all during the subsequent apnea. This turns out to be an efficient way of increasing the effective alveolar ventilation for a given $\dot{V}_E$. By contrast, the nonapneic type of PB pattern is a very inefficient means of generating alveolar ventilation since the tidal volumes of the hypopneic breaths barely clear the dead space.

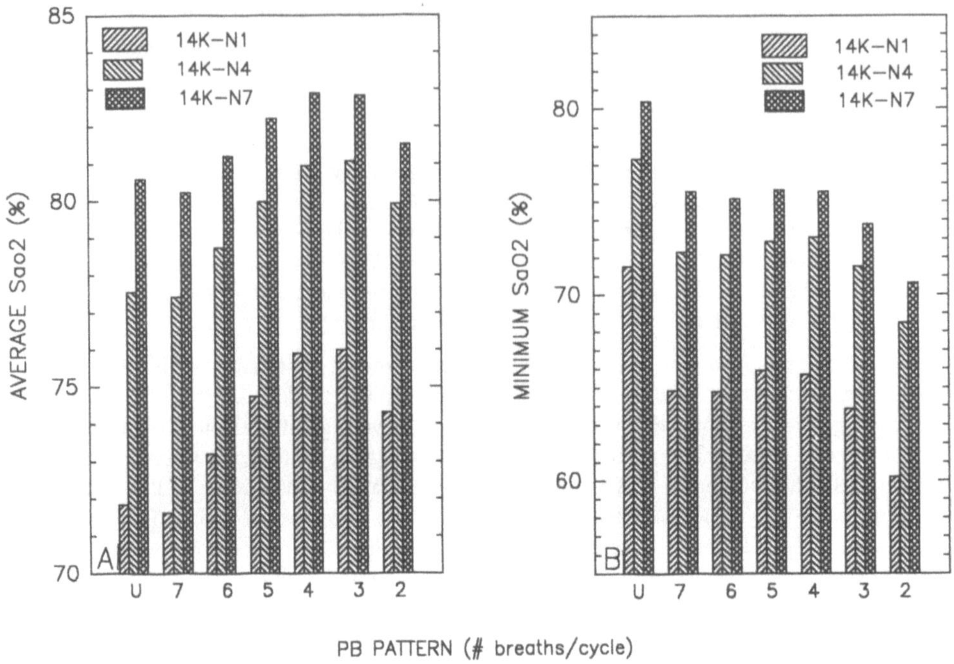

Fig.3. Effects of different ventilatory patterns on (A) average S_{aO_2}, and (B) minimum S_{aO_2}, on nights 1, 4 and 7 during acclimatization to 4300 m.

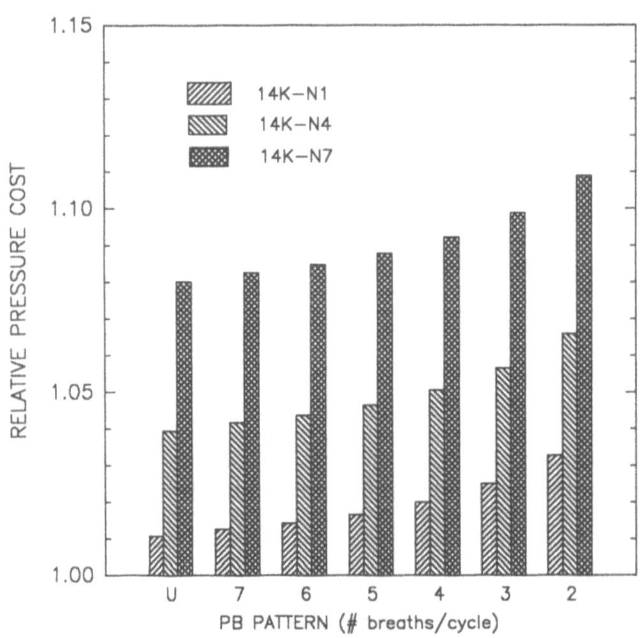

Fig.4. Effects of different ventilatory patterns on relative pressure cost on nights 1, 4 and 7 during acclimatization to 4300 m.

With PB, some savings in pressure cost accrue from the apneic phases. However, the increased pressure cost associated with the high inspiratory flow-rates of the hyperpneic breaths more than offset these savings. Thus, the net effect is a continual increase in pressure cost with increasing temporal nonuniformity of ventilation.

How do we interpret the results of this analysis in the light of what we know about altitude acclimatization? Upon ascent to high altitude, the immediate challenge encountered by the respiratory control system is hypoxemia. The chemoreflex-mediated increase in ventilatory drive acts to raise the level of oxygenation but this provides only partial compensation. At the expense of a small increase in pressure cost, ventilating with an apneic type of PB pattern enables the system to raise the average S_{aO_2} level by up to 4% beyond that which can be achieved by uniform tidal breathing at the same $\dot{V}_E$. Thus, a ventilatory pattern with a few large breaths followed by apnea appears to provide the best short-term solution to the problem of optimizing O_2 transport for a given level of ventilatory drive. Our analysis has shown that the PB pattern with N_B=4 achieves the highest mean S_{aO_2} and highest minimum S_{aO_2}. It is interesting to note the observation made by Reite et al.[2] that the typical PB pattern in their sleeping subjects at 4300 m consisted of "four or five rapid breaths (hyperpnea), followed by a 10-20 sec period of no breathing (apnea)". With the gradual increase in ventilatory drive that accompanies acclimatization, S_{aO_2} levels are also improved while pressure cost continues to increase. After a week of acclimatization (N7), the advantage of ventilating with a PB pattern of N_B=4 is significantly reduced: mean S_{aO_2} would only be increased by 2% beyond that achieved through uniform tidal breathing at the same $\dot{V}_E$. At the same time, ventilating with the PB pattern necessitates a further 1% increase in pressure cost. Thus, regularization of the breathing pattern becomes a more beneficial alternative, as this route of action would also lead to the elimination of fluctuating blood gas levels.

ACKNOWLEDGEMENTS

This work was supported in part by NIH Grants RR-01861 and an American Lung Association Career Investigator Award. The author is also the recipient of a NIH Research Career Development Award (HL-02536).

REFERENCES

1. Reite, M., D. Jackson, R.L. Cahoon and J.V. Weil. Sleep physiology at high altitude. Electroenceph. clin. Neurophysiol. 32:701-705, 1972.

2. Weil, J.V., M.H. Kryger and C.H. Scoggin. Sleep and breathing at high altitude. In: Sleep Apnea Syndromes, ed. C. Guilleminault and W.C. Dement, New York: Alan R. Liss, 1978, pp. 119-135.

3. Lahiri, S., K. Maret and M.G. Sherpa. Dependence of high altitude sleep apnea on ventilatory sensitivity to hypoxia. Respir. Physiol. 52:281-301, 1983.

4. White, D.P., K. Gleeson, C.K. Pickett, A.M. Rannels, A. Cymerman and J.V. Weil. Altitude acclimatization: influence on periodic breathing and chemoresponsiveness during sleep. J. Appl. Physiol. 63:401-412, 1987.

5. West, J.B., R.M. Peters Jr., G. Aksnes, K.L. Maret, J.S. Milledge and R.B. Schoene. Nocturnal periodic breathing at altitudes of 6,300 and 8050 m. J. Appl. Physiol. 61:280-287, 1986.

6. Anholm, J.D., A.C.P. Powles, R. Downey III, C.S. Houston, J.R. Sutton, M.H. Bonnet and A. Cymerman. Operation Everest II: Arterial oxygen saturation and sleep at extreme simulated altitude. Am. Rev. Respir. Dis. (in press).

7. Khoo, M.C.K. A model-based evaluation of the single-breath CO2 ventilatory response test. J. Appl. Physiol. 68:393-399, 1990.

8. Khoo, M.C.K. The noninvasive estimation of cardiopulmonary parameters. Ph.D. Dissertation, Boston, MA, 1981.

9. McGregor, M., and M. Becklake. The relationship of oxygen cost of breathing to respiratory mechanical work and respiratory force. J. Clin. Invest. 40:971-980, 1961.

10. Cotes, J.E. Ventilatory capacity at altitude and its relation to mask design. Proc. R. Soc. Lond. [Biol.] 143:32-39, 1954.

11. Milic-Emili, J. Work of breathing. In: The Lung: Scientific Foundations, ed. R.G. Crystal, J.B. West et al., pp.1065-1075, Raven Press, New York, 1991.

OPTIMAL CONTROL ANALYSIS OF THE BREATHING PATTERN BY USING THE WBREPAT COMPUTER SOFTWARE

Raimo P. Hämäläinen

Systems Analysis Laboratory
Helsinki University of Technology
02150 Espoo , Finland
email: raimo @ hut.fi

INTRODUCTION

Optimal control hypotheses about the control of breathing have been around for a long time. The early static models were relatively simple and they could be solved analytically. During the past twenty years we have been able to show that optimization concepts can be successfully used in more detailed dynamic descriptions of the respiratory system as well. However, the increasing need for complex computational solutions has left many physiologists out of the field. The models' solution algorithms and related software have traditionally not been easily accessible or transportable. This has been a major obstacle for wider experimental research aiming at the verification of the predictions and evaluation of the models' performance in different environments.

The aim of the **WBrePat**-Windows version of the **Breathing Pattern** Analysator software is to start a new era of distributable models and modelling software. It is also hoped that this revitalizes interest in optimal control modelling and in the different uses of these models in respiratory control research. WBrepat is available free of charge to interested researhers for noncommercial reseach and educational purposes.

THE GENERALIZED OPTIMAL CONTROL MODEL FOR INSPIRATION

The first version of the software is limited to the inspiratory phase of the hierarchical two-level model for the overall control of the mechanics of breathing. For a complete description of the overall model and a review of other related work see the references [1-6].

The software[7] is based on the modified criterion for the inspiratory period presented by Hämäläinen and Sipilä[5] which takes into account the decrease in the efficiency of muscular contraction. This is done by a weighting coefficient (multiplying the work rate term) which depends linearly on the instantaneous value of the driving pressure P(t):

$$J = \int_0^T [\ddot{V}^2(t) + \alpha(t) P(t) \dot{V}(t)]\, dt$$

$$\alpha(t) = \alpha_{11} + \alpha_{12} P(t)$$

Control of Breathing and Its Modeling Perspective, Edited by
Y. Honda *et al.*, Plenum Press, New York, 1992

After the substitution we get the following form for the criterion:

$$J = \int_0^T [\ddot{V}^2(t) + \alpha_{11} P(t)\dot{V}(t) + \alpha_{12}\dot{V}(t)P^2(t)]\,dt$$

In this software this criterion is for the first time solved for a general nonlinear system model, which also includes a nonlinear resistance in the airflow rate:

$$P(t) = KV(t) + R_1\dot{V}(t) + R_2\dot{V}^2(t)$$

We assume that a given fixed tidal volume, V_T, is inhaled starting from the initial FRC level, V_0. This gives the following natural boundary conditions:

$$V(0) = V_0 \qquad V(T) = V_T$$
$$\dot{V}(0) = 0 \qquad \dot{V}(T) = 0$$

The symbols used above are: P(t) is the total driving pressure; V(t) is the change in lung volume from the resting FRC level; the dot represents time derivative; V(t) with a dot above it is the airflow rate; V(t) with two dots represents the volume acceleration; V_O is the initial change from the resting FRC level; V_T is the tidal volume; K is the total elastance of the lung rib cage system; R_1 and R_2 are the linear and nonlinear components of the total resistance.

New minimun energy interpretation for the criterion

It is well known that if only the work rate term (pressure times flow rate) would be used in the integrand of the criterion, the solution would be a square shaped airflow pattern with an instantaneous initial jump to a constant airflow rate followed by a similar instantaneous final return to zero flow. In the author's previous papers the volume acceleration term was introduced to avoid these unnatural jumps. This improved the resulting predictions essentially. Earlier the new term was given the role of a penalty term which constraints the possible predictions to airflow patterns which do not have instantaneous jumps in the flow rate. However, more recently the author has been able to show (unpublished results) that a physiologically more relevant interpretation can be found if we use the present modified criterion together with a complete system model including also the mechanical inertia of the respiratory system (the coefficient of inertia being denoted by I):

$$P(t) = KV(t) + R_1\dot{V}(t) + R_2\dot{V}^2(t) + I\ddot{V}(t)$$

When the resulting optimal control model is solved one observes that the inertia term in the system equation and the square of the volume acceleration in the criterion create comparable terms into the equations defining the solutions. This means that we can, indeed, produce the same kind of airflow patterns even without having the volume acceleration term in the criterion. This does not happen, however, with the standard work rate model. You need the modified nonlinear work rate term too.

The essential new conclusion is that the model is, indeed, based on a minimum energy principle although it is mathematically implicitly formulated. The volume acceleration term in the criterion accounts for the energy losses related to overcoming the inertia in sudden changes in the flow rate. We need not describe it as a penalty term but a way of approximating these energy costs which are omitted due to the lacking inertia term in the system description. However, this observation needs further investigations. One needs to see what are the magnitudes and the weights of the related terms when realistic patterns are predicted and how do they correspond to the physiological data on the inertia of the respiratory system. This will be covered in a more complete article.

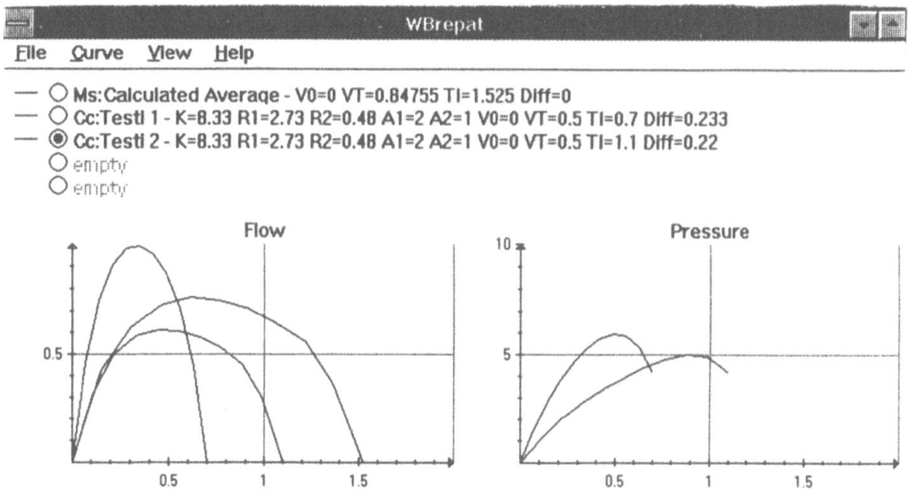

Figure 1. A WBrePat-window for comparing the model predictions and measurement data. The parameters and curves shown are only illustrative and do not represent the results of a real model fitting excercise.

INTERACTIVE MODELLING BY USING WBREPAT

The software[7] runs on DOS personal computers under the Windows 3.0 environment. It requires a VGA monitor and at least a 286 processor, the convenience of using the software can be greatly improved by a mathematics co-processor and a faster processor. A typical run time to solve the flow pattern for one set of parameters is five seconds on a 486 33MHz processor and the parameter estimation procedure takes some minutes to complete. These execution times are very fast when you consider the difficulty of the related mathematical problems. First to solve the optimal control problem we have a numerically very hard two-point boundary value problem for differential equations. It is solved iteratively by the so called multiple shooting method. In the estimation of parameters a Hooke-Jeeves search procedure is applied and the previous problem needs to be solved repeatedly at each step of the search.

User interface

The user interface is fully graphical and follows the Windows style, see Fig. 1. The main design philosophy has been to make WBrePat self guiding and so simple that any physiologist should be able to run it without difficulty. The printing facilities of the Windows environment are available, thus documents should be easy to create. The software has a context sensitive linked help feature. It allows the user to find more information and go across the topic levels from any help screen just by clicking on the related topic word in the text. The help system is also includes a complete description of the model theory and principles of the parameter estimation routines.

The basic modelling functions available in WBrePat are: analysis (averaging, grouping) and visualization of measurement data; generation and visualization of model predictions by any feasible combinations of respiratory parameters and model coefficients; estimation of the individual weighting coefficients giving the best fit to a specified set of measurements. The index of best fit is the integral of the difference between the curves (predicted-measured). You can also do the estimation so that one of the coefficients is restricted to a fixed value.

There are user friedly windows to change the parameters and to select different options for the presentation of the results. This should make the analysis fast and easy.

MODEL EVALUATION AND USE IN CLASSIFICATION AND TEACHING

One can directly see a number of different areas where WBrePat can be used. Of course the primary one is the possibility for researchers to independently test the predictions of the model under various environmental conditions. The ultimate goal is to test the predictions with the resistances and elastances estimated separately for each individual test subject. However, even comparisons based on estimated average values for the parameters are likely to be informative when the effects of changes in the environmental loading conditions are studied. Clearly, there is a wide range of most interesting experimental setups.

The principles of classifying breathing patterns is a question which can be approached in a new way by using this model and software. The early descriptive approaches have been based on external geometric properties of the airflow shapes like the relative rise times to reach peak flow. Another external way of describing the patterns is to use Fourier analysis and classify patterns by the Fourier coefficients[8]. The optimal control approach uses an internal approach as the differences in the system model can be eliminated from the description. The classification is done on the level of control criteria, more specifically in terms of the individual weighting parameters. The effects of environmental loads and individual differences in the mechanical structure can be eliminated by taking them into account already in the system model. Thus the classification indexes (if found to be constant) could possibly be reflecting a true deeper physiological constant the change of which could be a sign of a functional disorder and of potential diagnostic value. This is an interesting possibility which needs to be studied by extensive experimentation and model analysis. However, even the comparison of this method with the other ways of classifying breathing patterns is a relevant topic for future research.

The availability of a user friendly modelling software can also profit the teaching of modelling principles to physiologists and engineers. It can be used as a means of explaining the dynamics as well as the potential effects of changing the control principle. Of course one can also use the software as a general modelling or data analysis environment even without being devoted to the optimal control analysis.

REFERENCES

1. R. P. Hämäläinen and A. A. Viljanen, Modelling the respiratory airflow pattern by optimization criteria, *Biological Cybernetics* 29:143 (1978).
2. R. P. Hämäläinen and A. A. Viljanen, A hierarchical goal- seeking model of the control of breathing, Part I Model description, *Biological Cybernetics* 29:151 (1978).
3. R. P. Hämäläinen and A. A. Viljanen, A hierarchical goal- seeking model of the control of breathing, Part II Model performance, *Biological Cybernetics* 29:159 (1978).
4. R. P. Hämäläinen, Optimization of respiratory airflow, *in:* "Modeling and Control of breathing", B. J. Whipp and D. M. Wiberg, eds, Elsevier, New York (1983).
5. R. P. Hämäläinen and A. Sipilä, Optimal control of inspiratory airflow in breathing, *Optimal Control Applications and Methods* 5:177 (1984).
6. R. P. Hämäläinen, Respiratory systems:Optimal control, *in:* "Systems & Control Encyclopedia", M. G. Singh, ed., Pergamon Press, New York (1987).
7. R. P. Hämäläinen and H. Lauri, WBREPAT-Breathing Pattern Analysator Software ver. 1. 0, Systems Analysis Laboratory, Helsinki University of Technology, Espoo, Finland (1991)
8. G. Benchetrit, N. Blanc-Gras, B. Wuyam, T. Herve', and J. Demongeot, Influence of posture on the breathing pattern of adult humans at rest, paper presented at the 5th Oxford Meeting on Contol of Breathing, Fuji, (1991).

A MODEL FOR THE CONTRIBUTION OF LUNG RECEPTORS TO THE CONTROL OF BREATHING: POSITIVE AND NEGATIVE FEED-BACKS

C.P.M. van der Grinten and S.C.M. Luijendijk

Department of Pulmonology
University Hospital Maastricht
P.O. Box 5800
6202 AZ Maastricht
The Netherlands

INTRODUCTION

Experiments in dogs in which pulmonary and systemic circulations were controlled separately have shown that minute ventilation ($\dot{V}_E$) increased when the CO_2 content of pulmonary arterial blood was increased, while systemic arterial PCO_2 was kept constant. Similarly, $\dot{V}_E$ also increased when pulmonary blood flow was increased at constant PCO_2. These responses were abolished after bilateral vagal transection[10,20]. We present two positive feed-backs which may account for the increases in $\dot{V}_E$ observed in these experiments. Further, a negative feed-back is presented controlling the CO_2 related off-switch of inspiration.

THE FEED-BACKS

The Inspiratory Amplification Reflex

Loop A in figure 1 describes the positive feed-back which includes the inspiratory amplification reflex. The first relationship in this loop—increasing the output of inspiratory motoneurons leads to increased inspiratory flow ($\dot{V}_I$)—is trivial. The rest of the loop has generally been studied as a whole. In those studies, fully paralysed animals were ventilated mechanically using a pump triggered by the start of phrenic activity. Inflating dogs thus prepared with high $\dot{V}_I$ increased phrenic activity considerably[19]. We have shown[12] that in spontaneously breathing, anaesthetized cats the gain of this reflex loop is more or less constant over the entire range of ventilation from quiet breathing up to maximal ventilation. The gain was on average 1.08 for phrenic activity, which means that the inspiratory amplification reflex does not contribute much to phrenic activity. We were not able to quantify the amplification of intercostal activity properly. It may, however, have a much larger value[8].

In this feed-back we consider $\dot{V}_I$ as the appropriate stimulus, while rapidly adapting pulmonary stretch receptors (RAR) mediate the increase in $\dot{V}_I$. However, others[1,2,8] favour lung volume as the appropriate stimulus and slowly adapting pulmonary stretch receptors (SAR) as mediators of the response.

Control of Breathing and Its Modeling Perspective, Edited by
Y. Honda *et al.*, Plenum Press, New York, 1992

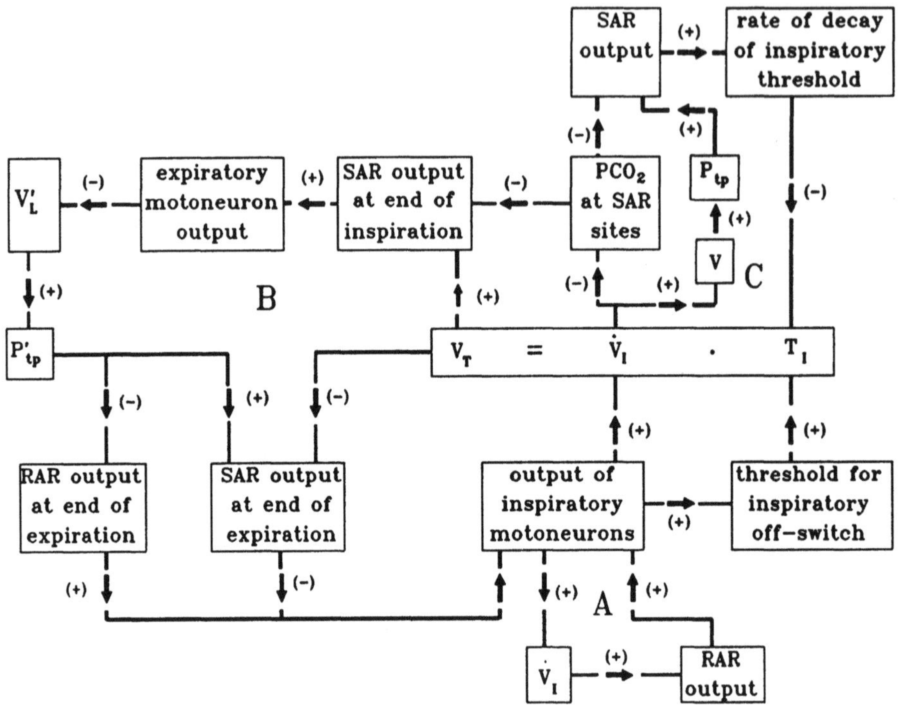

Figure 1. Diagram showing two positive feed-backs (A,B) and one negative feed-back (C). + and - signs indicate positive and negative correlation between two consecutive items, respectively

End-Expiratory Lung Volume and Inspiratory Activity

Feed-back B in figure 1 describes the relation between changes in end-expiratory lung volume (V'_L) and the inspiratory activity of the subsequent breath. Starting from the output of inspiratory motoneurons, the relation with $\dot{V}_I$ is again trivial. Increases in $\dot{V}_I$ are always accompanied by increases in tidal volume (V_T). This is the outcome of loop C, which will be discussed below.

The inverse relationship between V_T and V'_L depends on the respiratory stimulus used. In exercise V'_L decreases with increasing $\dot{V}_E$, whereas V'_L is nearly constant when breathing CO_2 enriched air[14]. A hypothesis how this difference might be caused is also indicated in loop B: increased V_T is accompanied by increased SAR activity. High SAR activity elicits expiratory activity in abdominal muscles[3], resulting in lowered V'_L[17]. The activities of SARs and RARs are reduced when PCO_2 in the airways is larger than 3 kPa[6]. Therefore, V'_L may be more reduced in exercise than during inspiratory CO_2 loading.

The direct, inverse relationship between V_T and end-expiratory SAR activity has been described by Mustafa and Purves[18]. Since these authors used artificially ventilated rabbits, the changes in end-expiratory SAR activity could not have been caused by changes in V'_L .

Changes in V'_L may be estimated from changes in end-expiratory transpulmonary pressure (P'_{tp}). They are proportional to each other, the ratio being lung compliance. We manipulated P'_{tp} and measured phrenic and intercostal activities, thus quantifying the remaining relations in this feed-back[11]. The resulting inverse relationships could be described by mono-exponential functions. E.g., $\ln(Phr_P/Phr_C)$ = Slope $\cdot \Delta P'_{tp}$ + Intercept, where Phr_P is phrenic activity following a change in P'_{tp} and Phr_C is the corresponding activity in control conditions. A similar equation described the response of intercostal activity. The (negative) slopes were steeper for intercostal than for phrenic activity, while intercepts were close to zero. An increase in P'_{tp} of 0.1 kPa (1 cm H_2O) resulted in a decrease in phrenic activity of about 12% in early inspiration, while the decrease in intercostal activity amounted to 42%, on average.

We have further shown[11] that the relationship between P'_{tp} and inspiratory activity is due to inhibition of this activity by SARs when P'_{tp} is increased, whereas with decreasing P'_{tp} excitation by RARs becomes more important.

The importance of the entire feed-back loop for hyperpneic breathing may be estimated by comparing the results of Henke et al.[14] obtained in humans with our results in cats, assuming that pressures are roughly comparable between species. They showed that $\dot{V}_I$ increased 3 times between rest and moderate exercise , while P'_{tp} decreased by 0.2 kPa. In cats, such a decrease in P'_{tp} would have increased phrenic activity by 35% and intercostal activity by 180% (on average). This means that this feed-back loop may contribute considerably to the sustenance of hyperpneic breathing.

The Influence of Airway CO_2 on the Inspiratory Off-Switch Mechanism

Feed-back C in figure 1 describes the dual relationship between inspiratory motoneuron activity and inspiratory duration (T_I), and the impact of PCO_2 at the SAR sites on this feed-back. Starting again from the inspiratory motoneuron activity and going upwards, the first relation with $\dot{V}_I$ is trivial. Increasing $\dot{V}_I$ increases inspired volume (V) and P_{tp} at a certain point in time during inspiration and so is the concomitant SAR output. Increased SAR output results in shortening of T_I . This part of feed-back C is in fact the classical Breuer-Hering inspiration-inhibitory reflex[15]. The addition of the PCO_2 at the SAR sites to this feed-back is derived from our experiments[13] and will be discussed below. The other part of feed-back C is derived from the work of von Euler and Trippenbach[9]. The hypothetical volume threshold has proved to be a very stimulating concept for research in this field. The threshold is thought to be set by central inspiratory activity and is proportional to the inspiratory motoneuron activity. This positive feed-back may explain why in long registrations of the flow pattern under constant conditions V_T is positively correlated with T_I , while mean inspiratory flow is almost constant[7]. Under such circumstances small breath-to-breath variations in the inspiratory threshold may cause these changes in T_I and V_T . However, V_T and T_I are negatively correlated when $\dot{V}_I$ is increased by exercise or hypercapnia[5].

In summary, an increase in $\dot{V}_I$ has two opposite effects on T_I. In practice, T_I decreases with increasing $\dot{V}_I$, but the relative decrease in T_I is less than the relative increase in $\dot{V}_I$, resulting in increased V_T .

The addition of the influence of airway CO_2 on the relations in feed-back C is new. In a series of experiments in spontaneously breathing, anaesthetized cats[13] we compared two loading techniques: conventional CO_2 loading (CCL) in which a constant level of CO_2 is inspired and a technique in which CO_2 is injected at the start of inspiration: slug CO_2 loading (SCL). It was demonstrated that T_I was longer for CCL than for SCL at the same arterial PCO_2 . This can be explained by assuming that for CCL the PCO_2 at the receptor sites is high during the entire inspiration, whereas it is low during the larger part of inspiration for SCL. Since CO_2 inhibits SAR activity[6], the decay of the inspiratory threshold will be slower when PCO_2 is high and as a result T_I is increased at the same $\dot{V}_I$. When $\dot{V}_I$ is increased by exercise, PCO_2 at the receptor sites will be decreased[16], which leads to shortening of T_I and increased breathing frequency. In our experiments a large difference in $\dot{V}_E$ was found: the slopes of the CO_2 response curves for SCL were—on average—two times larger than the slopes for the CCL response curves. This means that also $\dot{V}_I$ is higher during SCL compared to CCL at the same arterial PCO_2 . This finding may be explained by the connection between feed-backs B and C: the box "PCO_2 at the SAR sites". At the end of inspiration PCO_2 at the SAR sites is lower and, hence, SAR activity is higher during SCL than during CCL at the same V_T . Expiratory activity of abdominal muscles is increased when end-inspiratory SAR activity is increased[4]. This may cause a decrease in V'_L[17]. Following the rest of loop B, this will lead to an increase in $\dot{V}_I$.

CONCLUSIONS

Vagal lung receptors affect the control of the breathing in many aspects. They modulate the activities of inspiratory and expiratory muscles, the timing and amplitude of the breathing movements and the end-expiratory lung volume. The inverse CO_2 responsiveness of these receptors and the flow related CO_2 levels in the airways during inspiration imply, therefore, that these receptors may also contribute to the chemical control of breathing. In this respect, they may show not to be less important than central and peripheral chemoreceptors. The positive feed-backs presented may be important to sustain hyperpneic breathing, especially in exercise.

REFERENCES

1. Averill, D.B., W.E. Cameron and A.J. Berger, Monosynaptic excitation of dorsal medullary respiratory neurons by slowly adapting pulmonary stretch receptors. J. Neurophysiol. 52: 771-785 (1984).
2. Averill, D.B., W.E. Cameron and A.J. Berger, Neural elements subserving pulmonary stretch receptor-mediated facilitation of phrenic motoneurons. Brain Res. 346: 378-382 (1985).
3. Bishop, B., Abdominal muscles and diaphragm activities and cavity pressures in pressure breathing. J. Appl. Physiol. 18: 37-42 (1963).
4. Bishop, B. and H. Bachofen, Comparison of neural control of diaphragm and abdominal muscle activities in the cat. J. Appl. Physiol. 32: 798-805 (1972).
5. Clark, F.J. and C. von Euler, On the regulation of depth and rate of breathing. J. Physiol. 222: 267-295 (1972).
6. Coleridge, H.M., Coleridge, J.C.G. & Banzett, R.B., II. Effect of CO_2 on afferent vagal endings in the canine lung. Respir. Physiol. 34: 135-151 (1978).
7. Cunningham, D.J.C., Howson, M.G., Metias, E.F. & Petersen, E.S., Patterns of breathing in response to alternating patterns of alveolar carbon dioxide pressures in man. J. Physiol. 376: 31-45 (1986).
8. DiMarco, A.F., C. von Euler, J.R. Romaniuk and Y. Yamamoto, Positive feedback facilitation of external intercostal and phrenic inspiratory activity by pulmonary stretch receptors. Acta Physiol. Scand. 113: 375-386 (1981).
9. Euler, C. von and T. Trippenbach, Excitability changes of the inspiratory 'off-switch' mechanism tested by electrical stimulation in the nucleus parabrachialis in the cat. Acta Physiol. Scand. 97: 175-188 (1976).
10. Green, F.G. and M.I. Sheldon, Ventilatory changes associated with changes in pulmonary blood flow in dogs. J. Appl. Physiol.: Respirat. Environ. Exercise Physiol. 54: 997-1002 (1983).
11. Grinten, C.P.M. van der, W.R. de Vries, S.C.M. Luijendijk, Vagal modification of inspiratory activity during pressure breathing in cats. Europ. Respir. J. 2 Suppl 5: 311S (1989).
12. Grinten C.P.M. van der, W.R. de Vries, S.C.M. Luijendijk, Vagal amplification of phrenic activity at different levels of ventilation in spontaneously breathing cats. Europ. J. Appl. Physiol. 62: 49-55 (1991).
13. Grinten C.P.M. van der, E. Schoute, W.R. de Vries, S.C.M. Luijendijk, Conventional versus slug CO_2 loading and the control of breathing in anaesthetized cats. J. Physiology 445: 487-498 (1992).
14. Henke, K.G., M. Sharratt, D. Pegelow and J.A. Dempsey, Regulation of end-expiratory lung volume during exercise. J. Appl. Physiol. 64: 135-146 (1988)
15. Hering, E., Die Selbststeuerung der Athmung durch den Nervus vagus. Sitzungsber. Akad. Wiss. Wien 57: 672-677 (1868).
16. Luijendijk, S.C.M., Within-breath PCO_2 levels in the airways and at the pulmonary stretch receptor sites. J. Appl. Physiol.: Respirat., Environ. Exercise Physiol. 55: 1333-1340 (1983).
17. Martin, J.G. and A. De Troyer, The behavior of the abdominal muscles during inspiratory mechanical loading. Respir. Physiol. 50: 63-73 (1982).
18. Mustafa, M.E.K.Y. and M.J. Purves, The effect of CO_2 upon discharge from slowly adapting stretch receptors in the lungs of rabbits. Respir. Physiol. 16: 197-212 (1972).
19. Pack, A.I., R.G. DeLaney and A.P. Fishman, Augmentation of phrenic neural activity by increased rates of lung inflation. J. Appl. Physiol.: Respirat. Environ. Exercise Physiol. 50: 149-161 (1981).
20. Sheldon M.I. and F.G. Green, Evidence for pulmonary CO_2 chemosensitivity: effects on ventilation. J. Appl. Physiol.: Respirat. Environ. Exercise Physiol. 52: 1192-1197 (1982).

NEW RESULTS IN SYSTEM IDENTIFICATION

Donald M. Wiberg

Electrical Engineering Department
University of California
Los Angeles, CA 90024-1594

INTRODUCTION

This tutorial selectively presents significant results in system identification that have appeared in the last decade. Specifically, (1) transfer function bias, (2) system order estimation by predictive least squares, (3) near optimal recursive parameter estimation, (4) testing against a no-noise model, and (5) robustness theory for estimators are presented. No attempt is made at completeness, and the author's taste is the criterion for selection of specific results for presentation.

Many useful and insightful books have been published presenting techniques for system identification that appeared in the research literature prior to the last decade. The now classic text[1] by Ljung was published in 1987, and is presently the best available text on the subject. However, only 36 pages are devoted to recursive (on-line) methods, which are adequately covered in the 1983 book[2] by Ljung and Söderström. Supplementing Ljung's 1987 text[1] is the 1989 book by Söderström and Stoica[3], which has a number of good ideas not in Ljung's text[1]. Results not appearing or not emphasized in these books are presented here.

TRANSFER FUNCTION BIAS

Using least squares, maximum likelihood or other usual statistical criteria has resulted in bad data fits for some practical applications. In our laboratory, least squares estimation of the fundamental and first harmonic frequencies of vibration of a flexible beam resulted in a significant difference from the fundamental frequency observed during the natural mode of vibration. The first harmonic was accurately estimated. The false estimate of the fundamental is explained by transfer function bias.

Suppose the "true" system is autoregressive-exogenous (ARX) as

$$A_0(s)\, y(t) = B_0(s)\, u(t) + w(t) \quad , \tag{1}$$

where s is the differential operator d/dt (corresponding to the variable s if Laplace transforms are taken of (1)), and $w(t)$ is white noise with power spectral density

Control of Breathing and Its Modeling Perspective, Edited by
Y. Honda *et al.*, Plenum Press, New York, 1992

$\Phi_w(\omega)$. Suppose a parameter vector θ, such as $\theta = (\omega_f \; \omega_h)$ where ω_f is fundamental frequency and ω_h is harmonic frequency, is to be estimated. Then the model used to estimate θ is

$$A(s,\theta)\,\hat{y}(t,\theta) = B(s,\theta)\,u(t) \quad , \tag{2}$$

where $\hat{y}(t,\theta)$ is the estimated output. The error $\varepsilon(t,\theta)$ is then

$$\begin{aligned} \varepsilon(t,\theta) &= y(t) - \hat{y}(t,\theta) = \\ &= \left[B_0(s)\,A_0^{-1}(s) - B(s,\theta)\,A^{-1}(s,\theta)\right] u(t) + A_0^{-1}(s)\,w(t) \quad . \end{aligned} \tag{3}$$

The least squares cost criterion $J(\theta)$ can be interpreted in the frequency domain by using the Bessel equality

$$J(\theta) = \int_{-\infty}^{\infty} \varepsilon^2(t,\theta)dt = (2\pi)^{-1} \int_{-\infty}^{\infty} \Phi_\varepsilon(\omega,\theta)d\omega \quad , \tag{4}$$

where $\Phi_\varepsilon(\omega,\theta)$ is the power spectral density of the error $\varepsilon(t,\theta)$. But from (3),

$$\Phi_\varepsilon(\omega,\theta) = \left|B_0(j\omega)A_0^{-1}(j\omega) - B(j\omega,\theta)A^{-1}(j\omega,\theta)\right|^2 \Phi_u(\omega) + |\,A_0(j\omega)\,|^{-2}\Phi_w(\omega) \tag{5}$$

Clearly the integral in (4) is minimized for values θ^* such that

$$A(j\omega,\theta^*) = A_0(j\omega) \quad \text{and} \quad B(j\omega,\theta^*) = B_0(j\omega) \quad , \tag{6}$$

which is why least squares "works". But physiological and physical data come from processes in which $A_0(s)$ and $B_0(s)$ are very complex and are not true representations because of small nonlinearities. Consequently there exists no θ^* that satisfies (6) exactly, and even for $\Phi_w(\omega) = 0$, small differences must remain in (4) from which the estimate $\hat{\theta}$ is computed as the minimum. This is called **transfer function bias.** The right hand side of (4) can then be interpreted as a frequency weighting of θ. For example, if $\Phi_u = \Phi_w = 1$, A_0 is known exactly, and B_0 is known to be of the form

$$B(s) = 1 + s(\theta - \theta_0) \quad , \tag{7}$$

where θ_0 is the true value of θ, then (4) becomes

$$J(\theta) = (2\pi)^{-1} \int_{-\infty}^{\infty} |\,A_0(j\omega)\,|^{-2} \left[(\theta - \theta_0)^2\omega^2 + 1\right] d\omega \quad . \tag{8}$$

The ω^2 term in (8) above weights errors of θ from the true value θ_0 much more at high frequencies ω than at low frequencies ω. This is why our lab experiment could closely estimate the harmonic, yet have large deviations in the estimate of the fundamental frequency of vibration. The cure is to pass both $y(t)$ and $u(t)$ through a known filter shaped to offset the weighting in (4) before computing the estimate $\hat{\theta}$. Further explanation is given in the text[1], chapter 8, section 5, and in the article[4].

PREDICTIVE LEAST SQUARES

Given the remarks about no θ^* being able to satisfy (6), determination of model order from data takes on new significance. The model order n can be considered as part of the parameter vector θ, to be fit using some criterion such as (4). More significantly, estimates for such as Akaike's[5] AIC have been shown to be biased. An unbiased estimate for n that is conceptually satisfying and deceptively simple was published in 1986 by Rissanen,[6] which he calls **predictive least squares.**

Here, data $y(t)$ are collected for $t = 1, 2, \ldots, T$. Call this data sequence $y_{[1,T]}$. Any method of computing the estimate $\hat{\theta}$ is formed as a function f_n of the data, where f_n depends on the model order n, so that

$$\hat{\theta} = f_n(y_{[1,T]}) \quad . \tag{9}$$

The error $\varepsilon_n(T+1)$ at the next future time $T+1$ is found from (3) as

$$\varepsilon_n(T+1) = y(T+1) - \hat{y}(T+1\,,\; f_n(y_{[1,T]})\,) \quad . \tag{10}$$

The predictive least squares estimate $\hat{n}$ is the value of n that minimizes

$$J(n) = \sum_{T=0}^{N-1} \varepsilon_n^2(T+1) \quad , \tag{11}$$

where N is the total number of data collected in the experiment. Note that the estimate computed in (9) may be the least squares estimate based on T observations, so that (10) expresses the *prediction* error at the next step ahead, and that (11) is *not* the familiar least squares method.

NEAR OPTIMAL RECURSIVE PARAMETER ESTIMATION

The recursive parameter estimation problem can be re-formulated as a filtering problem as follows. The state equations for a linear model with noise can be written in vector-matrix form as

$$dx/dt = A(\theta)x + B(\theta)u + \Sigma(\theta)w \tag{12}$$

$$y = C(\theta)x + v \quad , \tag{13}$$

where $x(t)$ is the n-vector state variable, and $u(t)$, $w(t)$, $y(t)$ are as in (1). Here w and v are taken to be white noise. To estimate the constant parameter vector θ, differentiate to form the equation

$$d\theta/dt = 0 \quad . \tag{14}$$

Then the extended state $(x(t)\,\theta(t))$ can be estimated using optimal nonlinear filtering. Unfortunately, the partial differential equation for the nonlinear estimate is computationally insoluble except in the linear case, which is the Kalman filter. Previous approximations, such as the extended Kalman filter, often accumulate errors and are not globally convergent. At the expense of computing additionally a particular third order moment, a globally convergent **approximation of the optimal recursive parameter estimator** has been found.[7,8,9,10,11]

This globally convergent approximation has faster transient response (the error variance goes to zero faster) than extended least squares or recursive prediction error methods, which are not based on optimal filtering.

TESTING A NO-NOISE MODEL

Local minima are a difficulty encountered in the minimization of nonlinear functions of θ, such as (4), for which a global minimum is desired. Tests of whiteness of the residuals often do not uncover local minima for the same reason that no θ^* might be found to satisfy (6). Our laboratory simply plots the output of a **no-noise model** to compare with a plot of the data $y(t)$, to verify a global minimum. This is described further elsewhere.[12]

ROBUSTNESS THEORY FOR ESTIMATORS

Only in the past decade has the robustness of feedback control been of theoretical interest. Robustness in this case means that if a feedback controller has been designed on the basis of a simple model, will the controller work for the actual complicated process upon which the simple model is based? By duality, the same question has been asked in the field of filtering and the subfield of parameter estimation. Of recent interest is the so-called H^∞ theory, which is too complex to describe in the short space allotted here. An introduction to H^∞ as applied to control systems is provided in the paper[13], and to the theory in the book[14]. One approach for the use of H^∞ in parameter estimation is the article,[15] and a case study is given in the paper.[16]

REFERENCES

1. L. Ljung. "System Identification Theory for the User", Prentice Hall, Englewood Cliffs, NJ (1987).
2. L. Ljung and T. Söderström. "Theory and Practice of Recursive Identification", MIT Press, Cambridge, MA (1983).
3. T. Söderström and P. Stoica. "System Identification", Prentice Hall, Englewood Cliffs, NJ (1989).
4. B. Wahlberg and L. Ljung. Design variables for bias distribution in transfer function estimation, *IEEE Trans. Automatic Control*, **AC-31**, 134:144 (1986).
5. H. Akaike. A new look at statistical model identification, *IEEE Trans. Automatic Control*, **AC-19**, 716:723 (1974).
6. J. Rissanen. Order estimation by accumulated prediction errors, *in:* "Essays in Time Series and Allied Processes", J. Gani and M.B. Priestley, eds., Applied Probability Trust, Sheffield, England (1986).
7. D.M. Wiberg. Another approach to on-line parameter estimation, *Proc. 1987 American Control Conf.*, **1**, 418:423 (1987).
8. D.G. DeWolf and D.M. Wiberg. An ordinary differential equation technique for continuous time parameter estimation, *Proc. 1991 American Control Conf.*, **2**, 1390:1397 (1991). Also accepted, *IEEE Trans. Automatic Control.*
9. D.M. Wiberg and D.G. DeWolf. A convergent approximation of the optimal parameter estimator, *Proc. 30th Conf. on Decision and Control*, **2**, 2017:2023 (1991).
10. D.M. Wiberg and L.A. Campbell. A discrete-time convergent approximation of the optimal recursive parameter estimator, *Proc. 9th IFAC/IFORS Conf. on Identification and System Parameter Estimation*, **1**, 140:144 (1991).
11. D.M. Wiberg. The MIMO discrete-time convergent approximation of the optimal recursive parameter estimator, *Proc. 1992 American Control Conf.*, to appear.
12. D.S. Ward, J.I. Jensen, D.M. Wiberg, and J.W. Bellville. Noise models in respiration, *Proc. 1983 American Control Conf.*, **1**, 31:35 (1983).
13. W.S. Levine and R.T. Reichert. An introduction to H^∞ control system design, *Proc. 29th Conf. on Decision and Control*, **6**, 2966:2974 (1990).
14. T. Basar and P. Bernahrd. "H^∞-Optimal Control and Related Minimax Design Problems", Birkhäuser, Boston (1991).
15. A.J. Helmicki, C.A. Jacobsen, and C.N. Nett. Control oriented system identification: a worst case/deterministic approach in H^∞, *IEEE Trans. on Automatic Control*, **36**, 1163:1176 (1991).
16. K.M. Eveker and C.N. Nett. Model development for active surge control/rotating stall avoidance in aircraft gas turbine engines, *Proc. 1991 American Control Conf.*, **3**, 3166:3172 (1991).

EFFECTS OF CHANGES IN BREATHING PATTERN ON THE SENSATION OF DYSPNEA DURING INSPIRATORY LOADED BREATHING

Y. Kikuchi, M. Sakurai, W. Hida, S. Okabe, Y. Chung,
C. Shindoh, T. Chonan, H. Kurosawa and T. Takishima

The First Department of Internal Medicine
Tohoku University School of Medicine, Sendai, 980 Japan

INTRODUCTION

The ventilatory responses to external resistive and elastic loadings of the respiratory system have been extensively studied in a variety of experimental animals as well as also humans. It has been known that there are marked differences in the pattern of breathing during sustained inspiratory loadings between awake and anesthetized subjects (Margaria et al.,1973; Cherniack and Altose,1981; Milic-Emili and Zin,1986). In awake subjects sustained external elastic loading generally leads to increased breathing frequency and decreased tidal volume, whereas inspiratory resistive loading causes decreased frequency and increased tidal volume. By contrast in anesthetized subjects the changes in respiratory frequency are small or absent and independent of the type of external loading (Milic-Emili and Zin,1986). Because the different responses between the awake and anesthetized subjects cannot be explained as a simple reflex (Cherniack and Altose,1981; Milic-Emili and Zin,1986), it is assumed that conscious awareness of breathing or behavioral response may be responsible for the different breathing pattern observed during inspiratory resistive and elastic loadings. However, the nature of the behavioral control has not yet been evaluated. Thus, the aim of the present study was to examine the hypothesis that the breathing pattern during inspiratory loading might be optimized through cortical response so that the sensation of dyspnea would be minimized.

METHODS

Subject : Six healthy male subjects ranging in age from 25 to 40 yr were studied. Five of them were not familiar with the general purpose of the study.

Apparatus : All measurements were performed while the subject was seated on a chair and was breathing through a unidirectional Hans Rudolph valve with or without an inspiratory load. On the expiratory side of the valve there was a heated pneumotachograph with a Validyne differential pressure transducer, the signal of which gave expiratory flow and it was integrated to give tidal volume (Vt) and minute ventilation (Ve). The lung volume change was measured with Respitrace belts, which was calibrated using the isovolume technique while seated on the chair. Mouth pressure was measured with a Validyne differential pressure transducer. Expiratory fractions of CO_2 ($F_{ET}CO_2$) and O_2 ($F_{ET}O_2$) were monitored at the valve with a mass-spectrometer. The load was applied on the

Control of Breathing and Its Modeling Perspective, Edited by
Y. Honda *et al.*, Plenum Press, New York, 1992

inspiratory side of the valve. The inspiratory elastic load was applied by having the subject inspire from a rigid airtight container, which was vented to the atmosphere via a three way valve during each expiration. The magnitude of the elastic load was 33 cmH_2O/l. The inspiratory resistance was 30.7 $cmH_2O/l/sec$, measured at a flow rate of 0.5 l/sec. Breathing pattern was created as sine waves by an oscillator. For convenience, four types of breathing patterns, in which the frequency was increased by 4, 8, and 12 from the lowest with reciprocal changes in Vt such that the Ve values were equal among them, were created by four oscillators, simultaneously, and recorded on four separate channels of a Sony magnetic tape.

Protocols and Procedures: At first, the subject breathed freely with and without an inspiratory resistive or an elastic load and control respiratory parameters including Ve, Vt and breathing frequency (f) was recorded. It took three minute for each condition. Then the sensation of dyspnea was measured while the subject was copying the breathing pattern through either an inspiratory resistive or an elastic load. To match the breathing pattern, the breathing pattern previously recorded on a magnetic tape was displayed together with the actual volume signal measured with the Respitrace belts on a Tektronix dual-beam storage oscilloscope. The subjects attempted to keep the two signals superimposed.

Four breathing patterns were chosen for each load on the basis of the control breathing frequency (f_0) and minute ventilation during resting ventilation without load. For resistive loading breathing patterns of which the frequencies were f_0+4, f_0, f_0-4 and f_0-8 with reciprocal changes in tidal volume to maintain the control level of minute ventilation were chosen. Similarly, for elastic loading breathing patterns of which the frequencies were f_0+8, f_0+4, f_0, and f_0-4 were chosen. The subject was not informed about the characteristics of the pattern displayed, and the order of the pattern displayed on an oscilloscope was random. After copying each pattern for one minute, subjects rated the intensity of the sensation of difficulty in breathing (dyspnea) by marking on a visual analog scale (VAS) consisting of a line 15 cm in length. The two ends of the scale were designated "none at all" and "most intense imaginable", respectively. Subjects were instructed to avoid rating non-respiratory sensations. After each series of measurement consisting of four breathing patterns for each load, the subject was asked to choose the one of the four patterns which he felt the most easy to breath. Four or three series of the measurement were repeated for each load and the values were averaged. Values were expressed as means±SE. Statistical analysis was done using the paired t-test.

RESULTS

Table 1 shows control ventilatory parameters with and without load in six subjects. Vt increased and f decreased significantly during inspiratory resistive loading compared to those during resting breathing. In contrast Vt decreased and f increased significantly during elastic loading. These Vt and f changes during loading were compatible to those reported previously by several investigators (Freedman and Campbell,1970; Cherniack and Altose,1981; Milic-Emili and Zin, 1986; Kikuchi et al., 1991) in awake humans.

Table 1 . Breathing pattern during resting ventilation with and without load

	Control	Resistive Load	Elastic Load
Vc (l/min)	9.64±0.52	9.51±0.37	8.12±0.87
f (/min)	14.4±0.6	12.5±0.6**	16.7±1.4*
Vt (l)	0.67±0.02	0.77±0.03*	0.48±0.01**
FetCO2 (%)	5.58±0.19	5.70±0.17	5.80±0.22

* , ** significantly different from control; * P<0.05 ** P<0.01.

When the breathing frequency was changed by 4 or 8 from the f_0 with a reciprocal change in tidal volume to maintain control Ve, the intensity of the sensation of dyspnea expressed by VAS scores changed from the control level during both types of loaded breathing. The VAS score tended to be larger at the f_0+4 than VAS at f_0 and smaller at f_0-4 or f_0-8 during resistive loaded breathing. The mean values (±SE) of the VAS score against breathing frequency during copied breathing with the resistance in six subjects are shown in Table 2. The mean VAS score at f_0-4 was significantly smaller than that at f_0 and the mean VAS score at f_0+4 was significantly larger than that at f_0. The rate of incidence of the breathing frequency chosen by the subjects as the breathing pattern to be the easiest to breath during resistive loading is also shown in Table 2. The subjects most frequently selected the slow breathing pattern rather than the control pattern.

Table 2 also shows the mean VAS values against breathing frequency during elastic loaded breathing. The mean VAS value was significantly larger at f_0-4 and smaller at f_0+4 and f_0+8 than that at f_0. The rate of incidence of the breathing frequency chosen by the subjects as the breathing pattern to be the easiest to breath during elastic loading is also shown in Table 2. The subjects most frequently selected the rapid breathing pattern rather than the control pattern.

Table 2 . Sensation of dyspnea and rate of incidence chosen by the subjects during copied resistive and elastic loaded breathing.

	Resistive load				Elastic load			
	f_0-8	f_0-4	f_0	f_0+4	f_0-4	f_0	f_0+4	f_0+8
VAS (cm)	4.1±0.9	3.3±0.9*	4.3±0.9	5.2±1.2*	9.8±0.9**	8.6±1.0	6.8±1.0**	7.0±1.0*
Incidence (%)	6	64	26	4	0	0	53	47

VAS; visual analog scale. *, ** significantly different from that at f_0 ; *P<0.05, **P<0.01.

DISCUSSION

We found that, at constant levels of minute ventilation and chemical drive, the intensity of dyspneic sensation changes during either inspiratory resistive and elastic loaded breathing when the breathing frequency (and hence breathing pattern) was changed from the load-free level. The sensation of dyspnea was minimum at the 4 smaller frequency than control frequency during resistive loading, and at the 4 larger frequency during elastic loading. These breathing frequencies (patterns) with the minimum dyspnea sensation had a tendency to correspond to those adopted spontaneously during inspiratory resistive and elastic loaded breathing, respectively.

Several investigators (Cherniack and Altose,1981; Milic-Emili and Zin, 1986) suggest that higher brain centers, and in particular the cortex, is involved in changes in the breathing pattern observed during inspiratory resistive and elastic loadings in awake human subjects. However, how the breathing pattern or breathing frequency is determined through cortical responses during loaded breathing has not yet been clarified. Otis et al.(1950) pointed out that a particular frequency of breathing would be the least costly in terms of respiratory work. However, the difficulty with this idea is that it is unclear how the controller is able to sense a rate of work, which would require sensing and appropriate manipulation of three variables-force, distance, and time. Later Mead (1960) proposed that the respiratory frequency corresponds more closely to one at which the average force or tension developed by the respiratory muscles, rather than the work done by them on the respiratory system, is minimum. Recently Poon (1983) proposed a respiratory control model based on the hypothesis that the ventilatory output is set by the respiratory controller so as to minimize the net operating cost made up of the conflicting requirements to maintain arterial PCO_2 and to decrease ventilation (and hence work of breathing) to a minimum. Poon (1985) suggested that changes in the work of breathing may be sensed as changes in the sense of effort. In the present study we did not measure the work of breathing, however it seems theoretically clear that, at a given ventilation level, slow deep breaths will necessitate more work against an elastic load, whereas rapid shallow breaths will result in more energy loss overcoming resistance to gas flow. Thus, the particular breathing frequency that represents the minimum rate of work would be lower than the control load-free breathing frequency during resistive loading, and that would be higher during elastic loading. Therefore, these particular frequencies with minimum rate of respiratory work during resistive and elastic loadings would be similar to the particular frequencies obtained as the minimum dyspnea sensation in the present study. Hence, our findings that breathing frequency (pattern) with minimum dyspnea sensation during voluntarily copied breathing with each load had a tendency to correspond to the spontaneously adopted breathing frequency during either resistive or elastic loaded breathing may be explained by Poon's optimization model, and our findings are consistent with the possibility that the breathing pattern during loaded breathing might be optimized, at least in part, to minimize the sensation of dyspnea.

That such an optimizing principle may be at work not only during loaded breathing but also during load-free resting breathing has been demonstrated by Chonan et al.(1990), who have shown that at a constant level of minute ventilation respiratory sensation grows when the breathing frequency is either voluntarily increased or decreased from the spontaneously adopted level. It also has been shown that ventilation level (Chonan et al.,1990) and end-expiratory lung volume during inspiratory loading (Kikuchi et al.,1991) would be regulated so as to minimize the sensation of dyspnea. Therefore, we speculate that minimization of dyspnea is an important optimizing principle employed by the respiratory control system and that several respiratory responses, which had been believed to be due to cortical responses, might be accounted for by the optimization principle.

REFERENCES

Cherniack,N.S., and Altose,M.D., 1981, Respiratory responses to ventilatory loading, in :"Lung Biology in Health and Disease", Hornbein, T.F. ed., Vol.17, Regulation of Breathing, pt.II, p.905-964, Dekker, New York.

Chonan, T., Mullholland,M.B., Altose, M.D., and Cherniack, N.S., 1990, Effects of changes in level and pattern of breathing on the sensation of dyspnea, J. Appl. Physiol. 69:1290.

Freedman,S., and Campbell, E.J.M., 1970, The ability of normal subjects to tolerate added inspiratory loads, Respir. Physiol. 10:213.

Kikuchi,Y., Hida,H., Chonan,T., Shindoh,C., Sasaki,H., and Takishima,T., 1991, Decrease in functional residual capacity during inspiratory loading and the sensation of dyspnea, J. Appl. Physiol. 71:1787.

Margaria, C.E., Iscoe, S., Pengelly, L.D., Couture, J., Don H., and Milic-Emili, J., 1973, Immediate ventilatory responses to elastic loads and positive pressure in man, Respir. Physiol. 18:347.

Milic-Emili, J., and Zin ,W. A., 1986, Breathing responses to imposed mechanical loads, in: "Handbook of Physiology", Cherniack, N.S.,and Widdicombe, J.G., ed., Sect.3, The Respiratory System, Vol. II, p. 395-429, Am. Physiol. Soc., Bethesda, MD.

Mead, J., 1960, Control of respiratory frequency, J. Appl. Physiol. 15:325.

Otis, A.B., Fenn, W.O., and Rhan, H., 1950, Mechanics of breathing in man, J. Appl. Physiol. 2:592.

Poon, C.S., 1983, Optimal control of ventilation in hypoxia, hypercapnia and exercise, in :"Modeling and Control of Breathing", Whipp,B.J, and . Wiberg, D.M., ed., Elsevier, New York.

Poon C. S., and Greene,J.G., 1985, Control of exercise hyperpnea during hypercapnia in humans, J. Appl. Physiol. 59:792.

INTRODUCTION: ROLE OF POTASSIUM IN EXERCISE HYPERPNOEA

David J. Paterson, Peter A. Robbins and Piers C.G. Nye

University Laboratory of Physiology
Parks Road, Oxford UK

It is generally accepted that ventilation ($\dot{V}_E$) in exercise is regulated by a combination of neural and chemical drives, but the exact nature and relative contributions of the controlling signals are not agreed upon. Current thinking suggests that this control is based on a high degree of redundancy so that no single factor is responsible [1]. Recently, attention has focused on the idea that a substance that is released from exercising muscle [2,3] may contribute significantly to this control. This chapter examines evidence [4,5] which suggests that potassium fulfils the criteria of being a work substance and, that it stimulates $\dot{V}_E$ in exercise by increasing the sensitivity of the arterial chemoreflex.

RELEASE OF POTASSIUM FROM WORKING MUSCLE

It is well established that potassium (K^+) is lost from exercising muscle [6,7]. In exercising man this loss raises the concentration of arterial plasma K^+ in a manner related to metabolic rate [8,9] and values as high as 7-8mM have been recorded at exhaustion [6,10]. The rise in plasma K^+ occurs predominantly because the K^+ loss through delayed rectifier K^+ channels activated during the repolarisation phase of the muscle cell action potential [11] is incompletely taken up by the Na^+/K^+ pump. Other studies have also implicated calcium activated K^+ channels [12] and ATP activated K^+ channels [13,14] in the release of K^+ during muscle stimulation.

STIMULATION OF THE CAROTID BODIES AND VENTILATION BY K^+

Euler [15] suggested that $\dot{V}_E$ could be increased by KCl stimulating the carotid body, although no experimental evidence was produced to support this hypothesis. Comroe & Schmidt [16] later observed that $\dot{V}_E$ in the dog increased when KCl was injected into the femoral artery. Because this response could be abolished when the femoral and sciatic nerves in the vascularly isolated dog hindlimb were cut [17], it was concluded that KCl stimulation of $\dot{V}_E$ may have been caused by potassium-induced depolarisation of sensory fibres. Jarisch et al. [18], showed directly that the discharge of arterial chemoreceptors is stimulated by high concentrations of KCl injected close to the carotid body and Linton & Band [19] later showed that physiological increases in arterial K^+ stimulated them and also increased $\dot{V}_E$ (fig 1). These responses were abolished by surgical denervation of the peripheral arterial chemoreceptors [20].

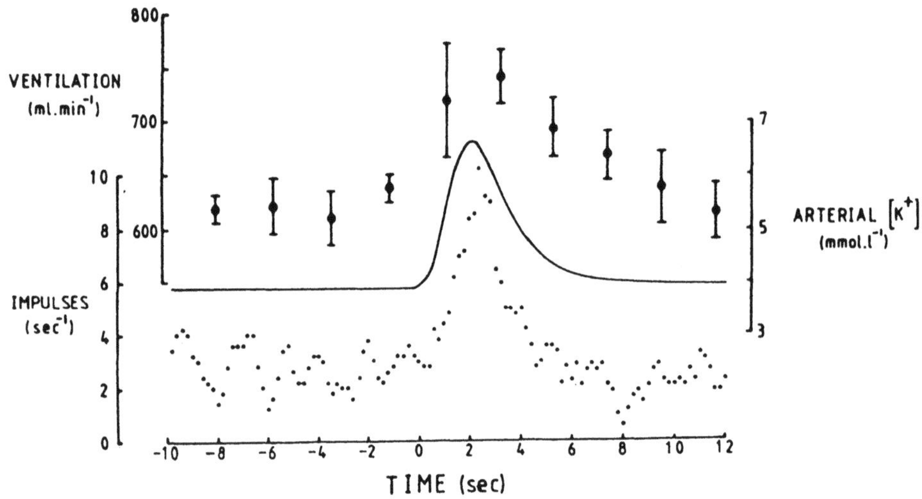

Figure 1. Response of ventilation and chemoreceptor discharge to an intravenous injection of KCl into the anaesthetized cat. Note the rise in both discharge and V_E as the K^+ electrode responds to the injection [from Linton and Band [19] with permission].

If $[K^+]_a$ is raised quickly to ca. 6mM and then held at this level chemoreceptor discharge shows a rapid increase in mean firing followed by a period of adaptation, although discharge remains above control [21](fig 2). If $[K^+]_a$ is raised further, chemoreceptor discharge becomes more sensitive to steady-state levels of K^+. Thus the stimulus response curve of K^+-excited chemoreceptor discharge is not straight, it curves upwards [22].

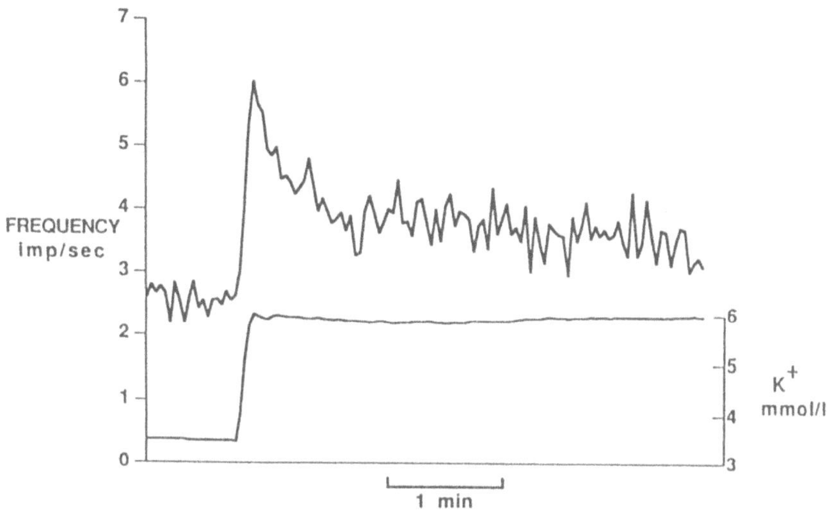

Figure 2. Chemoreceptor response to a steady-state infusion of KCl which raised $[K^+]_a$ to 6mM. Note that discharge increased and then adapted within 1 min after the start of the infusion [from Band & Linton [21] with permission].

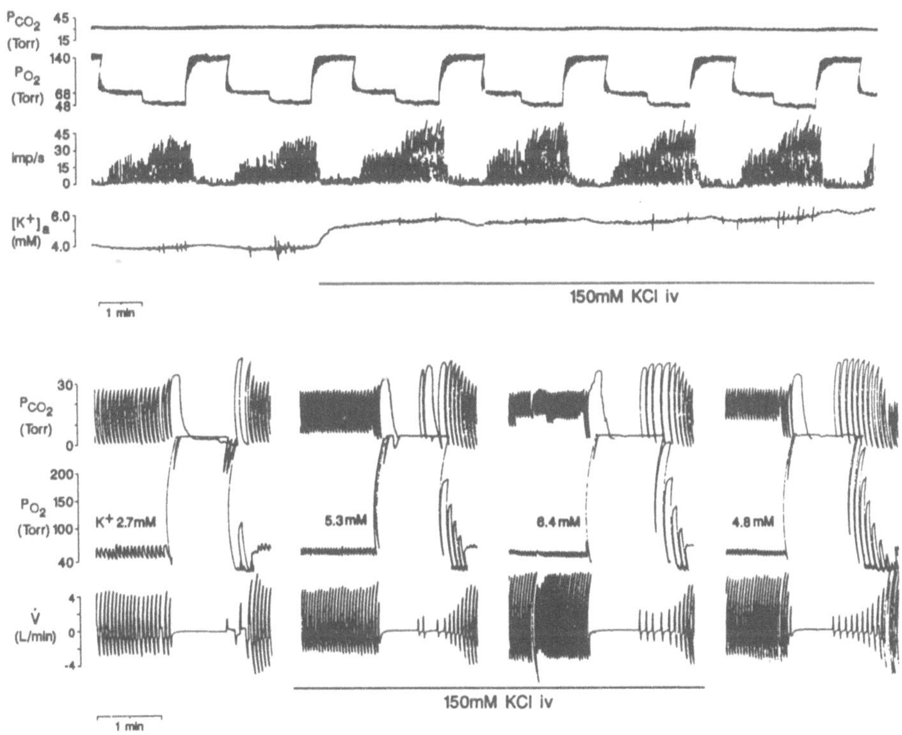

Figure 3. Top panel: excitation of chemoreceptor discharge in the anaesthetized cat in hypoxia is enhanced by K^+ but a higher PO_2 virtually abolishes this effect [from Burger et al. [23] with permission]. Bottom panel: $\dot{V}_E$ in the decerebrate cat is excited by K^+ and 100% O_2 abolishes it [from Paterson et al. [24] with permission].

INTERACTION AMONG K^+, OXYGEN, CHEMORECEPTOR ACTIVITY AND VENTILATION

Burger et al. [23] showed that the effect of hyperkalaemia on chemoreceptor discharge in the cat was enhanced by hypoxia and virtually abolished by a switch to a higher level of inspired PO_2 (fig 3 top panel). They also observed that steady-state hypercapnia did not enhance, but usually reduced, excitation by K^+. Furthermore, in the spontaneously breathing, decerebrate, hypoxic cat, K^+ excited $\dot{V}_E$ was abolished by an abrupt switch to 100% oxygen [24] (fig 3 bottom panel) and, long infusions of KCl into the sedated and analgesic-treated rhesus monkey ($[K^+]_a$ 6-8mM) increased V_E by ca. 30% in euoxia and by as much as 250% in hypoxia. Abrupt switches to 100% inspired O_2 removed the K^+ excited component of breathing in euoxia and stopped breathing altogether in hyperkalaemic hypoxia [25]. These results show that K^+ excites $\dot{V}_E$ only via the arterial chemoreceptors and that hyperoxia abolishes this excitation.

TEMPORAL RELATIONSHIP BETWEEN K^+ AND $\dot{V}_E$ IN EXERCISE

Ventilation and $[K^+]_a$ are well correlated during light [26] and heavy exercise [9,26,27] and also during incremental exercise tests, [8,27] the rise in $[K^+]_a$ being directly proportional to the increase in carbon dioxide production [9]. There is also a close temporal relationship between $[K^+]_a$ and $\dot{V}_E$ during recovery from exercise [8,10,27]. Stimulation of the arterial chemoreceptors is thought to be the major factor underlying the non-linear increase in $\dot{V}_E$ (above the anaerobic threshold) during incremental exercise tests, because subjects without carotid bodies do not show this response [28]. Acidosis has traditionally been regarded as the stimulus to the arterial chemoreceptors that mediates this non-linearity, however, subjects who cannot make acid during exercise (McArdle's syndrome) also show a non-linear ventilatory response during incremental exercise [29] and this casts doubt on the idea that acid is the only stimulus causing this response. In subjects with McArdle's syndrome, as in normals, $\dot{V}_E$ and $[K^+]_a$ are closely related during all stages of an incremental exercise test (fig 4). In addition, a recent study using normal subjects [30] showed that the anaerobic threshold is not changed by glycogen depletion although by reducing the availability of substrate, this alters the relation between work rate and arterial pH. In the same study a close temporal relationship was observed between $\dot{V}_E$ and $[K^+]_a$ before and after glycogen depletion and after glycogen repletion.

HOW MUCH OF THE CAROTID BODY DRIVE COMES FROM K^+ DURING EXERCISE?

This question is difficult to answer because mechanisms controlling the increase in sensitivity of the arterial chemoreflex loop in exercise are not fully understood. We know that the increase in chemoreceptor sensitivity is well correlated with increases in carbon dioxide production [1,3], and that the non-linear ventilatory response seen in exercise at higher work rates is mediated through the arterial chemoreflex, since patients without carotid bodies fail to show this ventilatory response [28]. Studies of subjects with McArdle's syndrome and of normal subjects following glycogen depletion suggest that acid is not the only stimulus contributing to this response. It is possible therefore that some of the increase in

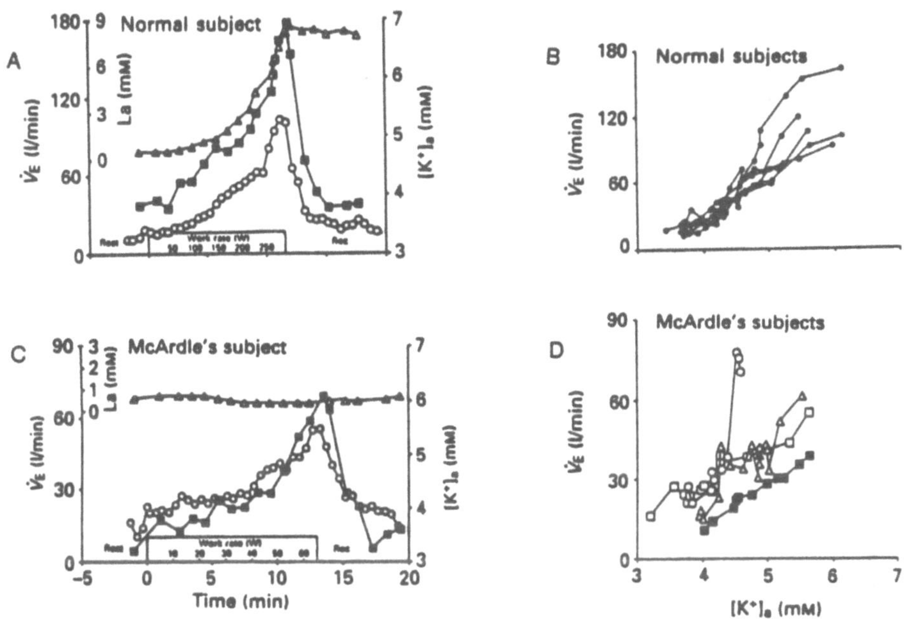

Figure 4. Changes in $[K^+]_a$ (squares) and $\dot{V}_E$ (circles) in normal subjects and subjects with McArdle's syndrome during an incremental exercise test. Panels A and C show that $\dot{V}_E$ and $[K^+]_a$ follow a similar time course during exercise and recovery and panels B and D show that the two variables are almost linearly related [from Paterson et al. [8] with permission].

chemosensitivity comes from K^+ depolarizing the sensory nerve endings and/or type I cells of the carotid body. It is also conceivable that K^+ interacts with an increased sympathetic drive since noradrenaline can excite the carotid body and increase $\dot{V}_E$ in the cat, monkey and man. This effect is also enhanced by hypoxia. Furthermore, beta blockade reduces the discharge of the carotid bodies and their sensitivity to hypoxia [31] and K^+ [22].

IS THE VENTILATORY RESPONSE TO K^+ MEDIATED BY PAIN AFFERENTS ?

Potassium is known to stimulate peripheral sensory fibres which reflexly results in increased $\dot{V}_E$, HR and blood pressure [17,32]. These fibres respond predominately to noxious stimuli and they terminate as free nerve endings within the muscle where they are in an ideal position to be excited by the products of metabolism. It is conceivable therefore that during heavy exercise, which is commonly associated with discomfort and muscle pain, K^+ efflux from muscle may excite $\dot{V}_E$ by depolarizing these sensory fibres. Although this is an attractive hypothesis, the animal studies cited above (cat and monkey), albeit in anaesthetized preparations, clearly show that the stimulatory effects of modest rises of K^+ (2-3mM) are mediated exclusively by excitation of arterial chemoreceptors. Moreover, during volitional incremental exercise in normal subjects $\dot{V}_E$ was unchanged by blockade of the afferents from working muscle by epidural anaesthesia, suggesting that stimulation of peripheral sensory fibres by external factors including K^+ may not play a major role in regulating $\dot{V}_E$ during exercise. Blood pressure was however, reduced [28]. It seems, therefore, that physiological rises in K^+ reflexly increase ventilation and do so only by exciting the arterial chemoreceptors.

ROLE OF K^+ IN THE OVERALL CONTROL OF EXERCISE HYPERPNOEA

Potassium fulfils all the major criteria of being the 'work substance' suggested by Asmussen and Nielsen,[3] it is (1) released by working muscle, (2) stimulates the arterial chemoreceptors and (3) appears and disappears from the blood more rapidly than lactic acid. It is therefore tempting to suppose that K^+ is their work substance and that its effects are enhanced at the arterial chemoreceptors by hypoxia (as described by Cunningham et al. [33]) and virtually abolished by hyperoxia (as described by Asmussen and Nielsen [3]). Finally, K^+ may interact with other stimuli at the carotid body, specifically catecholamines, and in so doing modulate the efficacy of the exercise arterial chemoreflex. Although there is no direct proof that K^+ can increase $\dot{V}_E$ in exercise, circumstantial evidence strongly supports the idea that it may make a significant contribution to the regulation of breathing in exercise.

Acknowledgments

We thank the Wellcome Trust, Medical Research Council and the Nuffield Foundation who supported our work described in this introduction.

REFERENCES

1. D.J.C. Cunningham, Studies on arterial chemoreceptors in man, *J Physiol* 384:1-26 (1987).

2. J. Geppert and N. Zuntz, Ueber die Regulation der Atmung, *Arch Ges Physiol* 42:189-244 (1888).

3. E. Asmussen and M. Nielsen, Studies on the regulation of respiration in heavy work, *Acta Physiol Scand* 12:171-188 (1946).

4. D.J. Paterson, Potassium and ventilation in exercise, *J Appl Physiol* (1992).(in press)

5. R.A.F. Linton and D.M. Band, Potassium and breathing, *News Physiol Sci* 5:104-107 (1990).

6. J.L. Medbo and O.M. Sejersted, Plasma potassium and high intensity exercise, *J Physiol* 421:105-122 (1990).

7. W.O. Fenn and D.M. Cobb, Electrolyte changes in muscle during activity, *Am J Physiol* 115:345-356 (1936).

8. D.J. Paterson, J.S. Friedland, D.A. Bascom, I.D. Clement, D.A. Cunningham, R. Painter, and P.A. Robbins, Changes in arterial K and ventilation during exercise in normal subjects and subjects with McArdle's syndrome, *J Physiol* 429:339-348 (1990).

9. C.G. Newstead, G.C. Donaldson, and J.R. Sneyd, Potassium as a respiratory signal in humans, *J Appl Physiol* 69:1799-1803 (1990).

10. D.J. Paterson, P.A. Robbins, and J. Conway, Changes in arterial potassium and ventilation in response to exercise in humans, *Respir Physiol* 78:323-330 (1989).

11. T. Clausen, Regulation of active Na-K transport in skeletal muscle, *Physiol Rev* 66:542-580 (1986).

12. C. Juel, Is a Ca dependent K channel involved in the K loss from active muscle? *Acta Physiol Scand* 132:P26 (1988).(Abstract)

13. D.J. Paterson, N. Vejlstrup, D. Willford, and M.C. Hogan, Effect of a sulphonylurea on dog skeletal muscle performance during fatiguing work, *Acta Physiol Scand* 114:399-400 (1992).

14. N.A. Castle and D.G. Haylett, Effect of channel blockers on potassium efflux from metabolically exhausted frog skeletal muscle, *J Physiol* 383:31-45 (1987).

15. U.S. Von Euler, Reflektorische und zentrale Wirkung von Kaliumionen auf Blutdruck und Atmung, *Skand Arch Physiol* 80:94-123 (1938).

16. J.H. Comroe and C.F. Schmidt, Reflexes from the limbs as a factor in the hyperpnea of muscular exercise, *Am J Physiol* 138:536-547 (1943).

17. K. Wildenthal, D.S. Mierzniak, N.S. Skinner, and J.H. Mitchell, Potassium-induced cardiovascular and ventilatory reflexes from the dog hindlimb, *Am J Physiol* 215:542-548 (1968).

18. A. Jarisch, S. Landgren, E. Neil, and Y. Zotterman, Impulse activity in the carotid sinus nerve folloowing intra-carotid injection of potassium cholride, veratrine, sodium citrate, adenosinetriphosphate and alpha dinitrophenol, *Acta Physiol Scand* 25:195-211 (1952).

19. R.A.F. Linton and D.M. Band, The effect of potassium on carotid chemoreceptor activity and ventilation in the cat, *Respir Physiol* 59:65-70 (1985).

20. D.M. Band, R.A.F. Linton, R. Kent, and F.L. Kurer, The effect of peripheral chemodenervation on the ventilatory responses to potassium, *Respir Physiol* 60:217-225 (1985).

21. D.M. Band and R.A.F. Linton, The effect of potassium on carotid body chemoreceptor discharge in the anaesthetized cat, *J Physiol* 381:39-47 (1986).

22. D.J. Paterson and P.C.G. Nye, The effect of beta adrenergic blockade on carotid body chemoreceptors during hyperkalaemia in the cat, *Respir Physiol* 74:229-238 (1988).

23. R.E. Burger, J.A. Estavillo, P. Kumar, P.C.G. Nye, and D.J. Paterson, Effects of oxygen, carbon dioxide and potassium on steady-state discharge of cat carotid body chemoreceptors, *J Physiol* 401:519-531 (1988).

24. D.J. Paterson and P.C.G. Nye, Effect of oxygen on potassium-excited ventilation in the decerebrate cat, *Respir Physiol* 84:223-230 (1991).

25. D.J. Paterson, K.L. Dorrington, D.H. Begel, G. Kerr, R.C. Miall, J.F. Stein, and P.C.G. Nye, Effect of potassium on ventilation in the rhesus monkey, *Expt Physiol* 77:217-220 (1992).

26. D.J. Paterson, P.A. Robbins, and J. Conway, Changes in arterial plasma potassium and ventilation during exercise in man, *Respir Physiol* 78:323-330 (1989).

27. T. Yoshida, M. Chida, M. Ichioka, K. Makiguchi, J. Eguchi, and Masao. Udo, Relationship between ventilation and arterial potassium concentration during incremental exercise and recovery, *Euro J Appl Physiol* 61:193-196 (1990).

28. K. Wasserman, B.J. Whipp, S.N. Koyal, and M.G. Cleary, Effect of carotid body resection on ventilatory and acid-base control during exercise, *J Appl Physiol* 39:354-358 (1975).

29. J.M. Hagberg, E.F. Coyle, J.E. Carroll, J.M. Miller, W.H. Martin, and M.H. Brooke, Exercise hyperventilation in patients with McArdle's disease, *J Appl Physiol* 52:991-994 (1982).

30. M. Busse, N. Maassen, H. Konrad, and D. Boning, Interrelationship between pH, plasma potassium concentration and ventilation during intense continuous exercise in man, *Euro J Appl Physiol* 59:256-261 (1989).

31. H. Folgering, J. Ponte, and T. Sadig, Adrenergic mechanisms and chemoreception in the carotid body of the cat and rabbit, *J Physiol* 325:1-22 (1982).

32. J.H. Coote, S.M. Hilton, and J.F. Perez-Gonzalez, The reflex nature of the pressure responses to hypoxia and hyperoxia to muscular exercise, *J Physiol* 215:789-804 (1971).

33. D.J.C. Cunningham, B.B. Lloyd, and D. Spurr, Doubts about the anaerobic work substance as a stimulus to breathing in exercise, *J Physiol* 186:110P-111P (1966).

ACID-BASE REGULATION AND VENTILATORY CONTROL DURING EXERCISE AND RECOVERY IN MAN

Karlman Wasserman
William Stringer
Richard Casaburi

Harbor-UCLA Medical Center
1000 West Carson Street
Torrance, CA 90509 U.S.A.

INTRODUCTION

At rest, 9 millimoles per minute of acid equivalents in the form of CO_2 are produced in a typical man. This rate increases to 45 mmol/min during moderate pace walking and up to 200 mmol/min during heavy exercise in a fit man. The primary mechanism by which an overwhelming body acidosis is avoided is by increasing alveolar ventilation (V_A) in proportion to the metabolic rate. If V_A failed to excrete only 10% of metabolically produced CO_2, the pH would fall by approximately 0.03 pH units per minute and the pH would decrease to approximately 7.1 after 10 min of walking. Since so many physiological functions depend on the pH of the perfusing blood, a well-designed ventilatory control system would take into account the acid load, including the rate of volatile (carbonic) and non-volatile (e.g. lactic) acid production.

Animal models do not demonstrate the precise pH regulation observed in man (1). This is likely because the ventilatory system serves a dual role in animals. Ventilation provides both gas exchange and temperature regulation (2). Lower mammals grow furry coats enabling them to retain body heat at rest without protective clothing. But these animals are handicapped during exercise when a high rate of heat calories are produced. In contrast to man, who eliminates heat during exercise by convection through the skin and sweat evaporation, these are inefficient mechanisms in furry animals. Evaporation of water from the tongue and airway surfaces becomes a very important means of thermoregulation in these animals. Thus, temperature regulation competes with pH regulation in determining the breathing pattern.

Our failure, to date, to define the mechanisms underlying ventilatory control in man is likely due to the small change of potential stimuli as compared to the large ventilatory response, the failure of the system to behave as a proportional controller and our willingness to rely on animal models to ascertain the ventilatory control mechanism of man.

The objective of this study was to determine the precision with which pH is dynamically regulated in response to exercise and recovery in man. The purpose of

Control of Breathing and Its Modeling Perspective, Edited by
Y. Honda *et al.*, Plenum Press, New York, 1992

the recovery component of the study was: a) to document the acid-base status in the non-exercising state without the changes that might result from anticipating exercise, and b) to determine the rapidity of recovery of the acid-base disturbance and the importance of the exercise lactic acidosis on the overall ventilatory control mechanism.

METHODS

The detailed methods and primary data for this study were previously reported (3). Eight reasonably fit male subjects (age = 26 ± 6.1, height = 178 ± 6.1 cm, weight = 72.5 ± 4.4 kg) volunteered for this study after it was explained to them. Signed informed consent was obtained. Their VO_2 max and lactic acidosis thresholds were 3.6 ± 0.6 and 2.2 ± 0.6 L/min STPD, respectively.

On the first day, the subject was familiarized with the laboratory and laboratory procedure and his lactic acidosis threshold (LAT) was determined during a progressively increasing work rate test as previously described (3). The LAT was determined by the V-slope method of Beaver et al (4) as modified by Sue et al (5). The span of VO_2 between the LAT and VO_2max was defined as delta (Δ). Three work rates were selected for study, each to be performed for 6 min, one at 80% of the LAT (120 ± 43.6 watts), the second at a work rate which provided a VO_2 at 35%Δ (i.e. LAT + 35%Δ) (205 ± 55.2 watts) and the third which provided a VO_2 at 75%Δ (269 ± 53.6 watts). On the second and third days of the study (each separated by one week), a catheter was inserted percutaneously by Seldinger technique into the left brachial artery to monitor the changes in arterial blood pH, PCO_2 and lactate. On this day, the 80% LAT and 75%Δ work rates were performed, each separated by two hours. One week later, the catheter was reinserted and the 35%Δ constant work rate was performed, followed in two hours by a progressive incremental exercise test. Arterial blood was collected every 7½ sec during the first minute of rest and 3 min of exercise, using a computer controlled and activated sampling device (6). This was followed by manual sampling every 30 sec for the remaining 3 min of exercise and the first 5 min of recovery. After 5 min of recovery, blood was sampled every 5 min for the remaining 30 min.

Ventilation and gas exchange were measured breath-by-breath at rest and during exercise on a calibrated cycle ergometer at the work intensities indicated above. A turbine device was used to measure expiratory and inspiratory volume, breath-by-breath. A mass spectrometer was used to measure O_2, CO_2 and N_2 concentrations. VO_2, VCO_2, and V_E were determined, breath-by-breath, as previously described (7). Arterial blood samples were measured for pH and blood gases using an IL 1306 or Radiometer ABL-2 blood gas analyzer. Lactate was measured on the supernatant of iced perchlorate precipitated arterial blood from the enzymatic conversion of NAD^+ to NADH (8). The standard bicarbonate values were calculated using an equation (3) derived from the Siggard-Andersen nomogram (9).

Data from the 8 subjects for the three exercise intensities studied were analyzed for change in arterial pH, $PaCO_2$, Std-HCO_3^-, and lactate from resting values. The resting values for each variable for each subject were calculated by averaging the results from 2 blood samples obtained from the sampling manifold prior to the start of exercise. The data during the first three minutes of exercise were analyzed in 10 second time bins. The data, after three minutes, were averaged into 30 sec time bins until the completion of 5 min of recovery. The further recovery measurements were averaged at 5, 10, 15, 20, 25 and 30 min.

Statistical analysis was performed using repeated-measures analysis of variance. Significant change was determined to be at a P value of less than 0.05.

RESULTS

The change in acid-base variables and lactate concentration from resting values for the transition to three exercise intensities is shown in Figure 1. The absolute resting values are shown in the legend. There was no significant change in pH, $PaCO_2$, lactate and bicarbonate during the first 20 sec of exercise at any of the three work rates. pH is regulated at resting levels during the first 20 sec of exercise at all three work intensities. After 20 sec, pH systematically decreased for all three exercise intensities, initially due to the increase in $PaCO_2$, and subsequently as a result of decreasing bicarbonate due to the accumulation of lactic acid for the heavy and very heavy work intensities.

For work below the lactic acidosis threshold, $PaCO_2$ increased, on average, about 3 mmHg and stayed at this level during the remaining six minutes of exercise. There is an appropriate decrease in pH to reflect this mild respiratory acidosis. The slight respiratory acidosis (increase in $PaCO_2$) noted for all 3 work intensities after 20 sec, disappeared after about 2½ min of exercise at the 35% and 75%Δ work rates, as ventilatory compensation developed for the lactic acidosis. The hyperventilation from 2½ to 6 min was more marked for the 75%Δ work rate.

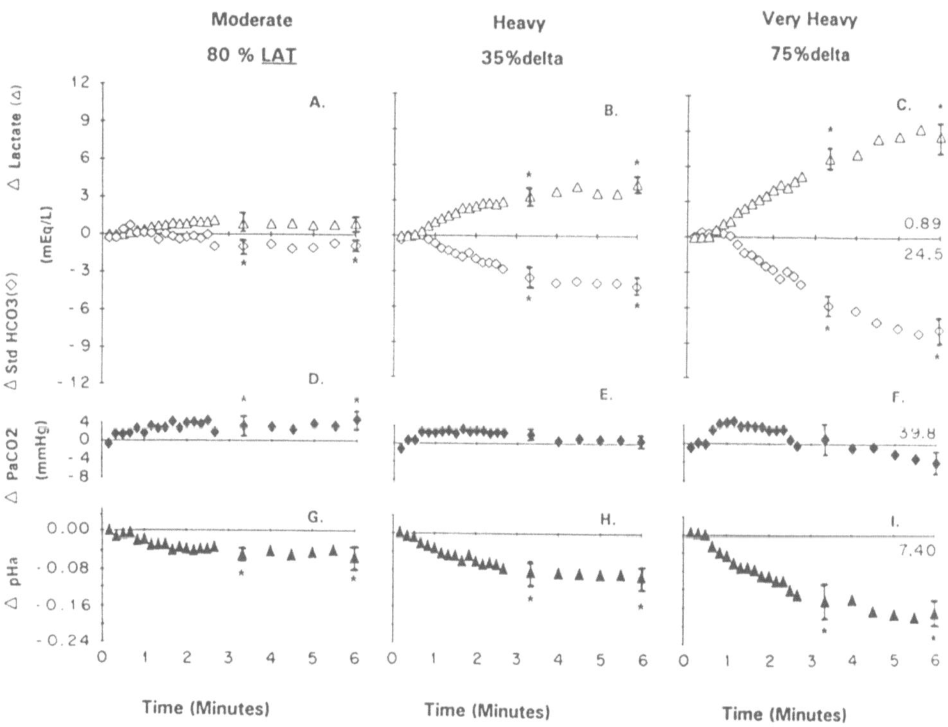

Figure 1. Average change in arterial lactate, standard bicarbonate, PCO_2 and pH as related to time for three 6-minute constant work rate tests for the 8 subjects studied. The three work rates performed were adjusted so that they were at moderate, heavy and very heavy work intensity, respectively, at 80% of the lactic acidosis threshold (LAT), and above the LAT at 35% and 75% of the difference between the LAT and VO_2 max. The mean and standard deviation of the control values are lactate = 0.89 ± .47 mEq/L, standard HCO_3^- = 24.5 ± 1.3 mEq/L, $PaCO_2$ = 39.8 ± 3.1 and pHa = 7.40 ± .03. The bars at 3 min and 6 min of exercise represent the standard errors of the mean. The asterisks indicate significant differences ($P < .05$). (From ref. 3)

All subjects at all three work levels developed an acidosis in response to exercise after the initial 20 sec period (Fig. 1). There were no instances of an increase in pH with exercise as has been reported in animals (10). For the 35% and 75%Δ work rates, the pH decreased in all subjects, being more striking for the 75%Δ work rate since the pH reaction was dominated by the development of lactic acidosis.

The changes in minute ventilation reflect the changes in acid-base status (Fig. 2). For the moderate work rate (below the LAT), V_E increased to a constant value by 3 min and remained constant throughout the remaining exercise period. However, for the 35%Δ work rate, V_E continued to rise and accounted for the reduction in $PaCO_2$ which took place after 2½ min of exercise. This phenomenon was reflected in a more striking manner for the 75%Δ work rate, where V_E continued to rise steeply and $PaCO_2$ continued to decrease even though work rate remained constant.

At about 40 sec, lactate started to increase and bicarbonate started to decrease for the work rates above the LAT (Fig. 1). This delay in development of lactic acidosis at suprathreshold work rates is, in part, due to a transport delay and, probably more importantly, the delay in the depletion of creatine phosphate to a critical level below which anaerobic metabolism is stimulated.

The finding of no systematic acid-base change during the first 15-20 sec of exercise is consistent with previous analyses of the gas exchange ratio (R) at the start of exercise (11). Figure 3 shows an example of the changes in R at the start of exercise that we measured in another study. The same work rate in the same healthy subject was repeated 6 times on 6 different days while gas exchange was measured breath-by-breath, and the results were interpolated to obtain values every second. Responses were time-aligned to the start of exercise. These studies show no change

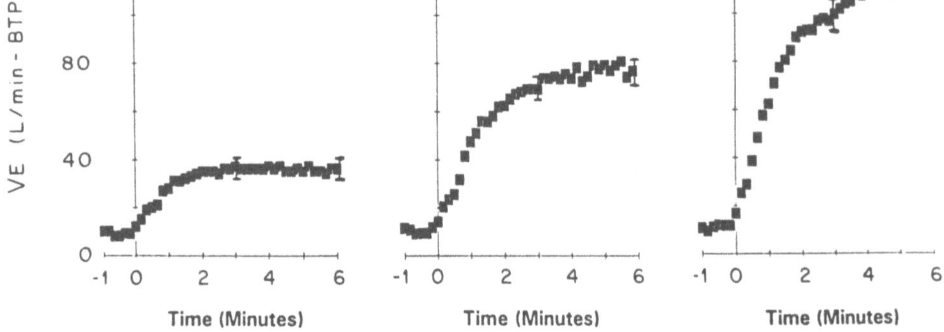

Figure 2. Average minute ventilation (V_E) in response to moderate (A), heavy (B) and very heavy (C) work rates for the 8 subjects whose acid-base data are shown in Figure 1. There was no difference for the V_E at 6 min compared to 3 min for the moderate work intensity. However, the V_E at 6 min was significantly higher than that at 3 min for the heavy and very heavy work intensities. It might be noted that the V_E for heavy and very heavy exercise increases more steeply at about 2½ to 3 min, the time at which ventilatory compensation starts for these work intensities, as evidenced by the decrease in $PaCO_2$ (see Figure 1). (From ref. 3)

in the gas exchange ratio (R).at the start of exercise through the first 15 to 20 sec despite abrupt increases in V_E. After 20 sec the gas exchange ratio decreased consequent to the more slow rise in VCO_2 relative to VO_2 before steady-state is reached. The constant R through the transition from rest to exercise suggests that the blood is neither hyperventilated (would cause an increase in R) or hypoventilated (would cause a decrease in R) through the transition from rest to exercise. The observations are consistent with the absence of a systematic change in $PaCO_2$ and pHa as observed in Figure 1 during the first 15-20 sec of exercise.

The decrease in R after the first 15 sec of exercise observed in Figure 3 can be accounted for by differential solubilities of O_2 and CO_2 in tissues. When exercise starts, O_2 is extracted from the blood in order to regenerate ATP, but the CO_2 formed from aerobic metabolism is partially retained in the tissues because of its relatively high solubility. Thus, relatively less CO_2 is added to the blood as it passes through the muscle capillary until the tissues generating the CO_2 come to a new equilibrium at a higher PCO_2. When total body R rises to its steady-state value, it represents the metabolic R.Q. of the body.

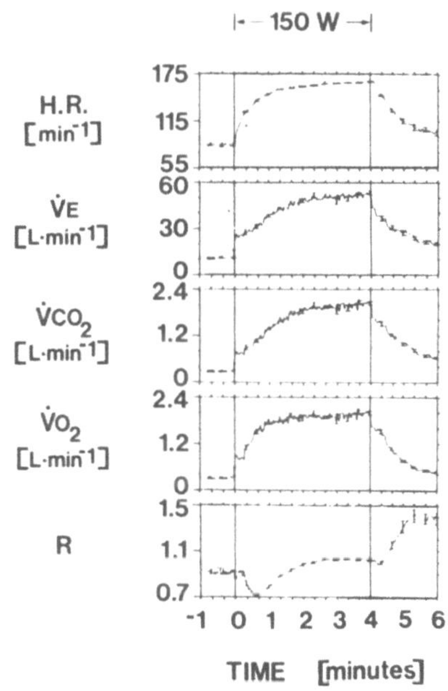

Figure 3. Change in heart rate, $\dot{V}_E$, $\dot{V}CO_2$, VO_2 and R at the start of exercise in a normal subject. This study is the average of six repetitions which have been time aligned to the start of exercise. Second-by-second values are obtained by interpolation. By averaging the time-aligned second-by-second data, the random noise is averaged out revealing the consistent physiological responses. The vertical bars on the records indicate the standard error of the measurements.

Recovery of ventilation and the restoration of resting steady-state acid-base status depends on the need to eliminate the CO_2 that was stored during exercise for moderate exercise intensity, and to catabolize lactate and regenerate HCO_3^- for heavy and very heavy exercise intensity. As shown in Figure 4, recovery for the work below the LAT was very rapid, with pH and $PaCO_2$ returning to the resting value within 2-3 min after exercise stopped. In contrast, recovery took a progressively longer time, the greater the lactic acidosis (Fig. 4). Bicarbonate could not be regenerated until lactate was catabolized. Thus, return of arterial pH, bicarbonate and PCO_2 are controlled by the lactate metabolism. If V_E returned to the resting value before the exercise-induced increase in lactate was catabolized and bicarbonate was regenerated, then pH would decrease because $PaCO_2$ would rise faster than HCO_3^-. However, $PaCO_2$ remained low in recovery and increased slowly back to the control value in concert with the rate of lactate catabolism and regeneration of bicarbonate.

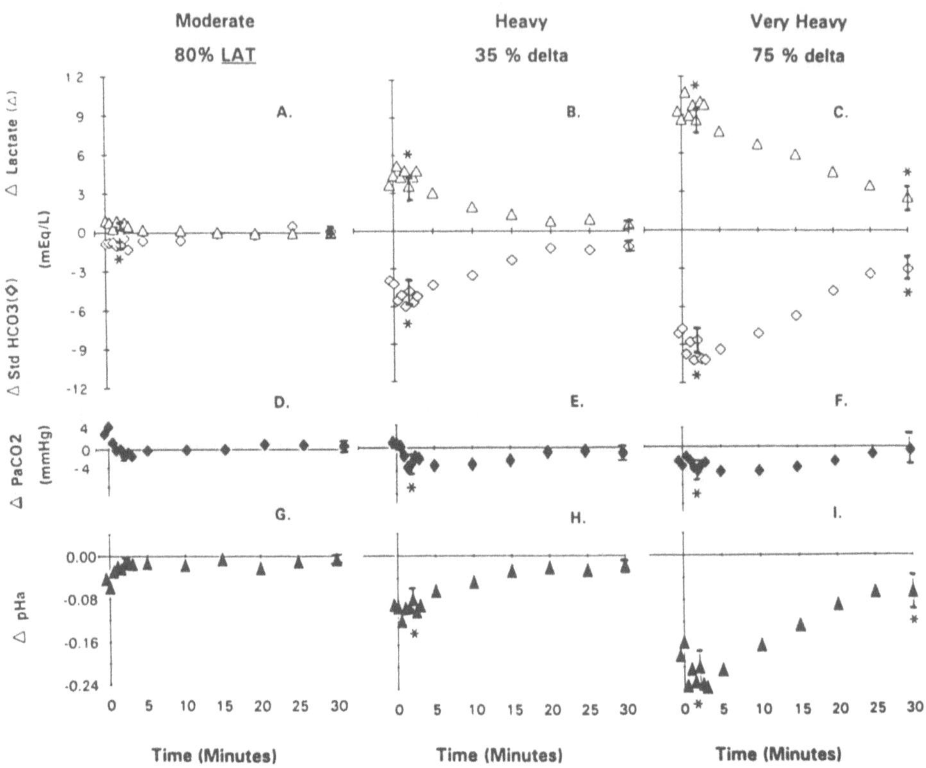

Figure 4. Change in arterial lactate, standard bicarbonate, PCO_2 and pH during the recovery from exercise for the 3 work intensities shown in Figure 1. Exercise stops and recovery begins at zero time. The zero ordinate value is the resting value shown in the legend of Figure 1. (From ref. 3)

DISCUSSION

The results of this study in normal man clearly shows that the pH reaction in response to exercise is characteristically either isohydric or acid. The respiratory alkalosis observed in animal models (10) is not the characteristic response in man, even transiently at the beginning of exercise.

In a number of studies, it had been demonstrated that $\dot{V}_E$ increases more slowly than VO_2 and VCO_2 in response to dynamic work rate forcings (12-15). The relative slowness of the increase in V_E correlated with but slightly lagged the CO_2 production (13,15). This was attributed to CO_2 providing a stimulus to breathing and the ventilatory control mechanism responding to this stimulus. How much pH error signal is needed to get a ventilatory response must vary from subject to subject and probably reflects the respiratory control sensitivity and work of breathing of the subject. Presumably, if the chemosensor is triggered by a certain change in pH, this change in pH will occur in a shorter time at a higher metabolic rate when the rate of acid production is increased as compared to that at a lower metabolic rate. Thus the ventilatory response should be more tightly linked to metabolic rate at high work rates.

pH changes in response to exercise are dependent on time and exercise intensity. Thus, the pH was found to be isohydric for the first 20 sec of exercise for the work rates below and above the lactic acidosis threshold. After 20 sec, a slight respiratory acidosis developed for all 3 work intensities. However, for work rates above the lactic acidosis threshold, the slight hypercapnia gave way to hypocapnia after 2½ to 3 min of exercise, most profoundly at the 75%Δ work rate. It is clear that the decrease in arterial PCO_2, as work duration progresses, depends on the severity of the lactic acidosis (3,15).

Recovery is rapid for work below the LAT, since it depends only on the need to reduce the CO_2 stores that were increased during exercise and return to the resting homeostatic state. Recovery of V_E took much longer for the work rates above the LAT. pH increased gradually to the control value except for a slight worsening of the lactic acidosis several minutes after recovery started. Hyperventilation persisted in these studies until arterial HCO_3^- returned to normal. Thus, V_E decreased in an orderly fashion as the metabolic acidosis was corrected.

In summary, the acid-base responses to exercise are predictable, in man. This probably relates to the fact that man does not need to use his ventilatory system for regulating body temperature. Ventilatory control in man appears to respond to the increased acid load induced by exercise thereby regulating arterial pH. For work rates below the LAT, volatile acid formation parallels that of VO_2; and V_E rises after start of exercise only subsequent to the appearance of CO_2 in the central circulation from metabolizing muscles. However, when lactic acid is produced in addition to the CO_2 acid, the acid load increases non-linearly and minute ventilation follows this same non-linear pattern. This disproportionate increase in V_E serves to constrain the decrease in pH, which otherwise would be profound. Thus when pH does not change, this means that alveolar ventilation is responding in exact proportion to the increased acid load and not to the increase in O_2 consumption or work rate. This suggests that the mechanism of the exercise hyperpnea is closely linked to H^+ production.

REFERENCES

1. K. Wasserman and R. Casaburi, Acid-base regulation during exercise in humans, in: "Exercise - Pulmonary Physiology and Pathophysiology," B.J. Whipp, ed., Marcel Dekker Inc. (1991).
2. M. Gleeson and J.H. Brackenbury, Effects of body temperature on ventilation, blood gases and acid-base balance in exercising fowl, Quart. J. of Exper. Physiol. 69:61-72 (1984).
3. W. Stringer, R. Casaburi and K. Wasserman, Acid-base regulation during exercise and recovery in man, J. Appl. Physiol. (In Press).
4. W.L. Beaver, K. Wasserman and B.J. Whipp, A new method for detecting the anaerobic threshold by gas exchange, J. Appl. Physiol. 60:2020-2027 (1986).
5. D.Y. Sue, K. Wasserman, R.B. Moricca and R. Casaburi, Metabolic acidosis during exercise in patients with chronic obstructive pulmonary disease, Chest 94:931-938 (1988).
6. R. Casaburi, J. Daly, J.E. Hansen and R.M. Effros, Abrupt changes in mixed venous blood gas composition after the onset of exercise, J. Appl. Physiol. 67:1106-1112 (1989).
7. W.L. Beaver, N. Lamarra and K. Wasserman, Breath-by-breath measurement of true alveolar gas exchange, J. Appl. Physiol. 51(6):1662-1675 (1981).
8. C. Olsen, An enzymatic fluorometric micro method for determination of acetoacetate B-hydroxybutyrate, pyruvate and lactate, Clin. Chem. Acta 33:293-300 (1971).
9. O. Siggard-Andersen, The pH-log PCO_2 blood acid-base nomogram revised, Scand. J. Clin. Lab. Invest. 14, 598-604 (1962).
10. H.G. Forster, L.G. Pan, G.E. Bisgard, R.P. Kaminski, S.M. Dorsey and M.A. Busch, Hyperpnea of exercise at various PIO_2 in normal and carotid body denervated ponies, J. Appl. Physiol. 54:1387-1393 (1983).
11. K. Wasserman, B.J. Whipp, R. Casaburi and W.L. Beaver, Carbon dioxide flow and exercise hyperpnea: Cause and effect? Am. Rev. Resp. Dis. 115:225-237 (1977).
12. L.B. Diamond, R. Casaburi, K. Wasserman and B.J. Whipp, Kinetics of gas exchange and ventilation in transitions from rest or prior exercise, J. Appl. Physiol. 43:704-708 (1977).
13. R. Casaburi, B.J. Whipp, W.L. Beaver, K. Wasserman and S. Koyal, Ventilatory and gas exchange

dynamics in response to sinusoidal work, J. Appl. Physiol. 42:300-311 (1977).
14. D. Linnarsson, Dynamics of pulmonary gas exchange and heart rate changes at start and end of exercise, Acta Physiol. Scand. Suppl. 425:1-68 (1974).
15. K. Wasserman, A. Van Kessel and G.G. Burton, Interaction of physiological mechanisms during exercise, J. Appl. Physiol. 22:71-85 (1967).
16. R. Casaburi, T.J. Barstow, T. Robinson and K. Wasserman, Influence of work rate on ventilatory and gas exchange kinetics, J. Appl. Physiol. 67:547-555 (1989).

INFLUENCE OF BODY CO_2 STORES ON VENTILATORY-METABOLIC COUPLING DURING EXERCISE

Susan A. Ward[1] and Brian J. Whipp[2]

[1]Department of Anesthesiology
[2]Department of Physiology
UCLA
Los Angeles, CA 90024, U.S.A.

INTRODUCTION

During the steady state of moderate exercise, ventilation ($\dot{V}_E$) is closely matched to pulmonary gas exchange rates ($\dot{V}O_2$, $\dot{V}CO_2$) and, therefore, to current metabolic demands. This maintains arterial PCO_2 ($PaCO_2$), pH (pHa) and PO_2 at, or close to, their resting levels. For the nonsteady state, however, the presence of intervening gas stores and circulatory delays between the sites of increased metabolic rate and the lungs transiently dissociates pulmonary and tissue gas exchange. The influence of the appreciable body CO_2 capacitance educes $\dot{V}CO_2$ kinetics which are substantially slower than for $\dot{V}O_2$ but similar to those of $\dot{V}_E$.[1,2,3,4] The close co-relation between $\dot{V}CO_2$ and $\dot{V}_E$ kinetics (with little change of $PaCO_2$) has led to the proposal of a CO_2-linked control of $\dot{V}_E$ during moderate exercise, although the precise mechanisms involved remain conjectural.

Above the lactate threshold (θ_L), the decrease in pHa is constrained by the degree of compensatory hyperventilation which, in turn, depends on the increase in supplemental $\dot{V}CO_2$ resulting from the rate of [bicarbonate] ($[HCO_3^-]$) decrease and the time course of the additional $\dot{V}_E$ response to the lactic acidosis. Despite augmented peripheral chemoreceptor (PC) stimulation,[5] however, $PaCO_2$ is typically elevated by some 3-4 mm Hg during the on-transient phase of constant-load supra-θ_L exercise.[6] As the time constant of the PC component is of the order of 10-15 s for the response to hypoxemia and hypercapnia in humans,[7,8] this suggests that the time course of the compensation for the metabolic acidosis of exercise is long, relative to that of PC responsiveness. Consequently, other mechanisms are likely to contribute to the compensation. The purpose of the present investigation was to characterize the influence of CO_2 stores on the time course of this compensatory hyperventilation.

Control of Breathing and Its Modeling Perspective, Edited by
Y. Honda *et al.*, Plenum Press, New York, 1992

METHODS

Ten normal subjects performed incremental exercise to exhaustion at incrementation rates of 50 W.min^{-1} ("fast"), 25 W.min^{-1} ("intermediate") and 6 W.min^{-1} ("slow"), on separate days. Respired volume was measured with a turbine transducer and gas concentrations by mass spectrometry, for on-line determination of ventilatory and gas exchange variables breath-by-breath. Arterialized venous blood was sampled at intervals throughout selected tests from an indewelling catheter in a superficial vein on the heated dorsum of the hand, and analyzed subsequently for blood-gas and acid-base variables. θ_L was estimated using conventional gas-exchange criteria for the demonstration of (i) hyperventilation relative to O_2 but not CO_2 and (ii) an acceleration in $\dot{V}CO_2$ relative to $\dot{V}O_2$ ("V-slope" technique)[9,10]. The parameters of the V-slope relationship were estimated as the least-squares linear regression coefficients for the sub- and supra-θ_L regions of interest (S1 and S2, respectively).[9,10]

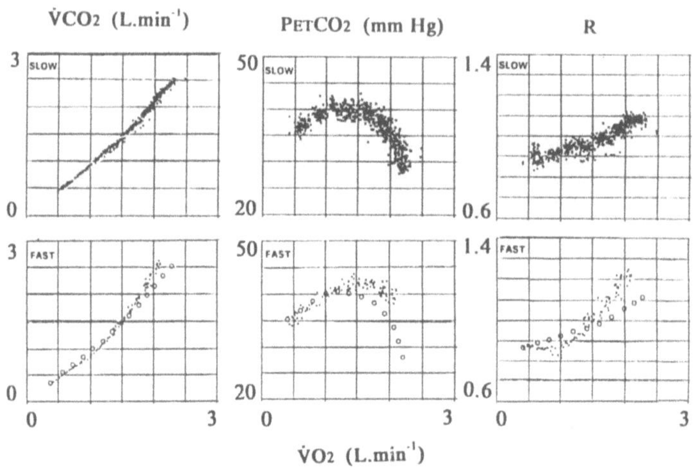

Figure 1. Responses of $\dot{V}CO_2$, $P_{ET}CO_2$ and R as a function of $\dot{V}O_2$ for the "slow" (**upper panels**) and "fast" (**lower panels**) ramp protocols, for a single subject. For the purposes of comparison, the "slow" response profiles (o) (averaging interval: 0.25 $\dot{V}O_2$) are also superimposed on the corresponding "fast" protocol.

RESULTS

Below θ_L, the slope of the $\dot{V}CO_2$-$\dot{V}O_2$ relationship (S1) was an *inverse* function of the ramp incrementation rate in every subject (Fig. 1), averaging 0.75 ± 0.17 for the "fast" tests, 0.88 ± 0.10 for the "intermediate" and 0.96 ± 0.07 for the "slow". For the "slow" protocol, the respiratory exchange ratio, R, increased progressively from ~0.85 towards 1.0 throughout the sub-θ_L range (Fig. 1); in contrast, a transient undershoot

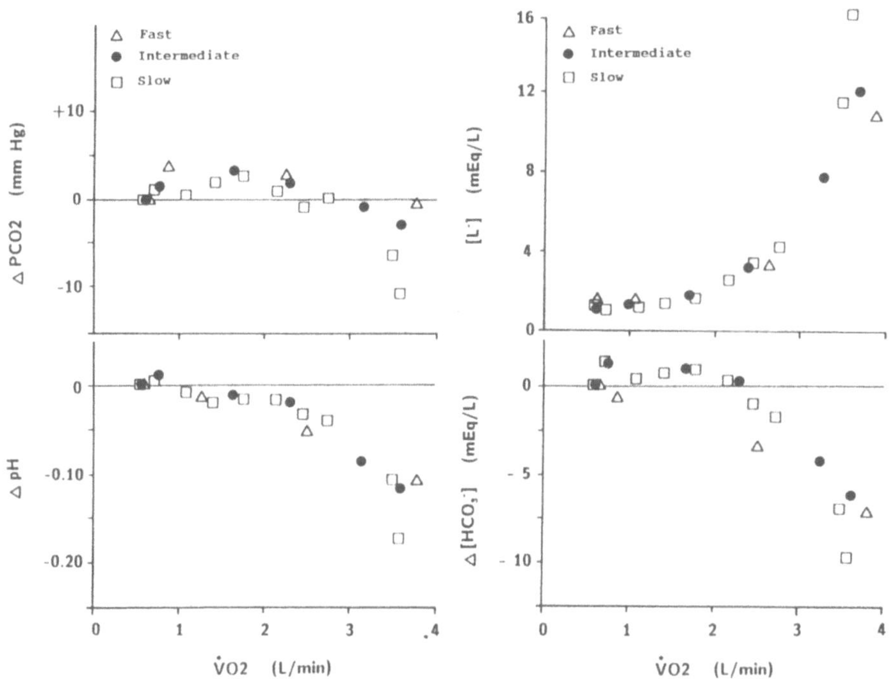

Figure 2. Responses of arterialized venous PCO2, pH, [lactate] and [HCO3⁻] as a function of V̇O2 for the "fast" (Δ), "intermediate" (●) and "slow" (□) ramp protocols, for a single subject. Responses are expressed as changes from the corresponding "0" W value.

was evident for the "fast" protocol (Fig. 1). Despite these differences, the profiles of both end-tidal PCO2 (PETCO2) and arterialized venous PCO2 (PavCO2) appeared similar for each protocol (Figs. 1 & 2) (although it should be noted that the sampling density for the "fast" protocol was sparse compared to the others).

Above θL, however, the V̇CO2-V̇O2 slope (S2) was a *direct* function of the incrementation rate (Fig. 1), being 1.68 $\pm$ 0.15 for the "fast" tests, 1.32 $\pm$ 0.10 for the "intermediate" and 1.19 $\pm$ 0.11 for the "slow". Both PavCO2 and PETCO2 were significantly lower on the "slow" protocol at any V̇O2 in this domain, relative to the faster tests (Figs. 1 & 2). At maximum exercise, PETCO2 averaged 30.9 $\pm$ 2.6 for the "slow" tests, 33.3 $\pm$ 1.0 for the "intermediate" and 38.1 $\pm$ 4.6 for the "fast", i.e. a difference of ~ 7 mm Hg between "slow" and "fast". In contrast, R at maximum was significantly higher for the "fast" tests (1.31 $\pm$0.08), relative to the "intermediate" (1.20 $\pm$ 0.08) and "slow" tests (1.07 $\pm$ 0.03). In the representative example shown in Figure 2, pHav fell progressively with increasing V̇O2 above θL, with the decline at maximum work rate tending to be greater for the "slow" protocol; this was consistent with a larger maximum increase in [lactate] at maximum for this protocol (Fig. 2).

On occasion, we observed what appeared to be a dissociation betweeen θL as estimated directly from the [lactate] profile and indirectly from the standard ventilatory and V-slope criteria. Figure 3 illustrates such a case where, for the "intermediate" protocol, a "threshold" V̇O2 was identified at which both the V̇CO2-V̇O2 relationship and the ventilatory equivalent for O2 (V̇E/V̇O2) evidenced sharp increases while the ventilatory equivalent for CO2 (V̇E/V̇CO2) had not yet started to

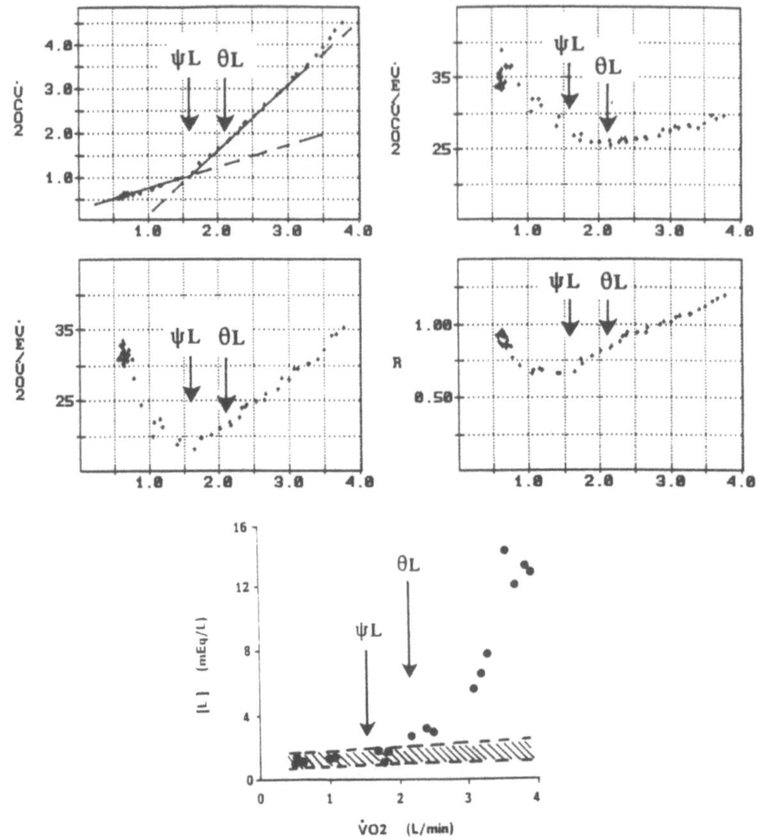

Figure 3. Responses of $\dot{V}CO_2$, $\dot{V}E/\dot{V}O_2$, $\dot{V}E/\dot{V}CO_2$, R and arterialized venous [lactate] as a function of $\dot{V}O_2$ for the "intermediate" ramp protocol, for a single subject. Note that a psuedothreshold (ψL) was evident, occurring at a lower $\dot{V}O_2$ than the directly-estimated lactate threshold (θL).

increase. However, the expected increase in [lactate]av was not evident until $\dot{V}O_2$ had increased further, by some 0.5 $L.min^{-1}$. We have termed this a "pseudothreshold" (ψL).[11]

DISCUSSION

The demonstration that the sub-θL component of the V-slope (S1) was less when the work rate was incremented rapidly, at a time when R was actually *declining* (Figs. 1 & 2), argues strongly for a component of the metabolically-produced CO_2 being washed into the body CO_2 stores, rather than being excreted at the lungs. As a corollary, therefore, the assumption that S1 consistently provides an index of the metabolic RQ for moderate exercise[9,10] must be used cautiously, unless it can be established that the response of R exactly reflects that of RQ (i.e., not necessarily the same, but changing at the same rate - the capacitative effect having become disappearingly small).

The phenomenon of a "pseudothreshold" (ψL) is, we believe, a further manifestation of the magnitude of the CO_2 capacitative effect that can occur during sub-θL exercise when work rate is incremented rapidly, as R falls progressively throughout this sub-ψ domain (Fig. 3). Whipp et al.[11] first demonstrated a pseudothreshold which met the required gas-exchange criteria *for* θL, but did not occur *at* θL. Under these conditions, the "threshold" behavior reflects an acceleration of $\dot{V}CO_2$ relative to $\dot{V}O_2$ that occurs as the CO_2 stores become charged up and the rate of CO_2 storage is reduced. As a result, the proportion of the metabolically-produced CO_2 reaching the lungs starts to increase towards the new steady-state level; it should not be confused with the acceleration of $\dot{V}CO_2$ that occurs above θL when buffering mechanisms augment $\dot{V}CO_2$ *beyond* the new steady-state expected on solely metabolic grounds.

Above θL, considerations of CO_2 exchange become further complicated by a washing-out of the body CO_2 stores by buffering-related HCO_3^- breakdown and compensatory hyperventilation for the lactic acidosis. That the supra-θL $\dot{V}CO_2$-$\dot{V}O_2$ slope (S2) is steepest for the "fast" protocol (Fig. 1) can be ascribed to the more-rapid

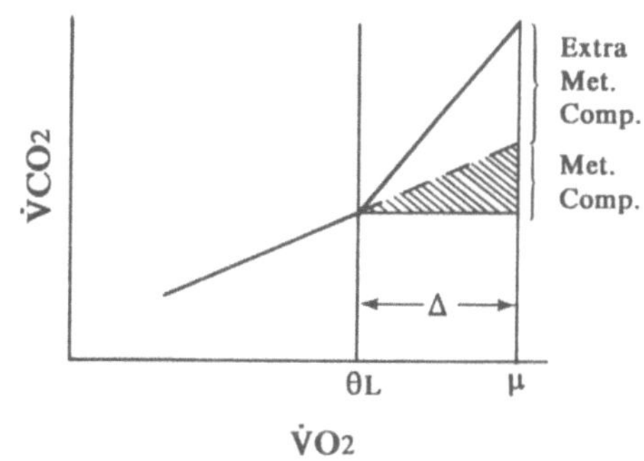

Figure 4. Schematic illustrating technique for estimating the "extra-metabolic" contribution to supra-θL CO_2 exchange. See text for details.

rate of $[HCO_3^-]$ decrease that releases volumes of CO_2 at a greater rate, rather than to compensatory hyperventilation for the lactic acidosis; i.e., in the "fast" test, $P_{ET}CO_2$ and $PavCO_2$ were both higher throughout the supra-θL range, compared to the slower protocols. It is possible to estimate this "extra-metabolic" contribution to the total CO_2 exchange above θL (Fig. 4). If it is assumed that the "metabolic" CO_2 contribution can be represented by a linear extrapolation of the sub-θL $\dot{V}CO_2$-$\dot{V}O_2$ relationship (shaded area), then the "extra-metabolic" component is defined by the region between this relationship and the observed $\dot{V}CO_2$-$\dot{V}O_2$ relationship (unshaded area). The volume of extra-metabolic CO_2 evolved throughout the entire supra-θL portion of each ramp test ($VemCO_2$) is therefore given by:

$$\Delta\dot{V}CO_2\ (>\theta L) * 0.5\Delta t\ (>\theta L)$$

and the average rate of its evolution ($\dot{V}emCO_2$) by:

$$VemCO_2/\Delta t\ (>\theta L)$$

For the "fast" protocol, $\dot{V}emCO_2$ averaged 0.63 ± 0.08 L.min^{-1}, compared with 0.21 ± 0.02 L.min^{-1} for the "slow"; i.e., a three-fold difference.

In contrast, $V_{em}CO_2$ averaged only 1.86 ±0.45 L for the "fast" protocol, compared with 4.63 ± 0.96 L for the "slow"; i.e., a three-fold greater *volume* of CO_2 was released from extra-metabolic sources when the work rate was incremented slowly. This additional extra-metabolic CO_2 provèd to be a consequence of the slowness of the development of the compensatory hyperventilation, consistent with the results of Rausch et al.[6] For example, $P_{ET}CO_2$ and $P_{av}CO_2$ were lower throughout the entire supra-θ_L domain in the "slow" protocol, compared with either the "intermediate" or the "fast" protocol. This reflects the unloading of appreciably more CO_2 from the arterially-perfused stores. In contrast, for the "fast" protocol, the compensatory phase for the lactic acidosis (i.e., in which $P_{av}CO_2$ was actually decreased) was delayed relative to θ_L.

Given the influence of ramp incrementation rate on the response profile of PCO_2 above θ_L, the assertion that S2 provides a useful index of the decrease in blood [HCO_3^-] that occurs above θ_L and, in turn, of the increase in blood [lactate][12] must be subject to careful scrutiny. This assertion is justified only if P_aCO_2 can be assumed (or, better, demonstrated) to remain unchanged throughout the entire supra-θ_L domain. It is clear that this condition becomes increasingly less likely as the ramp incrementation rate is reduced (Fig. 2).

In conclusion, therefore, considerations of ventilatory control below θ_L must take account of the dynamics of CO_2 stores wash-in. Above θ_L, these considerations become influenced by the complexities of CO_2 stores wash-out associated with differing degrees of bicarbonate breakdown and compensatory hyperventilation for the lactic acidosis. The mechanisms responsible for the relatively slow compensatory hyperventilation, and their interaction with the dynamics of the body CO_2 stores, remain to be elucidated.

REFERENCES

1. D. Linnarsson, D, Dynamics of pulmonary gas exchange and heart rate changes at start and end of exercise. *Acta Physiol. Scand.*, (suppl.) 415:1 (1974).
2. R. Casaburi, B.J. Whipp, K. Wasserman, and R.W. Stremel, Ventilatory control characteristics of the exercise hyperpnea as discerned from dynamic forcing techniques, *Chest* 73S:280 (1978).
3. B.J. Whipp, The control of exercise hyperpnea, *in*: "The Regulation of Breathing," T. Hornbein, ed., Dekker, New York (1981).
4. Y. Miyamoto, T. Hiura, T. Tamura, T. Nakamura, J. Higuchi, and T. Mikami, Dynamics of cardiac, respiratory and metabolic function in men in response to step work load, *J. Appl. Physiol.* 52:1198 (1982).
5. J.V. Weil and G.D. Swanson, Peripheral chemoreceptors and the control of breathing, *in*: "Pulmonary Physiology and Pathophysiology of Exercise," B.J. Whipp and K. Wasserman, eds., Dekker, New York (1991).
6. S.M. Rausch, B.J. Whipp, K. Wasserman, and A. Huszczuk, Role of the carotid bodies in the respiratory compensation for the metabolic acidosis of exercise in humans, *J. Physiol. (Lond.)* 444:567 (1991).

7. W.N. Gardner, The pattern of breathing following step changes of alveolar partial pressures of carbon dioxide and oxygen, *J. Physiol. (Lond.)* 300:55 (1980).
8. J.W. Bellville, B.J. Whipp, G.D. Swanson, and K.A. Aqleh, Dynamics of ventilatory responses to CO_2 in man: role of the carotid bodies, *J. Appl. Physiol.* 46:843 (1979).
9. W.L. Beaver, K. Wasserman, and B.J. Whipp, A new method for detecting the anaerobic threshold by gas exchange, *J. Appl. Physiol.* 60:2020 (1986).
10. C.B. Cooper, W.L. Beaver, D.M. Cooper, and K. Wasserman, Factors affecting the components of the alveolar CO2 output-O2 uptake relationship during incremental exercise in man, *Exp. Physiol.* 77:51 (1992).
11. B.J. Whipp, N. Lamarra, and S.A. Ward, Required characteristics of pulmonary gas exchange dynamics for non-invasive determination of the anaerobic threshold, *in*: "Concepts and Formulations in the Control of Breathing," G. Benchetrit, ed., Manchester University Press, Manchester (1987).
12. W.L. Beaver and K. Wasserman, Muscle RQ and lactate accumulation from analysis of the $\dot{V}CO2$-$\dot{V}$ relationship during exercise, *Clin. J. Sports Med.* 1:27 (1991).

RESPIRATION IN CHRONIC HYPOXIA AND HYPEROXIA:

ROLE OF PERIPHERAL CHEMORECEPTORS

S. Lahiri

Department of Physiology
University of Pennsylvania, School of Medicine
Philadelphia, PA 19104-6085, USA

INTRODUCTION

The peripheral chemoreceptors (carotid and aortic bodies) are environmental monitors - they monitor internal milieu, and external as well, to the extent that it influences the internal milieu. These are the only oxygen sensing organs which generate protective reflex respiratory and autonomic responses against the ambient hypoxia. In the absence of their neural linkage with the central nervous system or during their malfunction in disease, the organism cannot detect oxygen deprivation in the milieu and hence cannot respond appropriately. This "gateway" function of the peripheral chemoreceptors makes it indispensable for protection, particularly for the altitude sojourners and for the patients with hypoxemic disease. These organs work harder at higher altitudes. This paper will focus on the adaptive responses of the carotid body to hypoxia of high altitudes. For contrast, the observations on carotid body of animals which were exposed to high oxygen pressure will also be included. The theme, relating structure to function of the carotid body, will be the running thread in the paper.

RESULTS AND DISCUSSION

Hypoxia

Effects of chronic hypoxia on carotid bodies of sheep are shown in Figure 1. The control carotid bodies were obtained from sheep at sea level. For the effects of chronic hypoxia, carotid bodies were obtained from the sheep at 3800 m (inspired PO_2 = 91 Torr) in the Himalayas. It is quite clear that the glomus cells of the high altitude specimen were distinctly larger. They were also lighter. The ventilatory response to hypoxia was greater in these high altitude sheep (Lahiri, 1972).

The chemosensory responses were studied in the cat which were exposed to inspired PO_2 of 70 Torr for 28 days. The following characteristic changes occurred: (a) The chemosensory responses to hypoxia were augmented (Barnard et al., 1987). (b) The efferent inhibition on the chemosensory responses were

Control of Breathing and Its Modeling Perspective, Edited by
Y. Honda *et al.*, Plenum Press, New York, 1992

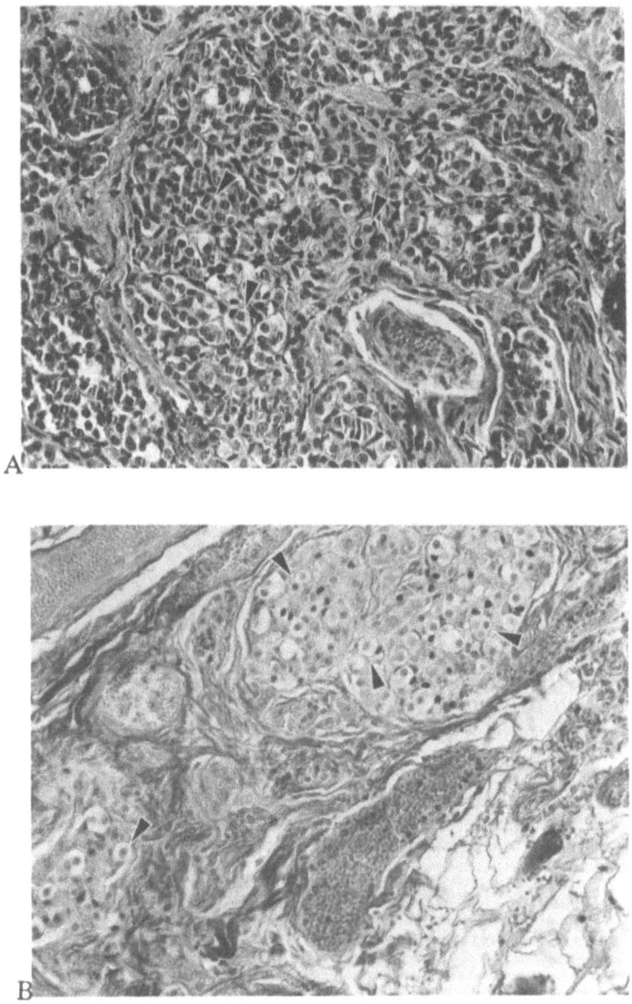

Figure 1. Sheep carotid bodies (x 250): A, upper panel inspired PO_2 = 150 Torr (Philadelphia, PA, USA); B, lower panel, inspired PO_2 = 91 Torr (Kunde, Nepal). Note that majority of the glomus (arrow-head) cells are significantly larger and lighter in chronic hypoxia at Kunde.

greater (Lahiri et al., 1983). This was unmasked by recording the responses before and after transecting the whole sinus nerve. (c) The inhibitory effect was dopaminergic because it was eliminated by dopamine receptor blocker (Lahiri et al., 1984). Chronic hypoxia also enhanced the release and concentrations of putative neurotransmitters in the carotid body (Hanbauer et al., 1981; Pequignot et al., 1987).

It appears then that during chronic hypoxia, the carotid body for a given arterial PO_2 releases a larger amount of neurotransmitters which generate a greater frequency of sensory discharge. The latter is expected to augment respiration and autonomic efferent activity as well, including those which influence function of the carotid body (O'Regan and Majcherczyk, 1983). This influence appears to be inhibitory, and conforms to the principle of negative feedback control. These efferent effects are summarized in Fig. 2. The responses were measured from a thin chemosensory filament before and after transection of the whole carotid sinus nerve. These results imply furthermore that an attenuation of the inhibitory activity could increase, and an augmentation could blunt the chemosensory response to hypoxia.

The incremental chemosensory activity could explain, in part, the ventilatory acclimatization to chronic hypoxia (Barnard et al., 1987; Bisgard et al., 1986 and Vizek et al., 1987). On the other hand, the blunting of the ventilatory chemoreflex response to hypoxia, as generally observed in the adult high altitude natives (Lahiri, 1984), requires either the chemosensory response to hypoxia is attenuated by the efferents or the ventilatory response to the sensory input is blocked. A central effect of chronic hypoxia seems to be a strong candidate (Pokorski and Lahiri, 1991), as originally proposed by Tenney and Ou (1977).

The central depressant effect of hypoxia is well manifested in ventilation (Lahiri, 1976) and in all types of medullary respiratory neuron activities (Richter et al., 1991), unlike the autonomic neural activity (Matsumoto et al., 1987a). The mechanisms of the differential effects of hypoxia in the two neuron pathways are not known. In the context of the observation that the respiratory neuron activity drives the sympathetic nerve activity (e.g., Huang and Lahiri, 1988), the persistent sympathetic excitatory response to hypoxia in the face of the phrenic failure in the chemodenervated cat (e.g., Matsumoto et al, 1987a) is intriguing. It is not due to possible hypotension but is likely to be due to a slow sympathetic chemosensing response (Matsumoto et al., 1987b).

Hyperoxia

Hyperoxia is an excess of oxygen as opposed to hypoxia. Accordingly, the effects of chronic hyperoxia on the chemoreceptors could be opposite to those of chronic hypoxia. However, the mechanisms of effects of low and high O_2 are unlikely to be the same. High O_2 may have dual effects - one is the prevention of hypoxia and the other is due to the production of oxygen related free-radicals which are biologically highly reactive and damaging to the membranes of cells and organelles and enzymes (see Fridovitch, 1988). It is well known that exposure of animals to 100 O_2 is fatal within a few days. However, all the tissues are not affected equally, perhaps because the tissue PO_2 increases are not equal, and the tissues differ in their antioxidant properties.

Carotid bodies are well known for their vascularity (see Fitzgerald and Lahiri, 1986, for review), and the balance of evidence suggests that their tissue blood flow is several fold higher than the nearest organs with high blood flow,

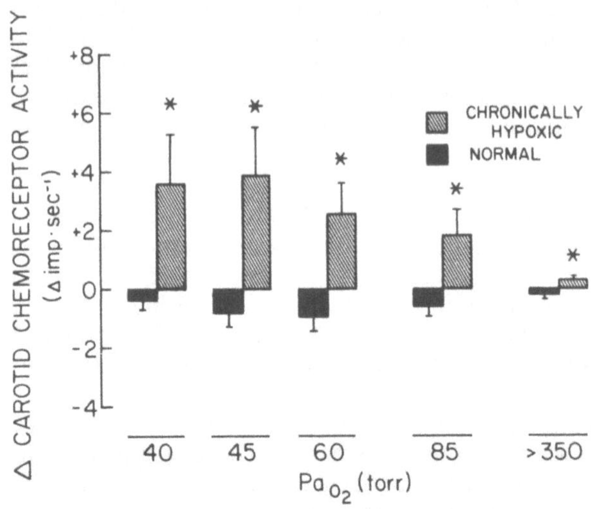

Figure 2. Augmentation of efferent effect during chronic hypoxia in the cat. The change of chemosensory activities (imp. sec^{-1}) after transecting the whole carotid sinus nerve are plotted against arterial PO_2.

e.g. brain and kidney (Barnett et al, 1988). Accordingly, it was predicted that the carotid bodies would be affected more than the central chemoreceptors and aortic body chemoreceptors during exposure of animals to 100% O_2.

The cats were exposed to 100% for 60-67 hr., and their ventilatory and carotid and aortic chemosensory responses were studied. Striking observations were that the subsequent hypoxic tests did not stimulate either ventilation or the carotid chemoreceptor activity (Lahiri et al.,1987), although aortic chemoreceptors were stimulated. Associated with the lack of hypoxic effect on the carotid chemoreceptors was the supersensitivity to hypercapnia. Moreover, the ventilatory response to CO_2 was not affected. These phenomena are illustrated in Figs. 3 and 4 for the same cat. Hypoxia was not stimulatory whereas hypercapnia was.

We found that the phenomena at the carotid body can be reproduced by the administration to the carotid body of reagents which prevent ATP synthesis coupled to oxidation (Mulligan et al., 1981). Accordingly, our interpretation is that chronic hyperoxia interferes with the energy metabolism in such a way that hypoxia cannot affect it anymore. Furthermore, the chemoreceptor cells are unable to maintain normal pHi regulation upon application of hypercapnia. From the normal chemosensory responses to hypercapnia (Lahiri and Delaney, 1975), it is reasonable to suggest that pHi in the chemoreceptor cells is not well regulated, and it parallels arterial pH. This relationship is further steepened as a result of chronic hyperoxia. Analogy with the effects of metabolic inhibitors suggests that inadequate metabolic energy is the key factor. It is well established that in most eukaryotic cells hydrogen ion concentration is maintained less than that dictated by the electrochemical gradient for H^+ (Roos and Boron, 1981). The energy is linked with the process which maintains Na^+ gradient across the cell membrane.

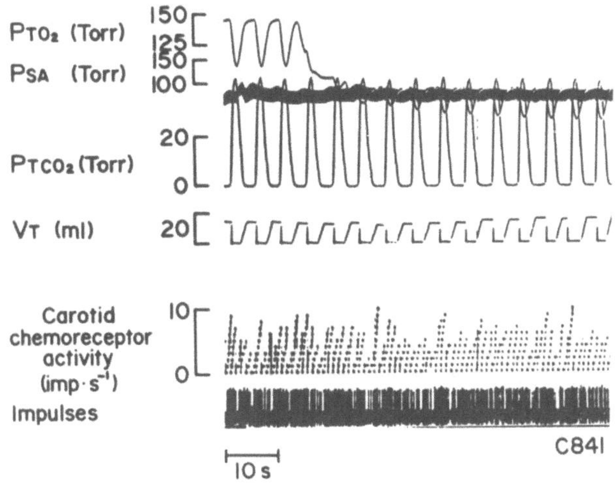

Figure 3. Effects of lowering tracheal PO_2 (PTO_2) on the carotid-chemosensory activity, tidal volume (VT), tracheal PCO_2 ($PTCO_2$) and arterial blood pressure in the chronically hyperoxic cat. Hypoxia did not stimulate carotid chemoreceptor activity. Consequently, ventilation was not stimulated and the end-tidal PCO_2 did not change.

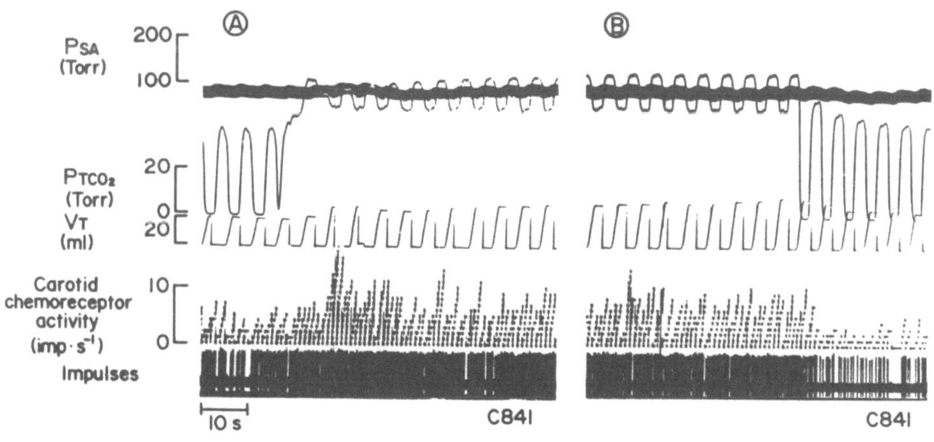

Figure 4. Effects of raising inspired and tracheal PCO_2 ($PTCO_2$) on carotid chemoreceptor activity, tidal volume (VT) and respiratory frequency and arterial blood pressure (PSA) in the same chronically hyperoxic cat as in Fig. 3.

CONCLUSION

Taken together, it is quite clear that carotid body as an oxygen sensing organ manifests plasticity of structure and function during the chronic changes in the oxygen environment. Augmented expressions for protein and enzyme synthesis are quite apparent during chronic hypoxia. The assumption that hypoxia increases synthesis and turnover rates of growth factors which, in turn, unleash cell growth and proliferation is reasonable. That the factor(s) is locally released in the carotid body is suggested by the fact that its response is not always associated with a stimulus which releases erythropoietin. For example, chronic inhalation of a trace of carbon monoxide leads to significant erythropoiesis without any responses of the carotid body (Sherpa et al., 1989). Elaboration of the local growth factor in the carotid body (including its capillary endothelium) is another significant issue in oxygen sensing in the carotid body.

The hyperoxic responses possibly envelope several processes: (a) adaptation to high PO_2 without the effects of free-radicals; (b) direct response to the reactive free- radicals, and (c) adaptive response to the free-radical injury. Although there is a hint that oxygen metabolism is the basis of oxygen chemoreception, it is curious that a mechanism which is selectively labile should be linked with a process so fundamental as the cellular oxidative metabolism. But then oxygen sensing is fundamental to the organisms for survival during impending hypoxia.

ACKNOWLEDGMENT

The paper owes much to many individuals who participated in the work which spanned over several years and to Suzanne Hyndman for her excellent secretarial assistance. The reference list is not exhaustive. Supported in part by grants HL-19737 and HL- 43413.

REFERENCES

Barnard P., Andronikou, S., Pokorski, M., Smatresk, N.J., Mokashi, A., and Lahiri, S., 1989, Time-dependent effect of hypoxia on carotid body chemosensory function. J. Appl. Physiol., 63:685-691.

Barnett, S., Mulligan, E., Wagerle, L. C., and Lahiri, S. 1988, Measurement of carotid body blood flow in the cat using radioactive microspheres, J. Appl. Physiol. 65:2486-2489.

Bisgard, G.E., Busch, M.A., and Forster, H.V., 1986, Ventilatory acclimatization to hypoxia is not dependent upon cerebral hypocapnic alkalosis, J. Appl. Physiol, 60:1011-1015.

Fitzgerald, R.S., and Lahiri, S., 1986, Reflex responses to chemoreceptor stimulation, in: Handbook of Physiology. The Respiratory System, Sec. 3, Vol. 2, Am. Physiol. Soc., Bethesda, pp. 313-362.

Fridovitch, I., 1988, The biology of oxygen radicals: general concepts, in: Oxygen Radicals and Tissue Injury, B. Halliwell, ed., FASEB, Bethesda, pp. 1-4.

Hanbauer, I., Karouson, F., Hellstrom, S., and Lahiri, S., 1981, Effects of long-term hypoxia on the catecholamine content in rat carotid body, Neuroscience, 6:81-86.

Lahiri, S., 1972, Unattenuated ventilatory hypoxic drive in ovine and bovine species native to high altitudes, J. Appl. Physiol. 32:95-102.

Lahiri, S. and Delaney, R.G., 1975, Stimulus interaction in the responses of carotid body chemoreceptor single afferent fibers, Respir. Physiol. 24:249-266.

Lahiri, S., 1976, Depressant effect of acute and chronic hypoxia on ventilation, in: Morphology and Mechanisms of Chemoreception, A.S. Paintal, ed., University of Delhi, Delhi, pp. 138-145.

Lahiri, S., Smatresk, N.J., Pokorski, M., Barnard, P., and Mokashi, A., 1983, Efferent inhibition of carotid body chemoreception in chronically hypoxic cats, Am. J Physiol. (Regulatory Integrative Comp. Physiol.14):R678-R683.

Lahiri, S., 1984, Respiratory control in Andean and Himalayan high altitude natives, in: High Altitude and Man, J.B. West and S. Lahiri, eds., Am. Physiol. Soc., pp. 147-162.

Lahiri, S., Smatresk, N.J., Pokorski, M., Barnard, P., Mokashi, A., and McGregor, K.H., 1984, Dopaminergic efferent inhibition of carotid body chemoreceptors in chronically hypoxic cats, Am. J. Physiol. (Regulatory Integrative Comp. Physiol. 17): R24-R28.

Lahiri, S., Mulligan, E., Andronikou, S., Shirahata, M., and Mokashi, A., 1987, Carotid body chemosensory function in prolonged normobaric hyperoxia in the cat, J. Appl. Physiol. 62:1924-1931.

Matsumoto, S., Mokashi, A., and Lahiri, S., 1987a, Cervical preganglionic sympathetic nerve activity and chemoreflexes in the cat, J. Appl. Physiol., 62:1713-1720.

Matsumoto, S., Mokashi, A., and Lahiri, S., 1987b, Ganglioglomerular nerves respond to moderate hypoxia independent of peripheral chemoreceptors in the cat, J. Autonom. Nerv. System 19:219-228.

Mokashi, A., Lahiri, S., 1991, Aortic and carotid body chemoreception in prolonged hyperoxia in the cat, Respir. Physiol. 186:233-243.

Mulligan, E., Lahiri, S., and Storey, B.T., 1981, Carotid body O_2 chemoreception and mitochondrial oxidative phosphorylation, J. Appl. Physiol. 51:438-446.

O'Regan, R.G., and Majcherczyk, S., 1983, Control of peripheral chemoreceptors by efferent nerves, in: "Physiology of the Peripheral Arterial Chemoreceptors,"H. Acker and R.G. O'Regan, eds., Elsevier, Amsterdam, pp. 257-298.

Pequignot, J.M., Cottet-Emard, J.M., Dalmaz, Y., and Peyrin, L., 1987, Dopamine and norepinephrine dynamics in rat carotid bodies during long-term hypoxia, J.Autonom. Nerv. System. 21:9-14.

Pokorski, M., and Lahiri, S., 1991, Endogenous opiates and ventilatory acclimatization to chronic hypoxia in the cat, Respir. Physiol. 83:211-222.

Richter, D. W., Bischoff, A., Anders, K., Bellingham, M., and Windhorst, U., 1991, Response of the medullary respiratory network of the cat to hypoxia, J. Physiol., 443:231-256.

Roos, A., and Boron, W.F., 1981, Intracellular pH. Physiol Rev. 61:296-433.

Sherpa, A.K., Albertine, K.H., Penney, D.G., Thompkins, B., Lahiri, S., 1989, Chronic CO exposure stimulates erythropoiesis but not glomus cell growth, J. Appl. Physiol. 67:1383-1387.

Tenney, S.M., and L.C. Ou, 1977, Hypoxic ventilatory response of cats to high altitude: an interpretation of blunting, Respir. Physiol. 30:185-189.

Vizek, M., Pickett, C.K., and Weil, J.V., 1987, Increased carotid body hypoxic sensitivity during acclimatization to hypobaric hypoxia, J.Appl. Physiol. 63:2403-2410.

CAROTID BODY CHEMOTRANSDUCTION

Robert S. Fitzgerald and Machiko Shirahata

Departments of Environmental Health Sciences, Physiology, Medicine
Anesthesiology/Critical Care Medicine
The Johns Hopkins Medical Institutions
Baltimore, Maryland 21205 USA

INTRODUCTION

Organisms have a fundamental dependence on their environment for continued survival. Most organisms can do without solid and liquid nutrition for several hours, sometimes days or even weeks. However, the other substrate needed to convert nutrition into biological energy is oxygen. Virtually all aerobic organisms cannot survive without environmental oxygen for longer than a few minutes. Diving mammals seem to be an exception to this dependence. In the course of evolution part of the machinery which these organisms have developed for delivering environmental oxygen to their tissues is the pulmonary and cardiovascular systems. Essentially each involves a pump, a series of collapsible tubes through which a fluid flows, and a gas exchanging surface. The control of these systems in the face of a challenge such as exercise or an unfavorable environment is of critical importance. The pumps might have to beat faster, the tubes may have to alter their resistance in general or regionally to redistribute the flow to various organs or parts of organs. The cardiopulmonary system may have to change overall or regional compliances to facilitate flow. These mechanical variables are controlled by local, humoral, and neural input. The neural control of the cardiopulmonary system is frequently described as the input/output of Receptors-Afferent Pathways-Central Nervous System-Efferent Pathways-Effectors. The carotid body is one of the Receptors in this model. Sampling the arterial blood, it increases its neural input into the nucleus tractus solitarius when the partial pressure of oxygen in the arterial blood falls, the carbon dioxide partial pressure or hydrogen ion concentration rises. Presently it is unclear how the carotid body converts hypoxia, hypercapnia, or acidosis into increased neural activity; that is, how it chemotransduces.

A working model of the carotid body's chemotransduction which is based on the morphology and histochemistry of the organ proposes that the stimuli somehow depolarize the Type I cell, or glomus cell. As a result intracellular calcium increases which promotes the release of an excitatory neurotransmitter. The neurotransmitter binds to a receptor on the apposed dendrite, depolarizing it and an action potential results. From such a model two questions arise: "What are the mechanisms responsible for depolarizing the Type I cell?" "What is (are) the excitatory neurotransmitter(s)?" The present report addresses the second question.

Control of Breathing and Its Modeling Perspective, Edited by
Y. Honda *et al.*, Plenum Press, New York, 1992

Specifically our hypothesis is that acetylcholine (ACh) is an excitatory neurotransmitter in the process of the carotid body's chemotransduction of hypoxia, hypercapnia, and acidosis. ACh was the first neurotransmitter thought to be involved in carotid body chemotransduction. Swedish investigators were particularly prolific in generating data supportive of an excitatory role for ACh. Subsequently, British investigators and others contested a role for ACh (for review see references 1 - 4). However, the opponents of the "Cholinergic Hypothesis" left open to a certain extent the possibility that ACh might be involved. Various contributions began to establish that virtually all the criteria that had been accepted as needing fulfillment in order to establish a substance as a neurotransmitter were being met in the case of ACh. Subsequent data generated by such investigators as Eyzaguirre, supportive of a role for ACh, were once again not broadly accepted as sufficiently persuasive. Currently, no single substance is broadly accepted as the excitatory neurotransmitter in carotid body chemotransduction.

METHODS

Two sets of experiments were performed. The animal preparation for each was the same. Cats were anesthetized with sodium pentobarbital (30mg/Kg) and fitted with tracheal cannula, arterial and venous catheters. The carotid body area was prepared as described previously[5,6]. In brief a loop fitted with a three-way stopcock was inserted into the common carotid artery. The carotid sinus nerve was prepared for whole nerve recording. In the first set of experiments the baroreceptors were removed by mechanical and thermal destruction. Hence, carotid sinus nerve output was exclusively or predominantly due to chemoreceptor activity. In the second set of experiments the baroreceptors were left intact.

The protocol for the first set of experiments called for the paralyzed animal to be ventilated on room air enriched with oxygen. This was followed by a mixture of 10% oxygen in nitrogen. Carotid body neural output rose to a maximum as hypoxic blood passing through the loop perfused the carotid body. After five minutes the stopcock was turned stopping the flow of blood, and the carotid body was perfused for two minutes with normoxic-normocapnic Krebs Ringer bicarbonate (KRB) solution. The stopcock was then returned to the original position so that hypoxic blood once again perfused the carotid body, elevating the neural output toward the original maximum. After the second five minutes of perfusion with hypoxic blood, the stopcock was adjusted for a second two minute perfusion of normoxic-normocapnic KRB. However, this time the KRB contained 2μM alpha-bungarotoxin, 134μM mecamylamine, and 314μM atropine. Then for the third time hypoxic blood was admitted into the carotid body.

The protocol for the second set of experiments called for the paralyzed animal to be ventilated on room air enriched with oxygen. The carotid body was first perfused by the animal with his own normoxic-normocapnic blood. Then the stopcock was adjusted so that the carotid body could be perfused with hypoxic KRB or with hypoxic KRB containing a mixture of the cholinergic blockers -- 2μM alpha bungarotoxin, 402μM mecamylamine, and 942μM atropine. The perfusion pump used in these experiments delivered the KRB in a pulsatile manner. Hence, the delivery had "systolic" and "diastolic" pressures.

RESULTS

The results of the first set of experiments (n = 4) are shown in Table 1.

The results from the first set of experiments showed that the response of the carotid body after a perfusion of normoxic-normocapnic KRB was somewhat similar to the original response to hypoxic blood. The response at 1 minute was significantly greater than the initial response at that point, but was 5% lower at the 3 minute mark. This was significantly lower. At the 5 minute mark there was no difference between the two responses to hypoxic blood. Hence, the overall effect was roughly the same although the pattern of response was somewhat different. The response to hypoxic blood after the KRB perfusion containing the blockers was significantly lower than the original response to hypoxic blood at the 3 and 5 minute marks. The response to nicotine paralleled these responses to hypoxic blood, being reduced after the blockers to 63% of the maximum response to nicotine administered before the blockers.

Table 1. Response to Hypoxic Blood (% of initial response at 5 minutes).

	CTL	1 MIN	3 MIN	5 MIN
Initial response to hypoxic blood	38±3	65±5	92±3	100
Response to hypoxic blood after KRB perfusion	--	76[a]±2	87[a]±1	96±1
Response to hypoxic blood after KRB plus blockers perfusion	--	63[b] ±3	76[ab] ±4	83[ab] ±3

[a] ($P<0.05$); different from Row 1 values; [b] from Row 2 values

The concentrations of blockers in the perfusates could be considered somewhat elevated, though it is highly unlikely that the tissues of the carotid body ever came into equilibrium with the perfusates. The purpose of the second series of experiments was to determine if the blocking action was due to non-specific blocking of neural transmission, and not a specific blocking of hypoxic chemotransduction activity. We assumed that if the effects we have reported were due to non-specific blocking, then the activity of the baroreceptors, at least as exposed to the perfusate as the chemoreceptors were, should decrease also. Table 2 shows clearly how during the hypoxic KRB perfusate the total integrated neural activity increased and exhibited a pulsatile activity having a mean neural "pulse pressure" of 0.60 μV. During the perfusion with the hypoxic KRB containing the blockers, the total integrated neural activity initially increased, but then decreased dramatically to about 35% of the maximum. However, the pulsatile response of the integrated activity continued to correlate with the pulsatile pressure generated by the perfusion pump, with no diminution whatsoever (mean neural "pulse pressure" was 0.68 μV).

Table 2

A. RESPONDING TO HYPOXIC PERFUSION	CTL	5 SEC	15 SEC	30 SEC	60 SEC	90 SEC	RECOV
P_{SP} (mmHg)	0	130	130	130	125	125	0
P_{DP} (mmHg)	0	85	80	80	75	77	0
PEAK INTEG. ACTIV. (μV)	1.1	3.4	6.8	9.4	9.1	9.1	0.4
TROUGH INTEG. ACTIV. (μV)	0.9	2.4	6.2	9.0	8.5	8.7	0.2
B. RESPONDING TO HYPOXIC PERFUSATE CONTAINING BLOCKERS							
P_{SP} (mmHg)	0	150	135	130	120	120	0
P_{DP} (mmHg)	0	95	90	85	75	80	0
PEAK INTEG. ACTIV. (μV)	0.7	2.8	6.1	5.4	3.8	3.3	0.7
TROUGH INTEG. ACTIV. (μV)	0.3	2.0	5.4	4.8	3.1	2.7	0.5

DISCUSSION

The data demonstrate that a cocktail of cholinergic blockers is capable of reducing the response of the carotid body to both hypoxic blood and to a hypoxic perfusate. Further, the data suggest that the reduction is not due to a non-specific blocking of nerves. It is worth calling attention to the fact that the concentration of the blockers in the second set of experiments was three times that in the first set of experiments.

How the Cholinergic Hypothesis is to be understood is not yet clear. The obvious understanding would be the one described in the model above. However, it is not at all clear that the presynaptic component, the Type I cell, releases ACh to bind to a receptor on the apposed dendrite. This model is the simplest, but the data presented here simply support an excitatory role for ACh in the process of chemotransduction in the carotid body.

Some investigators propose dopamine as the excitatory neurotransmitter. Dopamine is present in the carotid body in high concentrations. However, under those circumstances in which the exogenous administration of cholinergic agents evokes an increase in neural activity from the carotid body, the exogenous administration of dopamine ordinarily reduces the neural activity. But since high concentrations of dopamine under some conditions do increase neural activity, it is conceivable that if dopamine, released from the Type I cells, reached high enough concentrations in the synaptic cleft between the Type I cell and the apposed neuron, increased neural activity would result.

If this were the model, then our interpretation of our results would necessarily be different. Our interpretation would focus on the fact that ^{3}H QNB binding sites, putative muscarinic receptors, have been located on the Type I cell. If we postulate that hypoxia releases acetylcholine from the Type I cell and the acetylcholine proceeds to act on the QNB binding sites, functioning as autoreceptors, to release dopamine, then the action of our blockers would ultimately be to suppress the release of dopamine.

In summary we have shown that the administration of a cocktail of cholinergic blockers can reduce the response of the carotid body to hypoxic blood, and that the response of the carotid body to nicotine decreases also. Secondly we have shown that whereas the carotid body's response to a perfusion of the hypoxic KRB containing the blockers was significantly less than the carotid body's response to hypoxic KRB alone, the response of the carotid sinus baroreceptors was unaffected. These data support a role for acetylcholine in the response of the carotid body to hypoxia.

REFERENCES

1. W.W. Douglas. Is there chemical transmission at chemoreceptors? *Pharmacol. Rev.* 6:81-83 (1954).
2. C. Eyzaguirre, R.S. Fitzgerald, S. Lahiri, and P. Zapata. Arterial chemoreceptors, *in*: "Handbook of Physiology Section 2: The Cardiovascular System. Vol III. Peripheral Circulation and Organ Blood Flow," J.T. Sheperd and F.M. Abboud, eds., American Physiological Society, Bethesda, MD (1983).
3. S.J. Fidone and C. Gonzalez. Initiation and control of chemoreceptor activity in the carotid body, *in*: "Handbook of Physiology Respiration Vol. II. Part I. Control of Breathing," N.S. Cherniack and J.G. Widdicombe, eds., American Physiological Society, Bethesda, MD (1986).
4. G. Liljestrand. Transmission at chemoreceptors, *Pharmacol. Rev.* 6:73-78 (1954).
5. M. Shirahata and R.S. Fitzgerald. Dependency of hypoxic chemotransduction in cat carotid body on voltage-gated calcium channels, *J.Appl.Physiol.* 71:1062-1069 (1991).
6. M. Shirahata and R.S. Fitzgerald. The presence of CO_2/HCO_3^- is essential for hypoxic chemotransduction in the in vivo perfused carotid body, *Brain Res.* 545:297-300 (1991).

ALTITUDE ACCLIMATIZATION AND HYPOXIC VENTILATORY DEPRESSION : LOWLANDERS AND HIGHLANDERS

Shigeru Masuyama, Masashi Hayano, Akira Kojima, Kiyoshi Hasako, Takayuki Kuriyama, Atsuko Masuda*, Toshio Kobayashi*, Yoshikazu Sakakibara*, and Yoshiyuki Honda*

Department of Chest Medicine and Physiology*, School of Medicine, Chiba University, Chiba 280, Japan

INTRODUCTION

It has been well established that altitude acclimatization not only induces augmented ventilatory response to hypoxia[3] but also deteriorative effect on ventilation[1]. The latter is well known as hypoxic ventilatory depression (HVD) for patients with chronic mountain sickness[2,4]. As is presented in our other report in this proceedings[8], in early acclimatizing period we should take such HVD into account for comprehending serial change of ventilatory response. HVD in humans has been observed and studied by many scientists. In addition to biphasic ventilatory response to sustained mild hypoxia[5], paradoxical hyperpnea in response to oxygen administration in patients with chronic mountain sickness is also another important topics in current studies[7,10]. If a subject developed hyperpnea by oxygen breathing, it can be considered to be relieved from hypoxic ventilatory depression.

Our first question is whether the lowlander sojourners exhibited HVD at altitude or not? Secondly the same question wasexplored in healthy highlanders. Finally, whether or not HVD reveals the same characteristic features in both low- and highlanders.

METHODS

Subjects are eighteen young healthy Tibetan males and age matched 22 lowlander trekkers in Lhasa (3700m), Tibet. Their VE, VO_2, VCO_2 and ventilatory parameters were monitored continuously from breath by breath measurement using expiratory gas analyzer (MINATO MG 360, Tokyo, Japan) and a metabolic computer (MINATO RM300, Tokyo, Japan) at rest and during oxygen challenge and withdrawal test. Fig.1 illustrates the scheme of ventilatory response to oxygen inhalation in Lhasa (altitude 3700m ; barometric pressure, 470mmHg).

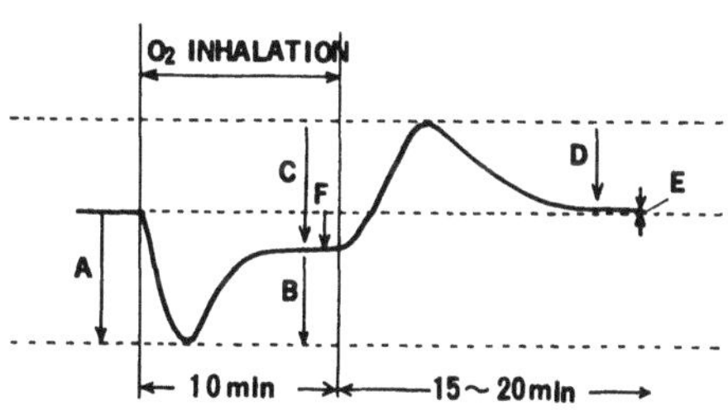

Fig.1 The scheme of ventilatory response to oxygen inhalation in Lhasa (3700M).

The subject was first maintained by inhaling room air for 5 minutes, then

Control of Breathing and Its Modeling Perspective, Edited by Y. Honda *et al.*, Plenum Press, New York, 1992

inspired humidified 100% O_2 for ten minutes from a Douglas bag and finally switched to room-air breathing again for longer than 15 minutes. We monitored continuously ventilatory parameters and calculated the magnitude of changes elicited by above experimental procedure. These data were analyzed in conjunction with hypoxic ventilatory response (HVR) by the isocapnic progressive hypoxia method performed on the same day.

RESULTS

Lowlanders

Fig.2-A illustrates typical time course of a lowlander trekker response to oxygen breathing. Immediately after oxygen inhalation, VE showed abrupt undershoot, which was followed by gradual recovery to the stable level that was smaller than that in room air breathing. $ETCO_2$ showed the mirror image of VE change. When returned to room air breathing, VE began to increase to the peak of ventilation then, in turn, VE began to fall and attained to the original level.

Fig.3 shows relationship among ventilatory changes due to oxygen and room air breathing. The upper figure demonstrates that initial increase in ventilation after room air

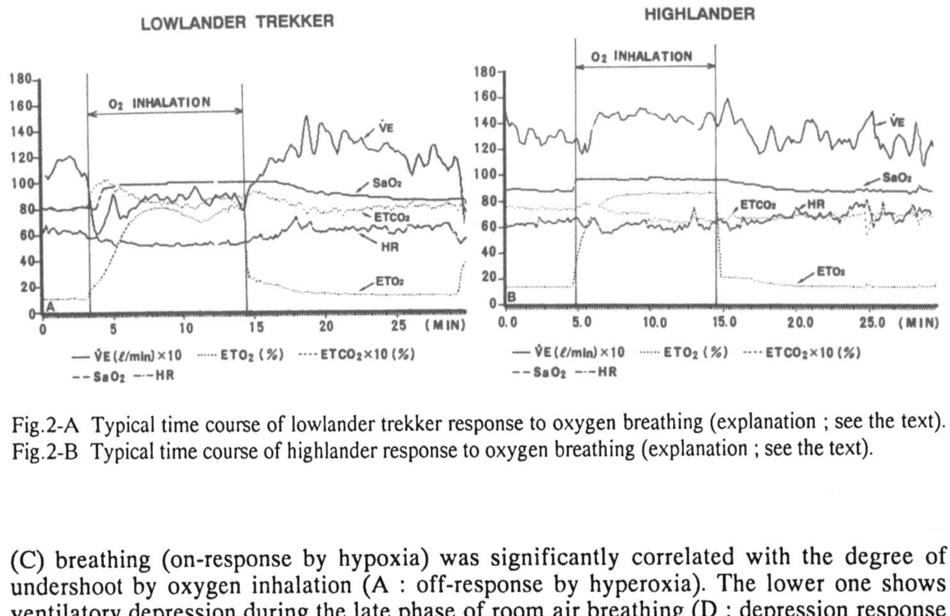

Fig.2-A Typical time course of lowlander trekker response to oxygen breathing (explanation ; see the text).
Fig.2-B Typical time course of highlander response to oxygen breathing (explanation ; see the text).

(C) breathing (on-response by hypoxia) was significantly correlated with the degree of undershoot by oxygen inhalation (A : off-response by hyperoxia). The lower one shows ventilatory depression during the late phase of room air breathing (D : depression response by hypoxia) was correlated with the ventilatory recovery during oxygen breathing (B : off-response).

Fig.4 shows the relationship between ventilatory parameters and HVR. The upper one shows that there is a significant correlation between initial increment in ventilation after room air (C) and the degree of HVR. Significant relationship between undershoot of off-response (A) and HVR is also detected.

Highlanders

In Fig.2-B, a typical time course of highlander response to oxygen breathing is shown. The initial undershoot was not comprehensive. VE increased to higher steady level than that in room air breathing. Though no apparent biphasic ventilatory response to hypoxia was seen, we can state that hypoxic ventilatory depression is still exist, because oxygen inhalation augmented ventilation.

As shown in Fig.5, no significant relationship was detected between paradoxical response represented onresponse (negative value of C) and HVR.

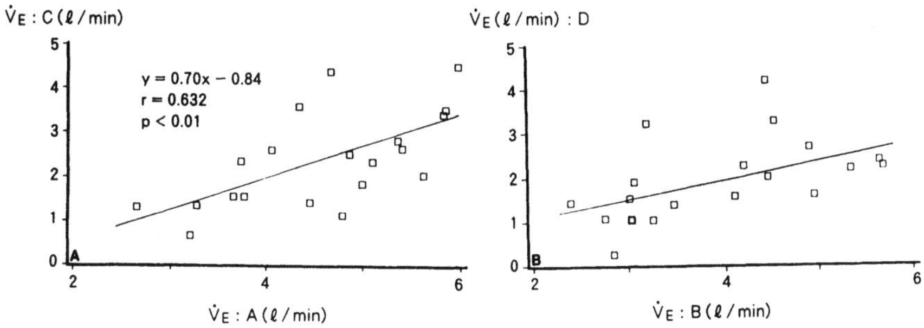

Fig.3 The initial increase in ventilation at on response (C) was significantly correlated with the degree of undershoot represented offresponse (A) (the upper). Ventilatory depression at onresponse (D) was correlated with the ventilatory recovery at offresponse (B) (the lower).

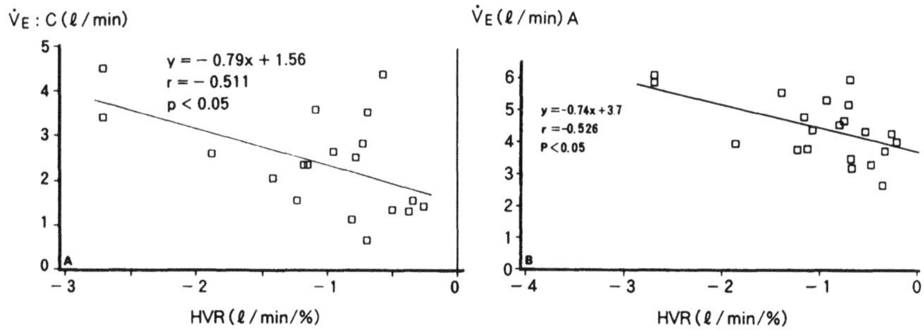

Fig.4 Ventilatory parameters and HVR (explanation ; see the text).

DISCUSSION

Fig.2-A shows clearly that resting ventilation in room air at high altitude in lowland sojourners exhibited hypoxic ventilatory depression. They showed the depression even in air breathing at 3700m.

Fig.3 suggests that ventilatory undershoot in off-response is induced mainly by disappearance of stimulation from the peripheral chemoreceptors and that its recovery within several minutes by oxygen might be induced centrally[1]. These speculation might be supported by Fig.4. It demonstrated that on and off responses to hypoxic stimulation (C and A) are well correlated with HVR. All of these findings in lowlanders suggest that ventilatory depression and stimulation during our experiment are explained on the basis of chemical control of ventilation.

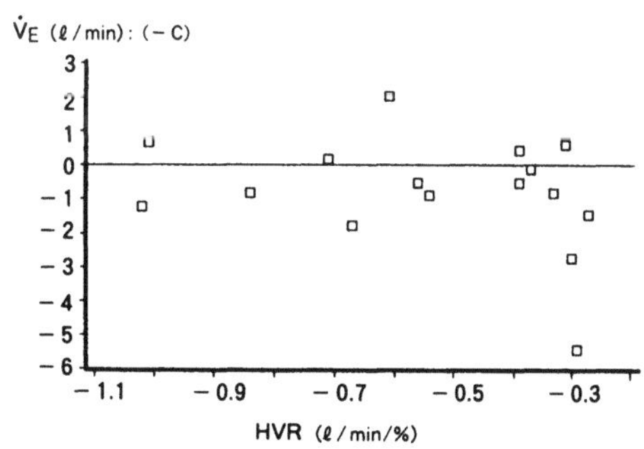

Fig.5 No significant relationship was detected between paradoxical hyperpnea response represented on-response (negative value of C) and HVR.

As for healthy highlanders. As shown in Fig.2-B, they exhibited the paradoxical hyperpnea-response to oxygen breathing as was found in CMS people[2,7]. Note that they are young and healthy highlander people. They showed larger VE, $PETO_2$ and SaO_2 than lowland trekker, though their HVR were relatively small[6]. Fig.5 does not find any relationship between HVR and the magnitude of paradoxical hyperpnea-response. These data suggest that the degree of blunting in hypoxic ventilatory response does not account for this paradoxical hyperpnea as had been claimed[2,7]. The characteristics of control of ventilation in highlanders, especially response to hypoxia, remains under debate[9,11]. Though we could find HVD in both lowland trekker and healthy highlanders, it was impossible to explain the depression of both groups on the same physiological basis. HVD in highlanders might be explained by other mechanics than we tried elucidate in this study.

SUMMARY

Though both highlander and lowlander people showed so called "hypoxic ventilatory depression", underlying mechanisms might be quite different between these depressions.

REFERENCES

1. P.A. Easton, L.J. Slykerman, N.R. Anthonisen, Ventilatory response to sustained hypoxia in normal adults. *J. Appl. Physiol.* 61 : 906-911, (1986).
2. P.H. Hackett, et. al. Control of breathing in Sherpas at low and high altitude. *J. Appl. Physiol.* 49 (3) : 374-379, (1980).
3. Y. Honda, N. Hata, Y. Sakakibara, et al. Central hypoxic-hypercapnic interaction in mild hypoxia in man. *Pfluegers. Arch.* 391 : 289, (1981).
4. A. Hurtado, Animals in high altitudes : resident man. *In*: Handbook Physiology. Adaptation of Environment. Washington D.C. : *Am. Physiol. Soc.*, p843-860, (1964).
5. S. Kagawa, et al. No effect of naloxone on hypoxia-induced ventilatory depression in adults. *J. Appl. Physiol.* 52 : 1030-1034, (1982).
6. T. Kobayashi, S. Masuyama, A. Masuda, Y. Sakakibara, M. Hayano, A. Kojima, K. Hasako, T. Kuriyama and Y. Honda, Control of Breathing and Metabolism in Tibetans. *in*: "High-Altitude Medical Science" G. Ueda et al., eds. Shinshu Univ., Matsumoto (1992)
7. Lahiri,S., J.S.Milledge et al. Respiration and heart rate of Sherpa highlanders during exercise. *J. Appl. Physiol.* 23 : 545-554, (1967).
8. A. Masuda, S. Masuyama, T. Kobayashi, Y. Sakakibara, M. Hayano, K. Hasako, A. Kojima, T. Kuriyama and Y. Honda, Serial Changes in Acute Hypoxic and Hypercapnic Ventilatory Responses during High Altitude Acclimatization. *in*: "Control of Breathing and Its Modeling Perspective" Y. Honda et al., eds. Plenum, New York (1992)
9. S. Masuyama, K. Hasako, A. Kojima, T. Kuriyama, Y. Honda, Do Nepalese Sherpas Maintain High Hypoxic Ventilatory Drive? *Jap. J. Mount. Med.* vol.10 : 81-90, (1990).
10. S.C. Severinghaus, C.R. Bainton, and A. Careln, Respiratory insensitivity to hypoxia in chronically hypoxic man. *Respir. Physiol.*, 1 : 308-334, (1966).
11. S. Sun, et al. Higher ventilatory drives in Tibetan than male residents of Lhasa (3685M). *Am. Rev. Respir. Dis.* 137 : A140, (1988).

SERIAL CHANGES IN ACUTE HYPOXIC AND HYPERCAPNIC VENTILATORY RESPONSES DURING HIGH ALTITIUDE ACCLIMATIZATION

Atsuko Masuda, Shigeru Masuyama*, Toshio Kobayashi, Yoshikazu Sakakibara, Masashi Hayano, Akira Kojima*, Kiyoshi hasako* Takayuki Kuriyama* and Yoshiyuki Honda

Department of Physiology and Department of Chest Medicine*,
School of Medicine
Chiba University
Chiba, 280 Japan

INTRODUCTION

Increased ventilation is important for persons going to high altitude and this is attributable to hypoxic stimulation of the peripheral chemoreceptors. However, the hypocapnia induced by hyperventilation and central hypoxic ventilatory depression substantailly attenuate the initial ventilatory response to hypoxia. Huang *et al.* (5) demonstrated that the combination of hypocapnia and sustained hypoxia might have blunted the ventilatory response on the first day after arrival on Pikes Peak (4300 m). Acute mountain sickness (AMS) occurs as acommon feature of ascent to high altitude usually within a few hours of ascent and lasting a few days. It has been recognized that ventilation is depressed in persons with symptoms of AMS (4). The reason for the hypoventilation in symptomatic persons is unknown. Moore *et al.* (8) suggest that hypoventilation in symptomatic compared to asymptomatic subjects is responsible for attenuated ventilatory response to acute hypoxia. In order to investigate the influence of altitude acclimatization on chemoresponsiveness, we consecutively examined hypoxic (HVR) and hypercapnic (HCVR) ventilatory responses in 7 lowlanders during sojourn in Lhasa (3700m), China.

METHODS

Seven normal healthy Japanese (age 24-42 yr) participated in the study. Although they live and spend daily life at sea level, they had no remarkable symptoms of AMS during sojourn in Lhasa. Hypoxic ventilatory response (HVR) by the progressive isocapnic hypoxia test (11) and hypercapnic ventilatory response (HCVR) by Read's rebreathing method (9) were examined consecutively. We measured ventilatory parameters by Minato Medical Expired Gas analyzer (Minato MG360) and Metabolic Computer (Minato RM300) at rest and during hypoxic and hypercapnic tests. Arterial oxygen saturation (Sa_{O_2}) and heart rate were monitored by pulse oximeter (Minolta Pulsox-7).

RESULTS

Fig. 1 illustrate typical changes in HVR and HCVR during sojourn in Lhasa. As shown in the upper panels, up to the 6th day after arrival at Lhasa, both response slopes decreased consecutively. On the other hand, following 6th day, the slope began to increase. In case of HVR, at 12th day, it exceeded the level of the first day and shifted to the right, showing the augmented oxygenation. The slope of HCVR also turned to increase and shifted to left, indicating developing of hypocapnia.

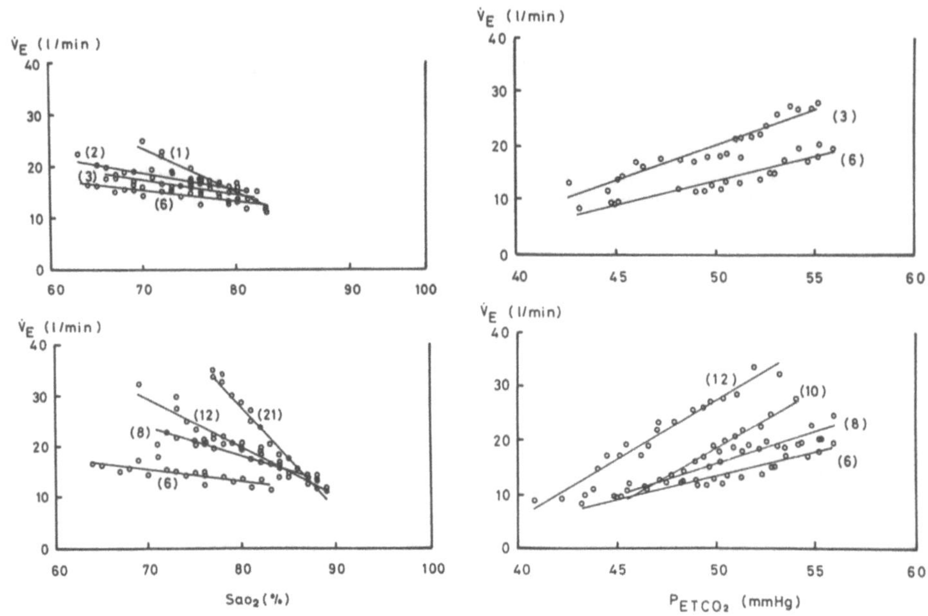

Fig. 1. Typical changes in HVR (left) and HCVR (right) during sojourn in Lhasa. The number in the parentheses indicate the day after arrival at Lhasa.

Serial changes in HVR and HCVR during sojourn in Lhasa of all subjects were summarized in left panels of Fig. 2. During the first several days after arrival at Lhasa, both response slopes decreased continuously, then converted to increase. Serial changes in Sao_2 and $\dot{V}_E$ at rest were shown in right panels of Fig. 2. For several days after arrival at Lhasa, Sao_2 stayed around 82 %, then increased up to about 88 %. On the other hand, $\dot{V}_E$ tended to augmented during initial several days, then turned to decrease gradually and attained the steady level.

DISCUSSION

The slope of both response curves, *i.e.*, HVR and HCVR, were changed with time. It gradually decreased during first several days, then turned to increase moderately up to about 10th day. Thereafter the slope progressively increased for more than 10 days, thus exceeding the value of the first day.

The influence of altitude acclimatization on ventilatory chemosensitivity has been a topic of this field for years. With regard to overall alteration of ventilatory responsiveness, our results except for downward shift of the response in early period agree with the findings of previous investigators.

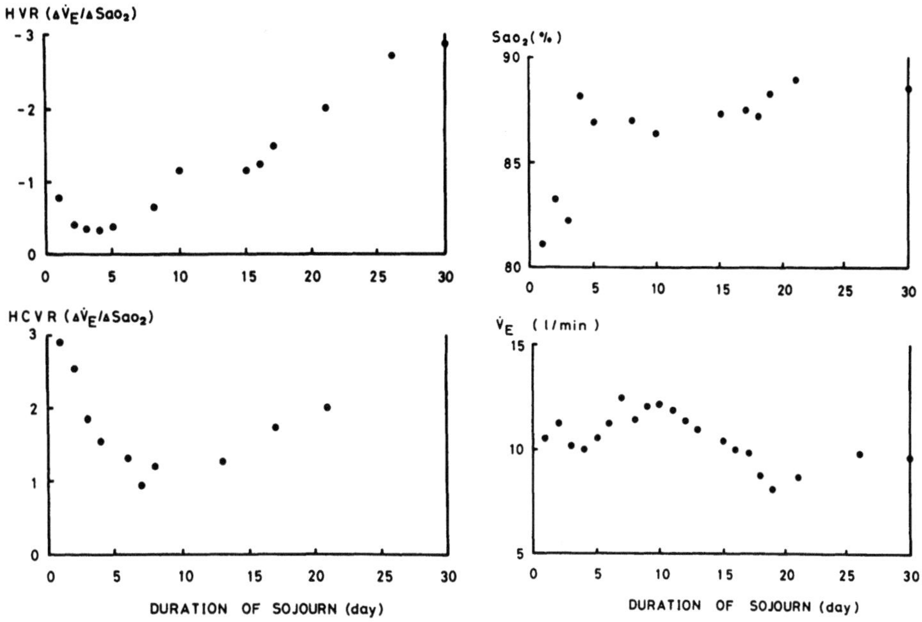

Fig. 2. Serial changes in HVR and HCVR (left) and Sao_2 and $\dot{V}_E$ at rest (right) during sojourn in Lhasa.

During acclimatization, there seems to develop a progressive increase of response. Forster *et al.* (3) observed that the hypoxic ventilatory response increased in the lowlanders during their sojourn at altitude 3000 m. White *et al.* (10) measured ventilatory response to acute isocapnic hypoxia at various period after ascent to the summit of Pikes Peak (4300 m). They reported that substantial and progressive steepening of the slope of the response despite a progressive decline in the basal Pco_2 at which the responses were measured. Weil (12) interpreted that such a progressive increase in hypoxic ventilatory chemosensitivity could be an important contributor to the progressive augmentation of ventilation seen during acclimatization.

Many studies have been conducted to see the influence of altitude acclimatization on ventilatory response to CO_2 as well as hypoxia (1,2,3,6,7). The consistent findings on the CO_2-ventilation response curve are as follows: 1) The response curve is shifted left so that magnitude of horizontal intercept or threshold for Pco_2 stimulus is diminished. 2) The slope or sensitivity of the CO_2 response curve is steepened. In this study, we also confirmed same tendency during later period sojourn in Lhasa.

On the other hand, we found that there was gradual diminution of chemosensitivity in early period after arrival at altitude and concurrently Sao_2 have been stayed at low level. Previous studies were performed every third or fourth days so that probably nobody noticed this phenomenon. Moore *et al.* (8) observed that in the measurements made prior to altitude exposure, ventilatory responsiveness to acute hypoxia was weaker in the subjects with histories of acute mountain sickness compared to the persons without such history. And they also found that symptomatic subjects hypoventilated at 4800 m relative to asymptomatic individuals accompanied by lower Sao_2. It may well be that the lowlanders susceptible to AMS may be suffered during the period of this depressed ventilatory activity with arterial desaturation. When the lowlanders recovered from this depression and augmented their ventilatory response more than the control level with improvement low Sao_2, danger for AMS will be less. We suggest that these serial changes in ventilatory chemosensitivity is an important physiological mechanism in high altitude acclimatization.

SUMMARY

We investigated the respiratory acclimatization to alitude. For this purpose, we have consecutively measured hypoxic ventilatory response (HVR) and hypercapnic ventilatory response (HCVR) in 7 lowlanders after arrival at Lhasa (3700 m). Both response curves showed biphasic change, *i.e.*, depressed for the first several days and then turned to increase. The initial attenuated responsiveness was accompanied by low Sao_2. This blunted chemosensitivity and desaturation might be accounted for possible risk factors of AMS. And danger for AMS will be less when the lowlanders recovered from this initial depression.

REFERENCES

1. J.C.Cruz,J.T.Reeves,R.F.Grover,J.T.Maher,R.E.McCullough,A.Cymerman,and J.C.Denniston,Ventilatory acclimatization to high altitude is prevented by CO_2 breathing,Respiration 39:121(1980).
2. E.J.Eger,Influence of CO_2 on ventilatory acclimatization to altitude,J.Appl.Physiol. 24:607(1968).
3. H.V.Forster,J.A.Dempsey,M.L.Birnbaum,W.G.Reddan,J.Thoden,R.F.Grover, and J.Rankin,Effect of chronic exposure to hypoxia on ventilatory response to CO_2 and hypoxia,J.Appl.Physiol.31:586(1971).
4. P.H.Hackett,D.Rennie,S.E.Hofmeister,R.F.Grover,E.B.Grover,and J.T.Reeves,Fluid retention and relative hypoventilation in acute mountain sickness,Respiration 43:321(1982).
5. S.Y.Huang,J.K.Alexander,R.F.Grover,J.T.Maher,R.E.McCullough, R.G.McCullough,L.G.Moore,J.B.Sampson,J.V.Weil, and J.T.Reeves, Hypocapnia and sustained hypoxia blunt ventilation on arrival at high altitude,J.Appl.Physiol,56:602(1984).
6. S.Lahiri,Dynamic aspects of regulation of ventilation in man during acclimatization,Respir.Physiol,16:245(1972).
7. C.C.Michel and J.S.Milledge,Respiratory regulation in man during acclimatization to high altitude,J.Physiol.(London),168:631(1963).
8. L.G.Moore,G.L.Harrison,R.E.McCullough,R.G.McCullough,A.J.Micco,A.Tucker, J.V.Weil, and J.T.Reeves, Low acute hypoxic ventilatory response and hypoxic depression in acute altitude sickness, J.Appl.Physiol.60: 1407(1986).
9. D.J.C.Read,A clinical method for assessing the ventilatory response to carbon dioxide,Australian Ann.Med.16:20(1967).
10. D.P.White,K.Gleeson,C.K.Pickett,A.M.Rannels,A.Cymerman,and J.V.White, Altitude acclimatization: influence on periodic breathing and chemoresponsiveness during sleep, J.Appl.Physiol.63:401(1987).
11. J.V.Weil,E.Brynne-Quinn,I.E.Sodal,W.O.Friesen,B.Underhill,G.F.Filley,and R.F.Grover,Hypoxic ventilatory drive in normal man, Clin. Invest. 49:1061(1970).
12. J.V.Weil,Control of ventilation in chronic hypoxia:Role of peripheral chemoreceptors,in:"Response and adaptation to hypoxia:Organ to organelle," S.Lahiri,N.S.Cherniack,and R.S.Fitzgerald,ed.,Oxford University Press,New York, Oxford,(1991).

ABNORMAL BREATHING PATTERN AND OXYGEN DESATURATION DURING SLEEP AT HIGH ALTITUDE

Koji Asano[1], Akio Sakai[1], Yasunori Yanagidaira[1], and Yukinori Matsuzawa[2]

[1]Dept. of Environmental Physiology and [2]1st Dept. of Medicine
Shinshu University School of Medicine, Matsumoto 390, Japan

INTRODUCTION

It is well known that healthy low-altitude residents often develop abnormal breathing pattern and oxygen desaturation during sleep at high altitude, although few studies of sleep during long-term sojourn at high altitude have been conducted. The purpose of this study was to investigate changes in breathing pattern and oxygen desaturation during sleep at high altitude, and their relation to pattern of ascent and descent, as well as to the acclimatization process in low-altitude residents.

METHODS

The present study was performed as part of a research expedition in Qinghai Plateau (China) in 1990. Eight Japanese males (23-47 yrs) who resided in regions below 610m altitude and who had no cardiopulmonary or sleep disorder were studied. The sleep study was

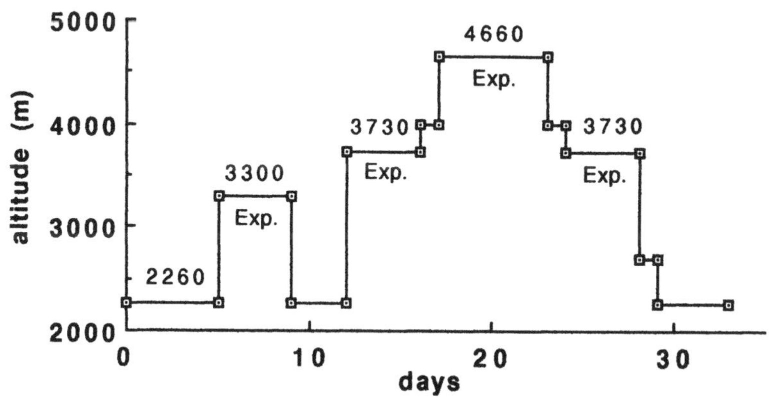

Figure 1. A profile of the ascent and descent in the research expedition. Subjects spent 4-5 nights at each altitude. Sleep was studied at the altitude indicated by "Exp." during the expedition.

Control of Breathing and Its Modeling Perspective, Edited by
Y. Honda *et al.*, Plenum Press, New York, 1992

carried out at the following altitudes: 3,300m (Qinghai Lake); 3,730m (Maqin, after brief descent to 2,260m); 4,660m (Base Camp of Mt. Amnemaqin); and 3,730m on descent. Fig. 1 shows a profile of ascent and descent during the expedition. Subjects stayed for 4-5 days at each altitude. No subject suffered any severe form of mountain sickness (i.e., high-altitude pulmonary edema or cerebral edema) during the expedition.

Arterial oxygen saturation (SaO_2) was monitored continuously during sleep with a pulse-oximetry (CSI 502, USA) in all subjects at each altitude. Abdominal respiratory movement, EEG, and eye movement were also monitored in 2-4 subjects. Data were recorded on a magnetic tape with a data-recorder (TEAC HR-30E, Japan). In addition, ventilatory response to isocapnic hypoxia (HVR) and to hypercapnia (HCVR) was measured at low altitude (610m, Matsumoto, Japan) in 7 of 8 subjects before and after the expedition.

Data are expressed as mean± SE, or individual values are given. Statistical analysis were performed with Students' t-test, and differences were considered as statistically significant when $p<0.05$.

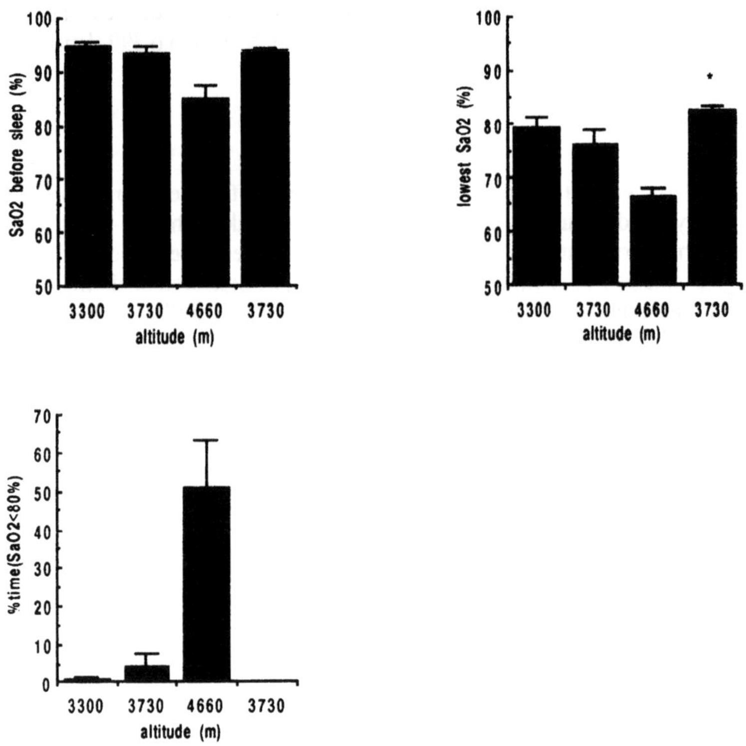

Figure 2. SaO_2 value before sleep (upper left), lowest value during sleep (upper right), and percentage of the time of $SaO_2<80\%$ in total recording time (lower) at each altitude in the expedition. Both before sleep and during sleep the most severe desaturation was observed at 4,660m. At 3,730m on descent, lowest SaO_2 during sleep was higher than that on ascent in contrast to no difference in the value before sleep, and SaO_2 never lowered below 80% during the sleep. Asterisk indicates $p<0.05$ vs. the value at 3,730m on ascent.

RESULTS

Fig. 2 shows the SaO_2 profile obtained during sleep study conducted at each altitude. With increasing altitude on ascent, SaO_2 both before sleep and the lowest value decreased, and the time that SaO_2 value was lower than 80% increased. At 3,730m on descent, the lowest SaO_2 value was significantly higher than that at the same altitude on ascent (82.5± 0.7% on descent vs. 76.2±2.8% on ascent: $p<0.05$), although the value before sleep did not differ from that on ascent (93.7±0.8% on descent vs. 93.5±1.3% on ascent). SaO_2 value never fell below 80% during sleep in any subject at 3,730m on descent.

Table 1 shows the frequency of periodic breathing (PB) during sleep at each altitude. PB showed large individual variation during the expedition; in subject K.Y., who never showed no PB, lowest SaO_2 during sleep at 3,730m increased considerably from 70% on ascent to 82% on descent, while the value before sleep on both ascent and descent was 94%.

Table 1. Frequency of periodic breathing (PB) during sleep at each altitude. (%)

Subject	3,300m	3,730m (a)	4,660m	3,730m (d)
K. A.	nd	nd	11.1	3.9
A. S.	3.9	9.9	18.2	nd
K. Y.	nd	0	1.7	0
T. F.	34.0	nd	16.9	nd

Data were expressed as the % of time that PB appeared in total recording time. nd: no data, a: ascent, d: descent

HVR ($\Delta \dot{V}_E/\Delta SaO_2$, l/min/%) after the expedition was significantly higher than that before (-0.89±0.17 vs. -0.71±0.17; $p<0.05$). HCVR ($\Delta \dot{V}_E/\Delta P_{ET}CO_2$, l/min/torr) after the expedition was also slightly higher than that before (1.50±0.27 vs. 1.30±0.24), although this difference was not significant.

DISCUSSION

In the present study, the primary finding was that desaturation during sleep at 3,730m was decreased on decent from 4,660m compared with that on ascent. This finding may suggest that the brief visit to the higher altitude facilitated the process of acclimatization with respect to oxygenation during sleep at the given high altitude.

There have been a few studies on the sequential relation between oxygen desaturation during sleep and acclimatization during long-term sojourns at high altitude. White et al[1]., in a study conducted during a sojourn at 14,110 ft (4,300m), observed that SaO_2 during sleep on the 7th night was increased compared with that on the 1st night. In this study, because the subjects stayed for less than 5 days and were examined for one times at each altitude, we can not exclude the possibility that the improvement of oxygenation observed in this study might have also occurred had the subjects simply remained for further days at 3,730m. Further investigation is required to examine this possibility.

The relation between PB and oxygen desaturation during sleep at high altitude has been argued. West et al[2]. reported that PB caused severe arterial hypoxemia during sleep at high altitude. On the other hand, Masuyama et al.[3] reported that the frequency of PB was negatively correlated with the degree of desaturation during sleep at high altitude. Unfortunately, in the present study respiratory movement during sleep was recorded only in 2-4 subjects because of logistics of the expedition precluded more extensive study. Thus so we were not able to clarify the relation between PB and desaturation, especially with respect to the improvement of oxygenation observed on descent at 3,730m. However, the findings in subject K.Y., who showed no either on ascent or descent, may provide indicate that the improvement observed at 3,730m was unrelated to the frequency of PB, since, at least in this subject, SaO_2 improved without any change of PB. Furthermore, Normand et al.[4] recently reported that mean SaO_2 value was not affected by PB. Thus we speculate that the improvement in SaO_2 is attributable to increase in total ventilation throughout sleep period, in both those with or without PB.

It has been generally accepted that strong HVR is beneficial to altitude-acclimatization not only in the awake state but also during sleep. In the study cited above, White et al. reported that HVR increased with time at high altitude, and that the increment was associated with increased minute ventilation and with improvement of oxygenation during sleep. Furthermore, several investigators reported that oxygenation during sleep improved when HVR was pharmacologically increased[5]. In this study, HVR after the expedition examined at low altitude was higher than that before. Although we could not examine HVR during the sojourn, the improvement of HVR was thought to due to factors occurring during the sojourn. In addition, the improvement of oxygenation at 3,730m on descent might be attributable to the increased HVR.

In summary, a brief visit to a higher altitude location may facilitate improvement of oxygenation during sleep at a given high altitude, which improvement may be due to increased HVR.

ACKNOWLEDGMENT

We thank Dr. Zhang Yanbo, Dr. Bai Zhiqin, and Dr. Wu Tianyi (Qinghai, China) for their great help with our research expedition in 1990.

REFERENCES

1. D.P. White et al., Altitude acclimatization: influence on periodic breathing and chemosensiveness during sleep, J. Appl. Physiol. 63:401(1987)
2. J.B. West et al., Nocturnal periodic breathing at altitude of 6,300 and 8,050 m, J. Appl. Physiol, 61:280 (1986)
3. S. Masuyama et al., Disordered breathing during sleep at high altitude and ventilatory chemosensitivities to hypoxia and hypercapnia: a study at high altitude (English abst.), Jpn. J. Mount. Med. 8:130(1988)
4. H.N.Normand et al., Periodic breathing and O2 saturation in relation to sleep stages at high altitude, Aviat. Space Environ. Med. 61:229(1990)
5. J.V. Weil et al., Sleep and breathing at high altitude, in:"Sleep Apnea Syndrome," Alan R. Liss, Inc., New York(1978)

INDEX

MIX
Papier aus verantwortungsvollen Quellen
Paper from responsible sources
FSC® C105338
www.fsc.org

If you have any concerns about our products,
you can contact us on
ProductSafety@springernature.com

In case Publisher is established outside the EU,
the EU authorized representative is:
Springer Nature Customer Service Center GmbH
Europaplatz 3, 69115 Heidelberg, Germany

Printed by Libri Plureos GmbH
in Hamburg, Germany